Y0-BXX-159

# New Techniques in Biophysics and Cell Biology

Volume 2

# New Techniques
# in
# Biophysics and Cell Biology

Volume 2


*Edited by*

R. H. Pain

*Department of Biochemistry,*
*University of Newcastle-upon-Tyne*

B. J. Smith

*Imperial Cancer Research Fund Medical Oncology Unit,*
*St. Bartholomew's Hospital, London*





A Wiley–Interscience Publication

JOHN WILEY & SONS

LONDON · NEW YORK · SYDNEY · TORONTO

***Library of Congress Cataloging in Publication Data*** (Revised):

Pain, Roger H.
New techniques in biophysics and cell biology.

'A Wiley–Interscience publication.'
Includes bibliographies.
1. Cytology—Technique—Collected works. 2. Biological physics—Technique—Collected works. 3. Molecular biology—Technique—Collected works. I. Smith, Basil J., joint author. II. Title. [DNLM: 1. Biophysics—Yearbooks. 2. Cytological technics—Yearbooks. 3. Cytology—Yearbooks. WLNE512N]

QH585.P34 574.8′028 72-8611
ISBN 0 471 65757 3 (v. 1)
ISBN 0 471 65758 1 (v. 2)

Made and printed in Great Britain by
The Garden City Press Limited,
Letchworth, Hertfordshire SG6 1JS

# Preface

'Of making many books there is no end'—and the Preacher is undoubtedly preaching today to the converted. It came, then, as something of a surprise to us to realize that there exists no collected volume of new techniques particularly suited to the needs of the young researcher entering the field of Molecular and Cell Biology. We were confirmed in this notion by the more than usual willingness of the authors to contribute to filling this gap. Established techniques are well documented. This series, however, is designed to give an account of new techniques which show every promise of being useful tools but most of which have, as yet, to become widely established. Molecular biology moves fast, so the aim has been to reduce publication time to the minimum.

Each technique is described in a way which should enable the research student and research worker to be able to assess its usefulness and applicability and sufficient theoretical and experimental background is provided so that he can quickly master the technique in practice. In addition, it is expected that the advanced undergraduate will find these chapters a useful introduction to the modern tools of the trade and thus broaden his concepts of modern biology as an experimental science.

We wish to thank the authors for their readiness to write and for their punctuality in providing manuscripts.

R. H. Pain

B. J. Smith

# Contributors to volume 2


| | |
|---|---|
| R. Cammack | *University of London King's College, Department of Plant Sciences, 68 Half Moon Lane, London SE24 9JF.* |
| E. C. Cocking | *Department of Botany, University of Nottingham, University Park, Nottingham NG7 2RD.* |
| P. K. Evans | *Department of Botany, University of Nottingham, University Park, Nottingham NG7 2RD.* |
| J. Feeney | *National Institute for Medical Research, Mill Hill, London NW7 1AA.* |
| J. M. Graham | *Imperial Cancer Research Fund Laboratories, PO Box 123, Lincoln's Inn Fields, London WC2A 3PX.* |
| Carl V. Hanson | *Department of Chemistry, University of California, Berkeley, California 94720, U.S.A.* |
| Ronald B. Herberman | *Cellular and Tumour Immunology Section, Laboratory of Cell Biology, National Cancer Institute, National Institutes of Health, Bethesda, Maryland 20014, U.S.A.* |
| Alan Suggett | *Unilever Research Laboratory, Sharnbrook, Bedfordshire, MK44 1LQ.* |
| C. C. Ting | *Cellular and Tumour Immunology Section, Laboratory of Cell Biology, National Cancer Institute, National Institutes of Health, Bethesda, Maryland 20014, U.S.A.* |
| Ingemar Wadsö | *Thermochemistry Laboratory, Chemical Centre, University of Lund, POB 740, S-22007 Lund 7, Sweden.* |
| Philippe Wahl | *Centre de Biophysique Moléculaire, C.N.R.S., 45045 Orleans Cedex, France.* |

# Contents

CHAPTER 1

# Cellular membrane fractionation

J. M. Graham
*Department of Tumour Virology,
Imperial Cancer Research Fund Laboratories,
P.O. Box 123, 44 Lincoln's Inn Fields,
London WC2A 3PX*

## I. INTRODUCTION

There exists today a vast number of published methods for the fractionation of a variety of tissues and isolated cells into their membraneous components. This seeming profusion of techniques conceals the fact that there are relatively few basic approaches to the problems of membrane fractionation, the majority of techniques being variations on these basic approaches.

A comprehensive survey, therefore, of every available membrane fractionation technique would be needlessly confusing and of little value. Naturally a number of these techniques will be discussed in order to emphasize the relevance of certain manipulations, but in general this review is directed more towards an evaluation of the problems involved in selecting and successfully executing a method whose end product satisfies the demands of the experimenter. The complete lack of a unified approach to membrane fractionation stems largely from an inability to reproduce operating conditions from laboratory to laboratory. This arises sometimes for obvious reasons, such as the use of different starting materials and the requirement for different homogenization techniques, sometimes for more obscure reasons such as the use of different centrifuge rotors. Regrettably these problems are frequently compounded by an inadequate description of the method.

The isolation and study of the membraneous components from a tissue or isolated cells involves a number of operations: (*a*) homogenization, (*b*) fractionation, (*c*) assessment of the efficacy of the fractionation, (*d*) storage of the products. These topics will be treated separately, although (*a*) and (*b*) are somewhat dependent on one another.

## II. MEMBRANE MARKERS

In order to assess the effects of the various steps in the homogenization and fractionation procedures employed to isolate surface membranes, assay methods (markers) are required which unambiguously identify each of the subcellular components. Only a brief survey of the most commonly used methods will be given here since an excellent review of this topic has been published recently (De Pierre and Karnovsky, 1973).

### A. Morphological Markers

Morphological markers can be used in a few restricted cases. Intact organelles such as nuclei and mitochondria are obviously recognizable by phase contrast and electron microscopy. Some other membrane elements such as rough endoplasmic reticulum can be identified by their ribosomal components: so can the plasma membrane in some instances. The lateral walls of liver parenchymal cells are connected through desmosomes; if their structures are maintained then any large sheets of membrane, particularly if the desmosomal junctions are intact, can be identified as plasma membrane (e.g. Neville, 1960; Emmelot *et al.*, 1964). The brush border plasma membrane from the small intestine is also instantly recognizable (Miller and Crane, 1961; Eicholz and Crane, 1965), as are the tissue-culture cell ghosts produced by the method of Warren *et al.* (1966). Small sheets of membrane or membrane vesicles, on the other hand, which can arise from various sources: plasma membrane, smooth endoplasmic reticulum, rough endoplasmic reticulum which has been stripped of its ribosomes, disrupted and/or fragmented organelles such as mitochondria or lysosomes, are not identifiable morphologically.

## B. Enzyme Markers for Plasma Membrane

The most widely used markers are those enzymes which are restricted to one subcellular component. The distribution of particular enzymes within the cell has been worked out either histochemically (e.g. Essner *et al.*, 1958; De Thé, 1968), or by correlation with a morphologically recognizable structure (Benedetti and Emmelot, 1968) or by correlation with a previously histochemically localized enzyme. The most commonly used surface membrane enzyme markers are: $Na^+/K^+$-stimulated $Mg^{2+}$-ATPase (Kamat and Wallach, 1966; Coleman *et al.*, 1967a; Evans, 1969; Gahmberg and Simons, 1970; Avruch and Wallach, 1971; Graham, 1972, 1973a, 1973b; Forte *et al.*, 1973; Hemminki and Suovaniemi, 1973; Lelievre, 1973) and 5′-nucleotidase (Coleman *et al.*, 1967a; Graham *et al.*, 1968; Berman *et al.*, 1969; Song *et al.*, 1969b; Weaver and Boyle, 1969; Allan and Crumpton, 1970; Gahmberg and Simons, 1970; Bingham and Burke, 1972; Hemminki and Suovaniemi, 1973; Lelievre, 1973). Other markers have also been used: leucine aminopeptidase (Hübscher *et al.*, 1965; Coleman and Finean, 1966; Rhodes *et al.*, 1967; Dod and Gray, 1968; Graham *et al.*, 1968; Evans, 1969), adenyl cyclase (Rodbell, 1967; McKeel and Jarett, 1970; Wolff and Jones, 1971; Combret and Laudat, 1972; Forte, 1972) and $K^+$-stimulated *p*-nitrophenyl phosphatase (Wolff and Jones, 1971; Forte *et al.*, 1973; Hemminki and Suovaniemi, 1973).

## C. Enzyme Markers for Contaminants

Contaminants of a plasma membrane fraction can also be detected enzymically. There are several enzymes normally used to characterize endoplasmic reticulum. One of these, glucose-6-phosphatase, has been detected histochemically in rat liver by Abrahamson *et al.* (1969) and Leskes *et al.* (1971). Glucose-6-phosphatase has been used as a marker by many workers: e.g. Coleman and Finean (1965); Graham *et al.* (1968); Rosenberg (1969); Song *et al.* (1969b); Weaver and Boyle (1969); Wolff and Jones (1971) and Forte *et al.* (1973). It should be pointed out, however, that glucose-6-phosphate serves as a substrate not only for glucose-6-phosphatase, but also alkaline phosphatase and acid phosphatase enzymes. Alkaline phosphatase has indeed been used as a plasma membrane marker (e.g. Bosmann *et al.*, 1968; Graham *et al.*, 1968). It is therefore advisable to include 4 m*M* EDTA and 1 m*M* KF to inhibit these two enzymes (Lauter *et al.*, 1972). Other marker enzymes for the endoplasmic reticulum include NADH diaphorase (Kamat and Wallach, 1966; Avruch and Wallach, 1971; Graham, 1972; Forte *et al.*, 1973), NADPH-cytochrome c reductase (Berman *et al.*, 1969; Dallman *et al.*, 1969, Wolff and Jones, 1971; Bingham and Burke, 1972) and NADH-cytochrome c reductase (McKeel and Jarett, 1970; Perdue and Sneider, 1970; Perdue *et al.*, 1973). Lelievre (1973) suggested UDPase was a good endoplasmic reticulum marker in spite of the fact that the specific activity of this enzyme in his plasma membrane fraction was nearly as high as that of the endoplasmic reticulum. Rough endoplasmic reticulum can also be detected by RNA assay. Cytochrome P450 is another marker for the endoplasmic reticulum—its presence in this membrane system has been demonstrated in rat liver (Gillette, 1966; Omura and Sato, 1966). Restriction of cytochrome P450 to the endoplasmic reticulum is not unreasonable since cytochrome-reducing enzymes are present in these membranes and not in the plasma membrane. This accounts for the yellow-brown colouring of endoplasmic reticulum. Mitochondria or inner mitochondrial membranes are routinely detected by assay for succinate dehydrogenase, and lysosomes by acid phosphatase. Golgi contamination is not often assessed but enzymes such as UDP-galactose *N*-acetylglucosamine galactosyl transferase (Fleischer and Fleischer, 1970; Leelavathi *et al.*, 1970; Sukeno *et al.*, 1972), UDP-*N*-acetylglucosamine glycoprotein *N*-acetylglucosamine transferase (Sukeno *et al.*, 1972) and thiamin pyrophosphatase (Leelavathi *et al.*, 1970) can be employed as markers. Thiamin pyrophosphatase however was not specifically concentrated in rat liver Golgi membranes (Fleischer and Fleischer, 1970).

The use of enzyme markers to establish the identity of particular membrane fractions is not however as reliable a method as one would wish. There are, for example, observations on the histochemical localization of 5′-nucleotidase (Widnell and Unkeless, 1968; Widnell, 1972) which suggest that this enzyme is present in the endoplasmic reticulum (particularly the rough membranes) as well as in the plasma membrane of rat liver. Song *et al.* (1969a) also suggested that this enzyme was present in a heavy microsomal subfraction of rat liver—presumably rough endoplasmic reticulum. To confuse the situation further Pletsch and Coffey (1972) indicated that 8% of the total 5′-nucleotidase from rat liver acted as a lysosomal marker and were able to show by chromatography on 'Sephadex' G-200 that the lysosome enzyme was not identical to the plasma membrane enzyme. There also appears to be some species specificity from a survey by Lauter *et al.* (1972): the plasma membranes from rat or guinea pig liver demonstrated a 16–19-fold enrichment in 5′-nucleotidase over the homogenate, whereas the fractions from rabbit or cat liver showed no enrichment whatsoever. The general level of 5′-nucleotidase in these two tissues is moreover very low and is entirely absent from rat adipose cell membranes (McKeel and Jarett, 1970). Both 5′-nucleotidase and glucose-6-phosphatase were detected in rat liver outer mitochondrial membrane preparations (Colbeau *et al.* 1971) although some contamination by smooth microsomes cannot be ruled out, indeed the relative concentration of monoamine oxidase in these membranes is greater than that of the other two enzymes. The $K^+$-stimulated *p*-nitrophenyl phosphatase is another unpredictable marker: Lelievre (1973), for example, was unable to demonstrate this enzyme in murine plasmacytoma surface membrane. Endoplasmic reticulum markers are not excluded from this uncertainty: Benedetti and Emmelot (1968) considered that 15–20% of the total NADH-cytochrome c reductase was intrinsic to the plasma membrane of rat liver.

## D. Other Intrinsic Plasma Membrane Markers

Another correlative parameter frequently used as a marker for surface membrane is the cholesterol content—in particular the cholesterol/phospholipid molar ratio which tends to be significantly higher in the surface membrane than in other membranes. Coleman and Finean (1966), compared surface membrane fractions from guinea pig liver, kidney and small intestine with other membrane systems of the same tissues: in all cases the surface membrane was enriched in cholesterol. Rat liver plasma membrane also demonstrates a higher cholesterol content (Benedetti and Emmelot, 1968). The absolute values reported for the cholesterol/phospholipid molar ratio in a particular membrane, vary alarmingly: the value for rat intestinal microvilli varies from 0·5 (Millington and Critchley, 1968) to 1·26 (Forstner *et al.*, 1968); the value for liver plasma membrane varies from 0·26 (Ashworth and Green, 1966; Evans, 1969), through 0·52 (Takeuchi and Terayama, 1965) to 0·81 (Skipski *et al.*, 1965). Values for plasma membranes from other sources are given in Table 1.1. The cholesterol/phospholipid molar ratio of the plasma membrane may vary from source to source but the variation in this ratio for liver plasma membrane prepared by different workers using a Neville (1960)-type procedure suggests that these discrepancies may also be a consequence of either subtle changes in the manipulation of the membranes or different cholesterol estimation techniques, or both. It is also pertinent to point out that the phospholipid/protein ratio is generally higher in the plasma membrane than in other subcellular structures (e.g. Forte *et al.*, 1973; Hemminki and Suovaniemi, 1973; Perdue *et al.*, 1973). Bosmann *et al.* (1968), however, recorded a lower value for HeLa plasma membrane than for other membrane fractions, although there was a significant enrichment in phospholipid over the homogenate. Cholesterol content should therefore be

**Table 1.1.** Cholesterol/phospholipid molar ratio in the plasma membrane from various cell types

| Source | Cholesterol/ phospholipid molar ratio | Workers |
|---|---|---|
| HeLa cells | 1·05 | Bosmann *et al.* (1968) |
| L cells | 0·69 | Weinstein *et al.* (1969) |
| Pig lymphocytes | 1·03 | Ferber *et al.* (1972) |
| Human lymphocytes | 0·69/0·75 | Demus (1973) |
| Chick embryo liver cells | 1·0 | Rosenberg (1969) |
| Calf thymocytes | 0·64 | Van Blitterswijk *et al.* (1973) |
| Chicken tumour cells | 0·72 | Perdue *et al.* (1973) |

expressed both in terms of phospholipid and protein. The cholesterol/phospholipid molar ratio of chick tumour cell surface membrane was 0·72 (against 0·57 for the homogenate), whilst figures for the cholesterol/protein weight ratio were 0·273 and 0·063 respectively (Perdue *et al.*, 1973). A plasma membrane fraction from human lymphocytes contained 263 μg cholesterol/mg protein, while the homogenate contained 33 μg (Demus, 1973).

Another possible marker is sialic acid. Many workers have demonstrated that this component is present on the surface of the cell and moreover that this represents at least 50% of the total sialic acid of the cell (Wallach and Eylar, 1961; Lichtman and Weed, 1970; Noseworthy *et al.*, 1972; De Pierre and Karnovsky, 1973). The presence of this component in nuclei and mitochondria (Nordling and Mayhew, 1966; Glick *et al.*, 1970), however, requires that sialic acid estimates should be accompanied by independent assessments of mitochondrial and nuclear contamination. Perdue *et al.* (1973) obtained a 3-fold enrichment in sialic acid over the homogenate in chick tumour cells; Van Blitterswijk *et al.* (1973) obtained a 5-fold enrichment with calf thymocytes. In the latter instance, however, only 3% of the total sialic acid was recovered—surface sialopeptides may have been removed during the hypotonic washing of the microsome fraction.

The recovery of surface antigens is a potentially useful tool which has been relatively little used. An antibody to a cell surface antigen is prepared and the various fractions tested for adsorption of this antibody. Wallach and Kamat (1964, 1966) recovered about 90% of the total surface antigen in their plasma membrane fraction, and significant enrichments in antigen content over the homogenate have been demonstrated in plasma membrane fractions by Gahmberg and Simons (1970) and Allan and Crumpton (1970) for baby hamster kidney cells and pig lymphocytes respectively. Boone *et al.* (1969) on the other hand, labelled intact HeLa cells with $^{125}$I-labelled antibody, fractionated the cells and followed the specific activity of the $^{125}$I. The peak of activity coincided with that of the $Na^{+}/K^{+}$-stimulated ATPase and represented a 50-fold purification. Regrettably, however, these workers provided no enzyme data for the homogenate so that a comparison of the enzyme and antigen methods is not possible.

## E. Extrinsic Plasma Membrane Markers

A generalization of the technique of Boone *et al.* (1969) is to label intact cells with a reagent which does not penetrate the cell membrane, and which remains firmly attached to some surface membrane component during fractionation. Most of these reagents, e.g. formyl methionyl sulphone methyl phosphate (Bretscher, 1971a, 1971b), diazonium salts of sulphanilic acid (Berg, 1969) and radioactive iodide (Phillips and Morrison, 1971) have been employed to study the arrangement of protein molecules in the erythrocyte membrane, using the rationale that only those proteins whose peptide chains are exposed at the external surface of the membrane will be labelled. The iodination technique has the merit of simplicity in so far as the entity binding to the protein is an iodine atom rather than a larger organic molecule. The reacting species is generated from iodide, hydrogen peroxide and lactoperoxidase (Marchalonis *et al.*, 1971; Phillips and Morrison, 1971): or to overcome the problem of hydrogen peroxide entering the cell, the peroxide may be generated at the cell surface by use of glucose and glucose oxidase (Hubbard and Cohn, 1972; Hynes, 1973). The technique has also been used for normal and neoplastic lymphocytes (Marchalonis *et al.*, 1971), platelets (Phillips, 1972) NIL, virus-transformed NIL and Lx cells (Hynes, 1973). Demonstration of the specificity of the surface labelling has been shown by protease treatment of the labelled material (Hubbard and Cohn, 1972; Phillips, 1972). Preliminary experiments with NIL cells (Hynes and Graham, unpublished observations) indicate that the majority of the iodine label fractionates with the plasma membrane. These externally applied markers provide an excellent criterion for membrane purity so long as they are specific, do not influence the efficacy of the fractionation and do not deleteriously affect the membrane's functional properties.

## III. HOMOGENIZATION

There are essentially four types of homogenization which rely on different forces to disrupt the cellular material: these are (*a*) mechanical shear, (*b*) liquid shear, (*c*) gaseous shear and (*d*) osmotic lysis.

### A. Mechanical Shear

Numerous instruments which operate on the domestic liquidizer principle and which embody a system of rotating metal blades, are available commercially. Those instruments which are driven by a motor mounted below the sample chamber generally require rather large volumes of material: e.g. the Kenwood is designed for volumes of 100 ml to 1 litre while the MSE Automix can be used for volumes from 50 ml to 1 litre. More commonly mechanical shear homogenizers are driven by an overhead motor. The earlier forms of these instruments also required sample volumes in excess of 50 ml; recent developments, however, enable the use of volumes as low as 2 ml. Instruments such as the MSE homogenizer (5 ml to 100 ml), the Townson and Mercer (5 ml to 1 litre), the Silverson range (2 ml to 4 litres) and the Polytron (2 ml to 10 litres) are currently available. The more sophisticated of these homogenizers (e.g. the Silverson type) involves (*a*) the drawing of the sample, by suction, into the working head, (*b*) mixing by the rotating blades within the working head and (*c*) shearing of the material by the blades during expulsion of the sample from the working head. These instruments are, however, used rather infrequently in the preparation of membrane fractions, only in a few special cases do such instruments find an application. Berezney *et al.* (1970) used a Waring blender to obtain nuclear membranes from bovine liver and both Morré *et al.* (1970) and Leelavathi *et al.* (1970) used a Polytron homogenizer to obtain Golgi membranes from rat liver. These machines are usually restricted to the disruption of large masses of tissue such as liver, although Zentgraf *et al.* (1971) did use a 'high-speed rotating knives homogenizer' to obtain both nuclear and plasma membranes from hen erythrocytes, and Wolff and Jones (1971) purified a plasma membrane fraction from bovine thyroid using a Polytron homogenizer to disrupt the tissue.

Another form of mechanical shear can be obtained by subjecting the material to a number of freeze–thaw cycles, in which the disruptive force is provided by the rapid growth and collapse of ice crystals. This method has found relatively little application except in rather special circumstances: Kjaerheim (1965) used this technique to remove the microvillus-free surface membrane from intact mouse jejunum, leaving the remainder of the cells remarkably intact. Rapid magnetic stirring in a medium containing 1 m$M$ $NaHCO_3$ and 2 m$M$ $Ca^{2+}$ (pH 8·0) was used by Lelievre (1973) to disrupt murine plasmacytoma cells.

## B. Liquid Shear

The most widely used method of homogenization employs liquid shear as the disruptive force. Practically, this is commonly achieved by forcing the cell- or tissue-containing medium through the narrow space between a moving pestle and a glass-containing vessel. The Dounce homogenizer uses a glass ball as the pestle which is forced through the cell suspension manually. In modern forms of the Potter–Elvehjem homogenizer the pestle is made from Teflon and shaped to fit the containing vessel which may have a rounded or conical bottom. The pestle may be driven either manually or mechanically.

The efficiency of the homogenization process and the composition of the resulting homogenate depends upon a number of factors: (*a*) the number of strokes of the pestle, (*b*) the thrust of the pestle, (*c*) the speed of rotation of the pestle (if mechanically driven), (*d*) the clearance between pestle and containing vessel, (*e*) the concentration of the cell material suspension and (*f*) the 'state' of the cellular material. Of these, only factor (*a*) is invariably well defined. Deplorably, factor (*c*) is often described in terms of some arbitrary rheostat setting; factor (*d*) is frequently given in terms of a 'loose' or 'tight-fitting' pestle and factor (*e*) is sometimes neglected altogether. Factor (*b*) is difficult to quantify since it depends upon the operator moving the containing vessel relative to the pestle, and the force required depends upon factors (*c*) to (*f*). Although factor (*b*) could be defined with respect to 'time per stroke', it is usually described in nebulous terms such as 'vigorous' or 'gentle' strokes. The unfortunate outcome of this lack of information is that it is often impossible to reproduce a set of operating conditions exactly.

Factor (*f*) is important when the operator is applying a method worked out for one type of cellular material to another type. The surface membrane from one cell may possess different physical properties from that of another cell so that conditions which effectively rupture one cell type will be less suitable for another. Parameters such as the size of the cell and its ability to swell in hypotonic media, may have a bearing on this problem. Large aggregates of cells may be ruptured by mechanical shear rather than liquid shear and intercellular adhesions can drastically modify the form of the end product. This will be discussed more fully later on, in relation to both the homogenization method and the homogenization medium, since the latter can significantly influence the efficacy of the process.

Other liquid shear methods are less common and are usually designed to suit rather special situations. Avruch and Wallach (1971) disrupted adipocytes by forcing them from a syringe through a stainless-steel screen (aperture size *c.* 100 $\mu$) contained within a Swinny filter holder. They found that nitrogen cavitation caused nuclear disruption whereas Potter–Elvehjem or Dounce homogenization resulted in loss of membraneous material into the fat cake which forms when the adipocytes release their contents.

### C. Gaseous Shear

Hunter and Commerford (1961) introduced nitrogen cavitation as a means of cell rupture and this method was subsequently used by Kamat and Wallach (1965) for Ehrlich Ascites cells. Nitrogen cavitation involves equilibrating a stirred cell suspension with oxygen-free nitrogen at pressures between 500 and 800 lb/in$^2$ for periods of 15–20 min. Cell disruption occurs upon sudden release of the pressure, either by a 'bursting balloon' process due to gas expanding within the cell or by the shearing forces of the rapidly forming bubbles of gas in the liquid phase. Practically, the method requires a stainless-steel pressure vessel. Two such vessels are available commercially; one is manufactured by Artisan Industries, Waltham, Massachusetts, U.S.A.; the other by Baskerville and Lindsay Ltd., Manchester, England. The bottom part of the apparatus is essentially a sample chamber of internal dimensions, 15 cm deep and 7 cm in diameter (Artisan model) which also contains an inlet port for the nitrogen. The top part of the apparatus, which screws on to the lower part to form a gas-tight seal, carries two outlet ports (a delivery tube which reaches nearly to the bottom of the sample chamber and a venting tube) and a safety valve. Inlet and outlet ports can be closed by needle valves. For large volumes (above 100 ml) the sample may be placed directly in the sample chamber; for smaller volumes (5–100 ml) the sample may be contained within a small plastic bottle centralized within a perspex collar. The Baskerville and Lindsay model also incorporates a mechanical stirring device, whilst with the Artisan model the cell suspension is agitated magnetically.

After equilibration of the sample with nitrogen at the appropriate pressure, the needle valve on the delivery tube is opened; the sample is forced up through the delivery tube and as soon as it passes the valve it is exposed to atmospheric pressure and cell disruption occurs.

The actual pressure required to cause disruption must (*a*) be high enough to cause complete vesiculation of the surface membrane, (*b*) not cause damage to nuclei and (*c*) be completely reproducible. Although the optimum conditions for disruption vary from cell to cell, a reasonable starting schedule would be 700 lb/in$^2$ for 15 min, using a cell suspension volume of 30 ml.

There are a number of advantages of nitrogen cavitation over liquid shear methods: (*a*) the process occurs in an inert atmosphere, (*b*) the local heating due to friction in liquid shear methods is eliminated and (*c*) the process appears to be largely independent of cell concentration over the range $10^7$–$5 \times 10^9$ per 30 ml (Graham, unpublished observation). Its main disadvantage is that the rapid release of bubbles of gas in the liquid phase causes foaming. The foam must be allowed to subside by gently stirring the homogenate in a vessel of large surface area (Graham, 1972), otherwise material becomes trapped within the foam during centrifugation.

### D. Osmotic Lysis

This has only been used in special cases: Barber and Jamieson (1970) disrupted platelets by loading them with glycerol to a concentration of 4·3 *M*; lysis occurred when the platelets were transferred to a relatively hypotonic medium containing 0·25 *M* sucrose.

## IV. APPLICATION OF HOMOGENIZATION TECHNIQUES

### A. Hypotonic Media

One of the first membrane fractionation techniques was worked out for rat liver by Neville (1960). He used a Dounce homogenizer and a medium containing 1 m*M* $NaHCO_3$ (pH 7·4). Intact liver tissue is somewhat exceptional in so far as the membranes of the opposing lateral surface of adjacent liver cells are connected through numerous desmosomes. This lends a rigidity to these membranes such that they remain largely intact during homogenization. The internal membranes constituting the endoplasmic reticulum tend to fragment or vesiculate under most conditions and are thus easily separable from the large rapidly sedimenting sheets of plasma membrane. The hypotonic medium used originally by Neville (1960) and later by Emmelot *et al.* (1964), Evans (1969) and Song *et al.* (1969b) has the advantage that the swelling of the cells which it induces, renders the cell more susceptible to rupture by liquid shear so that the homogenization need be relatively mild. Neville (1960), for instance, used '25–30 vigorous strokes' of a 'loose-fitting' pestle. The more severe the homogenization required, the more likely is the surface membrane to fragment into smaller more slowly sedimenting pieces.

The disruption of tissue-culture cells poses greater problems, since they do not possess these fortuitous desmosomal connexions, they are even more liable to fragment during homogenization. Contributory to this problem is the 'continuous' nature of liquid shear methods such as Dounce or Potter–Elvehjem homogenization. Not all the cells in the suspension are ruptured during the first stroke of the pestle—indeed cell rupture is occurring continuously throughout the period of homogenization. At the same time those surface membranes which are released earlier will be fragmented to a greater extent than those from cells ruptured in the later stages. Indeed the resulting homogenate will contain a whole spectrum of surface membrane forms, from sheets to vesicles. Other cell components such as mitochondria and nuclei will be similarly exposed to liquid shear forces for varying lengths of time. The composition of the homogenizing medium and the homogenization conditions may be arranged so as to favour either the larger or the smaller membrane elements; it is not, however, possible to overcome this problem completely. Consequently a variable proportion of the surface membrane will not sediment at the low centrifugation speeds normally employed to harvest large sheets of membrane. The plasma membrane fraction isolated must therefore represent only a selected population of surface membrane fragments: this is reflected in the generally rather low yields of surface membranes.

Forte *et al.* (1973), for example, disrupted Ehrlich Ascites cells by 35–40 strokes of the pestle in a tight-fitting Dounce homogenizer in a hypotonic Tris–saline medium. They processed only the surface membrane which was sedimentable at 7200 *g* min and thus recovered only 0·1 % of the protein in the homogenate as plasma membrane. Although these workers obtained a relatively pure plasma membrane fraction (as judged by an 18-fold enrichment in $Na^+/K^+$-stimulated ATPase over the homogenate), the total amount of this enzyme in the 15,000 *g* mitochondrial pellet was ten times greater than that in the plasma membrane fraction. Demus (1973) on the other hand, harvested both the rapidly sedimenting larger surface membrane fragments (7000 *g* for 20 min) and the slowly sedimenting smaller fragments (37,000 *g* for 1 h) from a homogenate of human lymphocytes, thereby recovering about 1 % of the protein of the total homogenate as plasma membrane.

Warren *et al.* (1966) overcame this problem with tissue-culture cells by using chemical agents such as $Zn^{2+}$ (about 1 m*M*) + 0·05 % 'Tween' 20, fluorescein mercuric acetate (FMA, about 1·5 m*M*), acetic acid (0·1 *N*) or Tris/$Mg^{2+}$ (both 50 m*M*) to stabilize or strengthen the surface membrane against fragmentation during homogenization. Tissue-culture cell ghosts or large sheets of surface membrane were produced by Dounce homogenization which were sedimentable at low speeds. The methods were developed for L cells but the FMA method has also been applied to baby hamster kidney cells, rat and rabbit lymphocytes, thymocytes (Warren *et al.*, 1966), and the $Zn^{2+}$ and Tris methods to Chinese hamster ovary and HeLa cells (Warren and Glick, 1969) with minor modifications of the techniques, in particular, the elimination of detergent from the zinc ion method. Bingham and Burke (1972), however, were unable to obtain satisfactory surface membrane preparations from chick embryo fibroblasts using the Tris/$Mg^{2+}$ method.

There are, however, attendant problems to the use of hypotonic media—in particular the rupture of internal organelles such as mitochondria and lysosomes, and also of nuclei. This will have serious consequences on the separation of a relatively pure surface membrane fraction. Mitochondrial and lysosomal membranes could be released in the form of sheets or fragments (or both) and hence contaminate other membrane fractions containing similarly sized elements. Indeed, De Duve (1971) noted that even gentle Potter–Elvehjem homogenization of rat liver released the outer membrane from 10 % of the mitochondria and caused damage to 15 % of the lysosomes. In view of the ease with which rupture of these organelles cocurs, it is surprising to find that the surface membrane fractions produced from such homogenates are rather infrequently assayed for succinate dehydrogenase. It is certainly insufficient to assess mitochondrial contamination by electron microscopy: any mitochondrial impurities banding isopycnically with plasma membrane will be in the form of sheets or vesicles rather than intact organelles. In addition, soluble proteins would be released into the medium from the swollen or damaged organelles—these would include soluble proteins from mitochondria, which tend to adsorb to other membranes and, perhaps more seriously, degradative enzymes (proteases and lipases) from lysosomes. There is also the possibility of nuclear rupture and consequent liberation of DNA: this will increase the viscosity of the medium and cause aggregation of membraneous material on a large scale. Such aggregation is virtually impossible to reverse. Nigam *et al.* (1971) indeed observed that using 1 m*M* $NaHCO_3$ as the homogenization medium; gelatinization of the components occurred upon standing at 4°. A number of modifications have been employed to overcome these problems.

## B. Iso-osmotic Media

Coleman *et al.* (1967b) used an iso-osmotic homogenization medium containing 1 m*M* $NaHCO_3$ and 0·3 *M* sucrose for rat liver, and Perdue *et al.* (1973) used unbuffered 0·25 *M* sucrose for chicken tumour cells. Such iso-osmotic solutions however prevent the advantageous swelling of cells which occurs in hypotonic media; consequently the homogenization required to disrupt the cells is more severe and membrane fragmentation is more likely. This situation even holds for rat liver (Coleman *et al.*, 1967b) and recently Hawkins and Jacquez (1972) in comparing three modifications of the Neville (1960) method for fractionating rat liver found that the iso-osmotic homogenization medium was inferior to hypotonic media (with or without $Ca^{2+}$) in terms of yield of surface membrane. On the other hand, Rosenberg (1969) reported that the use of isotonic saline (Krebs or Hanks buffers) or isotonic sucrose (plus 1 m*M* $Mg^{2+}$) had little effect on the disruption and fractionation of chick embryo liver cells using Potter–Elvehjem homogenization, although the sucrose solution did cause greater fragmentation of the surface membrane. Such sucrose-containing media do, however, afford considerable protection to mitochondria and lysosomes and some protection to nuclei. The latter, however, are better protected by divalent cations such as $Mg^{2+}$ or $Ca^{2+}$. Indeed media containing 2 m*M* $Mg^{2+}$, 2 m*M* $Ca^{2+}$ and 1 m*M* $NaHCO_3$ (in the absence of sucrose) have been successfully used for the preparation of intact nuclei from liver and hepatomas (Price *et al.*, 1972) and from polyoma-transformed baby hamster kidney cells and hamster embryo fibroblasts (Graham and Macpherson, unpublished observations). Accordingly, Takeuchi and Terayama (1965) used a homogenizing medium of 0·25 *M* sucrose/

0·5 m*M* $Ca^{2+}$; Berman *et al.* (1969) employed the same medium buffered with 5 m*M* Tris (pH 7·4); whilst Stein *et al.* (1968) substituted $Mg^{2+}$ for $Ca^{2+}$.

Since iso-osmotic sucrose media usually favour fragmentation of the surface membrane, Berman *et al.* (1969) emphasized the need to use a loose-fitting Dounce homogenizer to minimize this problem. Forte *et al.* (1973) employed a hypotonic medium containing 25 m*M* $Na^{+}$, 18 m*M* Tris, 0·5 m*M* $Ca^{2+}$; their rather severe homogenization conditions, however, negated to some extent, the advantages of this medium. Lelievre (1973) used 1 m*M* $NaHCO_3$/2 m*M* $Ca^{2+}$ (pH 8·0) to disrupt murine plasmacytoma cells, Ray (1970) used 1 m*M* $NaHCO_3$/0·5 m*M* $Ca^{2+}$ for rat liver and a medium containing 1 m*M* $Mg^{2+}$ and 10 m*M* Tris (pH 7·4) was employed by Boone *et al.* (1969) for HeLa cells.

Addition of divalent cations, while solving one problem, poses another in that these ions tend rather effectively to cross-link membranes—consequently aggregation occurs. This effect was deliberately exaggerated by Kamath and Rubin (1972) who found that addition of 8 m*M* $Ca^{2+}$ to a post-mitochondrial supernatant from rat liver enabled the microsomes to sediment at 30 *g* for 10 min. Nigam *et al.* (1971) therefore used iso-osmotic sucrose without divalent cations during the initial homogenization steps with rat liver. McKeel and Jarett (1970) used an iso-osmotic sucrose solution together with 1 m*M* EDTA for fat cells. Bosmann *et al.* (1968) went a stage further and disrupted HeLa cells in a medium containing 0·01 *M* EDTA/0·02 *M* Tris (pH 7·0), with a Dounce homogenizer. Hypotonic EDTA media have been used by Price *et al.* (1972) and Graham and Macpherson (unpublished results) to disrupt the nuclei from rat liver and polyoma-transformed baby hamster kidney cells respectively: it is most likely therefore that the conditions used by Bosmann *et al.* would result in nuclear rupture. Van Blitterswijk *et al.* (1973) found that addition of 6 m*M* EDTA to a thymocyte homogenate, generated by nitrogen cavitation, produced clumping of the material. This was probably due to release of DNA from the nucleus.

Perdue and Sneider (1970) were able to fractionate chick embryo fibroblasts using a modification of the $Zn^{2+}$ method of Warren *et al.* (1966). After disruption of the cells by Potter–Elvehjem homogenization, in a hypotonic medium containing about 1 m*M* $Zn^{2+}$ and 0·05% 'Tween' 20 the fractions containing surface membrane fragments were made 0·01 *M* with respect to EDTA. This was found necessary to disperse the membranes prior to centrifugation in sucrose density gradients. Perdue and Sneider (1970) also pointed out, however, that 0·01 *M* EDTA caused solubilization of 68% of the original protein. This is reflected in the rather low protein content (25%) of the membranes produced by this method: most other surface membranes contain 50–60% protein (Coleman and Finean, 1966; Benedetti and Emmelot, 1968; Weinstein, 1968). HeLa plasma membranes, however, exposed to the same concentration of EDTA (Bosmann *et al.*, 1968) were reported to contain about 60% protein.

Clearly, the frequently opposing influences of the composition of the suspending medium on the homogenization process itself and on the physical composition of the resulting homogenate, mean that an ideal system is impossible to realize. Consequently, compromise situations have been devised. Weaver and Boyle (1969), for instance, recognizing both the importance of a hypo-osmotic medium in facilitating the disruption of rat liver cells, and its destructive effect on cell organelles, used 0·08 *M* sucrose (pH 7·6). Boone *et al.* (1969) on the other hand, retained a hypotonic medium for the disruption of HeLa cells, but minimized its deleterious effects by adding sucrose immediately after homogenization so as to regain isotonicity. A rationale combining these two approaches was used by Hemminki and Suovaniemi (1973) who homogenized rat brain cells in a moderately hypotonic medium containing 90 m*M* sucrose/1 m*M* $Mg^{2+}$/40 m*M* Tris (pH 7·6) whose isotonicity was regained immediately afterwards by the addition of sucrose to 0·20 *M*. Gahmberg and Simons (1970) and Graham (1972, 1973a, 1973b) added 0·2 m*M* $Mg^{2+}$ to the medium so as to protect the nuclei of baby hamster kidney cells during nitrogen cavitation, but chelated it subsequently with 1 m*M* EDTA. Although this will prevent divalent cation-induced clumping, it also introduces the risk of EDTA-induced nuclear rupture and consequent clumping by DNA. The presence of 0·25 *M* sucrose should at least partially reverse this tendency, although Van Blitterswijk *et al.* (1973) using thymocytes, found this not to be the case. These workers disrupted calf thymocytes in Hanks solution using nitrogen cavitation; the addition of 6 m*M* EDTA to the homogenate caused clumping.

The effect of the pH of the homogenizing medium has been relatively little investigated. One of the few studies was carried out by Bell *et al.* (1971) using canine liver as the starting material. They found that Dounce homogenization which disrupted nearly 96% of the cells at pH 7·4 disrupted only 60% at pH 6·9. The majority of homogenization media are in fact buffered between pH 7·0 and 7·6, although both Forte *et al.* (1973) and Lelievre (1973) used a pH 8·0 medium to disrupt Ehrlich Ascites cells and murine plasmacytoma cells respectively. The rationale behind the selection of a particular pH is not usually made clear. Lelievre justified his use of a pH 8·0 medium on the observations of Steck *et al.* (1970) that such pH conditions should favour the separation of plasma membrane and endoplasmic reticulum vesicles. This situation, however, only holds when the density gradients for separating the two types of vesicles are of low osmotic activity (i.e. composed of high molecular weight ficoll or dextran). Lelievre (1973) on the other hand, used sucrose gradients for which the observations are not valid. This will be dealt with more fully later on.

It is clearly very difficult to make any definite recommendations regarding the adoption of a specific homogenization schedule using either a Dounce or Potter–Elvehjem homogenizer. Only the broadest of guidelines are feasible:

(*a*) To obtain large fragments of surface membrane from tissue-culture cells, use hypotonic media (1 m$M$ $NaHCO_3$) and 6–10 strokes of a loose-fitting Dounce.

(*b*) Hypotonic media should always include either 2 m$M$ $Ca^{2+}$ or 2 m$M$ $Mg^{2+}$ or both to protect nuclei.

(*c*) Hypotonic media should be rendered isotonic with sucrose as soon as possible.

(*d*) Small fragments or vesicles of surface membrane are best obtained using iso-osmotic sucrose and 20–30 strokes of a tight-fitting Dounce.

(*e*) The optimal clearance between pestle and containing vessel varies from cell to cell.

### C. Nitrogen Cavitation

Nitrogen cavitation solely depends upon a physical process, i.e. the sudden expansion of a gas previously dissolved in the homogenizing medium and/or the cell cytoplasm under pressure. Therefore, if all the possible variables: nitrogen pressure, equilibration time, medium composition and material concentration, are maintained constant, a high degree of reproducibility in the make-up of the homogenate should exist. Furthermore, the process is an essentially instantaneous event and as such, all cells and their components should be exposed to the same disruptive forces, in contrast to the 'continuous' nature of liquid shear methods. In the process, essentially all the plasma membrane is converted into vesicles: hence little surface membrane should be lost as rapidly sedimenting sheets or large fragments. There are, however, attendant problems, most important of which are (*a*) devising conditions such that nuclei do not rupture during nitrogen cavitation and (*b*) isolating the plasma membrane vesicles which possess very similar sedimentation coefficients and isopycnic densities (in sucrose) to endoplasmic reticulum vesicles. The latter will be dealt with in the section on Fractionation.

Kamat and Wallach (1965) disrupted Ehrlich Ascites cells with a nitrogen pressure of 900 lb/in$^2$ and an equilibration time of 20 min in a medium containing 0·25 $M$ sucrose and 0·2 m$M$ $Mg^{2+}$. Both Gahmberg and Simons (1970) and Graham (1972, 1973a) used the same medium, with nitrogen pressures of 800 lb/in$^2$ and equilibration times of 20 min for baby hamster kidney cells, although Graham (1973b) reduced the equilibration time to 15 min for both baby hamster kidney and NIL cells. The shorter equilibration time, together with the presence of $Mg^{2+}$ in the medium minimized nuclear rupture. Both Hunter and Commerford (1961) and Avis (1972) had recommended the use of $Mg^{2+}$-containing media to protect nuclei during nitrogen cavitation. On the other hand, Van Blitterswijk *et al.* (1973) found that 0·25 $M$ sucrose either on its own or together with 0·2 m*M* $Mg^{2+}$ or 2 m*M* $Ca^{2+}$ did not prevent rupture of calf thymocyte nuclei during nitrogen cavitation. These workers found that optimal results were obtained with Hanks solution, a nitrogen pressure of 500 lb/in² and an equilibration time of 15 min. In a study of the disruption of bovine thyroid tissue, by various homogenization techniques, Wolff and Jones (1971) observed that nitrogen cavitation at 900 lb/in² produced the smallest surface membrane fragments of all the methods: they were the most difficult to purify and possessed irregular functional properties.

## V. FRACTIONATION

As a general principle, fractionation of an homogenate involves two steps: (*a*) crude separation of the component(s) on the basis of differential sedimentation rates and (*b*) more complete separation of the component(s) in sucrose or polymer gradients on the basis of isopycnic densities. The development of a suitable differential centrifugation scheme has to be achieved largely by trial and error and in view of the wide variation in the size and integrity of the components in homogenates produced from different tissues by different techniques, it is likely that this part of the fractionation needs to be adapted to an individual's requirements. A mathematical approach to the task is difficult, mainly due to the complexity of the system. The rate of sedimentation of a particle in a centrifugal field depends upon a number of parameters including the size, shape and density of the particle and the density and viscosity of the suspending medium.

### A. Sheets of Plasma Membrane

#### *1. Isolation of crude fraction*

The advantage of producing large fragments or sheets of surface membrane is that these structures are easily separated from the endoplasmic reticulum: they will generally sediment with either the nuclear or the mitochondrial fraction, or both. So long as these organelles (nuclei and mitochondria) remain intact, their separation from the surface membranes should be achieved relatively easily in isopycnic sucrose gradients. The aim of most methods has been to produce a crude plasma membrane fraction by a differential sedimentation-rate centrifugation scheme which is then further purified. It should be emphasized yet again, however, that within the population of each subcellular component fraction there is a spectrum of particle sizes, and hence particle sedimentation rates: consequently only a fraction of these populations will be sampled.

The largest plasma membrane fragments have a sedimentation rate only slightly less than that of the nuclei. At low centrifugation speeds therefore

(1000–1500 *g* for 10 min) the nuclear pellet will be overlayed by a loosely packed surface membrane zone ('fluffy layers'), most of the mitochondria remaining in the supernatant. The sedimentation conditions are clearly critical: smaller membrane fragments would fail to sediment. Homogenates of liver (Neville, 1960; Emmelot *et al.*, 1964) and chick embryo liver cells (Rosenberg, 1969) have been fractionated in this manner. To remove any mitochondrial contamination, these 'fluffy coats' require resuspension and recentrifugation one or more times. The use of a non-sucrose homogenization medium may be advantageous in that the mitochondria will be swollen and less rapidly sedimenting than in the intact state. Indeed, Rosenberg (1969) observed that prior to final purification, the plasma membrane fraction from a sucrose solution homogenate contained about 50% more cytochrome oxidase than that from a saline solution homogenate. Coleman *et al.* (1967b) who also obtained large sheets of liver plasma membrane which co-sedimented with the nuclei, introduced an additional homogenization of the nuclear pellet. This allowed the nuclei to be separated from the membrane fragments which were produced.

The smaller fragments of liver plasma membrane from iso-osmotic sucrose homogenates (Berman *et al.*, 1969) were sedimented at 2000 *g* for 20 min. Boone *et al.* (1969) used similar centrifugation conditions for HeLa cells; whilst Bosmann *et al.* (1968) isolated surface membranes from a 4000 *g*/10 min supernatant. In this case large fragments of surface membrane are lost to the 4000 *g* sediment, indeed, approximately half of the total ATPase was found in this fraction. Hemminki and Suovaniemi (1973) on the other hand, isolated plasma membrane at 20,000 *g* for 20 min after removing the nuclei at 3500 *g* for 5 min. The variability in the response of different tissues to homogenization conditions is highlighted here, for in spite of using a slightly hypotonic medium and very mild liquid shearing forces to disrupt rat brain cells, less than 1% of the total $Na^+/K^+$ ATPase was recovered in the nuclear pellet. Perdue *et al.* (1973), after centrifuging the homogenate (chicken tumour tissue in isotonic sucrose) at 750 *g* for 10 min, sedimented the remaining material at 200,000 *g* for 1 h. The reason for such an intense centrifugal field is not at all clear. Presumably the 7500 *g* min supernatant would contain a significant number of mitochondria in a relatively intact state (in 0·25 *M* sucrose): these organelles would probably become partially disrupted and lose their outer membrane in the 200,000 *g* step (Wattiaux and Wattiaux-De Coninck, 1970).

As a prelude to the sedimentation of large membrane sheets Forte *et al.* (1973), centrifuged the homogenate at 200 *g* for 1 min to sediment whole cells and some of the nuclei. Although it may be more effective to sediment for a longer time, e.g. 150 *g* for 10 min (Berman *et al.*, 1969), the principle of removing larger material is a sound one, in that it will minimize the entrapment of membrane fragments by aggregating rapidly sedimenting particles at a later stage. Loss of material by physical trapping is indeed a very real problem in any differential centrifugation scheme. Extensive washing of the sedimented components can overcome this: Boone *et al.* (1969) washed a plasma membrane-containing pellet three times, Berman *et al.* (1969) washed a 2000 g pellet seven times. Such techniques are however time consuming, cause elution of membrane surface molecules and since the pellet requires resuspension by rehomogenization at each washing, result in further fragmentation of the membraneous components. An alternative approach is to minimize physical interactions between sedimenting particles by using high dilutions of the material (Rosenberg, 1969). Evans (1969, 1970) overcame this problem of repeated low speed centrifugation by sedimenting a crude liver nuclear fraction in 1 m*M* $NaHCO_3$ through a discontinuous sucrose gradient (6%–54% w/v) in a type 'A' zonal rotor. Zonal rotors and techniques of zonal centrifugation will be dealt with later on: suffice here to state that after centrifugation at 3900 rpm for 45 min three bands of material containing nuclei and cell debris, mitochondria (succinate dehydrogenase) and plasma membrane (5′-nucleotidase) were recovered.

*2. Purification of crude fraction*

In the majority of methods further purification of the surface membrane fraction is achieved on one or two sucrose gradients: these are usually, but not always, isopycnic. Figure 1.1 relates the density of sucrose solutions to their concentration (% w/w) at 5°. Most gradients are in fact not preformed but are generated from originally stepped gradients (Figure 1.2) which under the forces of diffusion and gravity become continuous but not necessarily linear. Generally speaking it may be assumed that for centrifugation times less than 3 h there will be a retention of the stepped nature to some extent: hence sharp increases of density will occur across the original interfaces. If the density of the membranes falls within these interfacial ranges, they will band sharply—this is clearly advantageous. An attendant problem, however, is that material moving through the gradient tends to build up in these interfacial regions and aggregation can occur with consequent loss of resolution within the gradient. The severity of this aggregation depends upon the particular gradient system used. In a sedimentation-rate density gradient the sample, of necessity, must be applied to the top of the gradient. If, however, the gradient separation is an isopycnic one, then the sample may be applied either (*a*) in low-density sucrose at the top of the gradient or (*b*) in high-density sucrose at the bottom of the gradient. In (*a*) all the material moves down through the gradient and the chances of aggregation at the first interfacial region are higher than in (*b*) where only the less dense material moves rapidly in a centripetal direction. Hemminki and Suovaniemi (1973),

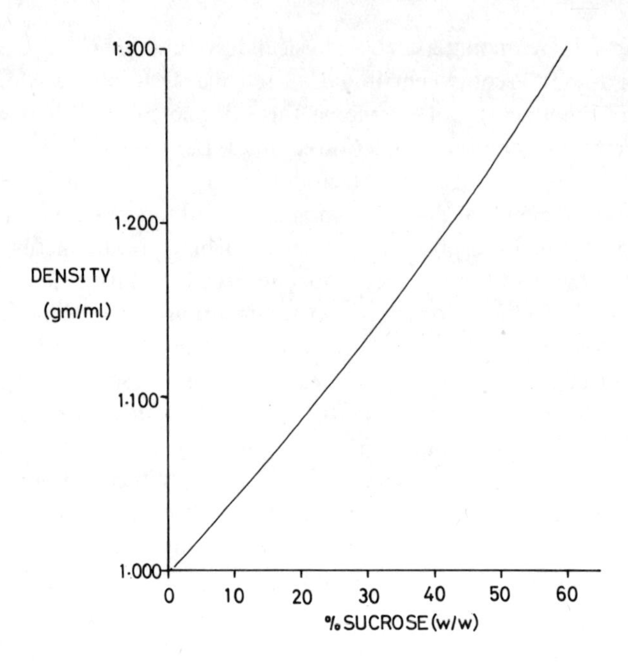


**Figure 1.1.** Relation between concentration (% w/w) and density of sucrose solutions at 5°

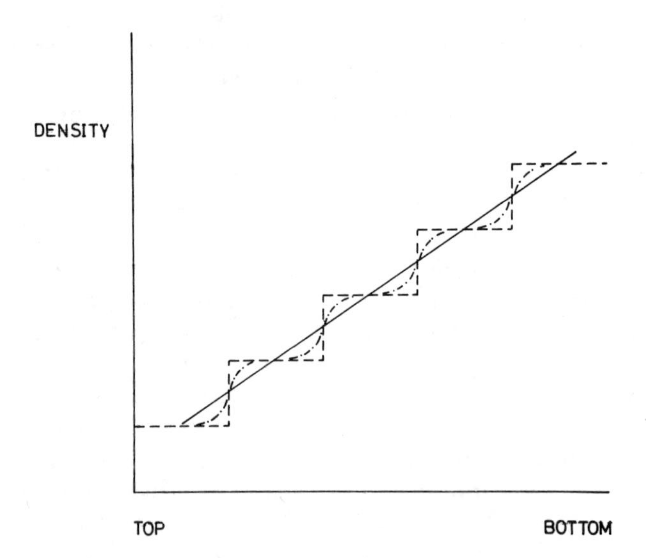


**Figure 1.2.** Generation of a linear gradient by centrifugation of a stepped gradient. – – – – – – – gradient at zero time, – · – · – · – · gradient at intermediate time, ——— gradient at end of run

put the sample in the middle of the gradient so that the plasma membrane and its denser contaminants moved in opposite directions.

The precise nature of the sucrose gradient varies from method to method. Table 1.2 gives the possible range of densities in sucrose for different mem-

**Table 1.2.** Density of membrane fractions in sucrose gradients

| Fraction | Density (g/ml) |
|---|---|
| Plasma membrane | 1·07–1·17 |
| Endoplasmic reticulum (smooth) | 1·15–1·17 |
| Endoplasmic reticulum (rough) | 1·18–1·22 |
| Mitochondria | approx. 1·22 |
| Nuclei | approx. 1·30 |

braneous components. The particular density gradient employed should therefore take account of the densities of the components to be separated.

Neville's original method (1960) included a simple two-step gradient of 37% and 40% (w/w) sucrose, the sample input being in the latter. To improve the separation from the remaining mitochondria, Emmelot *et al.* (1964) modified this by inserting an additional step of 41% (w/w) sucrose, thus producing three steps of density $\rho = 1{\cdot}16$, 1·18 and 1·22 (all densities are given in g/ml). Most sucrose gradients are a modification of this scheme: sometimes a lower density step ($\rho = 1{\cdot}13$ or 1·14) is added (Coleman *et al.*, 1967a; Rosenberg, 1969), sometimes a step at $\rho = 1{\cdot}20$ is added (e.g. Lelievre, 1973).

The separation of nuclei and mitochondria from the crude surface membrane fraction should pose little problem since they both possess considerably higher isopycnic densities than that of surface membranes: for mitochondria the value is approximately $\rho = 1{\cdot}22$ (de Duve *et al.*, 1959). However, if the mitochondria have been exposed to hypotonic conditions or excessive centrifugal fields, contamination of the plasma membrane by outer mitochondrial membranes becomes a real possibility. It is disconcerting therefore, that these membrane fractions are rarely, if ever, checked for either monoamine oxidase, kynurenine hydroxylase or rotenone-insensitive NADH-cytochrome c reductase which are generally accepted enzyme markers for the outer mitochondrial membrane (Sottocasa, 1967; Beattie, 1968; Greenawalt and Schnaitman, 1970; Racker and Proctor, 1970).

In a majority of cases the plasma membrane bands isopycnically around the $\rho = 1{\cdot}15/1{\cdot}17$ or $\rho = 1{\cdot}16/1{\cdot}18$ interfaces (Emmelot *et al.*, 1964; Berman *et al.*, 1969; Boone *et al.*, 1969; Rosenberg, 1969; Song *et al.*, 1969b; Forte *et al.*, 1973; Lelievre, 1973). However, in some cases this value was considerably lower: lymphocytes fractionated according to Demus (1973) had isopycnic densities of $\rho = 1{\cdot}05/1{\cdot}09$ and 1·09/1·12; for rat brain cell surface membrane (Hemminki and Suovaniemi, 1973) $\rho = 1{\cdot}03/1{\cdot}11$ and for chick embryo fibroblast surface membrane $\rho = 1{\cdot}06/1{\cdot}09$ and 1·11/1·12 (Perdue and Sneider, 1970; Perdue *et al.*, 1973). Any rough endoplasmic reticulum will band at a slightly higher density—usually around the 1·18/1·20 or 1·18/1·22 interfacial regions: mitochondria will either band in the same region as the rough endoplasmic reticulum or form a pellet and the nuclei will also pellet. Sometimes it is advantageous to remove the surface membrane material and recycle it through either a similar density gradient (e.g. Lelievre, 1973) or a more shallow gradient aimed at greater resolution (e.g. Perdue and Sneider, 1970; Perdue *et al.*, 1973). In some cases the increased purification resulting from recycling is worth the additional time spent in the operations, sometimes it is not. Lelievre (1973) for example achieved an increase in the specific activity of the $Na^+/K^+$-ATPase in the surface membrane (murine plasmacytoma cells) from 3·2 $\mu$moles ATP hydrolysed/h/mg protein to 14·0. At the same time there was a general decrease in the density of the fraction suggesting the removal of some adsorbed protein. On the other hand, Perdue and Sneider (1970) obtained only a marginal increase in the CTPase of their fraction.

## B. Plasma Membrane Vesicles

### 1. *Isolation of crude fraction*

Vesicles may be produced either by nitrogen cavitation (Kamat and Wallach, 1965; Gahmberg and Simons, 1970; Ferber *et al.*, 1972; Graham, 1972; Van Blitterswijk *et al.*, 1973) or by certain liquid shear methods (Graham *et al.*, 1968; Avruch and Wallach, 1971; Bingham and Burke, 1972). Separation of these vesicles from nuclei by differential centrifugation certainly possesses no problem: separation from mitochondria is, however, more difficult and should be studied carefully. Both Kamat and Wallach (1966) and Graham *et al.* (1968) used approximately 19,000 *g* for 15 min to remove the mitochondria from the post-nuclear supernatant, while Gahmberg and Simons (1970) and Van Blitterswijk *et al.* (1973) sedimented nuclei and mitochondria at 13,500 *g* for 15 min. Mitochondria normally possess a wide range of sedimenting properties. Rat liver mitochondria, for example, distribute themselves throughout a 16–48% (w/v) sucrose gradient during centrifugation (Loeb and Kimberg, 1970). If a baby hamster kidney cell post-nuclear supernatant is centrifuged at 15,000 *g* for 15 min some succinate-cytochrome c reductase activity is still detectable in the supernatant, whilst at 20,000 *g* for 15 min some of the heavier microsomes are lost to the pellet (Graham 1972). Chick embryo fibroblast mitochondria on the other hand appear to sediment far more easily: Bingham and Burke (1972), showed that at 4000 *g* for 10 min 92% of the mitochondria and 29% of the total endoplasmic reticulum pelleted, and although all of the mitochondria were sedimented at 9200 *g* for 10 min so was 58 % of the endoplasmic reticulum. To overcome this problem Avruch and Wallach (1971) separated a fat cell homogenate in a 27–54 % (w/v) sucrose gradient on a sedimentation-rate basis. 1 m*M* EDTA was included in the gradient to prevent aggregation of material. The microsomes which banded around $\rho = 1{\cdot}15$ contained 76 % of the 5′-nucleotidase and 100 % of the $Na^+/K^+$-ATPase and 3 % mitochondria. The latter banded around $\rho = 1{\cdot}18$. This method was unsuitable however for chick embryo fibroblasts (Bingham and Burke, 1972), since 48 % of the endoplasmic reticulum banded with the mitochondria. The authors do not state, however, how much of the surface membrane moved with the mitochondria. Graham (1972), however, was also unable to resolve adequately the microsomes and mitochondria from baby hamster kidney cells or rat embryo fibroblasts on linear 10–60 % (w/w) sucrose gradients, even utilizing the larger capacity (625 ml) of a zonal rotor. Only when a plateau of 30 % sucrose was inserted into the 10–60 % gradient were mitochondria and microsomes successfully resolved (Graham, 1972). The degree of resolution of these fractions on a particular gradient may depend to some extent on the material from which they are derived. Additionally, since the particles in the gradients are separated according to sedimentation rate, the distribution of microsomes within these gradients will depend on the size of the vesicles, hence the homogenization method will play a significant part in determining the efficacy of the gradient. Bingham and Burke (1972) also omitted EDTA from the gradient: this may have caused aggregation.

2. *Purification of crude fraction*

Vesicular microsomes cannot be adequately resolved into plasma membrane and endoplasmic reticulum on sucrose gradients. Resolution, however, can be achieved in magnesium-containing Ficoll (Kamat and Wallach, 1965) or dextran (Avruch and Wallach, 1971) gradients. These compounds have a high molecular weight (above 100,000); thus solutions of Ficoll or dextran possess an osmotic pressure some hundred times lower than a sucrose solution of the same concentration. The Kamat and Wallach (1965) method separates the water-filled vesicles, according to their buoyant density in low ionic strength Ficoll or dextran gradients. The latter is to be preferred because it can be obtained commercially in a more pure form and has a more well-defined molecular weight than Ficoll. The theory of the separation runs thus: inside each vesicle are a number of fixed negative charges, this will create an asymmetrical distribution of permeant ions across the membrane according to a Gibbs Donnan equilibrium. In a medium of low osmotic pressure and low ionic strength, the internal osmotic pressure created by the Donnan effect is significant and the vesicle swells. Low concentrations of $Mg^{2+}$ will titrate some of these negative charges; the Donnan effect will diminish; the vesicle will shrink, and its buoyant density will rise. The resolution of plasma membrane and endoplasmic reticulum vesicles in magnesium-containing Ficoll or dextran gradients relies on the quantitative difference in the rise in buoyant density when the fixed charges are neutralized by $Mg^{2+}$, (Steck *et al.*, 1970; Avruch and Wallach, 1971).

The water compartment within the vesicle contributes significantly towards the density of the whole particle. Thus the isopycnic density of these membrane vesicles is much lower than that of the membranes themselves. The dextran gradients normally employed have a density range $\rho = 1{\cdot}01–1{\cdot}10$ and are most effective at pH 8·6, irrespective of the material source. On the other hand the optimal $Mg^{2+}$ concentration varies. Kamat and Wallach (1966), Graham *et al.* (1968), Gahmberg and Simons (1970) and Van Blitterswijk *et al.* (1973) used 1 m*M* for Ehrlich Ascites cells, rat liver, baby hamster kidney cells and calf thymocytes respectively; Avruch and Wallach (1971) and Bingham and Burke (1972) used 0·5 m*M* for adipocytes and chick embryo fibroblasts and Graham (1972) used 2 m*M* for baby hamster kidney cells and rat embryo fibroblasts. Ferber *et al.* (1972) used 1 m*M* $Mg^{2+}$ at pH 8·2 for lymphocytes. As an alternative to a continuous dextran gradient it is also feasible to use a dextran barrier of the appropriate density which allows the endoplasmic reticulum to pellet and the plasma membrane to remain at or above the sample/dextran interface. The isopycnic density of the plasma membrane vesicles is 1·01–1·02 whilst that of the endoplasmic reticulum varies from 1·05 to 1·10.

The Kamat and Wallach (1965) method appears ideally suited to chick embryo fibroblasts, for Bingham and Burke (1972) obtained a plasma membrane fraction enriched in $Na^+/K^+$-$Mg^{2+}$-ATPase some 39× over the homogenate with no detectable endoplasmic reticulum (NADPH-cytochrome c reductase) contamination. The endoplasmic reticulum band represented a 9-fold enrichment over the homogenate with no contamination from plasma membrane. Twenty-seven per cent of the latter was recovered from the homogenate. Rat liver also fractionates rather well in this system with an approximately 19-fold increase in 5′-nucleotidase in the plasma membrane over the homogenate and little contamination from endoplasmic reticulum or intact mitochondria (Graham *et al.*, 1968).

If NADH diaphorase (NADH oxidoreductase) is accepted as an endoplasmic reticulum marker, however, it appears that the Kamat and Wallach method is not quite as effective in removing endoplasmic reticulum contamination from the plasma membrane in the case of rat adipocytes (Avruch and Wallach, 1971), pig lymphocytes (Ferber *et al.*, 1972) or thymocytes (Van Blitterswijk *et al.*, 1973). The Kamat and Wallach method requires that the vesicular microsomes be washed successively with 10 m*M* Tris then 1 m*M* Tris (both pH 8·6), (although Ferber *et al.* (1972) replaced the Tris by HEPES and lowered the pH to 7·5), to remove soluble proteins trapped within the vesicles and then dialysed against an $Mg^{2+}$/Tris medium to titrate the fixed charges. Any slowly sedimenting mitochondria remaining in the microsomes will certainly rupture. So long as the mitochondrial membranes remain in a non-vesicular state they will maintain their density, certainly greater than $\rho = 1{\cdot}10$ (the densest part of the dextran gradient) and will therefore pellet. If the fractionation is performed on a dextran gradient there should be very little contamination of either the plasma membrane or the endoplasmic reticulum bands; if on the other hand a dextran barrier is used, contamination of the endoplasmic reticulum pellet may occur. Vesicularization of the mitochondrial membranes during the washing stages would result inevitably in contamination of either the plasma membrane or the endoplasmic reticulum. Van Blitterswijk *et al.* (1973) used a Ficoll barrier and detected succinate dehydrogenase in the pellet. Bingham and Burke (1972) used a dextran gradient and detected no activity in their endoplasmic reticulum band.

If care is taken to remove essentially all the mitochondrial contamination (Graham, 1972) then it is possible to get subfractionation of both the plasma membrane and endoplasmic reticulum bands in 2 m*M* $Mg^{2+}$-dextran gradients. Under these conditions two plasma membrane bands ($\rho = 1{\cdot}04$ and $1{\cdot}024$) and two endoplasmic reticulum bands ($\rho = 1{\cdot}062$ and $1{\cdot}081$) are obtained. The values given, are for baby hamster kidney cells but normal and spontaneously transformed rat embryo fibroblasts give similar bands (one or two of which have slightly different buoyant densities).

## C. Special Approaches

Two methods which do not fit into the general categories will now be described. (*a*) Brunette and Till (1971). These workers developed a rapid system of fractionation which was based on the work of Albertsson (1960) who separated particles according to differences in surface properties in aqueous two-phase polymer systems. The crude surface membrane preparation (a 1000 *g*/15 min pellet) from a homogenate produced in the presence of 1 m*M* $Zn^{2+}$ is introduced into a two-phase system generated from 20% (w/w) dextran (M. Wt. 500,000), 30% (w/w) polyethylene glycol, and phosphate-buffered $ZnCl_2$, mixed and centrifuged. The membranes which collect at the interface can be purified by repeating the process. L cells were enriched in $Na^+/K^+$-ATPase by a factor of 15 (average of five reported experiments) although the variation in this value, 5–43, is rather alarming. (*b*) Wallach *et al.* (1972). The second method has been dubbed 'affinity density perturbation'. Use is made of the ability of the surface of a cell to bind certain ligands such as concanavalin A (Con. A). The rationale is to bind Con. A covalently to some small dense entity—these workers used a coliphage K29; the Con. A–phage complex (which can be labelled with $^{125}I$ in the ligand) is then bound to surface membrane vesicles isolated from the cell under study (in this case pig lymphocytes). This increases the density of the membrane from 1·18 to 1·30–1·32 (Wallach *et al.*, 1972) in CsCl gradients making the membrane Con. A–phage complex readily separable from other membranes within the cell which do not possess the Con. A receptor. If internal membranes do possess the receptor, then binding of the ligand to the surface membrane could be accomplished prior to homogenization. Removal of the Con. A–phage complex from the membrane is achieved by presenting the material with trehalose—a sugar which competes for the receptor.

## D. Zonal Centrifugation

The use of zonal centrifugation for the large-scale fractionation of crude cell homogenates or post-nuclear fractions has not, as yet, achieved great popularity. Although the equipment is expensive and rather more work is required in setting up the gradient system than in generating gradients in tubes, the actual operations are quite simple.

In zonal centrifugation, the gradient fills the entire body of the rotor. Figure 1.3 shows a cross-section and plan view of a typical zonal rotor. There are two fluid channels within the core of the rotor, one which exits at the surface of the core (marked A) and one which continues through the vanes within the rotor, to exit at the wall of the rotor (marked B). While the rotor is spinning (at 2000 rpm in the case of the MSE B XIV) the gradient (low-density end first) is fed to the wall of the rotor. As the gradient continues to move into the rotor the lightest part is displaced towards the core of the rotor eventually to exit via the core surface channel. The sample is then

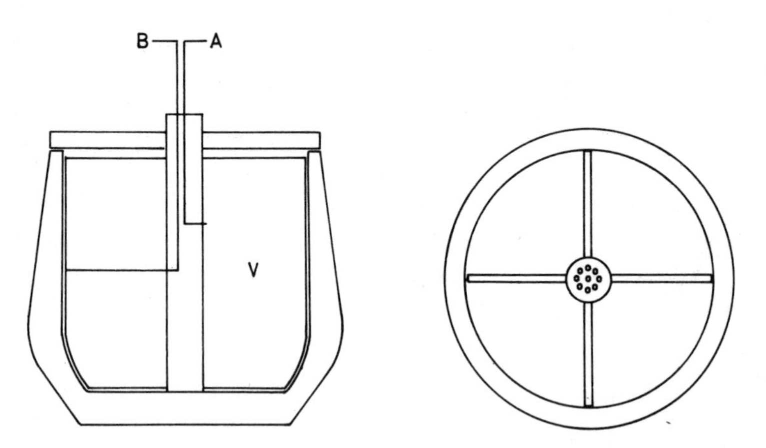


**Figure 1.3.** Cross-section (a) and plan view (b) of typical zonal rotor. V = vane

fed to the core of the rotor followed by an 'overlay' to move the sample into the centrifugal field. The rotor is accelerated to its running speed which is maintained for the appropriate time and then decelerated to 2000 rpm, when it is unloaded by passing high-density sucrose to the wall of the rotor, the gradient emerging from the centre. There are a number of zonal rotors currently available and these are listed in Table 1.3.

**Table 1.3.** Capacity and maximum speed of various zonal rotors

| Rotor | Capacity (ml) | Max. speed (rpm) |
|---|---|---|
| MSE B XIV (Aluminium) | 625 | 35,000 |
| MSE B XIV (Titanium) | 625 | 47,000 |
| MSE B XV (Aluminium) | 1650 | 25,000 |
| MSE B XV (Titanium) | 1650 | 35,000 |
| MSE HS | 700 | 10,000 |
| MSE A | 1300 | 5000 |

The advantages of the zonal system over a conventional swing-out rotor are considerable. Firstly there is the obvious increased capacity. Secondly, since the gradients can be loaded and unloaded dynamically (i.e. while the rotor is spinning), the gradient reorientation which occurs during the acceleration and deceleration of swing-out rotors, is eliminated. Thirdly, tube-wall effects are eliminated. These tube-wall effects contribute significantly to the broadening of the bands of membrane material and arise from the fact that whilst the direction of the centrifugal field everywhere is radial, the walls of the centrifuge tube are not. Figure 1.4 shows that only that part of the centrifugal field which passes through the middle of the tube is parallel to the wall of the tube. Any other line of centrifugal force drawn as a radius of the rotor, will make an angle with the tube wall. Sedimenting material follows the direction of the gravitational field, therefore a sedimenting particle within the input zone at a distance $x$ from the centre of the tube will be deflected towards the wall of the tube to an extent which is proportional to $x$ (see Figure 1.4). With a zonal rotor, on the other hand, such disturbances cannot exist since the vanes which divide up the rotor bowl into four or more chambers, are radial. An additional advantage of a zonal system is that the output from the rotor during unloading can be passed through a flow cell and its UV absorption monitored continuously, thus providing an immediate visualization of the material distribution.

Low-speed zonal centrifugation has been used to fractionate a nuclear pellet from rat liver by El-Aaser *et al.* (1966), Evans (1969, 1970) and Weaver and Boyle (1969). Both El-Aaser *et al.* and Evans employed an 'A'-type rotor, whilst Weaver and Boyle used a B XV rotor. The precise nature of the sucrose gradients and their generation varied: Weaver and Boyle and El-Aaser *et al.* used preformed gradients whilst Evans filled the rotor with steps of sucrose, and the composition of the gradient varied between 6–55% and 20–55% (w/w). After centrifugation at 3100 *g* for 60 min El-Aaser *et al.* (1966) obtained four well-defined peaks of material containing predominantly: (1) vesicles, (2) swollen mitochondria, (3) sheets of membrane and (4) nuclei (plus some membraneous material). Three peaks of 5′-nucleotidase activity coincided with (1), (3) and (4). The authors stated that peaks (3) and (4) had reached their isopycnic densities while (1) and (2) had not:

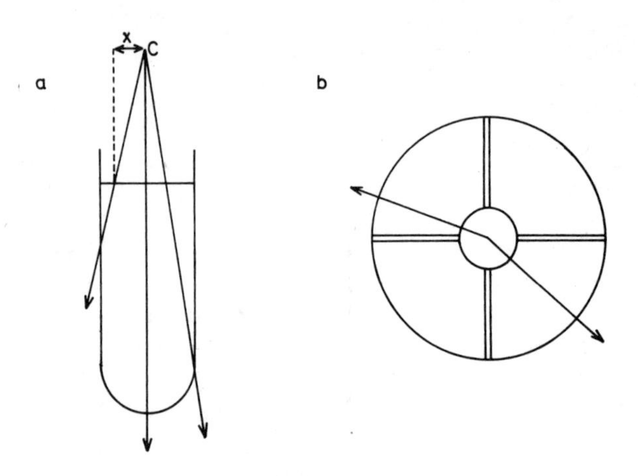


**Figure 1.4.** (a) Radial lines of centrifugal force (——→) passing through a tube from the centre (C) of a swing-out rotor. (b) Radial lines of centrifugal force (——→) in a zonal rotor

longer periods of centrifugation would therefore have caused an increasing overlap of peaks (2) and (3). Evans (1969, 1970) obtained very similar results on the distribution of 5′-nucleotidase. Weaver and Boyle (1969), on the other hand, using the same material, a rat liver nuclear pellet produced under apparently similar conditions, found that the bulk of the fast-moving plasma membrane overlapped the mitochondria, whilst the lighter surface membrane material overlapped the endoplasmic reticulum. These workers tried to capitalize on the more rapidly sedimenting property of the large plasma membrane sheets (compared to mitochondria) by reducing the speed of centrifugation and making the sucrose gradient more shallow (20–35% w/w). This enabled the membrane sheets to move away from the mitochondria and endoplasmic reticulum. Weaver and Boyle then unloaded the lighter part of the gradient (containing the contaminants) by introducing 55% sucrose at the wall of the rotor and recentrifuged the contents at 20,000 rpm for 1 h to band the plasma membrane isopycnically. This is a good example of the flexibility of zonal systems.

Prospero and Hinton (1973) used a similar system to isolate plasma membrane fragments from hepatoma. At 9000 rpm for 75 min a 10,000 *g*/10 min (mitochondrial) pellet was fractionated into a light microsomal fraction, a heavier plasma membrane/mitochondria band and an intermediate band of plasma membrane in a relatively uncontaminated state and containing 24% of the total 5′-nucleotidase. If the nuclei are not removed prior to zonal centrifugation, the intermediate fraction is not obtained. Further purification was obtained in an isopycnic sucrose zonal gradient by centrifugation in a B XIV rotor at 110,000 *g* for 2 h.

Endoplasmic reticulum and plasma membrane vesicles produced by nitrogen cavitation of tissue-culture cell suspensions are not completely separable on sucrose gradients. Graham (1973a, 1973b) therefore devised a gradient system in which a region of dextran (M. Wt. 40,000) was introduced into the sucrose gradient to enhance the slight separation which occurs in sucrose gradients. The gradient, in a B XIV rotor, was generated from, 15% and 19% sucrose, 22·5% dextran, 35% sucrose and 60% sucrose (all percentages in terms of w/w). After centrifugation at 105,000 *g* for 90 min it was observed that the bulk of the $Na^+/K^+$-ATPase activity (plasma membrane) failed to enter the dextran region, whereas most of the endoplasmic reticulum (NADH diaphorase) moved through the dextran region. Mitochondria banded around 40% sucrose. A dextran solution possesses a much higher viscosity and much lower osmotic pressure than a sucrose solution of the same concentration (% w/w); its density, however, is quite similar (Graham, 1973b). Such a solution may thus be interpolated into a sucrose gradient fairly readily, its high viscosity and molecular weight reducing diffusion. Its viscosity will retard the passage of membrane vesicles, while those vesicles which do enter the dextran region will gain water from the medium (due to its low osmotic activity), swell, and hence have their sedimenting properties modified. The efficacy of the process relies on the differential effect of these phenomena on the two types of vesicle. The actual concentration of the dextran is rather critical, reduction of the concentration to 20% (w/w) results in cross-contamination of both the plasma membrane and endoplasmic reticulum fractions from baby hamster kidney cells (Graham, 1973b). The precise nature of the gradient and the centrifugation conditions imposed may require modification from cell type to cell type. Since these vesicles contain trapped protein it is necessary to wash them in hypotonic Tris buffers (Kamat and Wallach, 1965) before they can be further purified by isopycnic banding in sucrose.

## E. Standardization of Centrifugation Conditions

### 1. *Temperature effects*

One important variable which should always be taken into account is the operating temperature of centrifugation. A whole spectrum of operating conditions exists from (*a*) rotors and tubes kept at 4°, rotor loading performed at 4° and centrifuge chamber precooled; to (*b*) rotors and tubes kept at room temperature, rotor loading at room temperature and centrifuge chamber not precooled. The former should be adhered to if possible, although not all centrifuge systems permit precooling of the centrifuge chamber. The temperature effect will be most noticeable during relatively short centrifugations which do not allow temperature equilibration between centrifuge chamber and tube contents. Since the rate of sedimentation of a particle through a medium is inversely proportional to the viscosity of the medium and since viscosity decreases with increasing temperature, then temperature will influence the rate of sedimentation. In a differential centrifugation scheme, the practical significance of this effect will depend upon the relative sedimentation velocities of those particles which sediment and those which remain in the supernatant. A limiting case of this may be encountered, i.e. a situation in which the suspension is fractionated into a supernatant, a tightly packed pellet and a loosely packed pellet. The material which constitutes the latter and overlays the tightly packed pellet must just sediment under the centrifugation conditions imposed, and is therefore influenced significantly by alterations in the sedimentation rate. This is part of the more general problem of sedimenting organelles such as mitochondria which possess a spectrum of sedimentation velocities; the lightest organelles which sediment under the particular operating conditions will be the most susceptible to small changes in sedimentation rate. Table 1.4 gives the viscosity of various

**Table 1.4.** Effect of temperature on viscosity (in centipoise) of sucrose solutions. [From the *Handbook of Biochemistry*, Anderson N. G., © The Chemical Rubber Co., 1970. Used by permission of the Chemical Rubber Co.]

| | Viscosity (centipoise) at temperature (°C) | | | | |
|---|---|---|---|---|---|
| % Sucrose (w/w) | 0 | 5 | 10 | 15 | 20 |
| 20 | 3·77 | 3·14 | 2·64 | 2·55 | 1·95 |
| 30 | 6·67 | 5·42 | 4·48 | 3·76 | 3·19 |
| 40 | 14·58 | 11·45 | 9·16 | 7·46 | 6·16 |
| 50 | 44·74 | 33·16 | 25·17 | 19·52 | 15·42 |

sucrose solutions at different temperatures. Whilst the precise temperature conditions are not usually described in published methods, the operator should at least be aware of the possibility that if a particular method is followed precisely, variation in the distribution of material in the fractionation scheme may result from centrifugation temperature irregularities.

*2. Rotor variation*

The precision with which a centrifugation step is described, may, in certain cases, determine the reproducibility of a particular fractionation method. Sometimes the only information given is the centrifugal force in the middle of the tube ($g_{av}$), sometimes the type of centrifuge rotor used and the speed in rpm are also given. In velocity sedimentation separations, and particularly in velocity sedimentation gradient separations, the operator should be wary of methods which merely give '$g_{av}$' values. Although the $g_{av}$ can be reproduced, $g_{min}$ and $g_{max}$ (centrifugal field at the top and bottom of the tube respectively) may be different if the centrifuge rotor is not identical to the one used by the author(s) of the method. Tables 1.5 and 1.6 list the $g_{min}$ values at a constant $g_{av}$ for various angle and swing-out rotors respectively.

**Table 1.5.** Comparison of $g_{min}$ values for various angle rotors when $g_{av} = 50 \times 10^3$

| Rotor type | $g_{min}$ ($\times 10^{-3}$) | Rotor type | $g_{min}$ ($\times 10^{-3}$) |
|---|---|---|---|
| MSE 10 × 10 | 33·4 | Beckman 12 × 38·5 | 32·2 |
| Beckman 12 × 13·5 | 32·2 | MSE 8 × 35 | 27·8 |
| MSE 8 × 50 | 26·8 | Beckman 8 × 38·5 | 29·0 |

**Table 1.6.** Comparison of $g_{min}$ values for various swing-out rotors when $g_{av} = 50 \times 10^3$

| Rotor type | $g_{min}$ ($\times 10^{-3}$) | Rotor type | $g_{min}$ ($\times 10^{-3}$) |
|---|---|---|---|
| MSE 6 × 14 | 21·2 | MSE 3 × 6·5 | 29·0 |
| Beckman 6 × 13·2 | 27·2 | Beckman 6 × 4·5 | 33·8 |
| Beckman 4 × 13·5 | 30·8 | Beckman 6 × 5 | 37·0 |
| MSE 3 × 25 | 30·6 | Beckman 3 × 5 | 32·2 |
| MSE 6 × 5 | 33·0 | MSE 3 × 70 | 29·0 |

In a velocity sedimentation gradient the sample, of necessity, is placed at the top of the gradient, so it will experience, initially $g_{min}$ rather than $g_{av}$. To take two extreme examples from Table 1.6 the $g_{min}$ experienced by the sample will be approximately 50% greater in the Beckman 4 × 13·5 than in the MSE 6 × 14. As soon as one starts to scale down or scale up the operation, the error may get greater: e.g. the $g_{min}$ of the Beckman 6 × 5 is 75% greater than the MSE 6 × 14. With angle rotors the differences between rotors, in the centrifugal gradient within the tube are less noticeable but nevertheless still exist (Table 1.5). As a general rule therefore, velocity sedimentation steps should be performed with rotors whose dimensions correspond as closely as possible to those used in the published method.

## VI. CHARACTERIZATION OF THE PRODUCTS

The intermediate and end products of a fractionation process are characterized by compositional and functional parameters—as described in the section on Membrane Markers. Because of the species and tissue variation in the amounts and distribution of membrane components, in the absence of some well-founded morphological criterion, this description should be as comprehensive as possible. It should include: protein, total lipid, phospholipid, cholesterol, RNA and DNA contents (both in absolute terms and as a percentage of the total homogenate), at least two accepted surface membrane enzyme markers, preferably including $Na^+/K^+$-stimulated $Mg^{2+}$-ATPase, together with glucose-6-phosphatase and/or NADH diaphorase, succinate dehydrogenase and monoamine oxidase (both in terms of specific activity and total amount). All other intermediate fractions should at least be tested for the two surface membrane enzymes and protein to account for loss of material.

Attempts to assess the purity of the membrane fractions produced frequently run into serious problems. Relative specific activity (RSA) figures (specific activity of enzyme in membrane/specific activity of enzyme in homogenate) emphasize the caution required in this assessment. A few examples with reference to 5′-nucleotidase and $Na^+/K^+$-stimulated $Mg^{2+}$-ATPase will serve to illustrate this point. A chick embryo fibroblast plasma membrane fraction prepared by Bingham and Burke (1972) exhibited an RSA of 5 for the 5′-nucleotidase and 39 for $Na^+/K^+$-ATPase; figures of 1·79 and 15·2 respectively were obtained by Gahmberg and Simons (1970) for a baby hamster kidney cell membrane fraction. On the other hand, these two parameters were of a similar value for rat liver (Berman *et al.*, 1969) and HeLa cells (Bosmann *et al.*, 1968). The RSAs for the two enzymes may not be in broad agreement for a variety of reasons: (*a*) differential loss or inactivation of one enzyme from the membrane during the procedure, (*b*) less specific distribution of one enzyme relative to the other or (*c*) selective loss of specific areas of the membrane in which the distribution of the two enzymes is non-random, e.g. the large membrane fragments which are processed may contain a different enzyme ratio than the smaller fragments which are not sampled in the fractionation process.

Frequently not enough information is given regarding the enzyme composition of the various fractions obtained during the fractionation, to be able to assess completely the distribution of subcellular components in each of these fractions. Only in a few cases is it possible to trace the source(s) of differential enzymic loss. For example, Avruch and Wallach (1971) showed that their adipocyte microsomal fraction contained only 76% of the 5′-nucleotidase but 100% of the $Na^+/K^+$-ATPase. Two enzymes which Lelievre

(1973) considered to be endoplasmic reticulum markers (for murine plasmocytoma cells)—alkaline phosphatase and glucose-6-phosphatase—were distributed in a completely opposite manner between a 27,000 *g*/20 min pellet and its supernatant. There are a number of examples of the recovery from gradients of two plasma membrane bands containing different concentrations of 5′-nucleotidase or $Na^+/K^+$-ATPase or CTPase (Graham, 1972; Lelievre, 1973; Perdue and Sneider, 1970). Whether this is due to a real subfractionation within the total population of plasma membrane fragments or whether it is the result of degradative or adsorption processes which occurred during the fractionation cannot be decided. More indicative of the existence of certain areas of specialization within the surface membrane are the results of Evans (1969) who was able to subfractionate a rat liver plasma membrane preparation, dispersed by forceful Dounce homogenization, into two subfractions on a sucrose gradient. Subfraction A contained the higher concentration of 5′-nucleotidase and leucine aminopeptidase while subfraction B contained most of the $Na^+/K^+$-$Mg^{2+}$-ATPase. Lelievre (1973) observed a similar segregation of activities in two plasma membrane fractions.

The actual concentration of these various marker enzymes within the plasma membrane varies quite astonishingly from preparation to preparation, so that it becomes virtually impossible to make valid comparisons between different preparations. The situation with $Na^+/K^+$-stimulated $Mg^{2+}$-ATPase is made more complicated by the fact that some workers quote activities for the $Mg^{2+}$-ATPase, some for the total $Na^+/K^+$-$Mg^{2+}$-ATPase and some for the $Na^+/K^+$-stimulated ATPase alone. $Na^+/K^+$-stimulated $Mg^{2+}$-ATPase in chick embryo fibroblast surface membrane varies from 4·2 μmoles ATP hydrolysed/h/mg protein to 64·1 (Bingham and Burke, 1972; Rosenberg, 1969 respectively); in HeLa cell surface membrane it varies from 1·5 to 35·8 (Boone *et al.*, 1969; Bosmann *et al.*, 1968 respectively). Figures for the 5′-nucleotidase in rat liver plasma membrane vary from 9·2 (Song *et al.*, 1969b) to 82·0 (Ray, 1970).

The contamination of surface membrane by endoplasmic reticulum is also subject to wide fluctuations, and the extent of this contamination may depend largely on the elimination of smaller endoplasmic reticulum fragments in the early stages of the fractionation. Berman *et al.* (1969) processed a 2000 *g*/10 min fraction and detected only 10% contamination of their plasma membrane fraction by NADPH-cytochrome c reductase or glucose-6-phosphatase. Perdue and Sneider (1970) on the other hand, processed a 200,000 *g*/20 min fraction and detected 25% contamination by NADH-cytochrome c reductase. On the other hand, Demus (1973) reported almost identical glucose-6-phosphatase contents for his light and heavy plasma membrane fractions.

It is unfortunately impossible to make any broad generalizations regarding the use of particular enzyme or compositional markers. The observed degree of purity of a membrane fraction is often dependent on the particular marker used to assess the purity. There are many cases of this problem: two examples will serve to illustrate the point. Song *et al.* (1969b) were unable to detect any cytochrome P450 in their rat liver plasma membrane preparation, yet the specific activity of glucose-6-phosphatase was half that in the microsomes. Both cytochrome P450 and the phosphatase enzyme are supposed to be markers for the endoplasmic reticulum. There are numerous interpretations of this result which depend upon the relative purity of the plasma membrane and microsome fractions. If they are pure then glucose-6-phosphatase is not an adequate marker for the endoplasmic reticulum in this situation; alternatively, the cytochrome P450 may have been washed off any microsomal contaminant of the plasma membrane. The second example is taken from Lelievre (1973): a plasma membrane fraction from murine plasmacytoma cells was enriched in 5′-nucleotidase and $Na^+/K^+$-ATPase but deficient in $K^+$-*p*-nitrophenyl phosphatase. The latter is present in the homogenate and has therefore either been inactivated during the preparation or it is localized at a site other than the plasma membrane.

It is, moreover, usually impossible to assess, in full, the efficacy of any fractionation method because of the lack of information. For example, it is impossible to assess adequately the endoplasmic reticulum contamination of the plasma membrane of Brunette and Till (1971) since NADH diaphorase activity is given only in terms of percentage recovery and not specific activity. Enzyme distribution should be estimated in terms of both of these parameters. Some methods even fail to report on the extent of contamination from any source—relying solely on plasma membrane markers to describe a fraction.

The extent of the variations both in the methods of fractionation and their products serves to emphasize the inadvisability of blindly applying one method worked out for one type of cell to another type of cell. Lauter *et al.* (1972) stressed the use of slightly different sucrose gradients for livers of different species. The homogenization should be monitored closely to determine the optimal conditions for each cell or tissue type. With Dounce homogenization the effects of altering all the possible variables described in the section on Homogenization, and with nitrogen cavitation—the pressure and time of equilibration, should be tested in terms of distribution of material in a differential centrifugation scheme. Phase contrast microscopy and DNA assay on a 1000 $g$/10 min pellet and supernatant will measure the removal of nuclei and their degree of breakage. 5′-nucleotidase and $Na^+/K^+$-stimulated ATPase estimations on pellets from centrifugation at 1000 $g$/10 min, 2500 $g$/10 min, 12,000 $g$/10 min and 100,000 $g$/30 min and the supernatant of the latter, will indicate the range of sizes of the membrane fragments obtained and the amount of the enzyme which is solubilized. Monoamine oxidase activities on these fractions would provide information regarding rupture of the outer mitochondrial membrane. During the purification of nuclear, mitochondrial or microsomal fractions, as many parameters as possible should be followed (at least two each for the plasma membrane and endoplasmic reticulum) in terms of both specific activity and percentage recovery. It should be constantly borne in mind that (*a*) a variable proportion of the plasma membrane will be lost; (*b*) specific components from the membrane may be lost or inactivated or both; (*c*) non-plasma membrane components may become adsorbed to the membrane and (*d*) the criteria of purity valid for one type of plasma membrane may be invalid for another.

## VII. MEMBRANE STORAGE CONDITIONS

An important sequel to the fractionation of plasma membranes is the conditions under which the membranes are stored. Certain membrane functions and physical characteristics are strongly influenced by these conditions. Although it is obviously preferable to make observations on freshly prepared membranes, this is not always feasible. Storage schedules are almost as varied as the fractionation techniques themselves. Membrane suspensions have been suspended in water and stored at 0° (Hawkins and Jacquez, 1972) or frozen (Rosenberg, 1969); suspended in buffered 0·25 *M* sucrose and stored at −20° (Demus, 1973) or −70° (Ferber *et al.*, 1972); suspended in dextran solutions and stored at −80° (Avruch and Wallach, 1971). Storage conditions should be thoroughly investigated with a view to achieving the greatest recovery of functional and structural integrity as possible. Lelievre (1973) protected the ATPase and 5′-nucleotidase enzymes of murine plasmacytoma plasma membrane by freezing them in a mixture containing buffered 50% glycerol, 1 m*M* AMP and 1 m*M* ATP. Parameters such as the concentration of membrane and the rapidity of freezing and thawing which also influence the recovery of functional and structural integrity, however, are usually overlooked. There has been no comprehensive investigation of these variables although some incomplete and unpublished observations made by Graham and Wallach on the effect of various storage conditions on the recovery of $Na^+/K^+$-ATPase activity and the maintenance of a ‘normal’ infrared spectrum in red cell membranes may be profitably included here. Generally the least favourable conditions were (*a*) suspension in water, (*b*) storage in an unfrozen state and (*c*) protein concentrations of more than 2 mg/ml, whilst the most favourable conditions were (*a*) suspension in buffered 0·25 *M* sucrose or 20% glycerol, (*b*) storage at −80° and (*c*) protein concentrations of approximately 1 mg/ml. Storage at −20° was significantly less successful in the recovery of ATPase activity. This difference between −80° and −20° storage becomes more marked if the membrane suspension is refrozen and rethawed. In fact, as a general principle, a membrane suspension should be frozen and thawed just once, and not subjected to deleterious repeated freeze–thaw cycles. Other beneficial storage practices include slow freezing to $-80°$ (achieved by placing the membrane sample within a polystyrene box) and relatively fast thawing by rapid agitation of the frozen sample in a 37° water-bath.

We also find that lyophilization, under carefully controlled conditions, is probably the best method for storing erythrocyte membranes. If (*a*) the membrane suspension (1 mg protein/ml) is shell-frozen on the internal surface of a rapidly rotating test-tube immersed in liquid nitrogen and (*b*) the lyophilized membrane is exposed to a water-saturated atmosphere prior to bulk resuspension in buffered sucrose, then these membranes may be stored almost indefinitely without significant loss of $Na^+/K^+$-stimulated $Mg^{2+}$-ATPase. It should be emphasized, however, that other membranes and other membrane functions may require different storage techniques.

## VIII. CONCLUSIONS

From the foregoing text it is clear that it is not possible to make any general recommendations for one particular membrane fractionation technique to the exclusion of others. The selection of a suitable preparative method depends upon the source of the material, the range of subcellular components required to be isolated, and the membrane parameters to be studied: it cannot be emphasized too strongly that the behaviour of a particular cell type and its components during homogenization and in a differential centrifugation scheme be exhaustively studied, as suggested at the end of Section VI. A few broad precepts can be set forth however to aid selection of an appropriate method. Although it is not feasible to relate these principles to specific tissues or cell types, one particular tissue, namely, rat liver, will be singled out because of its popularity as a source of subcellular components.

(1) The lateral surfaces of liver parenchymal cells possess a high tensile strength and are best isolated as large intact sheets by a procedure based on that of Neville (1960).

(2) The sinusoidal surface of the liver parenchymal cell fragments far more readily (Graham *et al.*, 1968) and is best isolated by a method based on that of Kamat and Wallach (1965).

(3) Tissue-culture cell surface membranes not required for functional studies are most easily isolated by one of the techniques developed by Warren *et al.* (1966). Some functional impairment may result from the action of the hardening agents upon the surface membrane.

(4) Requirements for the surface membrane alone may be satisfied by a method based on the Neville (1960) procedure, whilst requirements for both

plasma membrane and endoplasmic reticulum are probably best served by a method based on the Kamat and Wallach (1965) procedure.

(5) Where possible, a method employing an iso-osmotic isolation medium should be chosen to minimize functional inactivation or disruption of membraneous components. If hypotonic conditions are a prerequisite of the method, divalent cations should be included to prevent nuclear damage.

(6) For the future, the 'affinity density perturbation' method of Wallach *et al.* (1972) appears to offer the greatest potential for the isolation of specific membrane components.

## IX. REFERENCES

Abrahamson, D. E., J. L. Rigatuso and A. Lazarow (1969) 'Cytophotometric quantitation of glucose-6-phosphatase activity in rat liver', *J. Histochem. Cytochem.*, **17**, 107.

Albertsson, P. A. (1960) *Partition of Cell Particles and Macromolecules*, Wiley, New York.

Allan, D. and M. J. Crumpton (1970) 'Preparation and characterization of the plasma membrane of pig lymphocytes', *Biochem. J.*, **120**, 133.

Anderson, N. G. (1970) in *Handbook of Biochemistry*, 2nd ed. (Ed. H. A. Sober), The Chemical Rubber Co., Cleveland, Ohio, p. J288.

Ashworth, L. A. E. and C. Green (1966) 'Plasma membranes: phospholipid and sterol content', *Science*, **151**, 210.

Avis, P. J. G. (1972) 'Pressure homogenisation of mammalian cells', in *Subcellular Components—Preparation and Fractionation*, 2nd ed. (Ed. G. D. Birnie), Butterworths, London, p. 1.

Avruch, J. and D. F. H. Wallach (1971) 'Preparation and properties of plasma membrane and endoplasmic reticulum fragments from isolated rat fat cells', *Biochim. Biophys. Acta*, **233**, 334.

Barber, A. J. and G. A. Jamieson (1970) 'Isolation and characterization of plasma membranes from human blood platelets', *J. Biol. Chem.*, **245**, 6357.

Beattie, D. S. (1968) 'Enzyme localization in the inner and outer membranes of rat liver mitochondria', *Biochem. Biophys. Res. Comm.*, **31**, 901.

Bell, M. L., H. M. Lazarus, A. H. Herman, R. H. Egdahl and A. M. Rutenberg (1971) 'pH dependent changes in cell membrane stability', *Proc. Soc. Exp. Biol. Med.*, **136**, 298.

Benedetti, E. L. and P. Emmelot (1968) 'Structure and function of plasma membranes isolated from liver', in *The Membranes: Ultrastructure in Biological Systems*, Vol. 4 (Eds. A. J. Dalton and F. Haguenau), Academic Press Inc., New York, p. 33.

Berezney, R., L. K. Funk and F. L. Crane (1970) 'The isolation of nuclear membrane from a large scale preparation of bovine liver nuclei', *Biochim. Biophys. Acta*, **203**, 531.

Berg, H. C. (1969) 'Sulfanilic acid diazonium salt: a label for the outside of the human erythrocyte membrane', *Biochim. Biophys. Acta*, **183**, 65.

Berman, H. M., W. Gram and M. A. Spirtes (1969) 'An improved, reproducible method of preparing rat liver plasma cell membranes in buffered isotonic sucrose', *Biochim. Biophys. Acta*, **183**, 10.

Bingham, R. W. and D. C. Burke (1972) 'Isolation of plasma membrane and endoplasmic reticulum fragments from chick embryo fibroblasts', *Biochim. Biophys. Acta*, **274**, 348.

Boone, C. W., L. E. Ford, H. E. Bond, D. C. Stuart and D. Lorenz (1969) 'Isolation of plasma membrane fragments from HeLa cells', *J. Cell Biol.*, **41**, 378.

Bosmann, H. B., A. Hagopian and E. H. Eylar (1968) 'Cellular membranes; the isolation and characterization of the plasma and smooth membranes of HeLa cells', *Arch. Biochem. Biophys.*, **128**, 51.

Bretscher, M. S. (1971a) 'Human erythrocyte membranes: specific labelling of surface proteins', *J. Mol. Biol.*, **58**, 775.

Bretscher, M. S. (1971b) 'A major protein which spans the human erythrocyte membrane', *J. Mol. Biol.*, **59**, 351.

Brunette, D. M. and J. E. Till (1971) 'A rapid method for the isolation of L-cell surface membranes using an aqueous two-phase polymer system', *J. Membrane Biol.*, **5**, 215.

Colbeau, A., J. Nachbaur and P. M. Vignais (1971) 'Enzymic characterization and lipid composition of rat liver subcellular membranes', *Biochim. Biophys. Acta*, **249**, 462.

Coleman, R. and J. B. Finean (1965) 'Some properties of plasma membranes isolated from guinea-pig tissues', *Biochem. J.*, **97**, 39P.

Coleman, R. and J. B. Finean (1966) 'Preparation and properties of isolated plasma membranes from guinea-pig tissues', *Biochim. Biophys. Acta*, **125**, 197.

Coleman, R., R. H. Michell, J. B. Finean and J. N. Hawthorne (1967a) 'A purified plasma membrane fraction isolated from rat liver under isotonic conditions', *Biochim. Biophys. Acta*, **135**, 573.

Coleman, R., R. H. Michell, J. B. Finean and J. N. Hawthorne (1967b) 'A surface membrane preparation from rat liver prepared under isotonic conditions', *Biochem. J.*, **105**, 5P.

Combret, Y. and P. Laudat (1972) 'Adenyl cyclase activity in a plasma membrane fraction purified from "ghosts" of rat fat cells', *FEBS Letters*, **21**, 45.

Dallman, P. R., G. Dallner, A. Bergstrand and L. Ernster (1969) 'Heterogenous distribution of enzymes in submicrosomal membrane fragments', *J. Cell Biol.*, **41**, 357.

De Duve, C. (1971) 'Tissue fractionation. Past and present', *J. Cell Biol.*, **50**, 20D.

De Duve, C., J. Berthet and H. Beaufay (1959) 'Gradient centrifugation of cell particles: theory and applications', *Progr. Biophys. Chem.*, **9**, 325.

Demus, H. (1973) 'Subcellular fractionation of human lymphocytes. Isolation of two plasma membrane fractions and comparison of the protein components of the various lymphocyte organelles', *Biochim. Biophys. Acta*, **291**, 93.

De Pierre, J. W. and M. L. Karnovsky (1973) 'Plasma membranes of mammalian cells. A review of methods for their characterization and isolation', *J. Cell Biol.*, **56**, 275.

De Thé, G. (1968) 'Ultrastructural cytochemistry of the cellular membranes', in *The Membranes: Ultrastructure in Biological Systems*, Vol. 4 (Eds. A. J. Dalton and F. Haguenau), Academic Press Inc., New York, p. 121.

Dod, B. J. and G. M. Gray (1968) 'The lipid composition of rat liver plasma membranes', *Biochim. Biophys. Acta*, **150**, 397.

Eicholz, A. and R. K. Crane (1965) 'Studies on the organization of the brush border in intestinal epithelial cells. I. Tris disruption of isolated hamster brush borders and density gradient separation of fractions', *J. Cell Biol.*, **26**, 687.

El-Aaser, A. A., J. T. R. Fitzsimons, R. H. Hinton, E. Reid, E. Klucis and P.

Alexander (1966) 'Zonal centrifugation of crude nuclear fractions from rat liver', *Biochim. Biophys. Acta*, **127**, 553.

Emmelot, P., C. J. Bos, E. Benedetti and Ph. Rümke (1964) 'Studies of plasma membranes. I. Chemical composition and enzyme content of plasma membranes isolated from rat liver', *Biochim. Biophys. Acta*, **90**, 126.

Essner, E., A. B. Novikoff and B. Masek (1958) 'Adenosine triphosphatase and 5′-nucleotidase activities in the plasma membrane of liver cells as revealed by electron microscopy', *J. Biophys. Biochem. Cytol.*, **4**, 711.

Evans, W. H. (1969) 'Subfractionation of rat liver plasma membranes', *FEBS Letters*, **3**, 237.

Evans, W. H. (1970) 'Fractionation of liver plasma membranes prepared by zonal centrifugation', *Biochem. J.*, **116**, 833.

Ferber, E., K. Resch, D. F. H. Wallach and W. Imm (1972) 'Isolation and characterization of lymphocyte plasma membranes', *Biochim. Biophys. Acta*, **226**, 494.

Fleischer, B. and S. Fleischer (1970) 'Preparation and characterization of Golgi membranes from rat liver', *Biochim. Biophys. Acta*, **219**, 301.

Forstner, G. G., K. Tanaka and K. J. Isselbacher (1968) 'Lipid composition of the isolated rat intestinal microvillus membrane', *Biochem. J.*, **109**, 51.

Forte, L. R. (1972) 'Characterization of the adenyl cyclase of rat kidney plasma membranes', *Biochim. Biophys. Acta*, **266**, 524.

Forte, J. G., T. M. Forte and E. Heinz (1973) 'Isolation of plasma membranes from Ehrlich Ascites tumour cells. Influence of amino acids on $(Na^{+}+K^{+})$-ATPase and $K^{+}$-stimulated phosphatase', *Biochim. Biophys. Acta*, **298**, 827.

Gahmberg, C. G. and K. Simons (1970) 'Isolation of plasma membrane fragments from BHK21 cells', *Acta Pathol. Microbiol. Scand. Sect. B.*, **78**, 176.

Gillette, J. R. (1966) 'Biochemistry of drug oxidation and reduction by enzymes in the hepatic endoplasmic reticulum', *Advan. Pharmacol.*, **4**, 219.

Glick, M. C., C. Comstock and L. Warren (1970) 'Membranes of animal cells. VII. Carbohydrates of surface membranes and whole cells', *Biochim. Biophys. Acta*, **219**, 290.

Graham, J. M. (1972) 'Isolation and characterization of membranes from normal and transformed tissue culture cells', *Biochem. J.*, **130**, 1113.

Graham, J. M. (1973a) 'Isolation of surface membranes and internal membranes from baby hamster kidney cells and other tissue culture cells', in *Methodological Development in Biochemistry*, Vol. 3 (Ed. E. Reid), Longmans, London, p. 205.

Graham, J. M. (1973b) 'Fractionation of tissue culture cells into their membraneous components by rate-zonal centrifugation', *MSE Application Information*, A7/2/73, Measuring and Scientific Equipment Ltd., Manor Royal, Crawley, Sussex, England.

Graham, J. M., J. A. Higgins and C. Green (1968) 'The isolation of rat liver plasma membrane fragments', *Biochim. Biophys. Acta*, **150**, 303.

Greenawalt, J. W. and C. Schnaitman (1970) 'An appraisal of the use of monoamine oxidase as an enzyme marker for the outer membrane of rat liver mitochondria', *J. Cell Biol.*, **46**, 173.

Hawkins, C. F. and J. A. Jacquez (1972) 'Rat liver membrane preparations', *Anal. Biochem.*, **49**, 290.

Hemminki, K. and O. Suovaniemi (1973) 'Preparation of plasma membranes from isolated cells of newborn rat brain', *Biochim. Biophys. Acta*, **298**, 75.

Hubbard, A. L. and Z. A. Cohn (1972) 'The enzymatic iodination of the red cell membrane', *J. Cell Biol.*, **55**, 390.

Hübscher, G., G. R. West and D. N. Brindley (1965) 'Studies on the fractionation of mucosal homogenates from the small intestine', *Biochem. J.*, **97**, 629.
Hunter, M. J. and S. L. Commerford (1961) 'Pressure homogenisation of mammalian tissues', *Biochim. Biophys. Acta*, **47**, 580.
Hynes, R. O. (1973) 'Alteration of cell surface proteins by viral transformation and by proteolysis', *Proc. Natl. Acad. Sci. U.S.*, **70**, 3170.
Kamat, V. B. and D. F. H. Wallach (1965) 'Separation and partial purification of plasma membrane fragments from Ehrlich Ascites carcinoma microsomes', *Science*, **148**, 1343.
Kamath, S. A. and E. Rubin (1972) 'Interaction of calcium with microsomes: a modified method for the rapid isolation of rat liver microsomes', *Biochem. Biophys. Res. Comm.*, **49**, 52.
Kjaerheim, A. (1965) 'Attempts to isolate microvilli from mouse jejunal epithelium', *J. Ultrastruct. Res.*, **12**, 240.
Lauter, C. J., A. Solyom and E. G. Trams (1972) 'Comparative studies on enzyme markers of liver plasma membranes', *Biochim. Biophys. Acta*, **266**, 511.
Leelavathi, D. E., L. W. Estes, D. S. Feingold and B. Lombardi (1970) 'Isolation of a golgi-rich fraction from rat liver', *Biochim. Biophys. Acta*, **211**, 124.
Lelievre, L. (1973) 'Plasma membranes from fibroblastic cells in culture: isolation, morphological and enzymatic identification', *Biochim. Biophys. Acta*, **291**, 662.
Leskes, A., P. Siekevitz and G. E. Palade (1971) 'Differentiation of endoplasmic reticulum in hepatocytes', *J. Cell Biol.*, **49**, 264.
Lichtman, M. A. and R. I. Weed (1970) 'Electrophoretic mobility and *N*-acetyl neuraminic acid content of human normal and leukemic lymphocytes and granulocytes', *Blood*, **35**, 12.
Loeb, J. N. and D. V. Kimberg (1970) 'Sedimentation properties of rat liver mitochondria', *J. Cell Biol.*, **46**, 17.
Marchalonis, J. J., R. E. Cone and V. Santer (1971) 'Enzymic iodination. A probe for accessible surface proteins of normal and neoplastic lymphocytes', *Biochem. J.*, **124**, 921.
McKeel, D. W. and L. Jarett (1970) 'Preparation and characterization of a plasma membrane fraction from isolated fat cells', *J. Cell Biol.*, **44**, 417.
Miller, D. and R. K. Crane (1961) 'The digestive function of the epithelium of the small intestine. II. Localization of disaccharide hydrolysis in the isolated brush border portion of intestinal epithelial cells', *Biochim. Biophys. Acta*, **52**, 293.
Millington, P. F. and D. R. Critchley (1968) 'Lipid composition of the brush border of rat intestinal epithelial cells', *Life Sci.*, **7**, 839.
Morré, D. J., R. L. Hamilton, H. H. Mollenhauer, R. W. Mahley, W. P. Cunningham, R. D. Cheetham and V. S. Le Quire (1970) 'Isolation of a golgi apparatus-rich fraction from rat liver', *J. Cell Biol.*, **44**, 484.
Neville, D. M. (1960) 'The isolation of a cell membrane fraction from rat liver', *J. Biophys. Biochem. Cytol.*, **8**, 413.
Nigam, V. N., R. Morais and S. Karasaki (1971) 'A simple method for the isolation of rat liver cell plasma membranes in isotonic sucrose', *Biochim. Biophys. Acta*, **249**, 34.
Nordling, S. E. and E. Mayhew (1966) 'On the intracellular uptake of neuraminidase', *Exp. Cell Res.*, **44**, 552.
Noseworthy, J., H. Korchak and M. L. Karnovsky (1972) 'Phagocytosis and the sialic acid of the surface of polymorphonuclear leukocytes', *J. Cell Physiol.*, **79**, 91.
Omura, T. and R. Sato (1966) 'The carbon monoxide-binding pigment of liver microsomes', *J. Biol. Chem.*, **239**, 2370.

Perdue, J. F. and J. Sneider (1970) 'The isolation and characterization of the plasma membrane from chick embryo fibroblasts', *Biochim. Biophys. Acta*, **196**, 125.

Perdue, J. F., D. Warner and K. Miller (1973) 'The isolation and characterization of plasma membrane from cultured cells. V. The chemical composition of plasma membranes isolated from chicken tumors initiated with virus-transformed cells', *Biochim. Biophys. Acta*, **298**, 817.

Phillips, D. R. (1972) 'Effect of trypsin on the exposed polypeptides and glycoproteins in the human platelet membrane', *Biochemistry*, **11**, 4582.

Phillips, D. R. and M. Morrison (1971) 'Exposed protein on the intact human erythrocyte', *Biochemistry*, **10**, 1766.

Pletsch, Q. A. and J. W. Coffey (1972) 'Studies on 5′-nucleotidases of rat liver', *Biochim. Biophys. Acta*, **276**, 192.

Price, M. R., J. R. Harris and R. W. Baldwin (1972) 'A method for the isolation and purification of normal rat liver and hepatoma nuclear "ghosts" by zonal centrifugation', *J. Ultrastruct. Res.*, **40**, 178.

Prospero, T. D. and R. H. Hinton (1973) 'Isolation of plasma membrane fragments from hepatomas', in *Methodological Developments in Biochemistry*, Vol. 3 (Ed. E. Reid), Longmans, London, p. 171.

Racker, E. and H. Proctor (1970) 'Reconstitution of the outer mitochondrial membrane with monoamine oxidase', *Biochem. Biophys. Res. Comm.*, **39**, 1120.

Ray, T. K. (1970) 'A modified method for the isolation of the plasma membrane from rat liver', *Biochim. Biophys. Acta*, **196**, 1.

Rhodes, J. B., A. Eicholz and R. K. Crane (1967) 'Studies on the organisation of the brush border in intestinal epithelial cells. IV. Aminopeptidase activity in microvillus membranes of hamster intestinal brush borders', *Biochim. Biophys. Acta*, **135**, 959.

Rodbell, M. (1967) 'Metabolism of isolated fat cells. V. Preparation of "ghosts" and their properties; adenyl cyclase and other enzymes', *J. Biol. Chem.*, **242**, 5744.

Rosenberg, M. D. (1969) 'Plasma membranes of liver cells of the chick embryo. I. Isolation procedures', *Biochim. Biophys. Acta*, **173**, 11.

Skipski, V. P., M. Barclay, F. M. Archibald, O. Terebush-Kekish, E. S. Reichman and J. J. Good (1965) 'Lipid composition of rat liver cell membranes', *Life Sci.*, **4**, 1673.

Song, C. S., A. Kappas and O. Bodansky (1969a) '5′-nucleotidase of plasma membranes of the rat liver: studies on subcellular distribution', *Ann. N.Y. Acad. Sci.*, **166**, 565.

Song, C. S., W. Rubin, A. B. Rifkind and A. Kappas (1969b) 'Plasma membranes of the rat liver. Isolation and enzymatic characterization of a fraction rich in bile canaliculi', *J. Cell Biol.*, **41**, 124.

Sottocasa, G. L. (1967) 'Biochemical properties of inner and outer mitochondrial membranes of liver', *Biochem. J.*, **105**, 1P.

Steck, T. L., J. H. Straus and D. F. H. Wallach (1970) 'A model for the behaviour of vesicles in density gradients: implications for fractionation', *Biochim. Biophys. Acta*, **203**, 385.

Stein, Y., C. Widnell and O. Stein (1968) 'Acylation of lysophosphatides by plasma membrane fractions of rat liver', *J. Cell Biol.*, **39**, 185.

Sukeno, T., A. Herp and W. Pigman (1972) 'Enzymic characterization of golgi-rich fractions from rat submaxillary-sublingual glands', *Eur. J. Biochem.*, **27**, 419.

Takeuchi, M. and H. Terayama (1965) 'Preparation and chemical composition of rat liver cell membranes', *Exp. Cell Res.*, **40**, 32.

Van Blitterswijk, W. J., P. Emmelot and C. A. Feltkamp (1973) 'Studies on plasma membranes. XIX. Isolation and characterization of a plasma membrane fraction from calf thymocytes', *Biochim. Biophys. Acta*, **298**, 577.

Wallach, D. F. H. and E. H. Eylar (1961) 'Sialic acid in the cellular membranes of Ehrlich ascites-carcinoma cells', *Biochim. Biophys. Acta*, **52**, 594.

Wallach, D. F. H. and V. B. Kamat (1964) 'Plasma and cytoplasmic membrane fragments from Ehrlich ascites carcinoma', *Proc. Natl. Acad. Sci., U.S.*, **52**, 721.

Wallach, D. F. H. and V. B. Kamat (1966) 'Preparation of plasma membrane fragments from mouse ascites tumor cells', *Methods Enzymol.*, **8**, 164.

Wallach, D. F. H., B. Kranz, E. Ferber and H. Fischer (1972) 'Affinity density perturbation: a new fractionation principle and its illustration in a membrane separation', *FEBS Letters*, **21**, 29.

Warren, L., M. C. Glick and M. K. Nass (1966) 'Membranes of animal cells. I. Methods of isolation of the surface membrane', *J. Cell. Physiol.*, **68**, 269.

Warren, L. and M. C. Glick (1969) 'Isolation of surface membranes of tissue culture cells', in *Fundamental Techniques in Virology* (Eds. K. Habel and N. P. Salzman), Academic Press, New York, p. 66.

Wattiaux, R. and S. Wattiaux-De Coninck (1970) 'Distribution of mitochondrial enzymes after isopycnic centrifugation of a rat liver mitochondrial fraction in a sucrose gradient: influence of the speed of centrifugation', *Biochem. Biophys. Res. Comm.*, **40**, 1185.

Weaver, R. A. and W. Boyle (1969) 'Purification of plasma membranes of rat liver. Application of zonal centrifugation to isolation of cell membranes', *Biochim. Biophys. Acta*, **173**, 377.

Weinstein, D. B. (1968) 'The lipid composition of the surface membrane of the L cell', in *Biological Properties of the Mammalian Surface Membrane* (Ed. L. A. Manson), The Wistar Institute Press, Philadelphia, p. 17.

Weinstein, D. B., J. B. Marsh, M. C. Glick and L. Warren (1969) 'Membranes of animal cells. IV. Lipids of the L cell and its surface membrane', *J. Biol. Chem.*, **244**, 4103.

Widnell, C. C. (1972) 'Cytochemical localization of 5′-nucleotidase in subcellular fractions isolated from rat liver. I. The origin of 5′-nucleotidase activity in microsomes', *J. Cell Biol.*, **52**, 542.

Widnell, C. C. and J. C. Unkeless (1968) 'Partial purification of a lipoprotein with 5′-nucleotidase activity from membranes of rat liver cells', *Proc. Natl. Acad. Sci., U.S.*, **61**, 1050.

Wolff, J. and A. B. Jones (1971) 'The purification of bovine thyroid plasma membranes and the properties of membrane-bound adenyl cyclase', *J. Biol. Chem.*, **246**, 3939.

Zentgraf, H., B. Deumling, E-D. Jarasch and W. W. Franke (1971) 'Nuclear membranes and plasma membranes from hen erythrocytes', *J. Biol. Chem.*, **246**, 2986.

CHAPTER 2

# Techniques in the isolation and fractionation of eukaryotic chromosomes

Carl Veith Hanson
*Department of Chemistry,*
*University of California,*
*Berkeley, California 94720,*
*U.S.A.*

## I. GENERAL INTRODUCTION

Practical techniques for bulk chromosome isolation have been developed only over the past decade and until very recently have been restricted to the metaphase chromosomes of mammalian cells propagated *in vitro*. The availability of such isolated chromosomes in microgram to milligram quantities has not yet been very useful owing largely to a failure to separate the different chromosomes from each other. Indeed, the most likely route to success using the partial fractionations reviewed below will be by their application to organisms with karyotypes smaller and simpler than most of those previously studied.

The isolations discussed in detail here are confined to material from mitotic cells. These isolated chromosomes are useful for studying metaphase chromosome structure and the organization and functioning of genetic material in general. Condensed metaphase chromosomes are highly stable structures compared with interphase chromatin, and their distinct sizes and shapes enable not only their identification but also provide the basis for their fractionation.

One exception to the use of mitotic chromosomes is, however, described here: by employing isolation conditions involving high pH and other inhibitors of nucleases, Kavanoff and Zimm (1973) have detected unbroken chromosome-sized DNA from several eukaryotic sources. While not yet accomplished on a preparative scale, the existence of these procedures suggests the feasibility of using nucleic acid fractionation techniques to separate the pure DNA derived from different chromosomes.

## II. ISOLATION TECHNIQUES

### A. Cellular Sources of Chromosomes

Apart from the obvious requirement for the appropriate chemical conditions under which chromosomes will remain stable during isolation, many problems arise directly from the large-scale nature of the isolations being considered here. Since metaphase represents such a short proportion of the cell cycle there is also a need to accumulate cells in this part of the cycle. This is necessary to give a high yield, and also to minimize interphase contaminants. Techniques for obtaining metaphase cells both by selection of naturally occurring metaphases and by drug-induced mitotic arrest have been recently discussed by Mitchison (1971) and by Nias and Fox (1971). Of greatest popularity and practicality is the use of metaphase-arresting drugs such as colchicine, colcemid and vinblastine. The use of these or related agents requires in turn a source of cells which are both permeable to the chosen drug and growing fast. This has precluded the use of intact tissues and restricted most isolations to dispersed cells cultured *in vitro*. Exceptions include the use of rapidly dividing cells in mouse ascites tumour and the use of undifferentiated insect eggs treated to increase their permeability, which will be discussed in detail below.

So far, the most useful cells for chromosome isolations have been established lines of hamster, mouse and human (HeLa) cells. The more recent availability of an established cell line of *Drosophila melanogaster* (Schneider, 1972) has for the first time permitted the extension of large-scale chromosome isolation techniques to an invertebrate.

## B. Mammalian Chromosome Isolation

The steps involved in published bulk isolations of metaphase chromosomes are summarized in Table 2.1. A comparison of these techniques suggests the range of conditions which are capable of stabilizing eukaryotic chromosomes. For those who wish to reproduce these isolations, the descriptions in Table 2.1 and the following comments are, of course, not intended to substitute for the original published reports. Rather, an effort has been made to indicate unique or interesting aspects of each of the reported procedures. A discussion of purification techniques and of the properties of the isolated chromosomes is deferred to Section III below.

### *1. Early procedures*

With the exception of the pioneering attempts of Ris and Mirsky (1949) to obtain 'chromosomes' from interphase cells, the earliest reported successful bulk chromosome isolations were those of Chorazy *et al.* (1963a), Somers *et al.* (1963) and Lin and Chargaff (1964). While these early procedures provided quantities of material too small for chemical analysis, the resulting observations on the conditions required for chromosome stability have influenced all subsequent isolations. The early isolation techniques have in common the use of divalent cations and low pH. The use of non-ionic detergents, hypotonic preswelling of cells, and non-ionic additives including sucrose and formamide was also explored by these investigators. The presence of divalent cations in the range indicated in Table 2.1 was found in these studies to be an absolute necessity, while low pH was found to enhance chromosomal stability while increasing aggregation of non-chromosomal components. Hypotonic preswelling of cells and the presence of non-ionic detergent both aided the lysis of metaphase cells and the dispersal of individual chromosomes out of the native metaphase aggregate. The disaggregation as well as the stabilization of individual chromosomes was also judged by Somers and his coworkers to be

**Table 2.1.** A chronological tabulation of published bulk isolations of eukaryotic chromosomes. Values of pH and concentrations of ions and reagents refer to those present in the stabilizing media into which the chromosomes are dispersed by homogenization. In some isolations certain ingredients are added sequentially at various stages of the dispersal. Values for per cent DNA, RNA and protein were determined by the authors of the various isolations and refer to isolated chromosomes after all purification steps. Molecular weights of chromosomal DNA were determined by Wray *et al.* (1972) using Chinese hamster chromosomes isolated by reproducing several of the respective published procedures. These average molecular weight values are not absolute, but should be compared with a control value of $137 \times 10^6$ daltons determined for gently lysed mitotic cells which were unexposed to any chromosome isolation steps. Abbreviations used: 'PIPES', piperazine-*N*, *N'*-bis (2-ethane sulphonic acid) monosodium monohydrate; 'CAPS', cyclohexylamino-propane sulphonic acid; 'partial synch.' refers to partial synchronization of a culture prior to addition of the specified arresting agent

| Date | 1963a | 1963 | 1963 | 1964 | 1966 | 1966 | 1966 | 1967 |
|---|---|---|---|---|---|---|---|---|
| Authors | Chorazy *et al.* | Somers *et al.* | Somers *et al.* | Lin and Chargaff | Cantor and Hearst | Huberman and Attardi | Salzman *et al.* | Franceschini Giacomoni |
| Cell lines | Mouse tumor (*in vivo*) | Chinese hamster | Chinese hamster | HeLa | Mouse tumour (*in vivo*) | HeLa | HeLa | HeLa |
| Method of arrest | Colchicine 17 h | 0·06 μg/ml Colcemid 12–15 h | 0·06 μg/ml Colcemid 12–15 h | Partial synch. colchicine 24 h | Colchicine 18–24 h | Partial synch. colchicine 9–10 h | 1 μg/ml Vinblastine 10–14 h | Partial synch. + vinblastine |
| Hypotonic preswelling of cells | 0·003 *M* $CaCl_2$ 15′ | 0·1 *M* Sucrose 0·7 m*M* $Ca^{2+}$ 0·3 m*M* $Na^+$ 4°C, 3 min | 0·1 *M* Sucrose 0·7 m*M* $Ca^{2+}$ 0·3 m*M* $Na^+$ 4°C, 3 min | 10 m*M* Na hydrogen maleate | 0·25 × Hanks balanced salt solution | 0·1 *M* Sucrose 0·7 m*M* $Ca^{2+}$ 0·3 m*M* $Mg^{2+}$ | 0·25–1·0% Sodium citrate 30 min | 1% Na citrate 0·0005 *M* $Mg^{2+}$ 0·0005 *M* $Ca^{2+}$ |
| Method of homogenization | Shake with glass beads | 15 g needle | 15 g needle | Vortex | Virtis electric homogenizer | Glass 'Teflon' homogenizer | Shake by hand | Shake by hand 30 min |
| pH | 5·6 | Unbuffered | < 2 | 6·0 | 3·7 | 3·0 | 2·1 | 2·1 |
| $Mg^{2+}$ (*M*) | | 0·0005 | | 0·0025 | 0·001 | 0·0003 | | 0·0005 |
| $Ca^{2+}$ (*M*) | 0·003 | 0·0005 | | 0·0025 | | 0·0007 | | 0·0005 |
| Sucrose (*M*) | | 0·4 | | | | 0·1 | 0·1 | |
| Detergent | Digitonin | | | 0·17% 'Triton' X-100 | | | 0·1% 'Tween' 80 | |
| Buffer | 0·1 *M* Acetate HCl | | | 10 m*M* Maleate | Formate | | | |
| Other additives | 10% Formamide | | 30% Acetic acid | | | HCl to give pH 3·0 | 2·5% Citric acid | 2·5% Citric acid |
| DNA (%) | | | | ~3 | 14·1 | 16 | 20 | 25 |
| RNA (%) | 'High' | | | ~2 1/2 | 14·1 | 10 | 14 | 29 |
| Protein (%) | | | | ~94 | 71·7 | 74 | 66 | 46 |
| Relative DNA M. Wt. (Wray *et al.*, 1972) | | | $36 \times 10^6$ | | $16 \times 10^6$ | | | |
| Comments | | | 50% Acetic acid used in prep. for DNA M. Wt. | | DNA, RNA protein values normalized to 100% | | | |

| Date | 1967 | 1968 | 1968 | 1970 | 1970 | 1971 | 1972 | 1973 |
|---|---|---|---|---|---|---|---|---|
| Authors | Maio and Schildkraut | Mendelsohn *et al.* | Corry and Cole | Wray and Stubblefield | Burkholder and Mukherjee | Skinner and Ockey | Wray *et al.* | Hanson and Hearst |
| Cell lines | HeLa; mouse hamster | Chinese hamster | Chinese hamster | Chinese hamster | HeLa; mouse | *Microtis agrestis* | Chinese hamster | *Drosophila melanogaster* |
| Method of arrest | Vinblastine (various conc.) | 0·2 μg/ml Vinblastine 5 h | Colcemid; selective detachment | Colcemid; selective trypsinization | Colcemid; selective detachment | Colcemid; selective trypsinization | Colcemid; selective trypsinization | 2 μg/ml Vinblastine 20 h |
| Hypotonic preswelling of cells | Isolation medium saponin | 1% Na citrate 1 m$M$ $Mg^{2+}$ 1 m$M$ $Ca^{2+}$ 10 min | Isolation medium | 10 min in isolation buffer 37°C | 30 min in 0·25 × Hanks solution | Isolation medium | 10′ in isolation buffer 37°C | Homog. buffer minus NP40 |
| Method of homogenization | Dounce | Shake by hand | Dounce | 22 g needle | 18 g needle | Dounce or rapid decompression | 22 g needle | Dounce, or rapid decompression |
| pH | 7·0 | 3·0 | 9·6 | 6·5 | 2–3(?) | 3·2 | 10·5 | 10·0 |
| $Mg^{2+}$ ($M$) | 0·001 | 0·001 | 0·005 | | 0·001 | 0·001 | | |
| $Ca^{2+}$ ($M$) | 0·001 | 0·001 | | 0·0005 | 0·001 | 0·001 | 0·002 | 0·0013 |
| Sucrose ($M$) | | 0·1 | | | 0·1 | 0·1 | | 0·05 |
| Detergent | Saponin | | | | 0·01–0·5% 'Triton' X-100 | Saponin and 'Triton' X-100 | | 0·75% NP40 |
| Buffer | 0·02 $M$ Tris HCl | 0·1 $M$ Na acetate HCl | 0·005 $M$ Tris HCl | 0·0001 $M$ 'PIPES' | | 0·1 $M$ Na acetate HCl | 1 m$M$ 'CAPS' | 1 m$M$ 'CAPS' |
| Other additives | 0·001 $M$ $Zn^{2+}$ | | | 1·0 $M$ Hexylene glycol | 2% Citric acid | | 1·0 $M$ Hexylene glycol | 0·33 $M$ Hexylene glycol |
| DNA (%) | 16–17 | 9*, 28† | | 31 | | 24·6 | | |
| RNA (%) | 12–15 | 13*, 5† | | 0–5 | | 19·9 | | |
| Protein (%) | 68–72 | 78*, 67† | | 69 | | 55·5 | | |
| Relative DNA M. Wt. (Wray *et al.* 1972) | 64 × $10^6$ | | 82 × $10^6$ | 11 × $10^6$ | | | 84–104 × $10^6$ | |
| Comments | Value range for DNA, etc. is for 3 organisms | * Small chromosomes; † large chromosomes | | 20′ @ 4°C in growth medium before preswelling | Disaggregated in Virtis after needle shear | | 20′ @ 4°C in growth medium before preswelling | |

improved by the dehydrating effects of carefully controlled concentrations of sucrose.

A wide variety of methods of mechanical homogenization were explored by Chorazy *et al.* (1963a) and Somers *et al.* (1963). As confirmed in more recent work, a major challenge in the development of such isolation methods lies in finding a reproducible source of mechanical shear forces sufficient to disaggregate metaphase spreads without disrupting the structure of the individual chromosomes. Passage through a syringe needle or the use of various tissue homogenizers have proved most useful for this purpose, while such techniques as freeze–thaw or sonication have generally been found to be too harsh and unreproducible.

Refinements in more recently developed procedures allow greater yield with less chromosomal breakage, together with the possible advantages of the use of near-physiological pH. The observations of Somers *et al.* (1963) and of Chorazy and his colleagues (1963) are nevertheless of great value to anyone designing a new chromosome isolation system.

### 2. *Isolation from mouse ascites tumour at pH 3·7*

The isolation technique of Cantor and Hearst (1966) was among the first procedures to produce the milligram quantities of chromosomes required for accurate analysis of chemical composition and for physical studies of chromosome structure (Cantor and Hearst, 1969). This procedure shares with several others the limitations imposed by the possible extraction or fixation of chromosomal proteins at low pH. Of great interest in the development of this isolation is the discovery of a narrow pH range which permits chromosome dispersal under these solvent conditions. An increase or decrease of one half pH unit from the optimal pH 3·7 resulted in an irreversible aggregation of chromosomes in a gel-like matrix.

The merits of the Cantor and Hearst procedure include the simplicity of the isolation buffer which, for example, avoids the use of any detergent, and the high-speed blending with glass beads for 12 min in a Virtis homogenizer which provides optimal selective shearing forces in this system. While this method of homogenization is unusual among the published procedures (and itself precludes the use of some detergents due to foaming problems), its adaptability to large volumes could be important in many applications. Although not yet widely copied, the use in this study of non-centrifugal sedimentation for purification of chromosomes is also notable. Cantor and Hearst found that Brownian motion was sufficient to keep most free chromosomes suspended in an undisturbed vertical tube while most contaminating interphase nuclei settled to the bottom within 24 h.

### 3. *Isolation at very acidic pH from established cell lines*

The procedures for isolation of HeLa chromosomes at pH 3·0 (Huberman and Attardi, 1966) and at pH 2·0 (Salzman *et al.*, 1966) were developed simultaneously but independently and have in common with the previous method the use of simple isolation media involving very low pH, and the production of many milligrams of fairly pure intact chromosomes. As discussed in some detail in Sections IV and V below, these isolations are of particular importance since in both cases the authors have subsequently examined the chromosomes and partially fractionated them by velocity sedimentation.

As in the work of Cantor and Hearst, Huberman and Attardi found a narrow pH optimum in the acidic range. The different optima found by these two groups may have been influenced by the use of different organisms. The variety of conditions under which HeLa chromosomes have been isolated by other workers, however, suggests that synergistic effects between different levels of $Ca^{2+}$, $Mg^{2+}$, sucrose and pH are much more important.

The complete absence of divalent cations is unique to the procedure of Salzman *et al.* (1966), and its success may probably be attributed to a degree of protein fixation resulting from the extremely low pH employed. Their procedure, in common with approximately half of the other published isolations, involves the use of a non-ionic detergent—in this case a post-homogenization treatment with 0·1% 'Tween' 80 'to disrupt chromosome clumps and free chromosomes bound to membranes'. Perhaps the most important factor in this preparation is the successful use of porous stainless-steel filters for the efficient removal of contaminating interphase nuclei. This latter technique, which greatly facilitates high yields during purification, may nevertheless be limited in its applicability to other systems, as mentioned in Section III.A below.

This same group of investigators later modified their technique for application to Chinese hamster cells (Mendelsohn *et al.*, 1968) and in this latter form the technique was also applied to the unusual and very important system of banked human white blood cells (Schneider and Salzman, 1970). The most significant modification is the increase of pH to 3·0 in an attempt to preserve chromosomal histones. Although the authors do not elaborate on their addition of divalent cations, it is evident that the presence of such ions helps compensate for the relatively lower degree of stabilization at the slightly higher pH. A 3-fold reduction of swelling time in hypotonic medium was also deemed by Mendelsohn and his coworkers to be an important modification. It should be noted that carefully timed preswelling is in fact a feature of many of the reported procedures. The importance of avoiding prolonged cell swelling is not surprising considering that most hypotonic media used for the purpose are not in themselves capable of stabilizing chromosome structure.

Finally, it is useful to mention here the technique of Franceschini and Giacomoni (1967), and the more recent procedures of Burkholder and Mukherjee (1970a), Skinner and Ockey (1971), and Burki *et al.* (1973) to illustrate the continuing importance of chromosome isolation at acidic pH. The isolations performed by the first two groups may be summarized as modifications of that of Salzman *et al.* (1966), and the procedure followed by Burki *et al.* is closely related to that of Cantor and Hearst (1966). Although probably subject to acid extraction of some chromosomal protein, the preparation by Burkholder and Mukherjee proved suitable for studies of uptake of whole chromosomes by cultured cells (Burkholder and Mukherjee, 1970b) and a source of pulse-labelled chromosomal fibres for replication studies (Burkholder and Mukherjee, 1971).

The work of Skinner and Ockey (1971) illustrates a number of points: 1. The selection of an organism with a karyotype which facilitates chromosome fractionation. 2. The failure of the earlier isolation techniques when applied to a relatively unusual cell line. 3. The introduction of pressure homogenization. An established line of fibroblast cells from the field-mouse *Microtus agrestis* was selected by these investigators because this organism possesses sex chromosomes more than twice the size of the largest autosomes. By means of differential sedimentation, Skinner and Ockey were thus able to obtain a chromosome fraction, 75–85% of which was composed of these two large sex chromosomes.

The article by Skinner and Ockey is of special interest to those designing procedures for isolations from new systems. Both the earlier acidic procedures and the neutral pH technique of Wray and Stubblefield (1970) were found inadequate due to either incomplete removal of cytoplasmic debris or failure to disaggregate chromosomes. The authors speculate that an unusual level of collagen in their cell line may have precluded success with some of these techniques. The procedure evolved by Skinner and Ockey to overcome these complications may be described as one of the most complex of those reviewed here, and features the use of a very closely defined acidic pH value as in some of the early techniques.

Pressure homogenization (Avis, 1972) is particularly valuable for its applicability to large-scale preparations. Inert gas such as nitrogen is dissolved under a pressure of up to 1000 lb/in$^2$ in a suspension of swollen cells contained in a steel pressure-vessel or bomb. On slow release of the pressurized suspension through a valve, the explosive expansion of microscopic gas bubbles throughout each cell disrupts cellular structures to a degree proportional to the original pressure in the bomb. Advantages include the single exposure of each particle to the disrupting force, adiabatic cooling during decompression, and the ease of operating under sterile conditions when desired.

### 4. *Isolation at neutral or alkaline pH from established cell lines*

The isolation procedure of Maio and Schildkraut (1967) may be regarded as a major advance due to its successful use of near-physiological pH and the

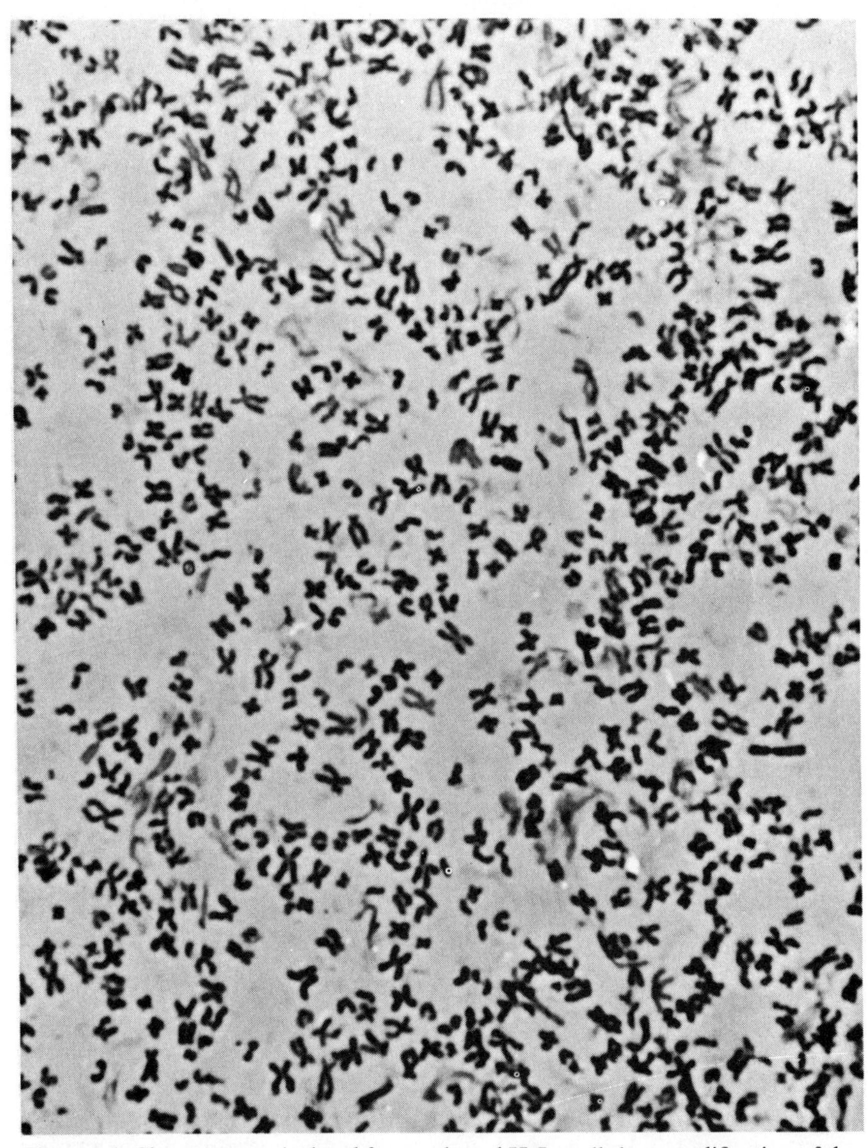

**Figure 2.1.** Chromosomes isolated from cultured HeLa cells by a modification of the procedure of Wray and Stubblefield (1970). HeLa cells were arrested by exposure for 18 h to 0·01 $\mu$g/ml of vinblastine sulphate, swollen in 0·2 × Earles balanced salt solution for 15 min, and disrupted in Wray and Stubblefield's pH 6·5 buffer by a single decompression after equilibration for 20 min with 150 lb/in$^2$ of argon (Avis, 1972). All isolation steps were at room temperature. Chromosomes were pelletted on to a glass slide (Lief *et al.*, 1971), stained with Giemsa and photographed with the aid of phase contrast optics. Isolation and staining were performed in the author's laboratory by R. Rubin

demonstrated applicability to a wide variety of mammalian cell lines. The unusual inclusion of $Zn^{2+}$ in this isolation medium results partly from the authors' observation that $Zn^{2+}$ and $Cd^{2+}$ are effective stabilizers of chromosome morphology (at neutral pH) at concentrations an order of magnitude lower than is required for stabilization by other common divalent metal ions. The Maio and Schildkraut procedure is a true bulk technique and, with only minor modifications in non-ionic detergent concentrations, has been used to produce milligrams of isolated chromosomes for a study of fractionation by velocity sedimentation (Maio and Schildkraut, 1969).

The more recent isolation near neutral pH by Wray and Stubblefield (1970) also possesses many advantages over the earlier techniques: The isolation medium is not only simple but is also gentle in the sense of avoiding both detergent and extremes of pH. Slightly modified, the technique can also be used to isolate the organelles most closely related to the metaphase chromosome, *viz.* the interphase nucleus and the intact mitotic spindle apparatus. In addition to its published application to Chinese hamster cells, the technique is well suited (with minor modification of the homogenizing method) to small- or large-scale preparation of HeLa chromosomes (see Figure 2.1). Central to the Wray and Stubblefield procedure is the stabilization of protein conformation by hexylene glycol, an approach introduced earlier by those making bulk isolations of the intact mitotic spindle apparatus (Kane, 1965; Sisken *et al.*, 1967). While the 12% organic solute content of this buffer is remote from physiological conditions, the resulting isolates satisfy several criteria for biological integrity. In particular, chromosome appearance in the electron microscope along with chemical composition and enzyme activity of both isolated chromosomes and isolated nuclei compare favourably with observations of material isolated by many other means. Wray and Stubblefield also point out the importance of using a non-metal-binding buffer such as 'PIPES.'

The range of conditions for successful isolation of chromosomes has recently been greatly extended by these same investigators (Wray *et al.*, 1972) in developing an alkaline isolation medium specifically for the purpose of maintaining high DNA molecular weight in isolated chromosomes. In this latter technique the presence of hexylene glycol and 0·002 *M* $Ca^{2+}$ appears to compensate for the destabilizing effects of high pH usually observed. The preservation of DNA molecular weight is presumed to be due to the inhibition of nucleases by alkaline pH. This aspect of the pH 10·5 isolation was partly inspired by the earlier work of Corry and Cole (1968) whose simple isolation procedure at pH 9·6 also produced high molecular weight DNA (see Table 2.1). Advantages of the techniques of Wray *et al.* over this earlier procedure include ease of handling and better morphology, while the slightly lower pH in the method of Corry and Cole may have advantages for the preservation of certain protein components.

In summary, procedures for isolating chromosomes in bulk from mammalian cell lines cover a wide range of pH and ionic conditions. The choice of a technique will depend upon the specific application. The discussion in Section III of the properties of chromosomes isolated by these various means should provide a guide to choosing the appropriate isolation method.

### C. *Drosophila* Chromosome Isolation

The availability of a variety of isolation procedures does not presently extend to invertebrate cells. Several laboratories have, however, been interested for many years in the bulk isolation of chromosomes from *Drosophila melanogaster* because of the genetic importance of this organism and because of the ease with which its chromosomes should be able to be fractionated as a result of its extremely small karyotype of diversely shaped chromosomes (Hearst and Botchan, 1970; Nishiura, 1972; Hanson and Hearst, 1973, and manuscript in preparation).

The requirement for drug-permeable, rapidly dividing cells suggests the use of undifferentiated embryos as a source of insect chromosomes. In the case of *Drosophila* the first dozen divisions of the preblastular nuclei occur synchronously every 10 min and are unaccompanied by cell wall formation. Depending on temperature, 3–8 % of such dividing nuclei in an asynchronous population of embryos are in metaphase at any given time, and the duration of metaphase is remarkably brief—from 0·3–0·7 min (Sonnenblick, 1965). Efficient chromosome isolation thus requires the introduction of a metaphase-arresting agent. The first step must involve removal of the chorionic layer from the eggs by immersion for 30 sec to 2 min in approximately 3 % sodium hypochlorite solution. The permeability of the remaining waxy coating and vitelline membranes must then be further enhanced. Nishiura (1972) found this could best be achieved by mechanical agitation of the eggs with ground glass. Alternatively, the chemical nature of the egg surface suggests brief extraction with an organic solvent such as toluene. We have also had some success treating with 0·1–0·3 % trypsin, as originally suggested by James Peacocke (personal communication).

After any of the above treatments the eggs may be incubated (with agitation, to prevent anoxia) in an aqueous solution of an arresting agent such as colchicine. We find that 90 % deuterated water ($D_2O$) produces efficient metaphase arrest, perhaps due to its high diffusion rate. $D_2O$ is thought to cause arrest by preventing anaphase spindle movement (Gross and Spindel, 1960) so that the chromosomes derived from such preparations might be of particular value for structural studies due to an absence of the usual colchicine-induced overcondensation.

A suitable medium for chromosome isolation from an established *Drosophila* cell line will be described next, but it should be emphasized that neither this nor any other medium has so far resulted in a high yield of chromosomes from *Drosophila* eggs. The obstacles are 2–fold: the difficulty of dispersing contaminating yolk granules, and perhaps more important, a general failure to disaggregate chromosomes from the metaphase plate or 'spread'. The above details for manipulation of the permeable egg system are nevertheless included here in the hope that they will be useful for further work with *Drosophila* and other invertebrates.

As mentioned in the introduction, an efficient technique for bulk isolation of chromosomes from an *in vitro* cell line of *Drosophila melanogaster* has recently become available (Hanson and Hearst, 1973, and manuscripts in preparation). Two cell lines originally established by Schneider (1972), have been grown in large-scale suspension culture where as much as 75% metaphase arrest is achieved by incubation for one generation period with relatively high doses of vinblastine sulphate (2–4 μg/ml). The composition of the preswelling and isolating media are summarized in Table 2.1. Preswelling occurs at room temperature during the 15 or so minutes required for two rinsings of the cells. A 3:1 dilution of the culture with isolation medium prior to the first rinse has given consistently better results than a sudden resuspension into buffer of cells pelleted directly out of growth medium. The temperature of the preparation is lowered to 0° and the detergent added just prior to homogenization. The conditions for disruption are critical, and may be achieved either by pressure homogenization or by 20–30 strokes in a siliconized glass Dounce homogenizer fitted with a size 'B' (tight) pestle. Approximately 250 lb/in$^2$ of argon has been found optimal in the pressure bomb, and, unlike the Dounce homogenizer, this technique is capable of producing acceptable homogenates in the absence of the detergent. At the time of this writing an analysis of the chemical properties of the isolated chromosomes has not been completed. The morphology of the isolated chromosomes as observed by light microscopy does however closely resemble that of chromosomes seen by staining the intact cell.

### D. Development of Isolation Procedures for New Sources

Chromosomes isolated by the methods summarized above vary in their chemical and physical properties (see the next section). Where the chromosomes of an organism can be isolated by more than one method, these properties provide criteria for choosing between isolation techniques. In developing a completely new isolation procedure, however, the properties of the isolated chromosomes may not in general be predicted in advance. The initial challenge is restricted to a search for conditions which will permit dispersal of the chromosomes in a stabilized state. The remainder of this section will consider how this may be approached using information from the above techniques as a guide, when none of the published procedures is found to give satisfactory results.

### *1. Chromosome sources*

When one wishes to isolate chromosomes from a particular species, the choice of tissue or type of cells can be critical. Ideally an established cell line should be available. Although a rapid division time for any such cultured cells will facilitate drug-induced metaphase arrest, the ability to synchronize the cells even in a slowly dividing cell line can have advantages. For with only partial synchrony the period of exposure to a metaphase-arresting agent may be reduced, thus minimizing such problems as pre-metaphase cell death or heterogeneous, drug-induced chromosome contraction. The ability to synchronize the cell cycles of newly fertilized sea urchin eggs has, for example, been exploited to isolate in bulk the intact mitotic spindle apparatus without the use of any arresting agent (Mazia and Dan, 1952; Mazia *et al.*, 1961; Kane, 1965). A potential disadvantage of cultured cells is however the fact that so many are aneuploid.

Alternatives to the use of established cell lines may be found among the procedures summarized above. Although the early chromosome isolations from mouse ascites tumour have not found recent popularity, this system remains potentially important. The application, for example, of any of the recent 'near-physiological' isolation techniques to these tumour cells has not been reported but this could possibly combine the advantages of the high yield obtained by Cantor and Hearst (1966) with the advantages of isolation under more gentle conditions.

The isolation of chromosomes from the peripheral lymphocytes of human blood by Schneider and Salzman (1970) is another notable departure from the use of established cell lines, and is unique in providing a source of chromosomes from normal euploid mammalian cells. Features of the technique include the introduction of phytohemagglutinin to stimulate cell division, and the use of vinblastine to achieve a final mitotic index of up to 20%. Possible extension of this procedure to other organisms would seem potentially exciting.

The use of newly fertilized eggs from *Drosophila*, or possibly from other organisms with acellular nuclear divisions in the early embryo, has been mentioned above. The major challenge here lies in the necessity of collecting large, approximately synchronous populations of eggs. In the case of *Drosophila*, large adult populations maintained with special concern for environmental stimuli are required to prevent the gross contamination of brief egg collections by older cellularized embryos.

Finally, techniques for bulk isolation of metaphase chromosomes have yet to be extended to simple eukaryotes such as yeast, or to plant cells. The restricted permeability of these types of cells is, of course, a potential difficulty in such efforts.

### 2. *Metaphase accumulation*

To obtain the highest yield of chromosomes the mitotic index (the percentage of metaphase cells) must be increased as much as possible. The appropriate types and optimal concentrations of arresting drugs for this purpose vary widely in different systems and must be determined empirically. In our laboratory vinblastine sulphate has consistently given higher levels of arrest in cultured *Drosophila* cells than has colchicine. On the other hand, since colchicine has only half the molecular weight and is chemically very different from vinblastine sulphate, situations might arise where permeability factors, for example, might favour the use of colchicine.

The mitotic spindle-arresting properties of $D_2O$ exploited with the *Drosophila* cells described in Section II.C may also be useful for cells which might be impermeable to larger molecules. The lack of specificity in the mechanism of this agent, however, means that the cell cycle may be disrupted elsewhere than at metaphase.

The advantage of selecting cells which are in natural metaphase has already been mentioned. There are two approaches to metaphase selection. The first of these is the synchronization of a population of cells with subsequent harvest at the time of mitosis, while the second involves the mechanical selection of metaphase cells out of a randomly dividing cell population. Both approaches have been discussed recently by Mitchison (1971).

### 3. *Isolation media and methods of homogenization*

It is important to discuss the composition of any chromosome isolation medium in relation to the choice of disruption procedures. Homogenization forces which are required successfully to disaggregate chromosomes in one system may in fact destroy chromosomes in another stabilizing medium. The aim in designing an isolation medium is to achieve a *selective destabilization* of all non-chromosomal components, such that some reproducible level of mechanical shearing forces will disrupt or solubilize only these components.

As we have seen above, divalent cations and other common additives are important factors in this respect. While many stabilizers of chromosome structure are known, most of them also stabilize the association of chromosomes within the metaphase chromosome spread. Under acidic isolation conditions, this aggregation has a highly specific pH dependence, while the integrity of individual chromosomes seems to remain unaffected over a wider pH range. The chemical reasons for this are not known and the isolation of metaphase chromosomes remains for the time being an empirical art.

Finally, in a discussion of homogenization conditions it is important to mention the requirements imposed by the presence of contaminating interphase nuclei. Unless conditions can be found under which cell breakage is confined to the more fragile metaphase cells (Corry and Cole, 1968), interphase nuclei will be present in any chromosome preparation. Such contamination would be a serious interference, for example, in isotopic labelling studies where much newly synthesized labelled nuclear RNA might be present. One goal during homogenization, therefore, is to keep such nuclei intact to facilitate their subsequent separation from chromosomes. This can be helped by the presence of appropriate divalent cations. Sucrose, which does not generally affect the structure of individual chromosomes (although affecting chromosome disaggregation), may also help to prevent rupture of nuclei by hypotonic shock. Finally, it should be pointed out that in preparations involving non-ionic detergents the resultant removal of the outer nuclear membrane renders contaminating nuclei more susceptible to hypotonic shock.

Hypotonic swelling not only serves to pull apart physically the metaphase spread within, but also gives time for the slow permeation of selective destabilizing elements into the nuclear region.

There are other, and as yet largely untried, approaches to bulk chromosome isolation. The first of these is an adaptation of the method of Kirsch *et al.* (1970) for the bulk isolation in non-aqueous media of nuclei from lyophilized cells. As applied to nuclei, the technique features ultrarapid freezing of whole cells followed by freeze drying and finally homogenization in water-free glycerol. When the nuclei are subsequently returned to aqueous media, the morphology and the labile and water-soluble constituents are found to be preserved remarkably well. We have made a preliminary attempt to isolate metaphase chromosomes from arrested cultured *Drosophila* cells frozen in this manner and though there was considerable chromosome aggregation we did see some stabilized free chromosomes of normal appearance (C. Hanson and T. Gurney, unpublished observations).

A second approach might use procedures employed for the isolation of the mitotic spindle apparatus. Such isolations in bulk have been achieved both in invertebrates (e.g. Mazia and Dan, 1952; Mazia *et al.*, 1961; Kane, 1965) and in vertebrates (e.g. Sisken *et al.*, 1967). The isolation of spindles on a microscopic scale has also been reported very recently from *Drosophila* eggs (Milsted and Cohen, 1973). Features of spindle stabilization techniques range from the early use of cold ethanol (Mazia and Dan, 1952) to that of hexylene glycol (Kane, Sisken) or disulphide stabilizers such as dithiodiglycol (Mazia *et al.*, 1961). The application of such techniques to chromosome isolation would require starting material which has not been metaphase arrested by spindle-destroying agents. The strategy implied here is the purification of the intact spindle apparatus away from cytoplasmic contaminants in the hope of subsequently dispersing intact chromosomes out of the purified apparatus.

Success with such an approach has not yet been reported, and it should be pointed out that in the development of the technique for cultured *Drosophila* cells promptness in shearing chromosome spreads after cell lysis was essential to successful dispersal. Yet another complication could arise from the difficulty of separating the spindles from similarly sized contaminating interphase nuclei.

## III. PROPERTIES OF ISOLATED CHROMOSOMES

Several recent works have reviewed the properties of metaphase chromosomes as revealed by microscopy and by analysis of bulk-isolated samples (Hearst and Botchan, 1970; Du Praw, 1970; Ris and Kubai, 1970; Huberman, 1973; Mendelsohn, 1973). The important question of the relationship of chromosomal properties *in vivo* to those in the isolated state has been approached by these reviews. This section will instead emphasize the differences in the properties of isolated chromosomes which result from different methods of preparation. Purification techniques which can be used with the various preparations are also summarized. These techniques are here considered as 'properties' of the isolated chromosomes.

### A. Purification Techniques

Immediately after homogenization it is usually desirable to separate the dispersed chromosomes from as much of the other material in the homogenate as is possible. Contaminants which are bigger than chromosomes include interphase nuclei, partially or wholly unbroken metaphase spreads and occasionally whole cells or cell fragments. Subchromosomal contaminants include chromosome fragments and soluble and insoluble cytoplasmic components. Perhaps the most difficult to separate are the occasional 'micronuclei'—abnormal structures often found in cells cultured for prolonged periods in the presence of metaphase-arresting agents. Most methods of purification involve the same separation principles as are applied to chromosome fractionation. Homogenates intended for chromosome fractionation rather than for chemical analysis of the chromosome mixture, will require less initial purification because of the purification which will occur during the subsequent fractionation.

The most effective technique for purification of chromosomes is differential sedimentation velocity. A wide range of centrifugal forces have been employed for separating chromosomes and contaminating nuclei. A force of 100 *g* for 5 min will sediment most nuclei and other large contaminants leaving most isolated mammalian chromosomes in the supernatant fraction. The conditions used to pellet chromosomes vary from 720 *g* for 15 min (Cantor and Hearst, 1966) to 2500 *g* for 30 min (Salzman *et al.*, 1966) and depend upon: 1. the average size of the smallest chromosomes in the organism employed, 2. the extent of swelling or condensation of the chromosomes by the isolation medium, 3. the viscosity of the medium and 4. the firmness of the pellet or the degree of separation from subchromosomal contaminants which is desired.

It is important to note that the small difference in sedimentation rates between chromosomes and some contaminants means that fractions must often be recycled as many as five times to obtain an acceptable purity. When greater efficiency or speed of processing is dictated by the need to preserve enzyme activities or by the unwieldiness of a very large-scale preparation, the use of zone sedimentation may greatly facilitate the purification. A bottom layer containing a few per cent sucrose to permit layer formation is used, its volume often approximating to that of the applied sample layer. The pellet is now uncontaminated by the more slowly sedimenting particles.

The use of a sucrose multistep gradient results in the separation of nuclei, chromosomes and cytoplasmic components through a combination of sedimentation velocity and buoyancy effects (e.g. Somers *et al.*, 1963; Skinner and Ockey, 1971). Very dense sedimentation media containing up to 2·2 *M* sucrose have been used. Chromosomes suspended in such media or applied as a zone have been successfully pelleted while most contaminants float to the top of the tube. A typical sedimentation force for such centrifugation is 50,000 *g* for 60 min (Maio and Schildkraut, 1967).

There are two general technical limitations to the sedimentation of chromosomes. The first is the aggregation of pelleted chromosomes, which can often be coped with by the rehomogenization of chromosome pellets. The severity of this phenomenon depends on the sedimentation force and must be carefully considered in any preparations where rehomogenization is undesirable. ‘Streaming’ of sedimenting particles presents a second limitation to chromosome sedimentation. This phenomenon is observed in the sedimentation of any large particles and involves the formation in the centrifuge tube of ‘streamer’ regions in which a local volume containing many particles sediments in bulk at an anomalously high rate. A discussion by Miller (1973) of the streaming encountered in the sedimentation of intact cells is equally relevant to streaming of chromosomes. Since there generally exists a particle concentration below which streaming does not occur (the streaming limit), the greatest restriction imposed by this phenomenon is that upon the processing of large quantities of material. The streaming limit observed by Huberman and Attardi (1967) for sedimentation of a zone of HeLa chromosomes through a glycerol–sucrose gradient is $10^9$/ml.

Selective filtration is, in a sense, the simplest technique for chromosome purification. The method is restricted, however, by the availability of filter materials which do not irreversibly adsorb chromosomes. Fibrous filters such as 'Millipore', for example, have not been found successful in this respect. Perhaps most successful is the use by Salzman *et al.* (1966) of stainless-steel filters for removal of interphase nuclei and other large contaminants from a homogenate of HeLa chromosomes. Filters of 5 μm mean pore size (porosity H, type FCH) from Pall Trinity Micro Corp., Cortland, N.Y., were used, and the homogenate of a 3 litre suspension culture could be accommodated on a single 9 cm diameter filter. Although the purity of the filtrate is impressive, it should be noted that the maximum recovery of chromosomes was 32%. Unfortunately, Cantor and Hearst (1966) found the use of such filters to be unsatisfactory with their homogenate of mouse ascites tumour chromosomes, partly due to even more severe losses of chromosomes on the filter. The success of the former group may be partly due to the use of chromosomes whose size or surface properties have been modified by isolation at pH 2·1.

Finally, Burkholder and Mukherjee (1970a) successfully removed clumped chromosomes and nuclei from a homogenate by suction filtration through 'Hysil' brand fine sintered-glass filters, porosity 20–30 μ.

Hanson and Hearst (1973) have recently shown the suitability of General Electric 'Nuclepore' polycarbonate filters for the processing of isolated *Drosophila* chromosomes. These filters, which are available in a variety of porosities, feature straight cylindrical pores which minimize surface interaction with the filtrate and greatly enhance the flow rate. These investigators have found the 3 μm filter removes contaminating interphase nuclei from homogenates containing *Drosophila* chromosomes. Difficulty in resuspending chromosomes trapped on 'Nuclepore' filters of smaller pore size has, however, precluded the use of this technique for the removal of subchromosomal contaminants. The further potential of 'Nuclepore' filters for chromosome fractionation will be outlined in Section IV.D. below. Finally, irreversible adsorption of chromosomes by many types of filters might perhaps be overcome by pretreatment of filters with silicone (see below) or related substances or by presaturation of chromosome-binding sites on the fibre with a protein or some other appropriate molecule.

Enzymic methods have been used for the 'purification' of isolated chromosomes. Chorazy *et al.* (1963a) found that treatment with a combination of ribonuclease and pepsin removed all the material which stains like RNA, although it causes severe morphological changes in many chromosomes. More recently, Mendelsohn *et al.* (1968) used a treatment of 150 μg RNase/ml for 1 h at 37°C to purify chromosomes which were subsequently used in successful chromosome fractionation experiments. In the absence of pepsin, this latter group found insignificant changes in chromosome morphology.

Finally, several general precautions deserve mention. Free chromosomes adsorb to glass, limiting the vessels and instruments which may be used in the isolation and the processing of chromosomes. In general, this may be prevented by pretreatment of glass surfaces with a hydrophobic coating such as

'Siliclad' (Clay Adams Co., Parsippany, N.J.). Temperature and speed of handling may also affect the properties of isolated chromosomes. While chromosome morphology in most of the media discussed above generally remains stable at room temperature, rapid purification at low temperature may be important in studies of chromosomal enzymatic activities. The non-centrifugal sedimentation procedure of Cantor and Hearst (see Section II.B.2), for example, is very valuable for efficient processing of large quantities, but the days required for recycling through several 24 h sedimentation periods severely limit its use in enzymic studies.

## B. Yield, Morphology and Stability

The quantities of purified chromosomes which may be prepared are limited by the overall yield and also by the numbers of cells which may be conveniently grown and arrested at one time. Authors of many of the isolation procedures have failed to state their yield of purified chromosomes. Among the early exceptions, however, are Somers and his coworkers (1963) who obtained a chromosome fraction containing 167 $\mu$g of DNA from $10^8$ cells with 50% mitotic index. Impressively large quantities of chromosomes were prepared by Cantor and Hearst (1966) in experiments using 5 to 7 mice with yields of 300–400 $\mu$g dry weight of purified chromosomes per mouse. The weight of chromosomes prepared by Huberman and Attardi (1966) is not stated, but one-third of the chromosomes originally in the arrested cells were reported to be retained throughout isolation and purification. Finally, Maio and Schildkraut (1967) were able to grow $1–2 \times 10^9$ HeLa or Syrian hamster cells, from which 60–75 $\mu$g dry weight of purified chromosomes were obtained. Approximately one-third of that yield was observed with Chinese hamster or L cells owing to lower mitotic indices.

In those instances where long term stability of chromosomes has been examined, storage at 0–4°C for several months has produced only minor effects on chromosome morphology as seen under the light microscope (Cantor and Hearst, 1966; Huberman and Attardi, 1966; Salzman *et al.*, 1966). This, of course, does not preclude a reduction with time of such properties as enzymic activities or molecular weight of chromosomal DNA. Huberman and Attardi (1966) further report the successful storage of isolated chromosomes at −70°C in a solution of millimolar HCl and divalent metal ions, and Skinner and Ockey (1971) have stored chromosomes for six months at −20°C in medium supplemented with 20% glycerol. The reader should note, on the other hand, the differences in electrofocusing behaviour of fresh and stored chromosomes as observed by Landel *et al.* (1972).

Studies of the morphology of isolated chromosomes in the light and electron microscope may be found in references cited at the beginning of Section III. Differences in chromosome ultrastructure resulting from various preparative procedures have not been very striking. Hearst and Botchan (1970) have observed in deliberately swollen isolated mouse ascites chromosomes the same size (230 Å) fibre as found in metaphase chromosomes *in situ* and in interphase nuclei. The reversibility of structural transitions occurring during isolation, however, remains an important question. The possibility exists of 'fixation' of chromosomal proteins by denaturation, extraction or cross-linking due to use of extreme pH conditions. It is fortunate that in spite of such possibilities, chromosomes isolated by a variety of methods are reported to respond to *in vitro* chemical manipulations of chromosomal condensation or compactness. In the only extensive physical study to date (Cantor and Hearst, 1969), a pH-dependent structural change which was reversible in the presence of $Mg^{2+}$ was reported. Earlier, the effect of enzymic treatment was explored by Chorazy *et al.* (1963a), who found an absence of morphological changes in the light microscope after treatment by pepsin or RNase alone but complete destruction of chromosome structure by DNase, trypsin or chymotrypsin. Reversible uncoiling of isolated chromosomes in 0·1 to 0·5 *M* NaCl has also been observed (Somers *et al.*, 1963). Finally, although not confirmed in other isolations, Salzman and his coworkers (1966), found that the overcondensation of chromosomes induced *in vivo* by vinblastine metaphase arrest was reversed upon exposure of the chromosomes to their isolation conditions.

## C. Chemical Properties

The chemical composition of chromosomes isolated prior to 1970 has been well reviewed by Hearst and Botchan (1970). The reported relative DNA, RNA and protein contents of chromosomes are listed in Table 2.1. Low molecular weight components or macromolecules other than protein or nucleic acid are generally believed to be absent or insignificant in metaphase chromosome structure. In one metal analysis of isolated chromosomes (Cantor and Hearst, 1966), only Cu and Mg ($\sim$0·005% of dry weight) were found. The possible presence of small amounts of lipid is suggested by observations of association between chromosomes and the nuclear membrane (Hearst and Botchan, 1970).

While the amount of protein in isolated metaphase chromosomes is generally higher than in isolated interphase chromatin (Sadgopal and Bonner, 1970), this comparison is complicated by the variable RNA content of the former. This variation is consistent with the identification of metaphase chromosomal RNA as a ribosomal RNA contaminant. The ribosomal nature of this material is inferred from the finding of equimolar amounts of 18 s and 28 s RNAs with the expected r-RNA base compositions (Salzman *et al.*, 1966; Huberman and Attardi, 1966; Maio and Schildkraut, 1967). More difficult questions include whether or not the RNA is present in the form of intact ribosomes and whether it is a natural constituent of chromosomes or a cytoplasmic contaminant acquired during homogenization. These questions have been studied by Salzman and his coworkers (1966) who found high levels of radioactivity in chromosomes isolated from a homogenate to which radioactive ribosomes had been added. Hearst and Botchan (1970) point out, however, that the results of this experiment may not be interpreted unambiguously. Perhaps the best evidence that RNA may not be an integral component of chromosome structure derives from the success of Wray and Stubblefield (1970) in isolating RNA-free chromosomes in their near-neutral medium. For many applications this observation alone could dictate the superiority of this unusual, hexylene glycol-containing medium. The chemical composition of chromosomes obtained in the other media involving hexylene glycol has not yet been reported.

When the chemical contents of the various preparations are recomputed by subtracting out the RNA component, most ratios for protein to DNA are in the region of 4:1. The exceptions are the preparations of Franceschini and Giacomoni (1967) and Skinner and Ockey (1971), which contain significantly higher amounts of RNA than any others, and that of Wray and Stubblefield (1970) which contain little or no RNA. The recomputed protein to DNA ratios for these latter three isolations are close to 2:1. More detailed analyses of metaphase chromosomal proteins, with regard to per cent acid solubility and histone v. non-histone content, have been reviewed by Hearst and Botchan (1970) for some of the earlier preparations. In brief summary, the fraction of protein soluble in 0·2 *N* HCl ranges from 26 % to 76 % depending on organism and isolation method and hence is much more variable than total protein content. Sadgopal and Bonner (1970) furthermore found that the content of acid-soluble non-histone protein was much greater than in interphase chromatin. Non-histones comprised 68 % of the protein soluble in 0·2 *N* HCl in chromosomes isolated at pH 3·0 by the method of Huberman and Attardi (1966) and at pH 7·0 by the method of Maio and Schildkraut (1967). Skinner and Ockey (1971), presenting the most complete data of this sort, confirmed in greater detail the similarity of histone patterns in *Microtus agrestis* chromosomes isolated at pH 3·2 and at pH 6·5 (Wray and Stubblefield, 1970). In both cases a significantly lower content of lysine-rich (f-1) histone was found than in interphase chromatin.

The molecular weight of DNA derived from isolated chromosomes was shown by Wray *et al.* (1972) to depend markedly on the isolation procedures. Their molecular weight data are included in Table 2.1, while in their article are further data obtained from non-bulk chromosome preparation procedures not listed in Table 2.1. Wray and his coworkers concluded that degradation of chromosomal DNA is enzymic and takes place during hypotonic swelling of cells whenever the pH is not alkaline. They attribute the excellent preservation of chromosomal DNA in their own buffer to its pH of 10·5.

Although their technique involves gently lysing cells directly on alkaline sucrose gradients, the possibility of mechanical or enzymic DNA breakage during this procedure still remains. Thus their data are consistent with the presence in alkali-isolated chromosomes of DNA with molecular weight even higher than that observed. Resolution of this interesting question would probably require application of the technique of Kavenoff and Zimm (1973) for measuring unbroken DNA, as mentioned in the Introduction to this chapter.

## IV. FRACTIONATION TECHNIQUES

### A. Introduction

Separation methods suitable for isolated chromosomes range from modifications of traditional techniques for fractionating soluble macromolecules to procedures developed more recently for handling whole cells or organelles such as nuclei. Partial fractionation on the basis of mass and overall dimensions has been accomplished by differential velocity sedimentation in preformed gradients and on a crude scale by selective filtration. Chromosomes have also been fractionated on the basis of their electric-charge properties by means of electrofocusing. Attempts to separate chromosomes according to density by isopycnic sedimentation in a gradient have just been reported. Finally, several techniques such as liquid-phase partition or countercurrent distribution remain to be exploited.

As mentioned in the General Introduction a barrier to complete chromosome fractionation is the presence in most karyotypes of different chromosomes of very similar size and shape. A potentially greater complication results from variations in the morphological state of a particular chromosome from different cells within a cell population. Such cell-to-cell variation can result in overlapping size ranges for populations of two chromosomes which are of distinctly different size within the karyotype of any single cell. This phenomenon is aggravated by commonly used metaphase-arresting agents. Vinblastine sulphate, for example, has been shown to condense chromosomes to a greater degree of compactness than in normal metaphase. Chromosomes in cells entering metaphase earliest within an arresting population, and hence arrested for the longest time, are thus the most condensed (Sasaki, 1961). Where partial synchrony of a cell culture is possible, the length of exposure to an arresting agent may be minimized by introducing the agent just prior to most of the mitoses of the synchronized cell population. The applications of partial synchrony to previous isolations are noted in Table 2.1 and have been most extensively discussed by Franceschini and Giacomoni (1967), who tabulated the dispersion in measured chromosome lengths in synchronized and non-synchronized cultures. A second common limitation on chromosome fractionation results from chromosome aggregation. This results in clumps of small chromosomes contaminating fractions containing large chromosomes.

The use of certain fractionation methods is precluded by the destructive effect of the supporting media involved. In particular, the use of dense media of high ionic strength for isopycnic sedimentation limits the application of this technique. As reported below a possible solution to this and related problems is the use of chromosomes which have been previously 'fixed' in a cytological sense. Such an approach might be appropriate, for example, in preparations to be used for studies depending only upon nucleotide sequences.

### B. Differential Sedimentation Velocity

The technique of differential sedimentation velocity is currently the most widely used method for chromosome fractionation. Most applications of the technique involve brief sedimentation through preformed sucrose gradients, with recycling of fractions. It should be noted that the streaming limit discussed in Section III.A applies equally to sedimentation-velocity fractionations.

Among the first reported sedimentation-velocity separations of chromosomes was that of Huberman and Attardi (1967) as shown in Figure 2.2. These investigators used HeLa chromosomes isolated in acidic medium (see Table 2.1) from unsynchronized cells which had been arrested for 15 h by vinblastine sulphate. The chromosomes were purified by pelleting through 2·2 *M* sucrose, 'Dounced' to disperse aggregates, and layered on sucrose gradients formed in 0·02 *M* Tris, pH 7, 2 m*M* $CaCl_2$ and 0·05% saponin. The authors report less chromosome aggregation during fractionation in this neutral buffer than in the acidic isolation medium. The 140 ml linear gradients were formed in 250 ml glass centrifuge bottles with 0–30% (w/w) sucrose and with a 30% (w/w) to 0% reverse gradient of glycerol. The purpose of the reverse glycerol gradient is to reduce the steepness of the strong viscosity gradient which accompanies the sucrose distribution. In the presence of the traditional uncompensated viscosity gradient the sedimentation of large particles farther down the tube is impeded more than that of small particles higher in the tube and the separation between large and small particles is thus reduced. While a slight viscosity gradient remained in Huberman and Attardi's sedimentation columns, this modification was reported to enhance greatly the difficult separation of HeLa chromosomes. Bearing in mind that the combination of the two gradient components must still maintain a good density gradient, it would be interesting in the future to determine the effect of even more extreme viscosity compensation.

These investigators report an additional artifact in velocity sedimentation

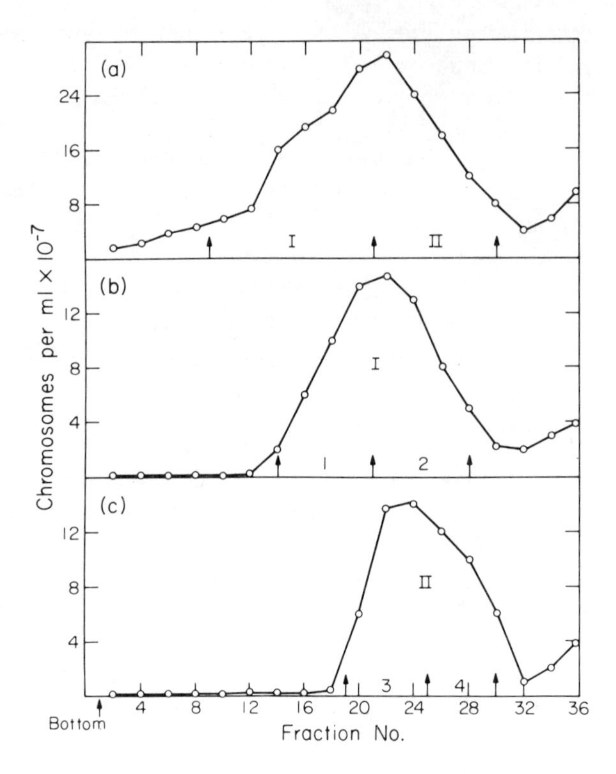


**Figure 2.2.** Distribution of isolated HeLa chromosomes after centrifugation through glycerol–sucrose gradients by the method of Huberman and Attardi (1967) as described in the text. (a) Chromosomes from approximately $1{\cdot}6 \times 10^9$ cells were sedimented through two gradients of 140 ml each, and fractions were assayed by counting in a bacterial counting chamber. Cut-off points for pooling the chromosome fractions into two classes are indicated by arrows. (b) the faster sedimenting chromosomes (I) from the two initial gradients were pooled and recentrifuged as in part a. (c) The slower sedimenting chromosomes (II) from the initial gradients were also pooled and recentrifuged as in part a. (Reproduced from Aloni *et al.* (1971) by permission of Academic Press Inc. (London) Limited)

experiments which results in the selective loss of large chromosomes: the sedimentation of particles through a cylindrical chamber in a radial force field results in the loss of some particles by impact on the wall of the vessel. The probability of this phenomenon is greatest for the largest chromosomes since they move farthest down the bottle.

In the experiments of Huberman and Attardi the gradients were centrifuged at 450 *g* for 40 min at 4°C with careful acceleration and deceleration. The result was a continuous distribution of chromosomes throughout the gradient as shown in Figure 2.2. The gradient was then divided into four arbitrary fractions. Photographs of chromosomes from these fractions were analysed to determine enrichment factors for the different chromosome types in the four gradient fractions. These factors differed up to an order of magnitude between remote gradient fractions for some of the chromosomes. This analysis was greatly complicated by clumping, morphological distortions and by overlap of chromosome sizes due to variable vinblastine-induced condensation. The results nevertheless show an unambiguous partial, or average, size fractionation of the chromosomes. The degree of fractionation was sufficient to allow the authors to demonstrate the localization of ribosomal RNA genes on the smaller chromosomes.

Franceschini and Giacomoni (1967) fractionated HeLa chromosomes on smaller scale gradients of 30–60% sucrose, centrifuging at 2500 *g* for 30 min at 5–10°C. This study convincingly demonstrated the importance of using chromosomes from synchronized, briefly arrested cells.

In another early study Mendelsohn *et al.* (1968) separated the chromosomes of Chinese hamster cells into three overlapping fractions by very low-speed sedimentation in a 10–40% linear sucrose gradient formed in their isolation medium (see Table 2.1). Chromosome suspensions were layered on to 24 ml gradients formed in siliconized nitrocellulose tubes and centrifuged at 50 *g* for 40 min at 25°C. Fractions were collected by tube puncture. Gradients of 180 ml in glass centrifuge bottles were also successfully employed for larger-scale preparations. Chinese hamster cells were chosen because they possess a greater size range of chromosomes and approximately one-third as many different chromosomes as in the karyotype of the more popular HeLa cells. The three fractions from the initial sedimentation were run in parallel on a second set of gradients and the yield in the three fractions pooled from the final gradients was 15% of the metaphase chromosomes initially present in the cell lysate.

Slowly sedimenting non-chromosomal material is found which cofractionates with the smallest chromosomes and accounts for their apparently higher protein and RNA content compared with large chromosomes. Mendelsohn and his colleagues used autoradiography of $^{14}$C-labelled preparations to demonstrate the absence of DNA and hence the non-chromosomal nature of this material. These observations obviously bear upon the question of identification of the intrinsic chemical constituents of metaphase chromosomes.

Maio and Schildkraut (1969) used an unusually brief, high-speed centrifugation. Neutral gradients of 0·1–1·8 *M* sucrose were first formed in 1″ × 3″ tubes and equilibrated overnight at 4°C. Chromosomes suspended by

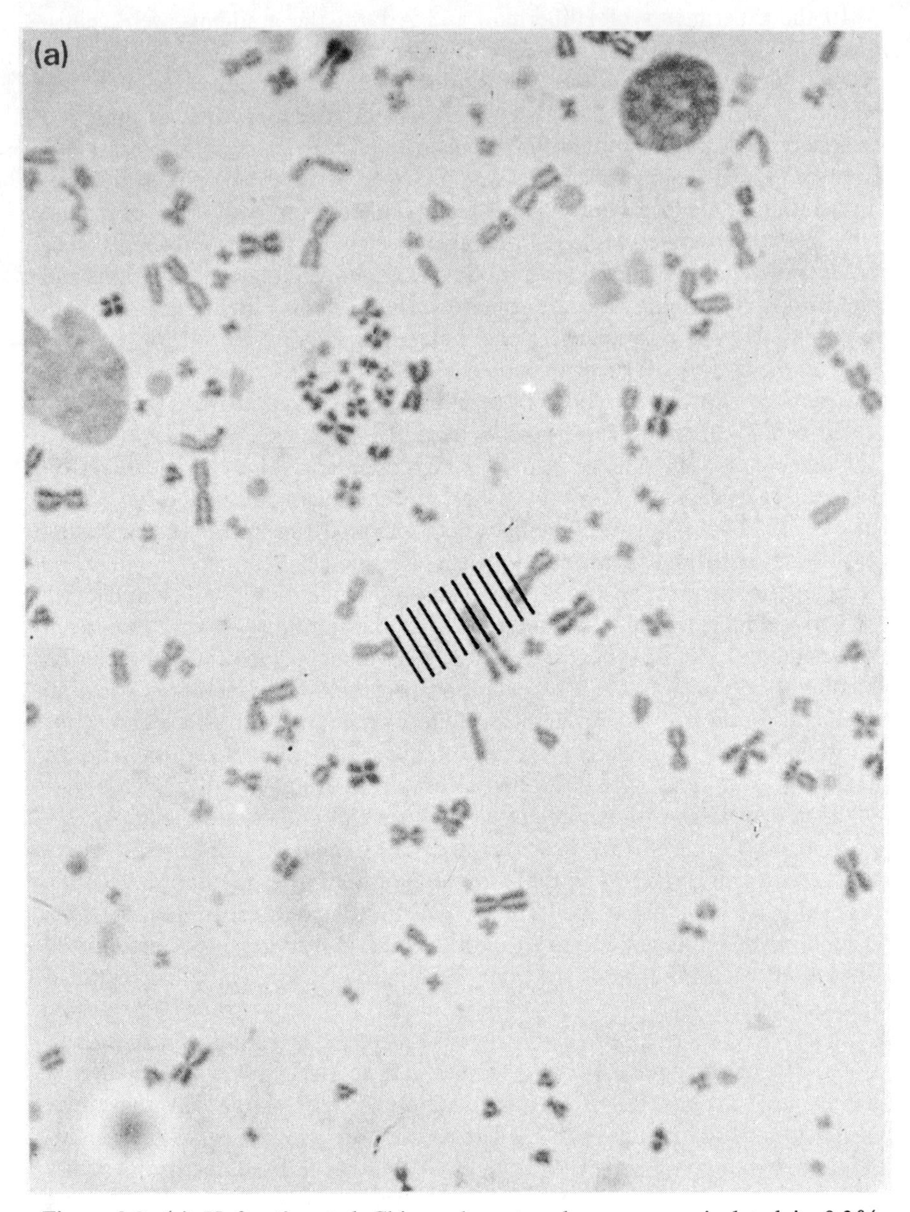


**Figure 2.3.** (a) Unfractionated Chinese hamster chromosomes isolated in 0·3% 'Triton' X-100, 1 m*M* $Mg^{2+}$, 1 m*M* $Ca^{2+}$, 1 m*M* KOH, 0·02 *N* formic acid, ph 3·7, according to the method of Burki *et al.* (1973). (b) $10^8$ to $10^9$ chromosomes at a time were fractionated by sedimentation for 30 min at 1000 rpm on a 5–30% sucrose gradient in an A12 zonal centrifuge rotor (International Equipment Co.). Selected fractions were then recycled by sedimentation through a second set of similar

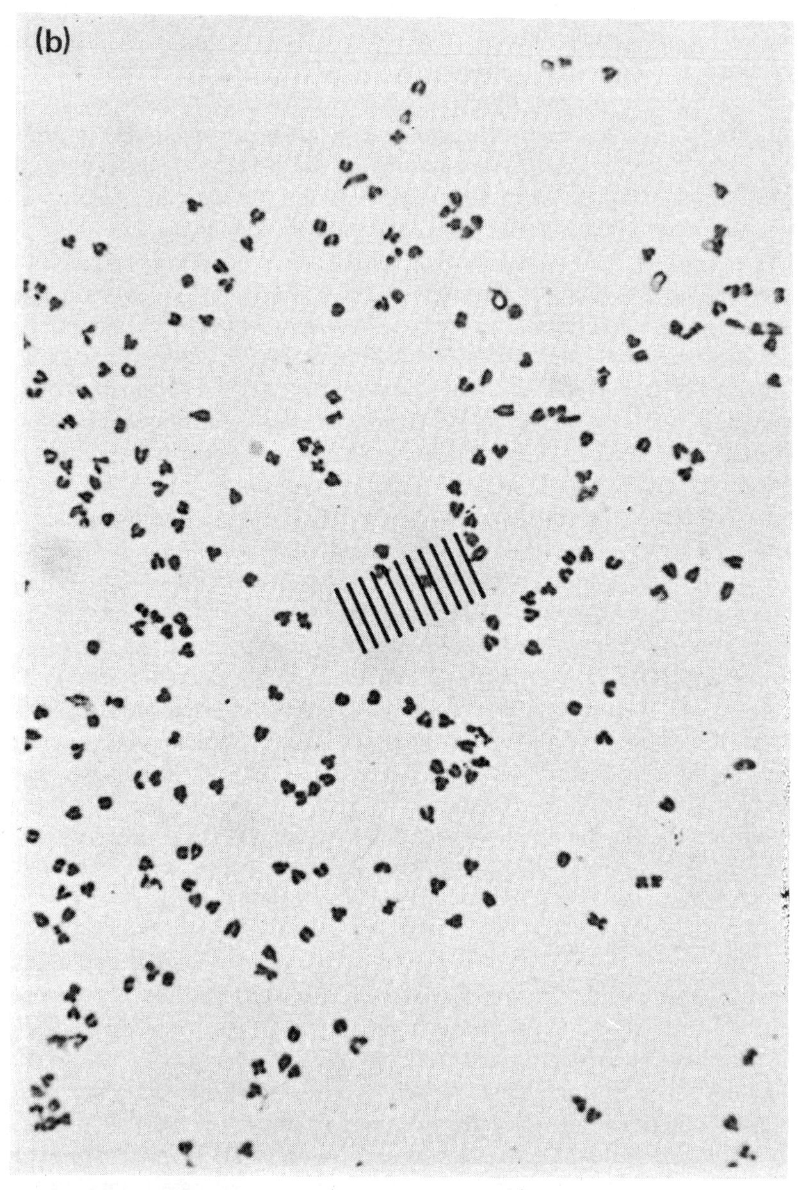


gradients (Regimbal, 1973). In part b is shown a fraction from the second cycle of sedimentation in which chromosome no. 9 comprises 95% of the material present. (Unpublished experiments by T. Regimbal, who kindly supplied the photographs and accompanying data.) Reticle lines in the centre of each field are spaced at 2 micron intervals

'Douncing' in gradient material removed from the uppermost portion of the tubes were layered back on to the remainder of the gradients, which were then centrifuged at 5000 rpm in a Beckman SW 25·1 rotor for 2 min at 4°C. The centrifuge brake is used for deceleration. The rapid acceleration provided by an ultracentrifuge was deemed by the authors to be an advantage in rapidly moving chromosomes away from the top of the tube where their high concentration might otherwise result in aggregation. The standard procedure of Maio and Schildkraut includes a second cycle of sedimentations.

The resolution and capacity of the zonal rotor have been exploited by Schneider and Salzman (1970) and by Burki *et al.*, (1973). Schneider and Salzman fractionated chromosomes isolated from human peripheral lymphocytes on an 1100 ml gradient of 15–40% sucrose underlayed with a 55% sucrose cushion. The gradient was formed in a type A-A12 zonal rotor and sedimentation was performed at 2000 rpm for 20 min. The first six of eighteen fractions collected contained fractionated chromosomes with subsequent fractions containing larger aggregates of chromosomes and nuclei. Microscopy and Denver group analysis of the first six fractions confirmed the success of the partial separation. In the recent work of Burki *et al.* (1973) and that of Stubblefield and Wray (1973) the advantages of the relatively simple karyotype of the Chinese hamster have also been explored by zonal rotor velocity sedimentation. The procedure of Burki *et al.* featured quantitative scoring of the fractionation results by means of automated scanning of photographed fraction material followed by computerized analysis with a pattern recognition program. In this research, the refractionation of selected initial fractions resulted in sharply reduced yields, but produced some nearly pure fractions of a single chromosome type (Regimbal, in preparation). Photographs of starting material and one such refined fraction are shown in Figure 2.3.

### C. Isopycnic Sedimentation

From the outset chromosome workers have been interested in the possibilities of applying the powerful technique of isopycnic, or density gradient equilibrium sedimentation to the fractionation of chromosomes. As in other applications of isopycnic sedimentation, yields may be increased by avoiding wasteful prepurification steps. A drawback, on the other hand, is the possibility of chromosomal aggregation due to concentration in narrow bands. The high sedimentation rates resulting from the extremely large size of chromosomes as compared to macromolecules permits banding at relatively low speeds or in a very short time span. This means that a wide choice of gradient materials and shapes can be used, since preformed gradients will be approximately maintained during the course of the experiment and the need for equilibrium gradients is avoided.

As mentioned above, the greatest limitation on the use of buoyant density gradients is the destructive nature of commonly used gradient materials possessing a sufficiently high density and low viscosity. Sucrose and related sugars are examples of water soluble agents which are free of this particular difficulty. In one early experiment, for example, Huberman and Attardi (1967) found isolated HeLa chromosomes to remain intact when banded in a sucrose gradient at pH 3. Unfortunately, there was no significant fractionation, with most of the chromosomes forming a single band at a density of 1·31 g/ml. This early unpromising result does not preclude fractionation in some other aqueous medium. In particular, the artificial chemical enhancement of slight differences in chromosomal properties might be exploited to generate a dispersion in chromosome densities, as discussed below in connexion with the use of non-aqueous gradients. Of possible interest in these regards are two recent reports of isopycnic banding of interphase chromatin. Raynaud and Ohlenbusch (1972) have successfully banded chromatin in a mixture of sucrose and glucose which provides higher densities than may be obtained with sucrose alone. The availability of such densities could be important if the densities of metaphase chromosomes could be selectively altered by pretreatments. Even higher densities have been attained by Hossainy *et al.* (1973) with aqueous gradients of choral hydrate in which interphase chromatin was also successfully banded in an intact state.

Gradients of certain water-soluble high molecular weight halogenated compounds such as sodium diatrizoate (3,5-diacetamido-2,4,6-triiodobenzoic acid sodium salt), have been used to resolve by isopycnic centrifugation various types of whole cells and cell components (Perper *et al.*, 1968, Ivarie and Péne, 1970). Although for chromosome work these substances have the disadvantage of being ionic, their extremely high molecular weights provide the possibility of achieving high-density solutions at relatively low molarities. Of even greater promise in this regard is the non-ionic substance metrizamide (2-(3-acetamido-5-*N*-methylacetamido-2,4,6-triiodobenzamido)-2-deoxy-D-glucose), which has very recently been introduced by Rickwood *et al.* (1973) for successful isopycnic banding of chromatin.

Perhaps the most promising attempt at isopycnic chromosome fractionation so far is the work of Stubblefield and Wray (1973), in which many of the above-mentioned problems have been avoided by the use of dehydrated chromosomes and non-aqueous gradients. A wide choice of dense organic substances is available for such gradients, and in this study a gradient of ethylene dichloride and carbon tetrachloride was used to generate a density range of 1·26–1·60. Although chromosomes dehydrated in ethanol remained free of clumping, rehydration could not easily be accomplished without chromosome damage. Successful reversible dehydration of chromosomes may be accomplished, however, by sedimentation of chromosomes from an aqueous sample layer through a layer of diethylene glycol on top of the non-aqueous gradient. Proper layer formation may be ensured by including sucrose in the diethylene glycol layer, and is further facilitated by the fact that the diethylene glycol is miscible with both the aqueous layer above it and the non-aqueous gradient below it.

Applying the above system to isolated Chinese hamster chromosomes, Stubblefield and Wray have obtained a single broad band with little fractionation. Chromosomes pretreated by controlled trypsinization, however, are found to band at increased densities due to preferential removal of protein from centromeric regions. Since centromeric regions comprise a relatively greater fraction of the mass of the smaller hamster chromosomes, such pretreatment provides a basis for selective density enhancement. As shown in Figure 2.4, a broad distribution of chromosomes with several peaks is thus obtained when the trypsinized chromosomes are sedimented in the above gradient system. It is perhaps unfortunate that this technique fails to provide a complement to velocity sedimentation fractionation, in that both techniques separate in the order of chromosomal masses.

### D. Selective Filtration

Chromosome fractionation by selective filtration is limited by four factors: (1) availability of filter materials which do not adsorb chromosomes; (2) capacity limitations, or tendency to clog; (3) availability of filters with a sufficient range of pore sizes; (4) uniformity of effective pore size in any particular filter. As a result of the latter two limitations, selective filtration is likely to be of use only for very crude fractionations or for selection of chromosomes of very unusual size.

The use of steel filters for chromosome purification (Salzman *et al.*, 1966) has been mentioned above, but for fractionation purposes this type of filter is limited by the inhomogeneity of its pores. Huberman and Attardi (1967) have reported that both 'Millipore' and 'Nuclepore' filters in the range of 2–10 $\mu$m pore size successfully retained chromosomes larger than pore size but were unsuitable for fractionation due to rapid clogging. More recently Hanson (MS. in preparation) has had preliminary success with partial fractionation of *Drosophila* chromosomes by means of passage through 0·4 or 0·6 $\mu$m 'Nuclepore' filters. In this case, the goal is to obtain a pure fraction of chromosome No. 4, which is the only chromosome small enough to pass through the filter. This fractionation is, in a sense, a special case since the No. 4 chromosome is approximately an order of magnitude smaller than the next largest in the karyotype. In spite of the reservations of Huberman and Attardi, 'Nuclepore' filters (General Electric Co., Pleasanton, California) seem to be the most suitable for chromosome studies on the basis of the four criteria above. Possibilities for further increasing the usefulness of filters by chemical pretreatments have been mentioned in Section III.A.

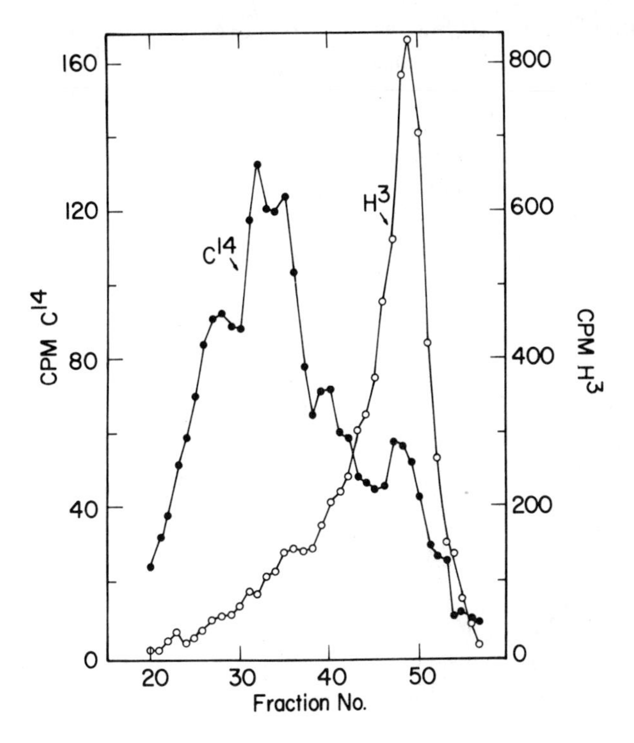


**Figure 2.4.** Profiles of dehydrated isolated Chinese hamster chromosomes banded isopycnically in a gradient of ethylene dichloride and carbon tetrachloride by the method of Stubblefield and Wray (1973) as described in the text. Density increases from right to left. In the chromosome mixture shown banded here, a control of chromosomes labelled with $^{3}H$ has been combined with trypsin-treated chromosomes labelled with $^{14}C$. Trypsinized chromosomes are seen to have variably increased densities, thus giving rise to a partial resolution of chromosomes throughout the gradient. (Reproduced from Stubblefield and Wray (1973) by permission of the Cold Spring Harbor Laboratory)

## E. Electrofocusing

The use of electrofocusing for the fractionation of isolated HeLa chromosomes has recently been reported by Landel *et al.* (1972). Electrofocusing, which is closely related to electrophoresis, involves the application of an electric potential difference between the ends of a column in which a pH gradient is established, with the anode at the acid end. Establishment of equilibrium pH gradients is achieved in practice by the electrophoretic movement of a mixture of low molecular weight compounds known as carrier ampholytes. The mixture contains numerous molecules each with a slightly different p*K* within the pH range of interest. Since movement of any compound ceases at the point in the column at which it is isoelectric, application of the electric field results in a short time in a distribution of ampholytes and hence a gradient of pH. Macromolecules or particles to be separated may be included throughout the gradient at the beginning of the run and after a period of hours will be found concentrated or 'focused' at their respective positions of isoelectric pH. A sucrose density gradient included in the column stabilizes the distribution and thus facilitates subsequent fraction collection from the bottom of the column. A simple measurement of the pH of each collected fraction conveniently determines the p*K* of substances found in that fraction.

Using both total HeLa chromosome mixtures and chromosomes previously size-fractionated by sucrose–glycerol gradient velocity sedimentation (Huberman and Attardi, 1967), these investigators were able to obtain broad electrofocused distributions with several reproducible peaks. Several methods were used to confirm that such distributions in fact represented a non-random separation of chromosomes. The chromosome fractions obtained in this case on the basis of charge differences were not correlated with size differences as in sedimentation velocity fractionation. One such electrofocusing pattern for a fraction of 'large' HeLa chromosomes obtained by a prior velocity sedimentation fractionation is shown in Figure 2.5.

Landel *et al.* found that some stripping of proteins from the chromosomes took place during prolonged exposure to the electric field. To the extent to which it is reproducible this may be regarded as a possible contributing factor in the fractionation. It was largely overcome by a modification of the technique in which the concentration of carrier ampholytes was decreased and in which electrofocusing times were greatly reduced by initially layering the chromosomes near their expected range of isoelectric positions. Chromosomes electrofocused by this modified procedure formed a continuous distribution in the pH range from 3·90 to 4·30 with some partially resolved groups which rebanded at the same pHs when reelectrofocused under identical conditions.

To demonstrate the potential usefulness of their new method Landel and coworkers studied the ribosomal RNA genes which were previously localized by Huberman and Attardi (1967) to smaller sized HeLa chromosomes as obtained from velocity sedimentation fractionation. In particular, hybridization experiments were used to localize these genes to the less-acid electrofocused subfractions of the previously velocity-fractionated smaller chromosomes. As pointed out by the authors, the resolution resulting from the combination of these two methods may find other important applications in the future.

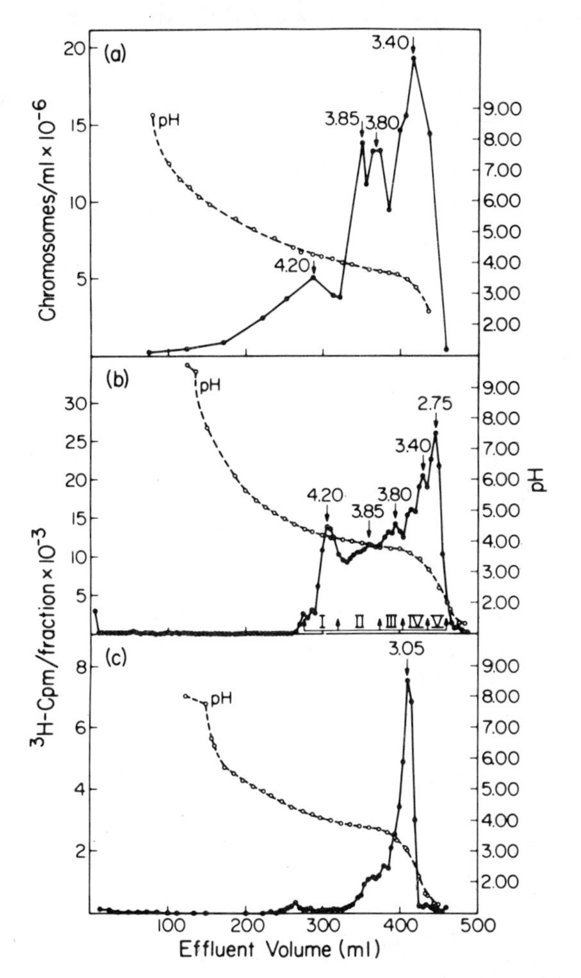


**Figure 2.5.** Electrofocusing profiles of large HeLa chromosomes isolated at pH 3 and previously fractionated by velocity sedimentation through a sucrose-glycerol gradient. (a) $10^9$ chromosomes were electrofocused 48 h in a 44 ml column by the standard procedure of Landel *et al.* (1972), and the number of chromosomes per ml in the collected fractions were determined by counting under a microscope. p*K* values for the peak fractions are indicated in the figure. (b) $^3$H-thymidine-labelled chromosomes were isolated, prefractionated and electrofocused as in part a. Chromosomal content of collected fractions was assayed by acid-precipitable radioactivity. (c) Fraction I from part b was re-electrofocused under the same conditions as in part b. Note the shift in p*K* manifested in the re-electrofocusing. (Reproduced from Landel *et al.* (1972) by permission of the American Chemical Society)

## F. Miscellaneous Techniques

Future progress in the fractionation of isolated chromosomes should result not only from refinement of the above techniques and the introduction of new ones, but in particular may depend upon the combined use of more than one type of technique along with careful selection of appropriate organisms. In these respects the artificial enhancement of slight differences in chromosomal properties may have the greatest impact on the refinement of existing techniques. The availability of isolated *Drosophila* chromosomes, on the other hand, may be the most important step towards facilitating a total separation of all the chromosomes of one organism.

Chromosome fractionation methods currently in use are largely based upon well-known methods for the separation of macromolecules. In addition to the chromosomal charge properties exploited in the electrofocusing study above, Marfey (1973) has detected differential binding of poly-L-lysine by isolated human chromosomes. This study involved *in situ* binding of radioactive poly-L-lysine followed by autoradiographic localization and counting of chromosomal grain contents. It was concluded that the differences were due to the different extents to which the nucleic acid in the chromosomes is exposed. The existence of this type of structural difference suggests, for example, the possible applicability of liquid-phase partition or countercurrent distribution (Albertsson, 1971) for chromosome fractionation. These techniques are sensitive to both total charge and surface properties of particles, and have already been successfully applied to fractionation of whole cells and certain organelles. Advantages of the techniques include gentle conditions involving low salt concentrations and no extremes of temperature or pH.

Another category of separation techniques is that in which the laminar flow of a liquid suspension is used to collect fractions containing particles from different portions of some physically induced distribution pattern. In one simple application of this principle, particles in a thin horizontally flowing sheet of liquid form a vertical distribution due to gravity and Brownian motion and are consequently resolved along the direction of flow according to their masses (Berg *et al.*, 1967). In a more sophisticated version of this approach known as the stable-flow free boundary (STAFLO) device, such properties as sedimentation velocity or electrophoretic behaviour are used to shift particles into regions of a liquid stream which are flowing at different rates (Mel, 1970). The possibility for future application of such devices to chromosome studies is suggested by previous successful use with particles of near chromosome size.

Finally, the 'cell-sorting' devices recently developed at Los Alamos (Fulwyler *et al.*, 1969) and at Stanford (Bonner *et al.*, 1972) could also prove useful to future chromosome workers. In this type of apparatus particles are fractionated one at a time on the basis of the physical properties of single particles contained within liquid microdrops and detected by ultrarapid mechanisms. The technique is limited both by capacity and by the low signal amplitudes predicted for particles of subcellular size. Of potential interest in chromosome work, however, is the capacity of the machine to fractionate on the basis of the fluorescence of stained particles.

In concluding this discussion of future prospects for chromosome fractionation it is important to emphasize the relevance of genetic manipulations. In general, two very similar chromosome types might be rendered resolvable by a sufficiently large deletion from either of them or by a translocation of material between them. Stubblefield and Wray (1973), for example, have suggested the usefulness of three such existing translocations in the Don-C line of Chinese hamster cells. Mutations of the sort exploited by Kavenoff and Zimm (1973) may also aid those seeking to fractionate isolated *Drosophila* chromosomes. Mutations having a gross effect on effective chromosome size, such as 'attached' chromosomes, are, of course, especially abundant in this latter organism. Hanson and Hearst (1973) have furthermore reported the presence of a neochromosome in an established *Drosophila* cell line. Since its intermediate size is distinct from any of the normal chromosomes, this presents the eventual possibility of isolating material which represents a subchromosomal segment of the genome.

## V. RESEARCH APPLICATIONS FOR ISOLATED CHROMOSOMES

Several of the applications to be discussed here have been mentioned above in connexion with the methods of isolation. The following brief summary is, however, to indicate how the availability of bulk isolated chromosomes may provide new approaches to certain biological questions.

### A. Unfractionated Chromosome Mixtures

Although separated chromosomes are highly desirable, much information has been obtained from the use of unfractionated mixtures of isolated chromosomes. This includes: (1) determinations of the chemical composition of metaphase chromosomes and comparison with corresponding interphase structures and (2) studies of the effects of pH, ions, detergents and enzymes on chromosomal structure. This information could only be obtained from large amounts of isolated chromosomes, but bulk isolation methods have also been an advantage in microscopic studies of chromosome morphology. This is because many parameters can be varied with the isolated chromosomes that cannot be when the chromosomes are examined *in situ*. Such conditions are also independent of interactions between chromosomes and adjacent structures. The comprehensive microscopy study by Stubblefield and Wray (1971) is one example of such an approach.

Finally, perhaps the most extensive use of unfractionated isolated chromosomes for physical studies is to be found in the work of Cantor and Hearst (1969). In this case the investigators exploited the availability of the milligrams of material which are required by such techniques as optical absorption, circular dichroism and electrometric titration. Their results included discovery of a dramatic pH- and $Mg^{2+}$-dependent structural transition as well as determination of the number of $Mg^{2+}$ binding sites and the corresponding binding constant.

At least two studies have taken advantage of the protected state of DNA in bulk isolated chromosomes for work in which metaphase structure was largely irrelevant. Corry and Cole (1968), for example, used isolated metaphase chromosomes as a source of mechanically undamaged DNA for studies of radiation-induced breaks in nucleic acid. Their alkaline isolation medium inhibited nuclease degradation of chromosomal DNA during chromosome isolation. Similarly, Burkholder and Mukherjee (1971) used metaphase chromosomes isolated from pulse-labelled cultures as a source of chromatin fibres for autoradiographic studies of replication. In their experiment isolated chromosomes were unravelled on to a glass slide with the aid of trypsin.

Finally, several groups have studied the spontaneous uptake of isolated chromosomes by cells in culture (Chorazy *et al.*, 1963b; Burkholder and Mukherjee, 1970a; Sekiguchi *et al.*, 1973). This phenomenon involves cellular uptake of intact chromosomes by processes related to pinocytosis or phagocytosis, and raises the intriguing possibility of a mechanism for genomic incorporation of heterologous genetic information. Early studies with radioactively labelled chromosomes indicated a rapid breakdown of such foreign chromosomes in the recipient cell. Sekiguchi and his coworkers, however, have recently reported a low-frequency incorporation of isolated rat chromosomes into the metaphase spreads of mouse cells. They partially attribute their success to the use of the 'near-physiological' chromosome isolation conditions of Maio and Schildkraut (1967).

### B. Fractionated Chromosomes

An important question is what protein or nucleic acid components are specific to a particular chromosome type. So far no significant differences have been found using the partially fractionated preparations. Huberman and Attardi (1967) found that DNA complementary to cytoplasmic messenger RNA is uniformly distributed in all HeLa chromosome size fractions. Maio and Schildkraut (1969) also found no differences in the CsCl density of total DNA or in the amount of satellite DNA in different size fractions of Chinese hamster chromosomes. More recently, Skinner and Ockey (1971) have shown a uniform distribution of histones among chromosome fractions from two different mammals. Stubblefield and Wray (1973) have, however, found indications of some enrichment of phenol-soluble non-histone protein in certain fractions of Chinese hamster chromosomes. Burki *et al.* (1973) also exploited fractionation of Chinese hamster chromosomes to show that the DNA content of chromatids is not proportional to chromatid length in this organism. In one study of DNA replication, the analysis of fractionated chromosomes from Chinese hamster cells which had been pulse-labelled with $^{3}$H-thymidine showed a selective synthesis of DNA in the smallest chromosomes late in S phase (Mendelsohn *et al.*, 1968).

The aim of several chromosome fractionation experiments has been the localization of certain genes by bulk hybridization of gene products to chromosome fractions. Huberman and Attardi (1967), for example, found ribosomal RNA genes localized on the smaller HeLa chromosomes, and Landel *et al.* (1972) later refined this localization to an even more restricted category of chromosomes obtained by electrofocusing. In an analogous study, Aloni *et al.* (1971) found transfer RNA genes and 5S RNA genes uniformly distributed in all size fractions of HeLa chromosomes. It should be noted that the approach to gene localization typified by these studies has to some extent been rendered obsolete by the elegant technique of *in situ* hybridization (Gall and Pardue, 1971; Jones, 1973). Bulk hybridization of components to chromosome fractions will remain, however, a valuable method for placing such studies on a quantitative basis. *In situ* hybridization is also insensitive to differences in the species which simultaneously hybridize to different chromosomes in experiments dealing with a heterogeneous mixture of hybridizing molecules. This limitation was overcome, on the other hand, in a study of the localization of rapidly hybridizing categories of heterogeneous nuclear RNA to DNA from fractionated large and small mammalian chromosomes (Pagoulatos and Darnell, 1970). After finding that such RNA hybridized to both chromosome fractions, the RNA which hybridized to each fraction was recovered after dissociation of the hybrids and then 'cross-hybridized' to DNA from the opposite chromosome fraction. The results showed an equal chromosomal distribution of the same rapidly hybridizing HnRNA sequences.

The availability of relatively homogeneous chromosome fractions may have an important impact on optical and other physical studies of chromosome structure. One possible example would be a comparison of the physical properties of a fraction of largely or wholly heterochromatic chromosomes *versus* a corresponding euchromatic fraction.

## VI. REFERENCES


Albertsson, P. (1971) *Partition of Cell Particles and Macromolecules*, 2nd ed., Wiley–Interscience, N.Y.

Aloni, Y., L. E. Hatlen and G. Attardi (1971) 'Studies of fractionated HeLa cell metaphase chromosomes. 2. Chromosomal distribution of sites for transfer RNA and 5 s RNA', *J. Mol. Biol.*, **56**, 555.

Avis, P. J. G. (1972) 'Pressure homogenisation of mammalian cells', in *Subcellular Components, Preparation and Fractionation*, 2nd ed. (Ed. G. D. Birnie), Butterworths, London.

Berg, H. C., E. M. Purcell and W. W. Stewart (1967) 'A method for separating according to mass a mixture of macromolecules or smaller particles suspended in a fluid. 2. Experiments in a gravitational field', *Proc. Natl. Acad. Sci. U.S.*, **58**, 1286.

Bonner, W. A., H. R. Hulett, R. G. Sweet and L. A. Herzenberg (1972) 'Fluorescence activated cell sorting', *Rev. Sci. Instrum.*, **43**, 404.

Burkholder, G. D. and B. B. Mukherjee (1970a) 'Uptake of isolated metaphase chromosomes by mammalian cells *in vitro*', *Exper. Cell Res.*, **61**, 413.

Burkholder, G. D. and B. B. Mukherjee (1970b) 'Fractionation of isolated mammalian metaphase chromosomes', *Exper. Cell Res.*, **63**, 213.

Burkholder, G. D. and B. B. Mukherjee (1971) 'Replication of chromosomal fibers from isolated metaphase chromosomes', *Exper. Cell Res.*, **64**, 470.

Burki, H. J., T. J. Regimbal, Jr. and H. C. Mel (1973) 'Zonal fractionation of mammalian chromosomes and determination of their DNA content', *Preparative Biochem.*, **3**, 157.

Cantor, K. P. and J. E. Hearst (1966) 'Isolation and partial characterization of metaphase chromosomes of a mouse ascites tumor', *Proc. Natl. Acad. Sci. U.S.*, **55**, 642.

Cantor, K. P. and J. E. Hearst (1969) 'The structure of metaphase chromosomes. 1. Electrometric titration, magnesium ion binding and circular dichroism', *J. Mol. Biol.*, **49**, 213.

Chorazy, M., A. Bendich, E. Borenfreund and D. J. Hutchison (1963a) 'Studies on the isolation of metaphase chromosomes', *J. Cell Biol.*, **19**, 59.

Chorazy, M., A. Bendich, E. Borenfreund, O. L. Ittensohn and D. J. Hutchison (1963b) 'Uptake of mammalian chromosomes by mammalian cells', *J. Cell Biol.*, **19**, 71.

Corry, P. M. and A. Cole (1968) 'Radiation-induced double-strand scission of the DNA of mammalian metaphase chromosomes', *Radiation Res.*, **36**, 528.

DuPraw, E. J. (1970) *DNA and Chromosomes*, Holt, Rinehart and Winston, Inc., N.Y.

Franceschini, P. and D. Giacomoni (1967) 'Isolation and fractionation of metaphase chromosomes from HeLa cells', *Atti Assoc. Genet. Ital.*, **12**, 248.

Fulwyler, M. J., R. B. Glascock and R. D. Hiebert (1969) 'Device which separates minute particles according to electronically sensed volume', *Rev. Sci. Instrum.*, **40**, 42.

Gall, J. G. and M. L. Pardue (1971) 'Nucleic acid hybridization in cytological preparations', *Methods Enzymol.*, **21**, 470.

Gross, P. R. and W. Spindel (1960) 'The inhibition of mitosis by deuterium', *Ann. N.Y. Acad. Sci.*, **84**, 745.

Hanson, C. V. and J. E. Hearst (1973) 'Bulk isolation of metaphase chromosomes

from an *in vitro* cell line of *Drosophila melanogaster*', *Cold Spring Harbor Symp. Quant. Biol.*, **38**, 341.
Hearst, J. E. and M. Botchan (1970) 'The eukaryotic chromosome', *Ann. Rev. of Biochemistry*, **39**, 151.
Hossainy, E., A. Zweidler and D. P. Bloch (1973) 'Isopycnic banding of chromatin in chloral hydrate gradients', *J. Mol. Biol.*, **74**, 283.
Huberman, J. A. (1973) 'Structure of chromosome fibers and chromosomes', *Ann. Rev. Biochem.*, **42**, 355.
Huberman, J. A. and G. Attardi (1966) 'Isolation of metaphase chromosomes from HeLa cells', *J. Cell Biol.*, **31**, 95.
Huberman, J. A. and G. Attardi (1967) 'Studies of fractionated HeLa cell metaphase chromosomes. 1. The chromosomal distribution of DNA complementary to 28 s and 18 s ribosomal RNA and to cytoplasmic messenger RNA', *J. Mol. Biol.*, **29**, 487.
Ivarie, R. D. and J. J. Pène (1970) 'Association of the *Bacillus subtilis* chromosome with the cell membrane: resolution of free and bound deoxyribonucleic acid on Renografin gradients', *J. Bacteriol.*, **104**, 839.
Jones, K. W. (1973) in *New Techniques in Biophysics and Cell Biology*, Vol. I (Eds. B. Smith and R. Pain), Wiley, London.
Kane, R. E. (1965) 'The mitotic apparatus. Physical–chemical factors controlling stability', *J. Cell Biol.*, **25**, 137.
Kavenoff, R. and B. Zimm (1973) 'Chromosome-sized DNA molecules from *Drosophila*', *Chromosoma* (*Berl.*), **41**, 1.
Kirsch, W., J. W. Leitner, M. Gainey, D. Schulz, R. Lasher and P. Nakane (1970) 'Bulk isolation in nonaqueous media of nuclei from lypholized cells', *Science, N.Y.*, **168**, 1592.
Landel, A., Y. Aloni, M. A. Raftery and G. Attardi (1972) 'Electrofocusing analysis of HeLa cell metaphase chromosomes', *Biochem.*, **11**, 1654.
Lief, R. C., H. N. Easter, Jr., R. L. Warters, R. A. Thomas, L. A. Dunlap and M. F. Austin (1971) 'Centrifugal cytology. 1. A quantitative technique for the preparation of glutaraldehyde-fixed cells for the light and scanning electron microscope', *J. Histochem. and Cytochem.*, **19**, 203.
Lin, H. J. and E. Chargaff (1964) 'Metaphase chromosomes as a source of DNA', *Biochim. Biophys. Acta*, **91**, 691.
Maio, J. J. and C. L. Schildkraut (1967) 'Isolated mammalian chromosomes. 1. General characteristics of nucleic acids and proteins', *J. Mol. Biol.*, **24**, 29.
Maio, J. and C. L. Schildkraut (1969) 'Isolated mammalian metaphase chromosomes. II. Fractionated chromosomes of mouse and Chinese hamster cells', *J. Mol. Biol.*, **40**, 203.
Marfey, P. (1973) 'Binding of $^{3}$H-poly-L-lysine to human metaphase chromosomes', *Biophys. Soc. Abstracts*, 1973, p. 145a.
Mazia, D. and K. Dan (1952) 'The isolation and biochemical characterization of the mitotic apparatus of dividing cells', *Proc. Natl. Acad. Sci. U.S.*, **38**, 826.
Mazia, D., J. Mitchison, H. Medina and P. Harris (1961) 'The direct isolation of the mitotic apparatus', *J. Biophys. Biochem. Cytol.*, **10**, 467.
Mel, H. C. (1970) 'Stable-flow free boundary cell fractionation as an approach to the study of hematopoietic disorders', in *Myeloproliferative Disorders of Animals and Man* (Eds. W. J. Clarke, E. B. Howard and P. L. Hackett), U.S. Atomic Energy Commission Div. of Tech. Info., Oak Ridge, Tennessee, p. 665.
Mendelsohn, J. (1973) in *The Cell Nucleus* (Ed. H. Busch), Academic Press, New York (in press).

Mendelsohn, J., D. E. Moore and N. P. Salzman (1968) 'Separation of isolated Chinese hamster metaphase chromosomes into three size-groups', *J. Mol. Biol.*, **32**, 101.

Miller, R. G. (1973) in *New Techniques in Biophysics and Cell Biology*, Vol. I (Eds. B. Smith and R. Pain), Wiley, London, p. 87.

Milsted, A. and W. D. Cohen (1973) 'Mitotic spindles from *Drosophila melanogaster* embryos', *Exper. Cell Res.*, **78**, 243.

Mitchison, J. M. (1971) *The Biology of the Cell Cycle*, Cambridge University Press, London, Chap. 3.

Nias, A. H. W. and M. Fox (1971) 'Synchronization of mammalian cells with reference to the mitotic cycle', *Cell Tissue Kinetics*, **4**, 351.

Nishiura, J. (1972) *Doctoral Dissertation*, University of Washington, Seattle.

Pagoulatos, G. N. and J. E. Darnell, Jr. (1970) 'Fractionation of heterogeneous nuclear RNA: rates of hybridization and chromosomal distribution of reiterated sequences', *J. Mol. Biol.*, **54**, 517.

Perper, R. J., T. W. Zee and M. M. Mickelson (1968) 'Purification of lymphocytes and platelets by gradient centrifugation', *J. Lab. Clin. Med.*, **72**, 842.

Raynaud, A. and H. H. Ohlenbusch (1972) 'Buoyant density of native chromatin', *J. Mol. Biol.*, **63**, 523.

Regimbal, T. J. (1973) 'Biophysics of mammalian chromosomes', *Ph.D. Thesis*, University of Califoronia, Berkeley.

Rickwood, D., A. Hell and G. D. Birnie (1973) 'Isopycnic centrifugation of sheared chromatin in metrizamide gradients', *Federation Exper. Biol.* (*Letters*), **33**, 221.

Ris, H. and A. E. Mirsky (1949) 'The state of the chromosomes in the interphase nucleus', *J. Gen. Physiol.*, **32**, 489.

Ris, H. and D. F. Kubai (1970) 'Chromosome structure', *Ann. Rev. Genetics*, **4**, 263.

Sadgopal, A. and J. Bonner (1970) 'Proteins of interphase and metaphase chromosomes compared', *Biochim. Biophys. Acta*, **207**, 227.

Salzman, N. P., D. E. Moore and J. Mendelsohn (1966) 'Isolation and characterization of human metaphase chromosomes', *Proc. Natl. Acad. Sci. U.S.*, **56**, 1449.

Sasaki, M. (1961) 'Observations on the modification in size and shape of chromosomes due to technical procedure', *Chromosoma* (*Berl.*), **11**, 514.

Schneider, E. L. and N. P. Salzman (1970) 'Isolation and zonal fractionation of metaphase chromosomes from human diploid cells', *Science, N.Y.*, **167**, 1141.

Schneider, I. (1972) 'Cell lines derived from late embryonic stages of *Drosophila melanogaster*', *J. Embryol. Exp. Morph.*, **27**, 353.

Sekiguchi, T., F. Sekiguchi and M.-A. Yamada (1973) 'Incorporation and replication of foreign metaphase chromosomes in cultured mammalian cells', *Exper. Cell Res.*, **80**, 223.

Sisken, J. E., E. Wilkes, G. M. Donnelly and T. Kakefuda (1967) 'The isolation of the mitotic apparatus from mammalian cells in culture', *J. Cell Biol.*, **32**, 212.

Skinner, L. G. and C. H. Ockey (1971) 'Isolation, fractionation and biochemical analysis of the metaphase chromosomes of *Microtus agrestis*', *Chromosoma* (*Berl.*), **35**, 125.

Somers, C. E., A. Cole and T. C. Hsu (1963) 'Isolation of chromosomes', *Exper. Cell Res. Suppl.*, **9**, 220.

Sonnenblick, B. P. (1965) in *Biology of* Drosophila (Ed. M. Demerec), Hafner, New York, p. 95.

Stubblefield, E. and W. Wray (1971) 'Architecture of the Chinese hamster metaphase chromosome', *Chromosoma* (*Berl.*), **32**, 262.

Stubblefield, E. and W. Wray (1973) 'Biochemical and morphological studies of partially purified Chinese hamster chromosomes', *Cold Spring Harbor Symp. Quant. Biol.*, **38** (in press).
Wray, W. and E. Stubblefield (1970) 'A new method for the rapid isolation of chromosomes, mitotic apparatus, or nuclei from mammalian fibroblasts at near neutral pH', *Exper. Cell Res.*, **59**, 469.
Wray, W., E. Stubblefield and R. Humphrey (1972) 'Mammalian metaphase chromosomes with high molecular weight DNA isolated at pH 10·5', *Nature New Biol.*, **238**, 237.

CHAPTER 3

# Microcalorimetry and its application in biological sciences

Ingemar Wadsö
*Thermochemistry Laboratory, Chemical Centre,*
*University of Lund, Sweden*

## I. INTRODUCTION

Calorimeters are instruments used for measurements of heat quantities or heat effects. Several calorimetric principles are in use and the instruments have a large number of different practical designs. If the instruments are very sensitive and require only small sample quantities they are usually called microcalorimeters. A modern microcalorimeter can, for example, be used for accurate measurements of mcal quantities of heat, which usually correspond to $\mu$mole quantities of substance. The heat evolution in biochemical experiments is often very small and it is therefore primarily microcalorimetric methods which are of interest in biocalorimetry.

In the past, and probably still at the present time, calorimetric measurements have been used mainly for thermodynamic investigations. In basic chemical and biological science, thermodynamic data are often required for the understanding of a process or for testing a new hypothesis. Thermodynamic data are further of immense importance in the chemical industry.

The most common property to be determined calorimetrically for a process is the enthalpy change, $\Delta H$. Enthalpy changes may also be derived from results of equilibrium measurements at different temperatures (van't Hoff enthalpies) but these latter values are usually less reliable. If $\Delta H$ is determined at different temperatures the change in heat capacity, $\Delta C_{\mathrm{p}}$, can be calculated. This property is attracting at present considerable interest in biochemical thermodynamics, in particular in connection with studies of medium effects, e.g. concerning 'hydrophobic bonding' (see, for example, Tanford, 1968, 1970).

If the equilibrium constant for a process is not too far from unity it is possible to determine by calorimetric experiments alone both the equilibrium constant and the enthalpy change and therefore also the entropy change. This method has recently been applied in several studies on biochemical systems, e.g. enzyme–inhibitor coupling reactions, cf. p. 115.

Practically every process, be it physical, chemical or biological, is accompanied by evolution of heat or heat absorption. The heat quantities are functions of the extent of the process and the heat evolved in a given process is proportional to the intensity or the rate of the process. It is thus obvious that calorimetry also forms a general analytical tool with potential applications in physics and in chemistry but perhaps in particular in the biological sciences.

Calorimetry is a very old science. Interestingly enough, some of the first calorimetric experiments had the character of biological thermodynamic experiments. About 1780, Lavoisier and Laplace performed measurements of heat evolution in connexion with animal respiration using a simple but ingenious ice calorimeter, see e.g. Armstrong (1964). However, it is not until the last few years that calorimetric techniques have started to be adopted by biochemists and biologists outside a few specialized laboratories.

One reason for this increased interest in biocalorimetry is due to the fact that microcalorimetric techniques have been greatly improved. Another reason is probably that bioscientists have a pronounced need for new analytical techniques to be applied in areas where few or no acceptable analytical methods exist. The need for more and reliable thermodynamic data has also been increasingly apparent, particularly in biochemistry.

The main cause of the present trend, however, is undoubtedly related to the fact that suitable calorimetric equipment is now commercially available. This does not mean that the present generation of these instruments is suitable for all biocalorimetric studies which can be envisaged. But with the present equipment, and with suitably modified instruments, there are large areas ready for experimental exploration. There are still other areas where application of a calorimetric technique would be important, but where suitable instrument designs do not yet exist.

The present chapter is meant to form a brief introduction to microcalorimetry as applied in biochemistry and biology. A comparatively broad treatment is given of the calorimetric principles and sources of errors in microcalorimetric instruments. Concerning practical designs, only those types of instruments are discussed which are used by most workers today. Specific applications have of necessity been treated rather briefly. No complete literature coverage has been aimed at; the intention has rather been to point out for the reader examples of the more recent work.

Among areas which possibly should have been given attention in this chapter are techniques and experiments in connexion with measurements of heat effects in tissues like muscles and nerves. However, two recent review articles, by Howarth (1970) and by Wooledge (1971), give an up-to-date coverage and these subjects have therefore not been treated here.

## II. MICROCALORIMETRY

Some general calorimetric principles and a few practical designs of microcalorimeters which have proved to be suitable for biochemical and biological work will be described. For a more comprehensive treatment of this subject the reader is referred to a recent review article (Wadsö, 1970, cf. also Skinner,

1969). Some important sources of systematic errors in microcalorimetric work will be discussed in a separate section.

## A. Calorimetric Principles

There are two main categories of calorimeter in use: adiabatic calorimeters and heat conduction calorimeters.

With an ideal adiabatic calorimeter there is no heat exchange between the calorimeter and its surroundings. The heat quantity evolved during the experiment, $Q$, is usually proportional to the observed temperature change, $\Delta T$, and to the heat capacity of the calorimetric system ('the calibration constant'), $\varepsilon$:

$$Q = \varepsilon \Delta T \tag{3.1}$$

When a heat quantity is determined in an experiment it is thus the temperature change which is measured.

With an ideal heat conduction calorimeter (also called heat leakage calorimeter) the heat evolved in the calorimetric vessel is quantitatively transferred to a body surrounding the calorimeter, the heat sink. Ideally, the heat capacity of the heat sink is infinitely large and the final temperature of the calorimetric vessel is the same as the starting temperature. With a heat conduction calorimeter, some property proportional to the heat flow between the calorimeter vessel and the heat sink is measured. The time integral for the heat flow, $\phi$, times a calibration constant, $\epsilon$, is proportional to the heat quantity evolved in the experiment

$$Q = \epsilon \int \phi \mathrm{d}t \tag{3.2}$$

Within these two extreme categories there are several types of calorimeters which have been adopted. Some of these will be described below.

### 1. *Isoperibol calorimeters**

The isoperibol calorimeter, also called 'constant temperature environment calorimeter' or 'isothermal jacket calorimeter', is probably the most common type of calorimeter in use. A schematic representation of an isoperibol calorimeter is shown in Figure 3.1. The calorimeter vessel (*a*) is separated by thermal insulation (*b*) from the surrounding thermostatic bath (*c*), which forms the 'isothermal jacket'. The insulation is usually vacuum (e.g. provided by a Dewar vessel) or air. In the calorimeter vessel there is a thermometer (*d*), usually a resistance thermometer, such as a thermistor, or a thermocouple. Depending on the purpose for which the calorimeter is designed, the vessel may be equipped with a number of additional details. The vessel may thus contain a calibration heater, some device for initiation of the process

* Isoperibol = constant radiation from the surroundings.

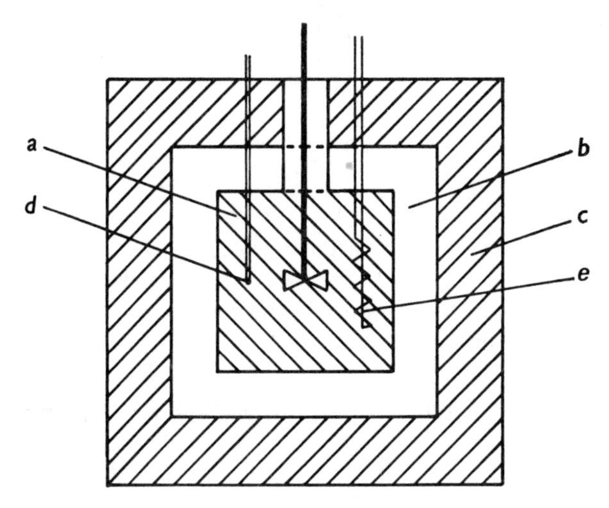


**Figure 3.1.** Isoperibol calorimeter. *a*, Calorimeter vessel; *b*, air gap or vacuum; *c*, thermostated bath or block; *d*, thermometer; *e*, heater. Reprinted from *Quart. Rev. Biophysics*, **3**, 383 (1970)

(e.g. an ampoule-breaking mechanism), stirrer, inlet tubes for gas or liquid, etc. If heat is evolved (exothermic process) or absorbed (endothermic process) in the calorimeter vessel, there will be a temperature change. If the insulation *b* had been perfect the calorimeter would have been a truly adiabatic calorimeter. In practice, there will always be some heat exchange between the vessel and the surroundings which cannot be neglected in accurate measurements. The isoperibol calorimeter can thus be characterized as a quasi-adiabatic calorimeter where a modest heat exchange is corrected for. This is done by applying Newton's cooling law, usually in the form of the Ragnault–Pfaundler relationship (see e.g. Coops *et al.*, 1956; Sturtevant, 1959; Skinner, 1969) or, for processes of very short duration, by using a graphical extrapolation procedure, see e.g. Skinner (1969).

Isoperibol calorimeters are in use both as very simple instruments but also, for comparatively fast processes, as very precise calorimeters. They are used, for example, as reaction and solution calorimeters and as combustion calorimeters, but isoperibol calorimeters have not yet found an extensive use as microcalorimeters built for biochemical or biological studies.

For slow processes (duration longer than about 30 min), the uncertainty introduced through the heat exchange with an isoperibol calorimeter will usually be inconveniently large. However, there will be no net heat exchange between the isothermal wall of the vessel and the surroundings if both have the same temperature. For an endothermic process this can in principle easily be arranged by introducing electrical energy into the calorimeter vessel in order to balance the heat absorbed in the process. In such a case an isoperibol calorimeter will be operated isothermally and the heat evolved in the process is equal to the electrical heat input. However, most processes suitable for calorimetric studies are exothermic.

### 2. *Adiabatic shield calorimeters*

For comparatively slow exothermic processes 'adiabatic shield calorimeters' have proved to be very useful. In these instruments, the reaction vessel is enclosed (or in practice nearly so) by a thin-walled metal envelope, an adiabatic shield. This is usually positioned in the vacuum or air space between the reaction vessel and the surrounding thermostatic bath. The temperature difference between the shield and the vessel is kept at zero during the experiment by the evolution of a suitable heat effect in a heater wound on the shield. Zero temperature difference is, for instance, indicated by thermocouples, and in modern calorimeters the shield effect is regulated automatically. Adiabatic shield calorimeters are used in some of the most important calorimetric applications, in particular for measurements of long duration and for measurements made at temperatures very different from room temperature. Important designs are found both as reaction calorimeters and as heat capacity calorimeters. Of particular interest in connexion with this chapter is, for example, the type of microreaction calorimeter described by Buzzel and Sturtevant (1951) and used in many biochemical studies. Recently, several sophisticated microversions of temperature-scanning heat capacity calorimeters have been described which employ adiabatic shields, see p. 110. These latter types of calorimeters have proved to be very useful for studies involving phase transitions in biochemical and biological samples.

### 3. *Thermoelectric heat pump calorimeters*

Heat evolved in an exothermic process can be actively transported out from the calorimeter vessel where the temperature is thus kept constant and equal to that of the surroundings. This can be achieved by use of the Peltier effect principle. A thermopile, preferably a semiconducting Peltier effect cooler, is positioned between the reaction vessel and the surrounding thermostatic bath which acts as a heat sink. The cooling effect, $-w$, produced at each cold junction is proportional to the current through the thermopile, $i$.

$$-w = \pi i \tag{3.3}$$

In practice, this cooling effect will be superimposed on the Joule heating effect, $Ri^2$, which is produced throughout the circuit. The actual cooling effect at each thermopile junction is thus given by

$$-w = \pi i - Ri^2 \tag{3.4}$$

where $R$ is an effective junction resistance.

In the well-known Calvet microcalorimeter (see p. 107), Peltier effect cooling is often used in conjunction with the 'thermopile heat conduction principle'. In several other designs which may not be called microcalorimeters, Peltier effect cooling has been successfully applied, see e.g. Christensen *et al.* (1968) and Becker and Kiefer (1969).

### 4. *Labyrinth flow calorimeters*

Heat may also be transported out from a calorimetric vessel by use of cooling liquid. This principle is used in 'labyrinth flow calorimeters'. By a heat exchange system ('a labyrinth') all heat evolved in the reaction vessel is absorbed by the liquid flow and the heat quantity evolved in the calorimeter is proportional to the temperature—time integral for the heat-transporting liquid. Naturally this principle can also be used with endothermic processes.

Sensitive labyrinth flow calorimeters suitable for very slow processes were developed by the Swietoslawski school (Swietoslawski, 1946; Swietoslawski and Zielenkiewicz, 1959). These calorimeters were generally called 'microcalorimeters'; but reaction volumes were rather large. Picker *et al.* (1969) have recently described a flow microcalorimeter where, in a similar manner, heat is transferred to a flow of thermostated liquid.

### 5. *Phase-change calorimeters*

A reaction system in a calorimetric vessel can be maintained at a constant temperature if the heat evolved or absorbed is compensated for by the presence of an auxiliary isothermal reaction system, e.g. a melting–crystallization process. This principle is used in so called 'phase-change calorimeters'. The most common type is the Bunsen ice calorimeter where the reaction vessel is enclosed by a hermetically sealed jacket and the space in between is completely filled with a pure liquid compound. A crust of crystals is allowed to freeze on the wall of the reaction vessel. At equilibrium, the two-phase system and the enclosed calorimeter vessel thus form an isothermal system where the melting temperature is equal to the melting point of the crystalline phase. The temperature of the surrounding thermostat is kept closely similar to that of the two-phase system and the calorimeter is thus operated nearly adiabatically.

The freezing-melting processes are accompanied by volume changes, which are proportional to the heat quantities evolved. Volume changes are usually measured by use of a mercury manometer. The most important working substance is not water, as in Bunsen's original calorimeter, but diphenyl oxide. This compound melts at 26·87°C which is close to the 'thermochemical reference temperature', 25°C. Bunsen calorimeters are quite sensitive and suitable for slow processes, but it is a limitation that the reaction temperature is fixed. Bunsen calorimeters do not seem to have been used in modern microcalorimetric designs.

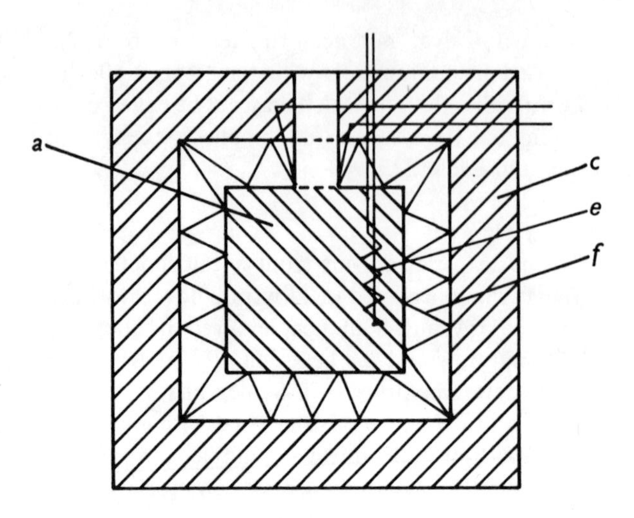


**Figure 3.2.** Thermopile heat conduction calorimeter. *a*, Calorimeter vessel; *c*, thermostated block; *e*, heater; *f*, thermopile. Reprinted from *Quart. Rev. Biophysics*, **3**, 383 (1970)

### 6. *Thermopile conduction calorimeters*

Thermopile heat conduction calorimeters have proved to be very suitable in microcalorimetric designs. Such calorimeters are also commercially available, which has made them the most common type of microcalorimeters in current use.

In all conduction calorimeters there is a controlled transfer of heat from the calorimetric vessel to a surrounding heat sink. Most simply this can be achieved by placing a thermopile wall (*f* in Figure 3.2) between the vessel and the surrounding heat sink. For each thermocouple, both voltage and heat flow are proportional to the temperature difference between the calorimeter wall and the heat sink. In the ideal case, all heat will be transported through the thermocouple and we will have the relationship

$$V = c\frac{\mathrm{d}Q}{\mathrm{d}t} \tag{3.5}$$

where $V$ is the thermopile voltage, $c$ is a proportionality constant and $\mathrm{d}Q/\mathrm{d}t$ is the total heat flow through the thermopile.

In practice, a significant part of the total heat flow does not pass through the thermocouple but takes place through air gaps, insulation material in the thermopile, mechanical supports between vessel and heat sink, leads to calibration heater, etc. However, heat flow can be expected to be proportional to

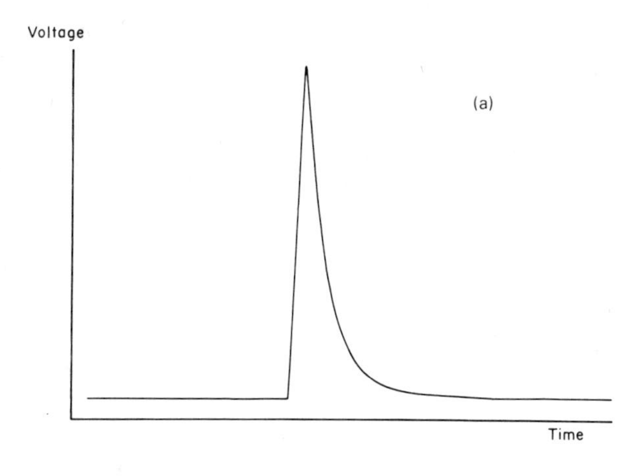


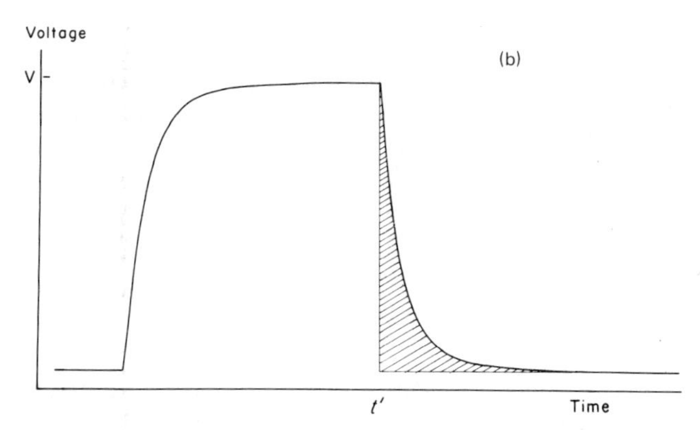


**Figure 3.3.** Voltage–time records for a thermopile conduction calorimeter. (a) Fast process. (b) Evolution of a constant amount of heat will result in a steady-state voltage value (calorimeter acts as a watt meter). Heating effect switched off at time $t$; the shaded area is proportional to the heat capacity of the reaction cell + contents. Reprinted from *Quart. Rev. Biophysics*, **3**, 383 (1970)

the temperature difference between the vessel and the heat sink at the particular site where it takes place. There should be only minor temperature differences between the vessel and the heat sink—as is the case in microcalorimetry—so that heat flow through convection can be neglected. Further, the thermocouples should be well distributed over the surface area of the vessel. The proportionality constant, $c$ (Equation 3.5), can be substituted by an 'effective value', $c'$.

$$V = c' \frac{\mathrm{d}Q}{\mathrm{d}t} \tag{3.6}$$

Usually it is not practicable to cover the entire surface area of the vessel but Equation (3.6) can still be valid to a good approximation. This is in particular expected to be the case if the wall of the calorimeter vessel has a high thermal conductivity so that the appearance of cold or hot spots is avoided. These could seriously interfere with the results if they occurred in a region of low thermocouple density.

Integration of Equation (3.6) will lead to

$$Q = \frac{1}{c'} \int V \mathrm{d}t \tag{3.7}$$

The evolved heat quantity is thus proportional to the surface area, $A$, under the voltage time curve,

$$Q = \epsilon A \tag{3.8}$$

where $\epsilon$ is a calibration constant.

In Figure 3.3a is shown schematically a typical voltage–time record for a thermopile heat conduction calorimeter where heat is evolved as a short pulse. The temperature of the vessel, and thus the thermopile voltage, will rapidly increase to a maximum value after which it will decrease comparatively slowly following an exponential curve. When the voltage has again reached zero, all heat has been conducted out to the heat sink provided the heat sink has an infinite heat capacity. In practice, the ratio between heat capacities of the vessel, including contents, and the heat sink may be of the order of $10^{-3}$, in which case 0·1% of the heat evolved will not leave the reaction vessel. However, the same fraction of the heat evolved will remain in the calorimeter vessel in a calibration experiment. It will thus automatically be corrected for and is included in the experimentally determined calibration constant.

If constant heat is evolved in the calorimeter, the thermopile voltage will attain a steady state value, Figure 3.3b, and baseline displacement ($V$) is directly proportional to the heat evolved in the reaction vessel, Equation (3.6).

Under such conditions and, approximately, for slow processes, the calorimeter clearly acts as a watt meter and can be used for kinetic studies. For fast processes there will be a considerable distortion of the voltage–time curve relative to the true kinetic curve (= the heat effect curve). In order to obtain precise kinetic information in addition to the thermodynamic result, it is then necessary to reconstruct the heat effect curve from the thermopile output data. Calculation schemes have been worked out for this purpose, see e.g. Calvet and Prat (1963).

Thermopile conduction calorimeters have primarily found application as reaction calorimeters, but they can also be used for the determination of heat capacities, $C_p$, cf. Figure 3.3b. Calvet microcalorimeters (see below) have been used rather extensively for this method of $C_p$ determinations but the attainable precision does not seem to meet the requirements for most $C_p$ determinations in biochemistry. A recent drop calorimetric method (Konicek, *et al.*, 1971), where a thermopile heat conduction calorimeter is used as a receiver, seems to be a much more promising approach for work on *e.g.* dilute solutions. However, in biochemical and biological connexions, as as well as in general thermochemistry, it is mainly adiabatic microcalorimeters which have been used for heat capacity work.

### 7. *Single and twin calorimeters*

A calorimeter can be in a single or a twin arrangement. The single arrangement is mechanically simpler but the twin calorimeter has advantages which often makes it attractive, in particular in microcalorimetry where twin arrangements are normal. If the two calorimetric vessels are arranged as perfect twins, it may be expected that disturbances from the surroundings will affect both vessels to the same extent. Therefore, if the differential signal (resistance change, voltage, etc.) from the two vessels is recorded, these disturbances will cancel.

With a twin calorimeter it is possible to carry out simultaneously a reaction in one of the vessels and, for instance, a dilution process in the other. A common type of biocalorimetric experiment is that where two buffer solutions, each containing a reaction component (e.g. enzyme + inhibitor), are mixed. It is in practice not unusual for the mixing process to be accompanied by a significant heat of dilution for one, and only one, of the reaction components. It may, in that case, be suitable simultaneously to mix this component in the reference cell with the same quantity of solvent, usually buffer. The differential signal from the calorimeter will then refer to the net reaction enthalpy value for the process.

The reference cell may also be used to balance out artifact heat effects such as those from wall adsorption and evaporation (see below).

A twin calorimeter may also be operated as a 'thermal balance'. An exothermal process in one of the vessels can be simulated by evolution of electrical energy in the other vessel in order to keep the temperature of the two vessels constant. Heat supplied to the twin vessel is, in the case of perfect twins, equal to the heat evolution in the reaction vessel and disturbances are at the same time expected to cancel.

For heat capacity determinations of biochemical compounds in dilute solutions, it is usual, or rather it has been found to be necessary, to use twin systems. If the solution is placed in one vessel and the pure solvent in the other, the small difference in heat capacity is directly measured. This is obviously a much more suitable procedure than measuring the heat capacity for the solution and for the solvent separately and then comparing these two nearly identical numbers.

*8. Batch and flow calorimeters. Mixing of reaction components*

When two or more components are brought together in a calorimeter it is often done by means of an ampoule or burette technique. Another method is to mix the contents of two separate compartments through rotation of the calorimeter. Such techniques are called batch procedures. It is also possible to mix the reaction components by means of a continuous-flow or a stopped-flow technique. These latter procedures can be of great convenience and are at present attracting a lot of interest in microcalorimetry, not least in connexion with the current promising trends in the analytical use of microcalorimeters.

With batch procedures, the method chosen for initiation of the process is very critical. In microcalorimetry this is of particular importance. Devices of the type used in 'macrocalorimetry' for separation of reaction components prior to the reaction will easily bring about heat effects which are too large and too poorly reproducible to be useful when only minute heat quantities are evolved. The methods used today which seem to create the smallest thermal disturbances are based on the use of open compartments where the contents are mixed by rotation of the calorimeter assembly. Here, however, the vessel will contain a rather large gas phase which can cause systematic errors due to evaporation effects (see below).

Use of mechanical devices such as ball mixers (Berger *et al.*, 1968) are efficient and can work without vapour space, but create comparatively large and not very reproducible heat effects.

In a flow calorimeter there is usually no gas phase present, which makes these calorimeters ideal for mixing and dilution experiments.

In a batch calorimetric experiment involving very dilute solutions, the possibility of adsorption effects (see below) must be considered. With a continuous-flow procedure, a steady state will be attained at which such surface effects should not be present.

Batch calorimeters may be suitable for reactions ranging from instantaneous processes to very slow processes. For a flow calorimeter it is normally required that the reaction must go to completion in the mixing chamber, i.e. the reaction time should be short compared to the retention time for the liquid in the flow cell. With a flow calorimeter it may also be possible to perform stopped-flow experiments. Here, the reaction time is not critical but the precision will inevitably be much lower than in a continuous-flow experiment.

For zero-order reaction systems and for very slow processes, a 'flow through procedure' may be used where reactions are preferably initiated outside the calorimetric system (Monk and Wadsö, 1969).

## B. Sources of Errors in Microcalorimetric Experiments

In the past, and perhaps still at the present time, calorimetry has had a solid reputation as a difficult experimental technique. But today, several calorimeters have been developed to a stage where they are very easy and convenient to operate. This is certainly the case for work in the biological and biochemical areas where the experimental technique can be made very simple indeed.

However, even if the instrument is well designed and the experiment is simple, there are, in calorimetric experiments, many possible ways of introducing systematic errors into the results. The risk of systematic errors is possibly greater in calorimetry than in any other scientific measurement technique since heat evolution or absorption takes place in all kinds of processes both chemical and biological and also in physical processes such as mechanical effects (friction etc.), evaporation, condensation and adsorption processes.

The risk is more serious in microcalorimetric measurements than in studies where comparatively large quantities of heat are evolved. Depending on what level of sensitivity or degree of accuracy is aimed at, these errors may be neglected or they may easily be controlled, but they may also make the results completely meaningless. It is thus important to be constantly aware of the various sources of systematic errors when designing an experiment as well as when the final results are being worked out.

We may note, however, that particularly in analytically oriented experiments, many errors will in practice be cancelled out by the procedures used for the standardization of the analytical method.

### *1. Mechanical effects*

When a liquid is agitated in a calorimetric vessel, for example, by a stirrer or by rotation of the vessel, heat will normally be generated by friction. In microcalorimetry, constant agitation of the medium throughout the experiment is usually avoided as the friction effects tend to be too large and not constant enough in comparison with the heat effects being studied. In practice, this kind of effect, together with mechanical effects in connexion with the initiation of a process, often limits the precision or the useful sensitivity of a microcalorimetric experiment. Constant stirring may normally be considered as a background effect and no correction is applied. But heat quantities evolved by, for example, one or a limited number of rotations of a reaction vessel must usually be corrected for—this is in practice true also when twin vessels are used and where thus only the differential heat quantity is observed.

In 'macro' reaction calorimetry, it is common to initiate a reaction by breaking an ampoule or opening a valve. Corresponding heat effects may be quite large, being typically of the order of 10 mcal or larger. For a macrocalorimetric experiment this can often be neglected. But it is inconveniently large and in general too poorly reproducible for most microcalorimetric experiments. Therefore, if any kind of mechanical device is used to start a reaction in a microcalorimetric experiment, it is very important to check frequently the heat quantities involved in its operation.

*2. Evaporation and condensation*

Water is invariably present in biochemical and biological experiments. The heat of vaporization of water is 10·5 kcal/mol (at 25°C), i.e. more than 0·5 mcal of heat is absorbed per $\mu$g of water evaporated. In some experiments the total heat evolution may be of the order of only 1 mcal or less. It is thus clear that *virtually no uncontrolled evaporation or condensation can be tolerated* with experiments at this sensitivity level.

Further, it is clear that the calorimetric vessel must be completely sealed—even a minute leakage will ruin the experiment. However, if there is leakage in the vessel of a microcalorimeter, the calorimeter signal is usually very erratic and leakage will thus rarely be recorded as systematic errors.

Evaporation or condensation processes inside the calorimetric vessel are more serious from the point of view of systematic errors. In order for such processes to take place there must be a gas phase in the calorimeter vessel. This is normally the case in batch microcalorimetric designs where the gas phase may be used as a means to facilitate efficient mixing. Typical examples are given by the reaction vessels used with the microcalorimeters shown in Figures 3.6 and 3.7. This category of mixing vessels seems to be the best from the point of view of a small and reproducible heat of friction at the mixing process. However, the possibility of distillation effects and of change of the vapour-phase composition must be considered (Wadsö, 1968).

If the composition of the two solutions is different, it is likely that the equilibrium solvent vapour pressure will also be different. This means that some distillation will be expected to take place between the two compartments before mixing takes place. There will thus be a cooling effect in the compartment containing the liquid with the higher vapour pressure and a heating effect of nearly the same magnitude in the other compartment. The only difference in the magnitude of the two heat effects will be related to dilution or concentration effects, which can normally be neglected.

The distillation process may cause a temperature gradient within the vessel. For a thermopile conduction calorimeter this is not expected to cause any systematic errors but it may possibly interfere with the temperature measurement in an unstirred isoperibolic or adiabatic shield calorimeter. It is usually more important to consider the change of the gas-phase composition which may accompany a mixing process. Prior to mixing, the gas-phase composition is not well defined since we do not have an equilibrium system. The composition will be different in different parts of the vessel and will depend on the distillation rate. This means that it is not possible to apply any rigorous corrections, even if the equilibrium vapour pressures are known for the two starting solutions, as well as for the final mixture. As an example we may consider an experiment where the gas phase is 3 ml and where a 1 $M$ aqueous solution is mixed with two parts of water (Wadsö, 1968). The most unfavourable situation would be if the vapour phase before the mixing were to be nearly in equilibrium with the pure water (distillation will occur from the solvent compartment to the solution). The maximum decrease in vapour pressure at mixing can be estimated from Raoults law to be approximately 0·15 mm Hg which corresponds to a condensation of $2{\cdot}5 \times 10^{-8}$ mole of water or a heat quantity equal to 0·3 mcal. Mixing of dilute aqueous solutions at moderate temperatures will not usually cause any significant change in the gas-phase composition. The same applies for mixing of strong but nearly identical aqueous solutions.

A mixing vessel with open compartments and a gas phase should never be used for mixing experiments involving volatile organic compounds where completely erroneous results may be obtained, cf. McGlashan (1962). In such cases, as well as in accurate experiments with two aqueous solutions having large differences in molar composition, experiments should be carried out using calorimetric vessels operating without a gas phase. Flow calorimeters seem to be the ideal instruments for such experiments.

Most biochemical and biological calorimetric studies are performed with reaction components at very low concentrations. The molar concentration of buffers and neutral salts may be high but their concentrations are normally kept identical for the two solutions. Therefore, in practice, most biocalorimetric experiments can be carried out without significant effects from the change of the gas-phase composition.

### 3. *Adsorption*

Heat effects due to adsorption or desorption of reagents from the walls of a calorimetric vessel can usually be neglected in a macrocalorimetric experiment. In microcalimetry, the ratio between wall surface area and the total heat quantity measured is more unfavourable and sorption phenomena can very seriously affect the results.

Benzinger (1969) has reported results from a series of NaOH–HCl neutralization experiments which were carried out in the micromolar range using a glass vessel. As well as the heat of neutralization, he noted a constant heat effect which was interpreted as due to the adsorption of $Cl^-$ ions on the glass wall. It was shown that the effect disappeared if small quantities of NaCl were added to the NaOH solution. From experiments with different quantities of NaCl present, the enthalpy of chloride adsorption was calculated to be —9·2 kcal/mol. When the experiments were carried out in a 'Teflon' vessel, the effect was reduced by a factor of two.

Similar effects have been observed in our laboratory between 18 carat gold vessels and various solutions of ionic compounds. If a clean reaction vessel of the type shown in Figure 3.7b was charged in both compartments with

0·5 $M$ KCl solution, significant exothermic heat effects were observed, about 1·5 mcal. For $K_2SO_4$ and $Na_2SO_4$, the effect was much smaller (*c.* 0·5 mcal) whereas the values for KBr were around 2·4 mcal. We interpret the observed effect as due to adsorption of ions to those parts of the vessel wall which had not been in contact with the salt solution before the mixing.

Similarly, if a clean and dry vessel is charged with pure water in both compartments, the first rotation will usually be accompanied by a significant heat evolution in excess of the differential heat of rotation. This effect will decrease with increasing equilibrium time. We believe it is caused by a slow attainment of the equilibrium between the aqueous phase and the non-wetted parts of the vessel.

These effects, due to adsorption of solvent components, are usually not completely cancelled even if the experiments are performed as twin experiments. This is not surprising since the properties of a surface may appreciably vary with its history. Further, it is not very likely that washing and drying procedures will lead to sufficiently reproducible results when adsorption of the order of $10^{-8}$ mole of material has to be considered.

In order to minimize errors due to adsorption effects, we work routinely in our laboratory with wetted cells in cases where small heat quantities are evolved ($\leqslant$ a few mcal). Before an experiment, the cleaned reaction vessel is washed with the reaction medium (usually a buffer solution). Likewise, the twin vessel is washed with the reference solution (usually identical to the reaction medium in the reaction vessel). The vessels are emptied and then charged again with the reaction and reference solutions. The vessels will thus contain an unknown but minute quantity of washing solution before charging, but corresponding mixing or dilution effects are usually insignificant. Sometimes it may be suitable to include the compounds reacting or those formed (cf. the experiment by Benzinger referred to above) in the washing solution.

In a continuous-flow experiment with a flow calorimeter no adsorption effects are to be expected. The walls of the flow cell will be saturated with the reaction mixture at the beginning of the experiment and when the instrument has reached a steady state there will be no net effects from adsorption processes.

### 4. *Gaseous reaction components*

If a gas is evolved in a biochemical or biological experiment it is always saturated with water vapour and there will be an evaporation effect if the gas is allowed to leave the reaction vessel. If 1 mole of a gas leaves an aqueous reaction mixture at 37°C this will be accompanied by an endothermic effect of 0·64 kcal ($p(H_2O) = 47$ mm Hg; $\Delta H_V(H_2O) = 10{\cdot}36$ kcal/mol). At 25°C, the corresponding effect is 0·33 kcal/mol ($p(H_2O) = 23{\cdot}8$ mm Hg; $\Delta H_V(H_2O) = 10{\cdot}52$ kcal/mol).

Similarly, if a more or less dry gas, such as air, is bubbled through an aqueous reaction system, there will be a large cooling effect. In fact, it is very difficult to perform a calorimetric experiment at a low heat level and at the same time add the air or oxygen which may be necessary in a biological experiment. Even if one tries carefully to saturate the gas with the same liquid as is used in the calorimeter vessel, at the same temperature, significant vaporization or condensation effects may take place.

In experiments where a mixed flow of liquid and gas can be pumped continuously through a flow calorimeter cell, the risk of such effects is greatly reduced (Eriksson and Wadsö, 1971; cf. p. 120).

If a gaseous compound is consumed in an experiment, the process is best defined if the compound is absorbed from the liquid phase without being replaced from a gas phase present in the reaction vessel. Microcalorimetric experiments of this kind have recently been described by Keyes *et al.* (1971) and by Langerman and Sturtevant (1971) who studied biochemical oxygenation reactions using flow calorimeters where no gas phase was present.

If dissolved gas takes part in the reaction but is replaced from a gas phase present in the reaction vessel, there will be a net consumption of gaseous compound. In practice, this pattern may not be followed to 100% in a calorimetric experiment, as during the experiment there might be a decrease in the amount of gas dissolved. The difference in $\Delta H$ for a reaction where a gaseous reaction component is taken from the gas phase and where it is taken from solution without being replaced equals the heat of solution of the gas. For oxygen in aqueous systems this value is $\Delta H_{soln} = -3{\cdot}8$ kcal/mol at 25°C. In quantitative work it is clearly important to have this kind of factor properly defined.

### 5. *Ionization reactions and other side reactions*

When a chemical reaction is to be followed calorimetrically it is usually desirable to correct the directly obtained value to an idealized process, e.g. a reaction performed at infinite dilution and under conditions where the reaction components have certain specified ionization states. From the directly obtained quantities the idealized reaction value may be obtained by performing auxiliary experiments such as dilution and ionization reactions or by the use of known data. It is particularly important to remember that if there is a release or an uptake of protons (or hydroxyl ions) in the reaction, the buffering medium will be involved in the gross process measured. Enthalpies of ionization are very different for different buffers. Carboxylic acid groups have $\Delta H_i$ values close to zero, for phosphate (second ionization) $\Delta H_i$ (25°C) is 1·13 kcal/mol, whereas $\Delta H_i$ values for amine buffers are strongly endothermic: $\Delta H_i$ (25°C) for Tris is +11·35 kcal/mol. A valuable tabulation of $\Delta H_i$ values for biochemical compounds and buffer substances is found in the compilation by Izatt and Christensen (1968). Examples of typical correction schemes for buffer ionization processes are discussed by Sturtevant (1962) and Wadsö (1969). The differences in $\Delta H_i$ values for different buffers can be utilized for determination of the number of protons involved in a given process, see e.g. Langerman and Sturtevant (1971).

Acid–base equilibria are always involved in reactions with biochemical compounds or biological systems. When two such systems are mixed in a calorimetric experiment, there may be significant proton transfer processes taking place if pH values for the two systems are not very nearly identical. (Cf. the fact that the number of buffer molecules or the number of acid or base groups present in the biological material may be very large compared to the number of reaction sites involved in the nominal reaction.) An effective protection against effects of this kind is to dialyse macromolecular reagents and to use buffer solutions which have been equilibrated with the biological material.

It is also important to realize that buffer substances, which are often present at high concentrations, may be bound to differing extents to the starting material and to the reaction products of a biochemical reaction. This can be expected to be the case especially when dramatic changes take place, e.g. with denaturation processes. Such binding would not be observed by most experimental techniques but calorimetric values are likely to be affected. It is therefore desirable to perform such calorimetric experiments in different buffer systems in order to investigate the possibility of this kind of effect.

### 6. *Incomplete mixing*

When a reaction is carried out in a batch experiment there should be no systematic error due to incomplete mixing. In a microcalorimetric experiment, the mixing procedure is usually repeated after the reaction has taken place in order to determine a correction for possible friction effects. If these heat effects are normal, or at least constant, this is a proof that the initial mixing was adequate. When reagents are mixed in a flow calorimeter, the risk of systematic error is greater. The presence of stirrers or other mechanical devices in the mixing cell is usually avoided and the mixing efficiency may therefore be comparatively low. The steady-state signal will look normal even if the reaction components are incompletely mixed when leaving the calorimeter.

It is therefore desirable to test the efficiency of the flow mixing cell if work is done with systems which may be expected to be difficult to mix, e.g. viscous solutions, cf. Monk and Wadsö (1968). This can be done by experiments where the molar ratio of the reaction components is changed. If a titration experiment is performed with a reaction system where the equilibrium position is strongly on the product side, there should be a sharp break at the expected equivalence point if the mixing is efficient.

### *7. Slow reaction*

In batch experiments, serious systematic errors can be introduced for slow processes if corrections or compensation procedures for heat exchange are in error. This applies in particular to isoperibolic calorimeters but usually not to thermopile conduction calorimeters where errors due to slow reactions will tend to be random.

In a continuous-flow experiment it is of course important that the process actually takes place in the anticipated 'reaction zone' of the flow system. For instance, in a flow mixing cell of a thermopile conduction calorimeter, the reaction must be complete well before the liquid reaches the exit of the cell. Otherwise, it is likely that too much heat, compared to that in the calibration experiment, will be transported out with the liquid flow.

When unknown or slow processes are run in a flow mixing cell, it is therefore desirable to carry out experiments with different flow rates which will give different retention times for the reaction mixture. It should then be noted that the calibration constant for the calorimeter may vary strongly with the flow rate, see e.g. Monk and Wadsö (1968).

### *8. Modification of instrument design*

It is not unusual for a standard instrument to be modified in order to make it more suitable for a certain experiment. For many kinds of physicochemical instruments it is quite clear if the alterations will or will not affect the calibrated response of the instrument. With a calorimeter, however, it is easy to be mistaken on this point.

When a calorimeter is developed it has to be shown by test experiments that the instrument, within stated uncertainty limits, does give correct results. It is important to realize that results of such test experiments hold true only for the particular design which was used for the test experiments. Even when small changes have been made on a calorimeter it is therefore advisable to test the modified instrument carefully if it is going to be used for accurate quantitative experiments. Electrical calibration experiments (e.g. with different heater positions) are most valuable for the evaluation of the effect of a design change, cf. Monk and Wadsö (1968), Kusano *et al.* (1972). Where chemical test experiments can be performed, it is as well to do so. Preferably they should involve well-known processes which are of the same type as those which will be studied with the modified instrument.

## C. Some Microcalorimeter Designs

In this section a few microcalorimeter designs will be briefly described. The examples have been chosen from among designs which have proved useful in biochemical and biological work. Preference has been given to those types which are commercially available and which therefore at present are

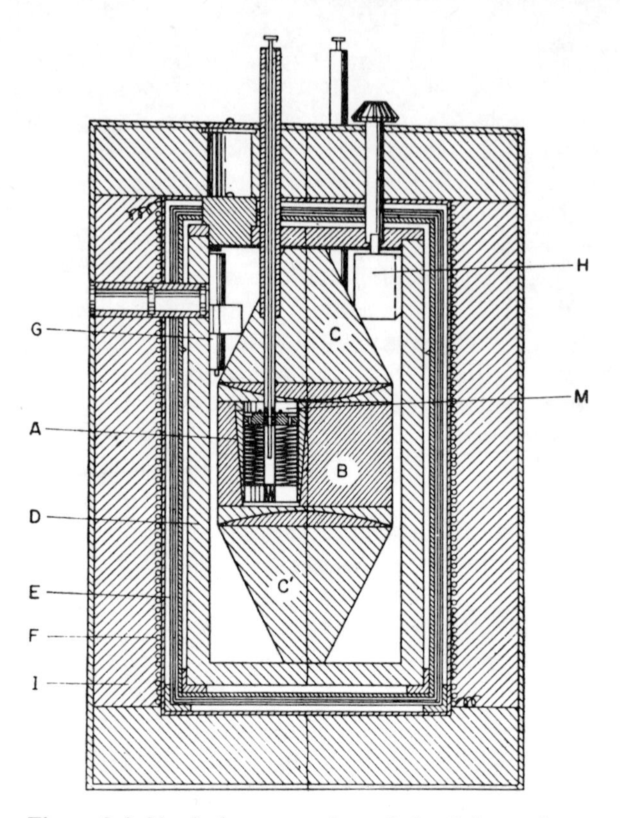


**Figure 3.4.** Vertical cross-section of the Calvet microcalorimeter. *A*, Conical socket cut into the block *B*; *B*, the metal block; *C* and *C′*, metallic cones; *D*, thick metal cylinder; *E*, thermostat consisting of several metal canisters; *F*, electrical heater; *G*, galvanometer; *H*, switch; *I*, thermal insulation; *M*, microcalorimeter element. Reprinted from *Recent Progress in Microcalorimetry*, Pergamon Press

those which are most widely used. For a more extended discussion the reader is referred to Wadsö, (1970) and Spink and Wadsö (1975).

### *1. Reaction calorimeters*

As noted earlier, calorimeters of the isoperibol type are in common use and several of them are commercially available, e.g. from LKB-Produkter, Stockholm-Bromma, Sweden (in particular reaction, solution and titration calorimeters), Tronac, Provo, Utah (titration calorimeters) and from American Instruments Co., Silver Spring, Md. (titration calorimeter). None of these instruments seem to qualify as microcalorimeters in the sense the term

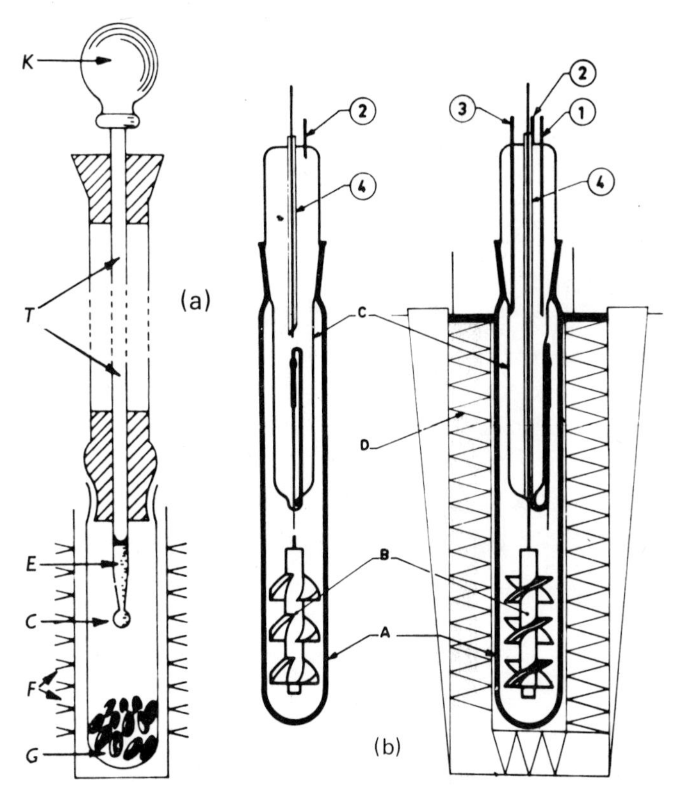


**Figure 3.5.** Example of calorimeter cells for a Calvet calorimeter. (a) Cell used for the study of germination of seeds (Prat, 1969). *G*, seeds; *F*, thermopile; *C*, vaseline plug; *E*, tube containing 1 ml of water; *T*, glass tube; *K*, rubber bulb. Reprinted from *Biochemical Microcalorimetry*, Academic Press. (b) Cell used for mixing of two solutions (Belaich and Sari, 1969). *A*, Pyrex calorimeter cell; *B*, helix stirrer in 'Teflon'; *C*, siphon: (1) tube for reactant introduction; (2) tube for the equilibration of pressure between the siphon and the exterior; (3) tube for the equilibration of pressure between the calorimeter cell and the exterior; and (4) access through the siphon for the stirrer axis; *D*, thermocouples surrounding the calorimeter cell. Reprinted from *Proc. Natl. Acad. Sci.*, **64**, 763 (1969)

is used here. But calorimeters of this sort are no doubt useful for many biochemical applications, in particular in experiments on model compounds and with inexpensive biochemical compounds.

Among microcalorimeters which are used today for the investigation of biochemical and biological processes, the thermopile heat conduction type is the most popular.

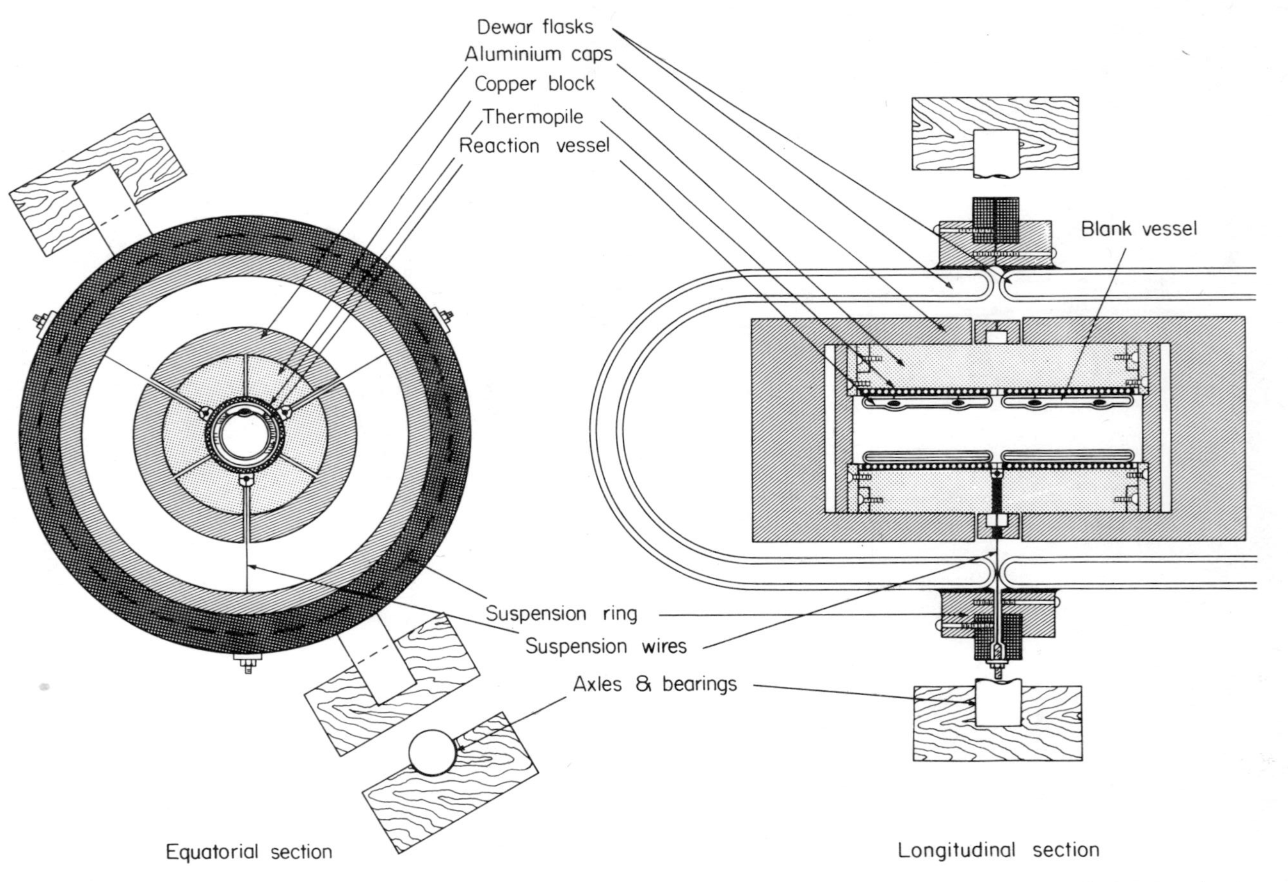


**Figure 3.6.** Thermopile heat conduction calorimeter (Kitzinger and Benzinger, 1960). External insulating shells are not shown except for the Dewar vessels. Reprinted from *Methods of Biochemical Analysis*, Vol. VIII, Interscience, 1960

A well-known instrument of this kind is the Calvet calorimeter which has been discussed in detail in several publications, see e.g. Calvet and Prat (1963).

The Calvet calorimeter is a twin calorimeter using the thermopile heat conduction principle. However, the heat conductivity of the thermopile is rather low and means are provided therefore for the use of Peltier effect cooling. A cross-section of this instrument is shown in Figure 3.4. In the centre of the calorimeter units are tube-shaped reaction vessels, each surrounded by two thermopiles. One of these is used for Peltier effect cooling and the other for temperature indication and heat-flow measurement. Peltier effect cooling is not used for total compensation but for balancing the major part of the heat produced.

Initiation of a process is started by, for example, breaking an ampoule, by pipette or by siphon techniques, see Figure 3.5 and 3.10. In recent modifications, it is possible to mix the components by rotation of the block (Evans, 1969; Brown, 1969).

Studies carried out with Calvet calorimeters include thermogenesis of plants, microorganisms, insects and small animals. For literature references in these areas see e.g. the recent reviews by Forrest (1969) and Prat (1969).

The Calvet calorimeter is made commercially by Setaram, Lyon, France.

In Figure 3.6, the Benzinger 'heat burst' calorimeter is shown (Kitzinger and Benzinger, 1960; Benzinger, 1969). This construction is a typical twin thermopile heat conduction calorimeter. In a metal block, which acts as a heat sink, there is a hole which contains two cylindrical thermopiles connected in opposition and which enclose the two reaction vessels. The thermopiles are made from a constant spiral which has been copper plated on every half turn to form a larger number of copper–constantan thermocouple junctions connected in series.

The reaction cells shown in the figure are formed by two cylinders joined together at the ends to form vessels with an annular cross-section. The inner cylinder has dimpled recesses which house one of the reactants prior to mixing. The other is contained in the annular space. Bicompartmented vessels for mixing equal volumes have also been described, see Benzinger (1969). Mixing is achieved by rotation of the calorimeter assembly.

A Benzinger type of microcalorimeter has been marketed by Beckman Instruments (Palo Alto, Calif.).

A flow version of the Beckman calorimeter has also been described (Sturtevant, 1969) which very recently has been used for a number of biochemical thermodynamic studies, in particular by Sturtevant and coworkers.

Figure 3.7 shows another thermopile heat conduction calorimeter in a twin arrangement (Wadsö, 1968). In this calorimeter, as in the Benzinger–Beckman calorimeter, heat conduction takes place rapidly through the thermopile and there is no need for Peltier effect cooling.

The reaction vessels consist of rectangular or squared cans of glass or

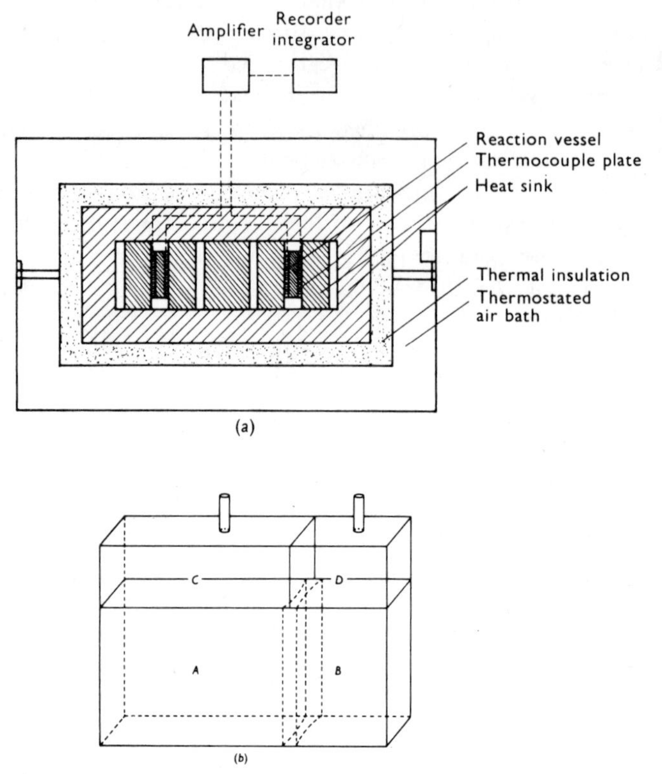


**Figure 3.7.** Thermopile heat conduction calorimeter with semiconducting thermocouple plates (Wadsö, 1968). (a) calorimeter block and air thermostat; (b) reaction vessel. Normal liquid volumes are 4 + 2 ml. Reprinted from *Acta Chem. Scand.*, **22**, 927 (1968)

gold which are divided by a partition wall into two compartments, Figure 3.7b. A heater used for electrical calibration is positioned in the partition wall. The reaction cells are on both sides in thermal contact with thermocouple plates to form a sandwich-like construction. The thermocouple plates are semiconducting 'thermoelectric coolers' which have a high heat conductivity and a very low electrical resistance. The calorimeter block is suspended in a thermostated ($\pm 0{\cdot}01\,°K$) air bath. Mixing of reagents originally kept in spaces $A$ and $B$ is achieved by rotation of the calorimeter assembly.

Like other thermopile heat conduction calorimeters in a twin arrangement, this calorimeter is suitable for measurements of both fast and slow (hours or days) processes.

Recently, this type of microcalorimeter has been used as a receiver in a

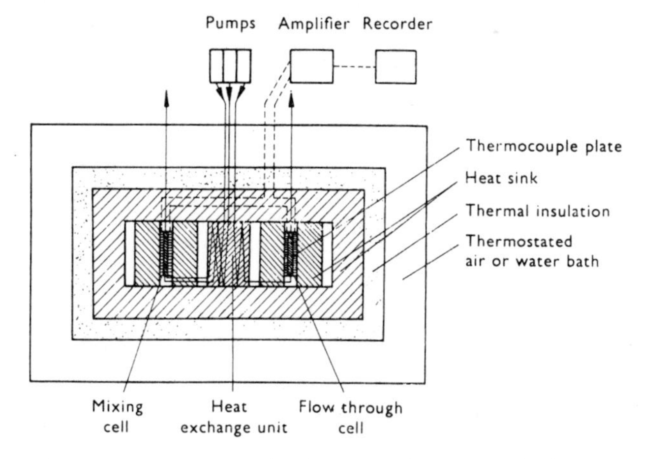


**Figure 3.8.** Thermopile heat conduction calorimeter. Flow version of the batch calorimeter shown in Figure 3.7 (Monk and Wadsö, 1968). Reprinted from *Acta Chem. Scand.*, **22**, 1842 (1968)

'drop calorimeter' for precise heat capacity determinations of small liquid or solid samples (Konicek *et al.*, 1971).

Figure 3.8 shows a flow version of the batch calorimeter just described (Monk and Wadsö, 1968). The solutions are first pumped through a heat exchange unit positioned in the centre of the calorimeter block. The calorimeter usually contains two different flow cells. One is a heat of mixing cell and the other is a flow-through cell (Monk and Wadsö, 1969). The former is used for mixing processes and for measurements of fast processes. The flow-through cell is used for slow processes which are preferably initiated outside the calorimeter, e.g. in enzyme assay experiments involving substrate-saturated enzyme solutions (Monk and Wadsö, 1969) or in microbiological growth experiments (Delin *et al.*, 1969; Eriksson and Wadsö, 1971; Poole *et al.*, 1973).

Commercial versions of both the batch and the flow type of this instrument are marketed by LKB-Produkter, Stockholm-Bromma, Sweden.

*2. Calorimeters used for determination of heat capacities and thermally induced transitions*

Heat capacity determinations of biochemically important compounds in solution form a very important area of investigations in biothermochemistry. However, such measurements are often difficult to perform. For a 1 % aqueous solution about 99·6 % of the heat capacity is due to the water. In order to determine the apparent heat capacity for the solute with the rather moderate precision of 1 %, it is thus necessary to perform the calorimetric measurement with a precision of 0·004 %, or better if the heat capacity of the calorimetric vessel is significant compared to that of the solution. This precision can be achieved by the most precise reaction calorimeters of the isoperibol type (see Biltonen *et al.*, 1971). However, it is then necessary to use rather large quantities of substance—of an order of 100 ml of a 1 % solution—and such studies can therefore only be performed on inexpensive compounds.

The new type of microdrop heat capacity calorimeter referred to above seems promising for work on small samples of dilute solutions as well as for solid materials.

Most macromolecules in solution undergo thermal transitions when heated to higher temperatures. Calorimetric investigations of such phenomena are preferably made by heat capacity calorimeters designed for temperature-scanning operation, see e.g. Ackermann (1969). During the last few years, several adiabatic shield calorimeters suitable for temperature-scanning studies on small samples of dilute solutions have been reported (Privalov *et al.*, 1964; Danforth *et al.*, 1967; Clem *et al.*, 1969; Bakradze *et al.*, 1971). They all use a differential arrangement, i.e. the heat capacity of the sample solution is compared with that of a reference solution.

This type of calorimeter has recently been used for a substantial number of important studies on biochemical compounds in solution (for a review see Wadsö, 1972) as well as on biological material (see e.g. Andronikashvili *et al.*, 1970). However, these calorimeters are quite intricate in their design and so far none of them has been made commercially available.

Instruments in the same category are the 'Differential Scanning Calorimeters' (DSC) marketed by Perkin-Elmer, Norwalk, Conn. These are useful for transition studies and heat capacity determinations on small solid or liquid samples and over a wide temperature range. Sample requirements are in the mg range. It has been demonstrated that the 'DSC' is useful for temperature-scanning studies on fairly concentrated biochemical solutions (*c.* 5 %), see e.g. work by Crescenzi and Delben (1971), and on solid biopolymers or biological materials, see e.g. work by Berlin *et al.* (1970), Reinert and Steim (1970), Chapman and Urbina (1971) and Chapman and Chen (1972). Hasl and Pauly (1971) have recently studied the thermal properties of water in concentrated protein solutions at temperatures below 0°C using a modified Perkin-Elmer DSC-1B instrument. In particular the new model DSC-2 seems to offer significant opportunities for use on biochemical and biological systems.

## III. APPLICATIONS OF MICROCALORIMETRY IN BIOLOGICAL SCIENCES

Biocalorimetry is conducted in areas extending from general chemistry into the field of biology and it includes thermodynamic work as well as analytical applications. In the next few paragraphs, various kinds of applications will be commented upon and a few specific examples will be discussed.

## A. Thermodynamic Studies

Thermodynamic data for a process are directly related to the breaking and formation of all types of chemical bonds and to changes in molecular interactions. For simple chemical systems it is often possible to make a detailed analysis of the origin of the various contributions to determined thermodynamic values. For biochemical systems this is generally not yet the case but with an increasing amount of data available, including some for model systems, we are moving towards a clearer understanding of the quantities determined and thereby of the systems themselves. Naturally, our ability to interpret thermodynamic quantities of complex systems will increase the more other pieces of quantitative information are available for the particular system. All the physicochemical data taken together form a quantitative description of the system which can be utilized for basic scientific discussions as well as for various empirical correlations or classifications.

The primary goal of a thermochemical study is usually to determine an enthalpy change, $\Delta H$. If the measurements are made at different temperatures its temperature coefficient, $\Delta C_p$, can be evaluated

$$\frac{d\Delta H}{dT} = \Delta C_p \tag{3.9}$$

$\Delta C_p$, the change in the heat capacity of the system, is a most important thermodynamic function in biochemical connexions, but its determination requires very precise calorimetric work.

As will be discussed below, it is sometimes possible to derive the enthalpy change as well as the equilibrium constant, $K$, from results of one series of calorimetric experiments. From the latter, derived calorimetrically or by other methods, the standard free energy change, $\Delta G^0$, can be calculated:

$$\Delta G^0 = -RT\ln K \tag{3.10}$$

By combination of the values for free energy and enthalpy changes, the entropy is obtained

$$\Delta G = \Delta H - T\Delta S \tag{3.11}$$

### *1. Model systems*

It is not an easy task to discuss on a molecular level the results from calorimetric studies performed on complex biochemical systems. One useful approach towards a better understanding of biothermochemical results is to study appropriate model systems. Within a series of model compounds the structure can be varied systematically and it is then possible to make correlations between thermodynamic quantities and structural parameters.

One area of significant importance in current calorimetric work on model compounds deals with the interaction between water and simple compounds and with the transfer of model compounds between water and other media. Such studies are mainly performed as heat of solution measurements, giving $\Delta H$ and $\Delta C_p$ values.

If $\Delta C_p$ for a solution process ($\Delta C_{p_{\text{soln}}}$) and the heat capacity for the pure compound ($C_p{}^0$) are known, one can calculate the very important value for the (apparent) partial molar heat capacity of the compound in solution, $C_{p_2}$

$$C_{p_2} = C_p{}^0 + \Delta C_{p_{\text{soln}}} \tag{3.12}$$

In work with model compounds, it is often possible to perform precise heat of solution measurements and this approach will then usually lead to more precise $C_{p_2}$ values for dilute solutions than the comparison between directly determined heat capacities for solvent and solution, respectively. Calorimetric studies on biochemical models are often best performed by the use of precise 'macrocalorimeters' where work is done on the mmole scale, cf. p. 104.

### 2. *Biochemical systems*

For simple biochemical compounds the accuracy of thermochemical measurements may approach that obtained with model compounds. With complex biochemical compounds, however, it is usually not possible to define the state of the samples to better than a few per cent. But it is still often desirable to perform the calorimetric experiment with a very high precision. This is certainly true for heat capacity experiments with dilute solutions, but also, for example, in specific binding reactions.

Studies can be based on a comparison between a series of well-defined compounds interacting with the biopolymer, or conditions such as pH, buffer system and temperature may be varied systematically. The state of the less well-defined biochemical compound is constant throughout the series of experiments and will thus serve as an internal reference. Even if a high precision is obtained, and can be utilized, in this kind of work, it should be remembered that the results may not be very accurate in an absolute sense.

Current calorimetric–thermodynamic work on biochemical model compounds and on well-defined biochemical systems has recently been reviewed (Wadsö, 1972, cf. also Brown, 1971). Among studies in this latter category there is a rapidly increasing number of studies performed on the specific binding of low molecular weight compounds to proteins, see e.g. Ross and Ginsburg (1969), Bjurulf *et al.* (1970), Shiao (1970), Hinz *et al.* (1971), Bolen *et al.* (1971), Velick *et al.*, (1971), Maurer *et al.* (1971) and Barisas *et al.* (1972).

Another active area is formed by the studies of specific interactions between biopolymers, see e.g. Ross and Scruggs (1969), Sjöquist and Wadsö (1971), Stauffer *et al.* (1970), Hearn *et al.* (1971).

A significant field involves investigations of unfolding processes for biopolymers. The unfolding processes may be induced chemically by changing the pH or by addition of denaturants, see, for example, work by Rialdi and Profumo (1968), Biltonen *et al.* (1971), Lapanje and Wadsö (1971), Atha and Ackers (1971). Temperature-scanning heat capacity calorimeters have been used in several studies involving thermally induced unfolding processes for proteins, polypeptides, nucleic acids and polynucleotides, see e.g. Jackson and Brandts (1970), Tsong *et al.* (1970), Privalov and Tiktopulo (1970), Privalov *et al.* (1971) and Khechinachivili *et al.* (1973). As pointed out earlier, this kind of study usually requires rather intricate calorimetric equipment which is not yet commercially available, although it has been shown (see e.g. Crescenzi and Delben, 1971) that the Perkin-Elmer 'DSC' can be used for studies of thermal unfolding in comparatively concentrated solutions of biopolymers.

### 3. *Biological systems*

Biological systems with more or less intact life functions are very poorly defined from a physicochemical point of view. Calorimetric investigations on such systems will therefore not provide thermodynamic results which can be discussed on a molecular level in the same sense as those determined for purified systems. Another kind of thermodynamic work involves studies of chemical potentials in metabolic processes. Such studies, mainly performed as non-calorimetric work, are not judged to be of major current interest. We seem to have reasonable information about the various energy steps in the important metabolic reactions, and it is doubtful if a substantial increase in the amount or quality of data in this field can be of much value at the present time.

Still another reason for making thermodynamic investigations on biological systems is to determine energy balances, undertaken to show the degree of utilization of nutritional substances.

## B. Analytical Studies

The heat quantity evolved or absorbed in a process is quantitatively related to the extent and to the intensity of the process. It is thus clear that calorimetry forms a very general analytical principle which is of a potential interest for all fields of biochemical and biological analysis.

At the moment, however, calorimetric analytical methods are still of only marginal importance in the biological sciences. This can be attributed to several factors: calorimetric methods have a low specificity, suggested experimental methods have usually been time consuming and experimentally difficult and the sensitivity has often not been adequate.

However, due to the instrumental developments during recent years, the picture is now very much more interesting. Modern microcalorimeters are usually very simple to operate and their sensitivities are high (even if there are many applications which call for an as yet unattainable sensitivity).

For measurements on biochemical systems, the inherent low specificity of all calorimetric methods may not be a decisive disadvantage as these systems are often sufficiently specific to allow use of an unspecific analytical method.

For the biological field, we should further note the very important property that living processes are always accompanied by a heat effect. If the sensitivity is sufficient cellular reactions may always be recorded by a calorimetric method and it is not necessary to disturb the biological system by the addition of reagents or by infrared radiation.

Another factor of great importance to analytical investigations on biological systems is that calorimetric methods, in contrast to spectrophotometric methods, do not require optically clear objects, but can be used on non-transparent systems like animals, plants, tissues, soil or cell suspensions.

It has been demonstrated that flow microcalorimeters in particular can be used as convenient analytical instruments for enzyme assays in biochemical systems (Monk and Wadsö, 1969). However, it is likely that by far the most important analytical applications for calorimetry will be in various biological experiments where quantitative or qualitative measurements are needed for following properties like the growth and response of living cells and tissues. This field seems to include a vast number of practically important areas where no methods or only inadequate analytical methods are available at the present time.

## C. Some Specific Examples of Microcalorimetric Studies

As an illustration of the use of modern microcalorimeters a few examples will be briefly discussed. The examples have been chosen from areas which, in the author's opinion, are those which at present are of greatest current interest: thermodynamic work on well-defined biochemical systems and analytical work in physiology and cell biology.

### *1. Calorimetric determination of* $\Delta$G, $\Delta$H *and* $\Delta$S

During recent years, many binding constants have been determined calorimetrically; in particular for acid–base equilibria and for metal–ligand binding reactions. This technique has also been adopted successfully for biochemical systems. As an example let us consider the formation of a simple 1:1 complex between an enzyme and an inhibitor

$$E + I \rightleftharpoons EI \tag{3.13}$$

where the binding constant $K_I$ may be expressed in terms of equilibrium concentrations (activity coefficients taken equal to unity).

$$K_I = \frac{[EI]}{[E][I]} \tag{3.14}$$

When $K_I$ is moderately large, a variation of the starting concentrations of E or I will result in a significant variation in the degree of binding. If the experiment is performed in a calorimeter, the heat quantities evolved can be related to the equilibrium concentrations. For a series of calorimetric experiments where the concentration of enzyme is kept constant but where the inhibitor concentration is varied we will have

$$K_I = \frac{[E_t]Q/Q_m}{([E_t]-[E_t]Q/Q_m)([I_t]-[E_t]Q/Q_m)} \cdot \tag{3.15}$$

$Q$ is the heat quantity, corrected for dilution processes, evolved in the calorimetric experiment and $Q_m$ is the corrected heat quantity for the hypothetical process where all the enzyme is complexed by inhibitor. $[E_t]$ and $[I_t]$ are the total concentrations of enzyme and inhibitor, respectively.

Figure 3.9 summarizes results from microcalorimetric experiments with lysozyme and two saccharide inhibitors, $(GlcNAc)_3$ and $(GlcNAc)_2$, (GlcNAc = *N*-acetylglucosamine) (Bjurulf *et al.*, 1970; cf. also the work by Vichutinskij *et al.*, 1969).

In these experiments the enzyme concentration was kept constant while the concentration of inhibitor was varied. For $(GlcNAc)_3$ (Figure 3.9a) the vertical line indicates the value for $[E_t]$. It is seen that the equivalence point is at the rather sharp bend between the nearly straight arms of the 'titration curve', indicating the formation of a comparatively strong 1:1 complex. In this case, an approximate value for $\Delta H$ is obtained from the nearly horizontal arm of the curve where $Q$ is very close to $Q_m$. Then, using the values of $Q$, $[I_t]$ points around the equivalence point, precise values for $K$ and $\Delta H$ can be calculated from Equation (3.15) by successive approximations.

$(GlcNAc)_2$ forms a weaker complex, as seen from the shape of the curve in Figure 3.9b. For the evaluation of $\Delta H$ and $K$ values in this case it is useful to convert Equation (3.15) to

$$\frac{1}{Q} = \frac{1}{Q_m} + \frac{1}{Q_m K_I} \cdot \frac{1}{[I_t]-[E_t]\dfrac{Q}{Q_m}} \tag{3.16}$$

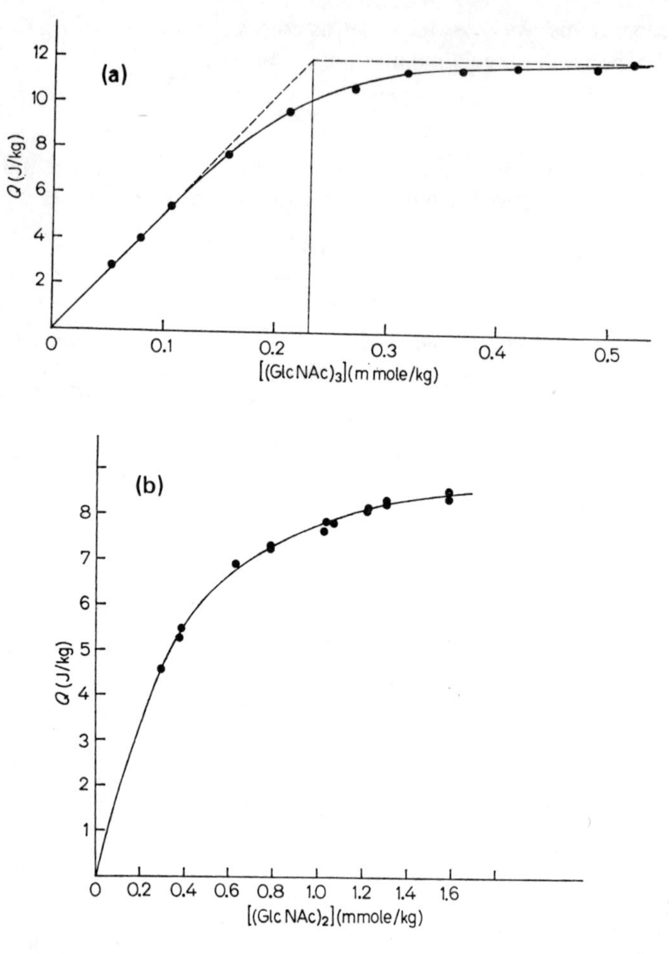


**Figure 3.9.** Results from microcalorimetric studies of binding saccharide inhibitors to lysozyme (Bjurulf *et al.*, 1970). Lysozyme concentration was kept constant at 231 μmole/kg. (a) $(GlcNAc)_3$. (b) $(GlcNAc)_2$. Reprinted from *Eur. J. Biochem.*, **14**, 47 (1970)

or

$$\frac{1}{Q} = \frac{1}{Q_m} + \frac{1}{Q_m K_I} \cdot \frac{1}{[I]} \tag{3.17}$$

where [I] is the concentration of free inhibitor in the final calorimetric solution and $K_I$ is the intrinsic inhibitor binding constant.

The unknown $Q_m$ and $K$ values can be solved by minimizing the error for the straight line described by Equation (3.17). $\Delta H$ is obtained from the intercept, $1/Q_m$, after which $K$ can be calculated from the slope of the line, $1/Q_m K_I$.

The results were in good agreement with data from corresponding equilibrium measurements supporting the view that 1:1 complexes are formed. From the values of the thermodynamic quantities ($\Delta G$, $\Delta H$, $\Delta S$ and $\Delta C_p$) for lysozyme–saccharide binding reactions (cf. also the study of Bjurulf and Wadsö, 1972), conclusions could be drawn concerning differences in the mode of binding between the various inhibitors. These conclusions are in agreement with those earlier obtained from the results of X–ray investigations on crystalline material. The resulting $\Delta C_p$ values were close to zero which supports the view that in the binding processes there is no significant net change in the exposure of hydrophobic groups to the bulk water.

In an important microcalorimetric study, Bolen *et al.* (1971) treated data on mononucleotide binding to ribonuclease in a more general manner. Assuming that there are $n$ identical and independent binding sites on the macromolecule, Equation (3.17) still holds true if the free inhibitor concentration [I] is given by

$$[\mathrm{I}] = [\mathrm{I_t}] - n\frac{Q}{Q_\mathrm{m}}[\mathrm{E_t}]. \tag{3.18}$$

Hence, that value for $n$ which gives the best straight line in a 'double reciprocal' plot according to Equation (3.17) should be the correct value. Bolen *et al.* pointed out, however, that a more stringent test for the value for $n$ is obtained if a second series of experiments is performed where inhibitor concentration is kept constant and the enzyme concentration is varied.

These experimental data can be represented by Equation (3.19)

$$\frac{1}{Q'} = \frac{1}{Q'_\mathrm{m}} + \frac{1}{Q'_\mathrm{m}\alpha[\mathrm{E}]} \tag{3.19}$$

where $Q'_\mathrm{m}$ is the heat quantity evolved when all the inhibitor is bound to the enzyme, i.e. $Q'_\mathrm{m} = Q_\mathrm{m}/n$,

$$[\mathrm{E}] = [\mathrm{E_t}] - \frac{Q'}{Q'_\mathrm{m}}\frac{[\mathrm{I_t}]}{n} \tag{3.20}$$

and $\alpha = nK_\mathrm{I}$.

From the two series of experiments, the best values for $Q_\mathrm{m}$ and $K_\mathrm{I}$ and for $Q'_\mathrm{m}$ and $\alpha$, respectively, were evaluated for assumed values of $n$. Comparison of the results leads to a value for $n$.

### 2. *Determination of insulin activity*

Boivinet *et al.* (1968) have reported results of calorimetric experiments where they investigated the metabolic effect of insulin added to rat epidydimal fat tissue. This kind of experiment represents an example of the use of

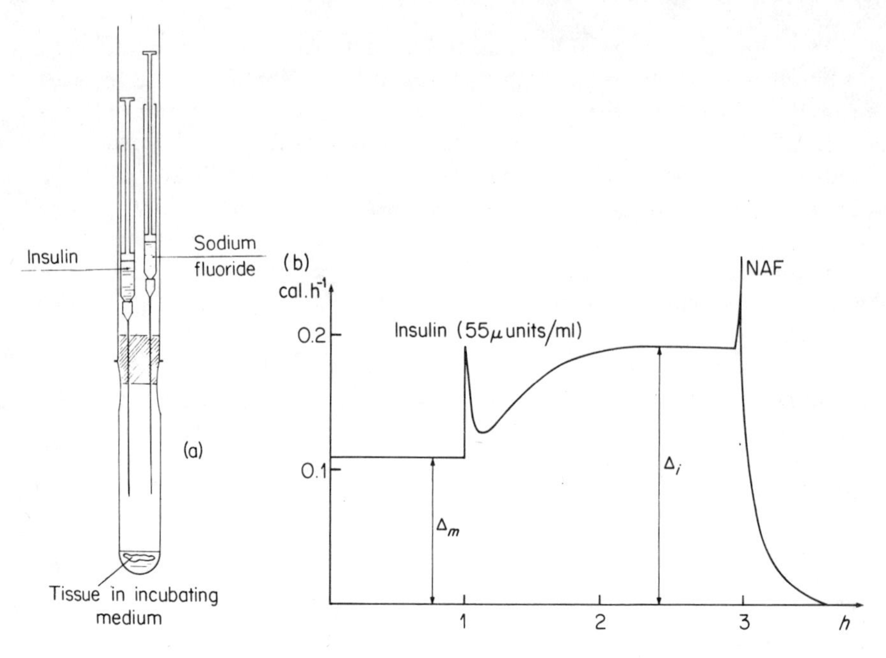


**Figure 3.10.** Microcalorimetric determination of insulin activity (Boivinet *et al.*, 1968). (a) Reaction tube in a Calvet microcalorimeter holding ca. 100 mg fat tissue kept in a nutrient solution. (b) Calorimetric curve showing an increased heat effect from the fat tissue as a result of added insulin. Reprinted from *C. R. Soc. Biol.*, **162**, 1770 (1968)

calorimetry for investigations of life processes and their response to agents such as activators or inhibitors.

Fat tissue (50–700 mg) was kept in about 1 ml of nutrient solution which was saturated with 95% $O_2$, 5% $CO_2$ in the reaction tube of a Calvet calorimeter. Solutions of insulin and of an agent which will stop the metabolic activity (e.g. NaF) were added through syringes positioned inside the calorimeter tube, Figure 3.10a. The result of a calorimetric experiment is shown in Figure 3.10b. The sharp peaks on the 'thermogram' accompanying the addition of insulin and of NaF represent heats of mixing. From $(\Delta_1 - \Delta_m)\Delta_m$, a relative measure for the action of the insulin preparation is obtained.

### 3. *Studies on human blood cells*

Levin (1971) has recently described flow calorimetric experiments where he measured the heat effect produced by leucocytes and thrombocytes. Results from one such series of experiments are shown in Figure 3.11. Samples were sucked at a constant flow rate of about 10 ml/h through the flow-through cell of an LKB flow microcalorimeter. The recorder line to the far left in the

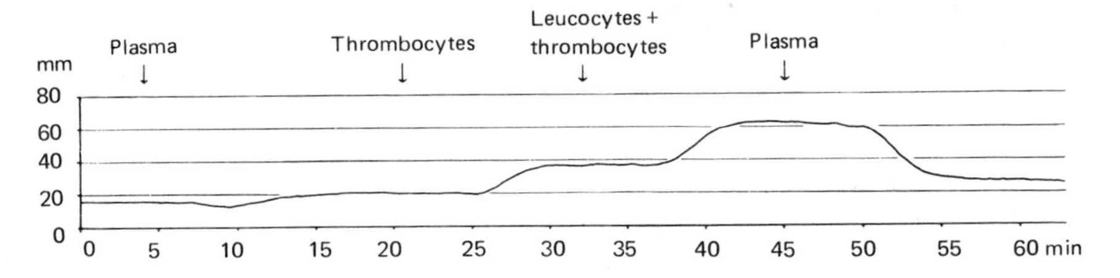


**Figure 3.11.** Flow calorimetric study on human blood cells (Levin, 1971). The arrows indicate the time of introduction of the various liquids or suspensions into the tube system. A calorimetric response of 1 mm corresponds to ca. 1 $\mu$W. Each fraction was ca. 3 ml. Reprinted from *Clin. Chim. Acta*, **32**, 87 (1971)

figure represents the instrument zero obtained when 0·9% NaCl was sucked through the cell. At the first arrow, plasma was introduced to the tube system of the calorimeter and reached the flow cell within 5 min. After a further 5 min, a steady-state value was reached. The difference between this value and the instrument zero line represents partly heat effects from increased viscous heating but is mainly due to reactions taking place in the plasma. At the second arrow, a suspension of thrombocytes in plasma (ca. 150,000/$mm^3$) was introduced into the calorimeter. A new steady-state level was reached after 10 min and the difference from that obtained for plasma thus represents the heat produced by the thrombocytes. At the third arrow, a mixture of thrombocytes (ca. 200,000/$mm^3$) and leucocytes (ca. 8000/$mm^3$) was sucked into the calorimeter, giving a still higher steady-state value.

At the time indicated by the last arrow in Figure 3.11, plasma was again sucked into the tube system and the calorimetric response returned to a low level. It may be noted that this final value is significantly higher than the initial steady-state value for plasma. This is explained by some adherence of polymorphonuclear leucocytes to the walls of the cell as verified by cell counts on the suspensions before and after the passage of the calorimeter.

Results from studies on human blood cells point to a direct clinical application for microcalorimetry in the near future, cf. Levin (1971). For example, the recent flow calorimetric study by Boyo and Ikomi-Kumm (1972) shows a significant difference in heat production between normal red blood cells and cells from patients suffering from sickle-cell anaemia.

### 4. *Aerobic bacterial growth*

Calorimeters have proved to be valuable as analytical instruments as well as thermodynamic tools for studies of bacterial growth processes, see e.g. Forrest (1969). In particular, flow microcalorimeters are widely used in this field. However, with aerobic growth of dense and fast-growing cultures, there are problems with the supply of oxygen to the bacteria in the calorimetric flow line. One recent solution to this problem involves pumping a mixed

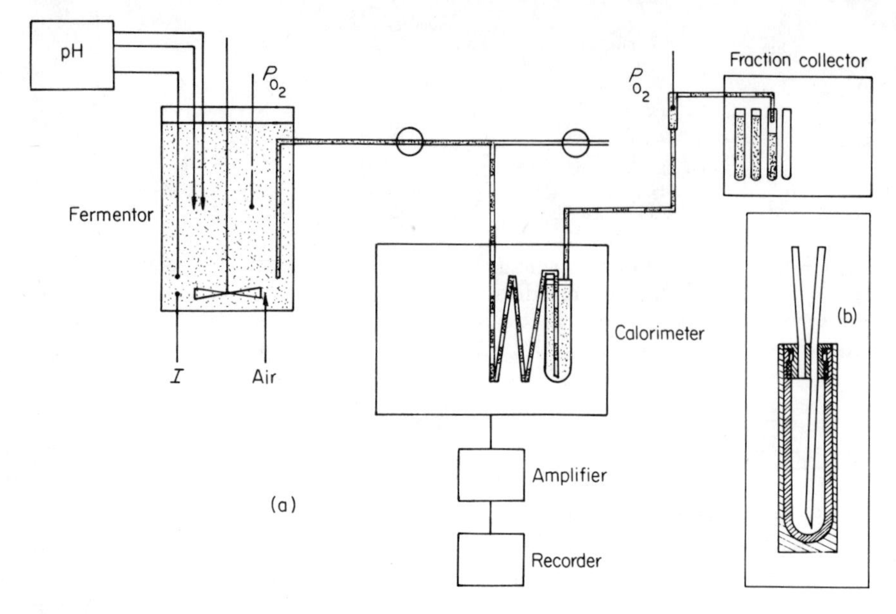


**Figure 3.12.** Equipment for flow calorimetric study of aerobic bacterial growth. (a) The bacteria are cultured in a fermentor with arrangements for aeration, temperature regulation ($T$), pH measurement and adjustment (pH) and recording of the oxygen potential ($P_{O_2}$). From the fermentor the bacterial suspension is pumped to a T-piece where it is met by a flow of gas with a suitable oxygen content. After passage through the calorimeter, the mixed liquid–gas flow passes a second oxygen electrode and is finally collected by a fraction collector and held at a temperature which is sufficiently low to prevent further growth of the culture. (b) Calorimetric flow cell positioned in a hole in an aluminium plate. This is surrounded by thermocouple plates, cf. Figure 3.8

flow of air and bacterial suspension through the calorimeter. Figure 3.12a shows schematically the experimental assembly used in our laboratory and in Figure 3.12b is shown a recent version of the flow-through cell used in these studies. It is positioned in one of the calorimetric units of a flow calorimeter similar to that shown in Figure 3.8. Immediately before entering the calorimetric cell, the mixed flow of liquid and gas passes through the heat exchange unit in the calorimeter. In addition to the temperature equilibration taking place, a proper saturation of the gas phase will be attained and vaporization in the calorimetric cell will thus be avoided.

Suitable chemical and biological analysis can be made on the collected fractions and the results can be correlated with the determined calorimetric curve.

Figure 3.13 shows results from a simple growth experiment with *E. coli* at 37°C on a medium containing 6 g glucose/l and salt medium (Eriksson and Wadsö, 1971). The suspension flow from the fermentor (40 ml/h) was aerated

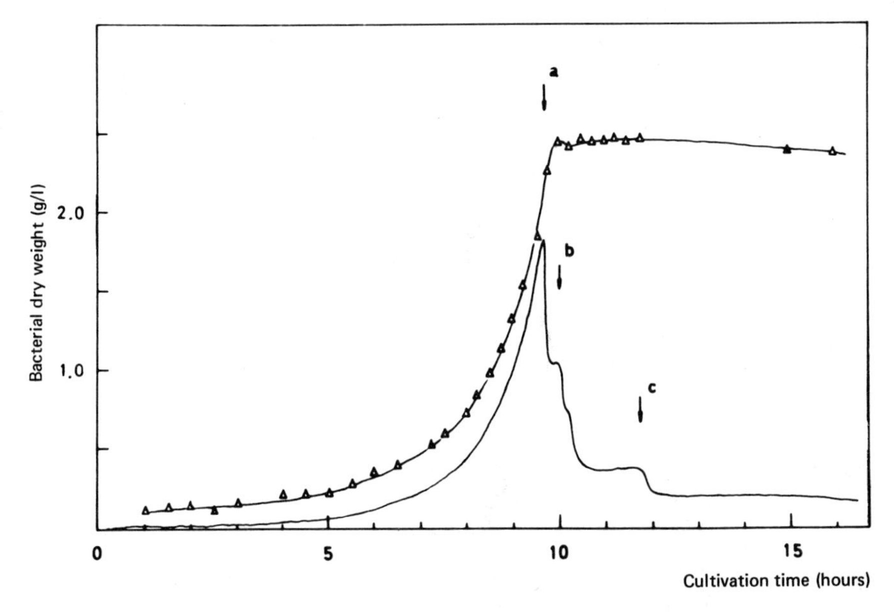


**Figure 3.13.** Results from an aerobic growth experiment with *E. coli* (Eriksson and Wadsö, 1971). ———, Thermogram; —— △ ——, dry weight curve. Reprinted from *First European Biophysics Congress* (1971), *Proceedings*, Vol. IV

by an equal volume of oxygen. Oxygen potentials measured in the fermentor and after the calorimeter showed that aerobic conditions prevailed during the whole experiment.

The thermogram in Figure 3.13 shows several distinct phases (a, b and c). From the results of glucose analysis, the first phase was identified as due to the exponential growth on glucose.

The origins of the two or three shoulders on the curve after the main peak have not yet been elucidated. In similar experiments performed at 27°C (4 g glucose/l), one distinct peak appeared after the major glucose peak. In that case, results from acetate analysis experiments suggested that the second phase was due to the consumption of acetate formed during the exponential growth (Eriksson and Wadsö, 1971).

Figure 3.13 illustrates the fact that a calorimetric curve can provide a valuable means of monitoring the progress of bacterial growth processes.

## D. Some Recent Applications

After this chapter was prepared a number of papers in the field of analytical biocalorimetry have appeared, *cf* Spink and Wadsö (1975). Some examples are referred to below.

Work on human blood cells has been reported by Monti and Wadsö (1973), Levin (1973a, 1973b) and by Ross *et al.* (1973), and on yeast cells in papers by Lamprecht and Meggers (1972), Schaarschmidt and Lamprecht (1973) and by Poole *et al.* (1973).

Measurements on bacterial activity in soil have been performed by Mortensen *et al.* (1973). Binford *et al.* (1973) have reported on a method for antibiotic sensitivity testing on clinically important bacteria and Boling *et al.* (1973) have described a microcalorimetric method for bacterial identification.

## IV. REFERENCES

Ackermann, T. (1969) 'The calorimeters: adiabatic calorimeters', in *Biochemical Microcalorimetry* (Ed. H. D. Brown), Academic Press, New York.

Andronikashvili, E. L., E. Y. Roinishivili and N. N. Khechinashivili (1970) 'Calorimetric investigation of muscle and liver tissues at low temperatures (on the possibility of the phase transformation existence in the cooled tissues of animals)', *Biofizika*, **15**, 484.

Armstrong, G. T. (1964) 'The calorimeter and its influence on the development of chemistry', *J. Chem. Educ.*, **41**, 297.

Atha, D. H. and G. K. Ackers (1971) 'Calorimetric determination of denaturation enthalpy for lysozyme in guanidine hydrochloride', *J. Biol. Chem.*, **246**, 5845.

Bakradze, N. G., D. R. Monaselidze, G. M. Mrevlishvili, A. D. Bibikova and L. L. Kisselev (1971) 'Microcalorimetric determination of tRNA hydration', *Biochim. Biophys. Acta*, **238**, 161.

Barisas, B. G., J. M. Sturtevant and S. J. Singer (1972) 'Thermodynamics of the binding of haptens to rabbit anti-2, 4-dinitrophenyl antibodies', *Biochemistry*, **11**, 2741.

Becker, F. and M. Kiefer (1969) 'Kontinuerliche Bestimmung von Mischungswärmen durch isotherme Enthalpietitration', *Z. Naturf.*, **24*a***, 7.

Belaich, J. P. and J. C. Sari (1969) 'Microcalorimetric studies on the formation of magnesium complexes of adenine nucleotides', *Proc. Natl. Acad. Sci.*, **64**, 763.

Benzinger, T. H. (1969) 'Ultrasensitive reaction calorimetry', in *A Laboratory Manual of Analytical Methods of Protein Chemistry* (Ed. P. Alexander and H. P. Lundgren), Pergamon Press, New York.

Berger, R. L., Y. Fok Chick and N. Davids (1968) 'Differential microcalorimeter for biochemical reaction studies', *Rev. Scient. Instrum.*, **39**, 362.

Berlin, E., P. G. Kliman and M. J. Pallansch (1970) 'Changes in state of water in proteinaceous systems', *J. Colloid Interface Sci.*, **34**, 488.

Biltonen, R. L., T. Schwartz and I. Wadsö (1971) 'Calorimetric studies of the chymotrypsinogen family of proteins', *Biochemistry*, **10**, 4136.

Binford, J. S., L. F. Binford and P. Adler (1973) 'A semiautomated microcalorimetric method of antibiotic sensitivity tests', *Am. J. Clin. Pathol.*, **59**, 86.

Bjurulf, C., J. Laynez and I. Wadsö (1970) 'Thermochemistry of lysozyme-inhibitor binding', *Eur. J. Biochem.*, **14**, 47.

Bjurulf, C. and I. Wadsö (1972) 'Thermochemical studies of lysozyme inhibitor binding', *Eur. J. Biochem.*, **31**, 95.

Boivinet, P., J. C. Garrigues and A. Grangetto (1968) 'Dosage microcalorimetrique de l'insuline', *C. R. Soc. Biol.*, **162**, 1770.

Bolen, D. W., M. Flögel and R. Biltonen (1971) 'Calorimetric studies of protein-inhibitor interaction. I. Binding of 3′-cytidine monophosphate to ribonuclease A at pH 5·5', *Biochemistry*, **10**, 4136.

Boling, E. A., G. C. Blanchard and W. J. Russel (1973) 'Bacterial identification by microcalorimetry', *Nature*, **241**, 472.

Boyo, A. E. and J. A. Ikomi-Kumm (1972) 'Increased metabolic heat production by erythrocytes in sickle-cell disease', *Lancet*, **1**, 1215.

Brown, H. D. (1969) 'Multiple calorimeters', in *Biochemical Microcalorimetry* (Ed. H. D. Brown), Academic Press, New York.

Brown, H. D. (1971) 'Protein topography by calorimetry', *J. Agr. Food Chem.*, **19**, 669.

Buzzel, A. and J. M. Sturtevant (1951) 'A new calorimetric method', *J. Am. Chem. Soc.*, **73**, 2454.

Calvet, E. and H. Prat (1963) '*Recent Progress in Microcalorimetry* (Ed. H. A. Skinner), Pergamon Press, London.

Chapman, D. and S. Chen (1972), 'Thermal and NMR spectroscopic studies of lipids and membranes' *Chem. Phys. Lipids*, **8**, 318.

Chapman, D. and J. Urbina (1971), 'Phase transitions and bilayer structure of Mycoplasma laidlawii B' *FEBS Letters*, **12**, 169.

Christensen, J. J., H. D. Johnston, H. Dee and R. M. Izatt (1968) 'Isothermal titration calorimeter', *Rev. Sci. Instr.*, **39**, 1546.

Clem, T. R., R. L. Berger and P. D. Ross (1969) 'A differential adiabatic microcalorimeter for the study of heats of transition in solution', *Rev. Scient. Instrum.*, **40**, 1273.

Coops, J., R. S. Jessup and K. van Ness (1956), 'Calibrations of Calorimeters for reactions in a bomb at constant volume', in *Experimental Thermochemistry* (Ed. F. D. Rossini), Interscience, New York.

Crescenzi, V. and F. Delben (1971) 'Application of differential enthalpy analysis to the study of biopolymer solutions', *Int. J. Protein Res.*, **3**, 57.

Danforth, R., H. Krakauer and M. Sturtevant (1967) 'Differential calorimetry of thermally induced processes in solution', *Rev. Scient. Instrum.*, **38**, 484.

Delin, S., P. Monk and I. Wadsö (1969) 'Flow microcalorimetry as an analytical tool in microbiology', *Sci. Tools*, **16**, 22.

Eriksson, R. and I. Wadsö (1971) 'Design and testing of a flow calorimeter for studies of aerobic bacterial growth', in *First European Biophysics Congress, Proceedings*, Vol. IV (Eds. E. Broda, A. Locker and H. Springer-Ledever), Wiener Medizinischen Akademie, Wien, p. 319.

Evans, W. J. (1969) 'The conduction-type microcalorimeter', in *Biochemical Microcalorimetry*, (Ed. H. D. Brown), Academic Press, New York.

Forrest, W. W. (1969) 'Bacterial calorimetry', in *Biochemical Microcalorimetry*, (Ed. H. D. Brown), Academic Press, New York.

Hasl, G. and H. Pauly (1971) 'Kalorische Eigenschaften des gebundenen Wassers in Proteinlösungen', *Biophysik*, **7**, 283.

Hearn, R. P., R. M. Richards, J. M. Sturtevant and G. D. Watt (1971) 'Thermodynamics of the binding of S-peptide to S-protein to form ribonuclease S', *Biochemistry*, **10**, 806.

Hinz, H. J., D. D. F. Shiao and J. M. Sturtevant (1971) 'Calorimetric investigation of inhibitor binding to rabbit muscle aldolase', *Biochemistry*, **10**, 1347.

Howarth, J. V. (1970) 'The technique of thermal measurements in excitable tissues', *Quart. Rev. Biophys.*, **3**, 429.

Izatt, R. M. and J. J. Christensen (1968) 'Heats of proton ionization and related thermodynamic quantities', in *Handbook of Biochemistry* (Ed. H. A. Sober), J-49, The Chemical Rubber Co., Cleveland.

Jackson, W. M. and J. F. Brandts (1970) 'Thermodynamics of protein denaturation. A calorimetric study of the reversible denaturation of chymotrypsinogen and conclusions regarding the accuracy of the two-state approximation', *Biochemistry*, **9**, 2294.

Keyes, M. H., M. Falley and R. Lumry (1971) 'Studies of haem proteins. II. Preparation and thermodynamic properties of sperm whale myoglobin', *J. Am. Chem. Soc.*, **93**, 2035.

Khechinachivili, N. N., P. L. Privalov and E. J. Tiktipulo (1973), 'Calorimetric investigation of lysozyme thermal denaturation' *FEBS Letters*, **30**, 57.

Kitzinger, C. and T. H. Benzinger (1960) 'Principle and method of heat-burst microcalorimetry and the determination of free energy, enthalpy and entropy changes', in *Methods of Biochemical Analysis*, Vol. VIII (Ed. D. Glick). Wiley-Interscience, New York.

Konicek, J., J. Suurkuusk and I. Wadsö (1971) 'A precise drop heat capacity calorimeter for small samples', *Chemica Scripta*, **1**, 217.

Kusano, K., B. Nelander and I. Wadsö (1972) 'A calorimeter for studies of adsorption of gases and liquids on solids', *Chemica Scripta*, **1**, 211.

Lamprecht, I. and C. Meggers (1972) 'Mikrokalorimetrische Untersuchungen zum Stoffwechsel von Hefen. II. Wachstum auf festen Nährböden', *Biophysik*, **8**, 316.

Langerman, N. and J. M. Sturtevant (1971) 'Calorimetric studies of quaternary structure and ligand-binding of hemerythrin', *Biochemistry*, **10**, 2809.

Lapanje, S. and I. Wadsö (1971) 'A calorimetric study of the denaturation of lysozyme by guanidine hydrochloride and hydrochloric acid', *Eur. J. Biochem.*, **22**, 345.

Levin, K. (1971) 'Heat production by leucocytes and thrombocytes measured with a flow microcalorimeter in normal man and during thyroid dysfunction', *Clin. Chim. Acta*, **32**, 87.

Levin, K. (1973a) 'Determination of heat production from erythrocytes in normal man and in anaemic patients with flow microcalorimetry', *Scand. J. Clin. Lab. Invest.*, **32**, 55.

Levin, K. (1973b) 'A modified flow calorimeter adopted for the study of human leucocyte phagocytosis', *Scand. J. Clin. Lab. Invest*, **32**, 67.

Maurer, W., W. Haar and H. Rüterjans (1971) 'Calorimetric investigations of the interaction of inhibitors and substrates with ribonuclease A and ribonuclease T', in *First European Biophysics Congress Proceedings*, Vol. IV (Eds. E. Broda, A. Locker and H. Springer-Lederer), Wiener Medizinischen Akademie, Wien, p. 375.

McGlashan, M. L. (1962) 'Heats of mixing', in *Experimental Thermochemistry*, Vol. II (Ed. H. A. Skinner), Interscience Publ., New York.

Monk, P. and I. Wadsö (1968) 'A flow micro reaction calorimeter', *Acta Chem. Scand*, **22**, 1842.

Monk, P. and I. Wadsö (1969) 'Flow microcalorimetry as an analytical tool in biochemistry and related areas', *Acta Chem. Scand.*, **23**, 29.

Monti, M. and I. Wadsö (1973) 'Microcalorimetric measurements of heat production in human erythrocytes', *Scand. J. Clin. Lab. Invest.*, **32**, 47.

Mortensen, U., B. Norén and I. Wadsö (1973) 'Microcalorimetry in the study of the activity of microorganisms, *Bull Ecol. Res. Comm.* (*Stockholm*), **17**, 189.

Picker, P., C. Jolicoeur and J. E. Desnoyers (1969) 'Steady state and composition scanning differential flow microcalorimeters', *J. Chem. Thermodyn.*, **1**, 485.

Poole, R. K., D. Lloyd and R. B. Kemp (1973), 'Respiratory oscillations and heat evolution in synchronously dividing cultures of the fission yeast Schizosaccharomyces pombe 972 h' *J. Gen. Microbiol.*, **77**, 209.

Prat, H. (1969) 'Calorimetry of higher organisms', in *Biochemical Microcalorimetry* (Ed. H. D. Brown), Academic Press, New York.

Privalov, P. C., G. M. Monaselidze, G. M. Mrevlishvili and V. A. Mageldadze (1964) 'Intramolecular heat of fusion of macromolecules', *J. Expt. Theor. Phys.* (*U.S.S.R.*), **47**, 2073.

Privalov, P. L. and E. I. Tiktopulo (1970) 'Thermal conformational transformation of tropocollagen. I. Calorimetric study', *Biopolymers*, **9**, 127.

Privalov, P. L., N. N. Khechinashvili and B. P. Atanasov (1971) 'Thermodynamic analysis of thermal transitions in globular proteins. I. Calorimetric study of chymotrypsinogen, ribonuclease and myoglobin', *Biopolymers*, **10**, 1865.

Reinert, J. C. and J. M. Steim (1970) 'Calorimetric detection of a membrane–lipid phase transition in living cells', *Science*, **168**, 1580.

Rialdi, G. and P. Profumo (1968) 'Calorimetric measurements on DNA helix-coil transition in 2·2 *M* urea', *Biopolymers*, **6**, 899.

Ross, P. D. and R. L. Scruggs (1969) 'Heat of the reaction between polyribocytidylic acid and polyriboinosinic acid', *J. Mol. Biol.*, **45**, 567.

Ross, P. D. and A. Ginsburg (1969) 'A calorimetric study of the binding of two feedback inhibitors to the glutamine synthetase from *Escherichia coli*', *Biochemistry*, **8**, 4690.

Ross, P. D., A. P. Fletcher and G. A. Jamieson (1973) 'Microcalorimetric study of isolated blood platelets in the presence of thrombin and other aggregating agents', *Biochim. Biophys. Acta*, **313**, 106.

Schaarschmidt, B. and I. Lamprecht (1973) 'Ultraviolet irradiation and measuring of the total optical density of microorganisms in a microcalorimeter', *Experienta*, **29**, 505.

Shiao, D. D. F. (1970) 'Calorimetric investigations of the binding of inhibitors to α-chymotrypsin. II. A systematic comparison of the thermodynamic functions of binding of a variety of inhibitors to α-chymotrypsin', *Biochemistry*, **9**, 1083.

Sjöquist, J. and I. Wadsö (1971) 'A thermochemical study of the reaction between protein A from *S. aureus* and fragment Fc from immunoglobulin G', *FEBS Letters*, **14**, 254.

Skinner, H. A. (1969) 'Theory, scope and accuracy of calorimetric measurements', in *Biochemical Calorimetry* (Ed. H. D. Brown), Academic Press, New York.

Spink, E. and Wadsö, I. (1975) 'Calorimetry as an analytical tool in biochemistry and biology', in *Methods of Biochemical Analysis*, Vol. 23 (Ed. D. Glick). Wiley-Interscience, New York.

Stauffer, H., S. Srinivasan and M. A. Lauffer (1970) 'Calorimetric studies on polymerisation–depolymerisation of tobacco mosaic virus protein', *Biochemistry*, **9**, 193.

Sturtevant, J. M. (1959) 'Calorimetry', in *Techniques of Organic Chemistry*, 3rd ed., Vol. I (Ed. A. Weissberger), part 1, John Wiley and Sons, New York.

Sturtevant, J. M. (1962) 'Heats of biochemical reactions', in *Experimental Thermochemistry*, Vol. II (Ed. H. A. Skinner), Interscience, London.

Sturtevant, J. M. (1969) 'Flow calorimetry', *Fractions*, 1.

Swietoslawski, W. (1946), *Microcalorimetry*, Reinhold Publ. Comp., New York.

Swietoslawski, W. and W. Zielenkiewicz (1959) 'On a new labyrinth flow calorimeter', *Bull. Acad. Polon. Sci.*, **7**, 101.

Tanford, C. (1968) 'Protein denaturation', *Adv. Protein Chem.*, **23**, 122.

Tanford, C. (1970) 'Protein denaturation. Part C. Theoretical models for the mechanism of denaturation', *Adv. Protein Chem.*, **24**, 1.

Tsong, T. Y., R. P. Hearn, D. P. Wrathall and J. M. Sturtevant (1970) 'A calorimetric study of thermally induced conformational transitions of ribonuclease A and certain of its derivatives', *Biochemistry*, **9**, 2666.

Velick, S. F., J. P. Baggott and J. M. Sturtevant (1971) 'Thermodynamics of nicotine-adenine dinucleotide addition to glyceraldehyde-3-phosphate dehydrogenases of yeast and of rabbit skeletal muscle. An equilibrium and calorimetric analysis over a range of temperatures', *Biochemistry*, **10**, 779.

Vichutinskij. A. A., B. J. Zaslowsky, A. L. Platonov, L. A. Timmerman and A. J. Khorlin (1969) 'Use of calorimetry to study an enzyme–inhibitor reaction', *Dokl. Akad. Nauk SSSR.*, **189**, 432.

Wadsö, I. (1968) 'Design and testing of a micro reaction calorimeter', *Acta Chem. Scand.*, **22**, 927.

Wadsö, I. (1969) 'Experimental approach and desired accuracy in biochemical thermochemistry', in *Biochemical Microcalorimetry* (Ed. H. D. Brown), Academic Press, New York.

Wadsö, I. (1970) 'Microcalorimeters', *Quart. Rev. Biophys.*, **3**, 383.

Wadsö, I. (1972) 'Biochemical thermochemistry', in *MTP International Rev. of Science. Physical Chemistry Series One.* (*Thermochemistry and Thermodynamics*), Vol. 10 (Ed. H. A. Skinner), Butterworth, London.

Wooledge, R. C. (1971) 'Heat production and chemical change in muscle', in *Progress in Biophysics and Molecular Biology*, Vol. 22 (Eds. J. A. V. Butler and D. Noble), Pergamon Press, Oxford.

CHAPTER 4

# The techniques of plant cell culture and somatic cell hybridization

P. K. Evans and E. C. Cocking
*Department of Botany,*
*University of Nottingham,*
*University Park,*
*Nottingham NG7 2RD*



## I. INTRODUCTION

The techniques of plant tissue and cell culture have been available to plant physiologists and biochemists for some time. Over the years various nutrient media have been developed, which now allow the successful culture of a wide range of plant species as callus tissue, a largely undifferentiated mass of cells growing on solidified nutrient medium. Originally it was necessary to add various complex nutrients such as coconut milk or yeast extract but now these have generally been replaced by completely defined media composed of a salt solution, various vitamins, plant hormones such as an auxin and a cytokinin, and a carbon source, usually sucrose.

The technique of cell suspension culture whereby plant cells grow as free cells or small groups of cells in an agitated liquid nutrient medium has been more recently developed. This method has now teached a considerable level of sophistication, allowing plant cells to be grown either as batch culture or as a continuous culture in the manner of the chemostat or turbidostat (Wilson, King and Street, 1971). These callus and suspension cultures normally develop as a result of cell division in excised plant organs and thus arise as a wound response of the tissue. In contrast, until recently, little success had been achieved with the isolation and culture free cells derived directly from the plant. But the development of new methods whereby free cells can be obtained in large quantities directly from the plant, either as cells or protoplasts (cells from which the cell wall has been removed), has led to an upsurge of interest in this area of plant cell culture. These isolated cells and protoplasts have some distinct advantages in certain experimental systems over cells from *in vitro* cell cultures. Large quantities of morphologically uniform cells can be obtained and also the genetic uniformity is likely to be higher than in *in vitro* cell cultures where chromosome aberrations are known to occur, the frequency of which increases with the time in culture (Sunderland, 1973). The protoplast as a naked cell opens up the possibility of somatic cell fusion and cell hydridization (Cocking, 1972; Tempé, 1973). This somatic cell fusion combined with the ability of single plant cells to develop into entire plants has led to the possibility of using such techniques for the production of hybrid plants which for various reasons could not be produced by conventional sexual means. Because of the possible implications of such methods for crop improvement, a considerable amount of research effort in laboratories around the world has been focused on plant protoplast isolation and culture, and in consequence some rapid advances have been made. The naked cell also offers an approach to plant cell modification which is either not possible or not so readily achieved with the walled cell. The uptake of viruses and their subsequent replication (Coutts, Cocking and Kassanis, 1972; Zaitlin and Beachy, 1973), the possible uptake of isolated nuclei (Potrykus and Hoffmann, 1973), bacteria (Davey and Cocking, 1972) and blue-green algae and the possible uptake of isolated chloroplasts (Carlson, 1973; Potrykus 1973) are all being actively investigated. These approaches, together with the modification of cells by the uptake of foreign DNA (Ohyama, Gamborg and Miller, 1972; Hoffmann and Hess, 1973) and the uptake of specialized transducing bacteriophage (Johnson, Grierson and Smith, 1973; Doy, Gresshoff and Rolfe, 1973), may perhaps lead, in the future, to the production of plants with novel and advantageous properties.

Although the successful culture of isolated cells and protoplasts owes much to the study of callus and suspension cell cultures (Street, 1973), the scope of this article will be restricted in the main, to the culture and isolation of cells obtained directly from the plant. Over the past few years we have been impressed by the number of workers in fields other than traditional plant cell and tissue culture who have expressed an interest in these new techniques and who have seriously considered that these methods could be of use in their own particular research area. It is these workers we have had in mind in attempting to bring together those methods which show particular promise for the isolation and culture of plant cells and their isolated protoplasts.

## II. ISOLATION OF CELLS

The concept of obtaining free isolated cells from the plant body and studying their growth and development is attributed to Haberlandt (1902, translated into English by Krikorian and Berquam, 1969). He was able to isolate palisade and spongy mesophyll cells mechanically from the bracts of the red dead nettle, *Lamium purpureum*. These cells were apparently viable and for several days would 'assimilate' in the light. Although these cultured cells remained alive in a simple nutrient medium for up to a month and during this time increased in size and often became round or pear shaped, they failed to divide. This failure to obtain division from isolated leaf cells was regarded by later workers to be due to the fact that the mesophyll cell is a highly differentiated photosynthetic system. Moreover, cells within the mature leaf are not destined to undergo any further cell division. For these reasons leaf tissue was regarded for many years as an unsuitable material for the initiation of cell cultures (White, 1963), and those interested in cell cultures turned their attention to the growth of excised tissues.

In spite of the fact that they could not be successfully induced to divide, isolated leaf cells were still an attractive system in some areas of investigation, especially the study of virus replication and photosynthesis. A preparation of soybean mesophyll cells capable of fixing $^{14}CO_2$ for a short period after isolation was obtained by Racusen and Aronoff (1953) by grinding leaves in a test-tube homogenizer in sucrose and phosphate buffer at 4°C. Whilst Zaitlin (1959) obtained a preparation of isolated tobacco leaf cells capable of supporting limited viral replication by incubating leaf pieces with 0·2% pectinase in sucrose and phosphate buffer. These two early methods, although of limited success, demonstrate the two basically different approaches to the isolation of plant cells, that of mechanical and enzymatic isolation.

### A. Enzymatic Isolation

Although both enzymatic and mechanical methods of protoplast isolation were developed in parallel we shall consider first the enzymatic isolation. Jyung, Wittwer and Bukovac (1965) using a modification of Zaitlin's pectinase method, were able to isolate cells from the pith and leaves of tobacco and the stem, root and leaves of bean and study the ion uptake of these isolated cells. Peptone, the presence of which Chayen (1952) had previously found to be beneficial to the isolation of root cells, was added to counteract the possible damaging effects of proteolytic enzymes present in the pectinase preparation. The presence of EDTA as well as pectinase was necessary for the isolation of cells from bean leaves. Cocking (1960a) observed that although treatments with EDTA and other chelating agents lead to cell separation in roots, they were always found to have an irreversible deleterious effect on the metabolism of the isolated cells. Takebe, Otsuki and Aoki (1968) further refined the technique of enzymatic isolation. The lower epidermis of tobacco leaves was removed and the leaf pieces treated with crude polygalacturonase under plasmolysing conditions in the presence of potassium dextran sulphate. These isolated cells proved highly viable and capable of supporting viral multiplication. The presence of potassium dextran sulphate leads to an improved yield of isolated cells, although the actual mechanism of action is not clear. Takebe *et al.* (1968) make the suggestion that since polyanions bind to basic proteins, potassium dextran sulphate binds to and inactivates some toxic basic proteins present in the pectinase. Indeed, the activity of RNAse present in the enzyme mixture used for protoplast isolation is greatly reduced by potassium dextran sulphate (Coutts, 1973). Ruesink and Thimann (1965) found that ribonuclease could lyse *Avena* coleoptile protoplasts. This lysis, however, was not dependent on enzyme activity as an inactive carboxymethylated derivative of bovine pancreatic ribonuclease was about as effective as active RNAse in lysing protoplasts. RNAse behaved, therefore, like ionic detergents which are known to destabilize the membrane. This lysis could be prevented by the presence of divalent cations, suggesting the presence of negatively charged sites on the protoplast surface (Ruesink, 1971). Potassium dextran sulphate presumably protects these sites and enhances membrane stability.

The isolation of these cells under hypertonic conditions is of considerable significance and was regarded by Zaitlin and Beachy (1973) as a major contributing factor to the successful viral multiplication in the isolated cells. There is no doubt that the activity of the crude polygalacturonase is not solely limited to the middle lamella but in all probability leads to a general weakening of the wall. This weakened wall under non-plasmolysing conditions could lead to local 'blow-outs' of the protoplasts causing plasmalemma rupture and cell damage. Indeed, Takebe *et al.* (1968) commented that 'at mannitol concentrations below 0·3 *M*, protoplasts appear to burst within the cell walls and the cells assumed a damaged appearance'. Using this isolation method, Otsuki and Takebe (1969) obtained leaf cells from a wide range of species, including monocotyledons and dicotyledons. The following schedule which is a modification of the original method of Takebe *et al.* (1968) yields large quantities of sterile viable tobacco mesophyll cells.

Fully expanded dark-green leaves, from plants 60–80 days old are selected. The leaves are surface sterilized by immersion in 70% ethanol for 30 sec, followed by 30 min in 3% sodium hypochlorite. A small quantity (0·05%) of 'Teepol' (BDH, Poole, U.K.) or 'Cetavlon' (I.C.I. Ltd., Macclesfield, U.K.) may be added as a wetting agent. The sterilant is removed by 3–4 washes in sterile distilled water. With the aid of sterile fine jeweller's forceps inserted into a junction of the midrib and a vein, areas of epidermis can be peeled away. With practice substantial areas of epidermis can be removed. The peeled areas of the leaf are excised with a sterile blade. Approximately 2 g of peeled leaf pieces are placed into a sterile 100 ml Erlenmeyer flask containing 20 ml of filter-sterilized (Sartorious membrane filter, GmbH, Gottingen, Germany) enzyme solution, containing 0·5% 'Macerozyme' (All Japan Biochemicals Co. Ltd., Nishinomiya, Japan) 0·8 *M* mannitol and 1% potassium dextran sulphate (molecular weight source dextran 560, sulphur content 17·3%, Meito Sangyo Co. Ltd., Japan) pH 5·8. The enzyme is infiltrated into the leaf by briefly evacuating the flask with a vacuum pump. The flask is then placed on a reciprocating shaker with a stroke of 4–5 cm at the rate of 120 cycles a minute. The temperature is maintained at 25°C. After 15 min shaking, the enzyme solution, which now contains broken cells and debris, is removed and discarded and replaced with a further 20 ml. The enzyme solution is replaced at 30 min intervals for 2 h. The enzyme solution removed after the first 30 min incubation contains predominantly spongy mesophyll cells, the second 30 min incubation had a mixed population of cells whilst the last two incubations contained palisade cells. These enzyme incubations are combined and the cells sedimented by centrifugation at 100 g for 1 min. The cells are washed twice with 0·7 *M* mannitol to remove residual enzyme and are ready for use.

Not all plants tested by Otsuki and Takebe (1969) yielded cells; maize and wheat were especially resistant. Unsuccessful attempts have also been made to isolate cells after treatment with pectinase from wheat, rye and barley leaves (Evans, unpublished results). The shape of the mesophyll cell and their arrangement inside the cereal leaf may be a key factor. These cells appear elongated with a number of constrictions and within the leaf they may form an interlocking structure preventing isolation.

## B. Mechanical Isolation

The method of mechanical isolation was used by Ball and Joshi (1965) who were interested in the culture of the cells isolated directly from the plant. They found that cells could be removed with a fine scalpel from the palisade layer of torn leaves of peanut, *Arachis hypogaea*. Many of these cells were viable and capable of dividing in culture. They tested a wide range of plants but found only a few had leaves which allowed the successful removal of cells in this way (Joshi and Ball, 1967). Similarly, viable cells were isolated from leaves of the plume poppy, *Macleaya cordata* (Kohlenbach, 1966). A cell suspension was obtained by shaking briefly, with sand, leaves which had previous been cut into very thin sections. A method which dispensed with the need to cut these sections was developed by Gnanam and Kulandaivelu (1969). A preparation of leaf cells capable of photosynthesis was obtained by grinding leaves in medium in a porcelain pestle, filtering the homogenate through fine muslin cloth and centrifuging the filtrate at 200 *g* for 30 sec. The supernatant contained the free cells. All steps were performed between 0 and 4°C. A substantial number of plant species yielded mesophyll cell suspensions by this method, including apparently, grasses, although these were not identified. Edwards and Black (1971), using a basically similar method in which leaf shreds were ground gently in a medium at 4°C with a pestle and mortar and subsequently filtered, were able to isolate separate suspensions of mesophyll and bundle sheath cells from crab grass, *Digitaria sanguinalis*, and mesophyll cells of spinach.

A mechanical isolation method which produces a high yield of viable cells from leaves of bindweed, *Calystegia sepium* has been developed by Rossini (1969). This method has also produced good yields of free cells from peanut (Jullien, 1970), *Asparagus* (Figure 4.1) and *Ipomea* leaves (Harada, Ohyama and Cheruel, 1972). Leaves are immersed in a filtered suspension of 7% calcium hypochlorite for 10 min and then rinsed at least three times with sterile water. Some find it advisable to seal cut surfaces with wax before sterilization. Approximately 2·5 g of leaf in 30 ml of salt solution (Lin and Staba, 1961) plus 1% sucrose are ground in a Potter–Elvehjem glass homogenizer until all the leaf pieces are broken up. The resulting cell suspension is passed through two sterile metal Tyler filters (W. S. Tyler Co. Cleveland, Ohio, U.S.A.); the upper filter with mesh diameter 0·061 mm and the lower 0·038 mm. The remaining fine debris can be removed either by slow-speed centrifugation (200 g) which sediments only the free cells or by retaining the isolated cells on a sintered-glass filter whilst the fine debris is removed. The isolated cells are then suspended in a volume of nutrient medium ready for culture. This procedure emphasizes the advantages of the mechanical method of isolation, the simplicity of technique, the speed with which cell suspensions can be obtained, the lack of exposure of the cells to hydrolytic enzymes and the need for plasmolysing conditions. But it is yet to be seen if it is as widely applicable as the enzymatic procedure.

## III. ISOLATION OF PROTOPLASTS

### A. Mechanical Isolation

We have already seen that in the mechanical and enzymatic methods we have two different approaches to the isolation of cells. Similarly, these two methods are used for protoplast isolation, but because of the close relationship of the cell wall and protoplast, mechanical isolation has definite limitations. However, with a few exceptions mechanical methods were the only way of isolating protoplasts prior to 1960. These procedures involved the incubation of the plant tissue in a hypertonic solution leading to the plasmolysis of the cells. The protoplasts of many cells will shrink away from the cell wall and take on a spherical shape and occupy a much reduced volume within the cell wall. Sectioning of the tissue at a thickness calculated to cut cells only once followed by slight deplasmolysis results in the extrusion of the plant protoplasts from the cut ends of the cells. This method naturally has the disadvantage of producing many damaged cells and the yield of intact protoplasts is always small. The method is also restricted to tissues which are highly vacuolate and can thus be readily plasmolysed and further, it is essential that as the cells plasmolyse they should contract away completely from the cell wall. The successful use of this method has therefore been restricted largely to parenchymatous cells of storage tissues such as beet (Whatley, 1956), although epidermal cell protoplasts from onion bulb scales (Chambers and Hofler, 1931) and radishes (Törnävä, 1939) have been isolated in this way.

### B. Enzymatic Isolation

Except for certain experiments in which exposure to the hydrolytic enzymes is regarded as detrimental to the cell system (Pilet, 1973) workers have now largely abandoned mechanical in favour of enzymatic isolation. Indeed, to obtain protoplasts from a wide range of tissues and in the large numbers which are essential for successful culture, one must make use of the enzymatic degradation of the cell wall to liberate the protoplasts. Initially it was necessary for workers to make their own preparations of enzymes. Cocking (1960b) used concentrated solutions of crude fungal cellulase from *Myrothecium verrucaria* to isolate protoplasts from tomato roots. Since 1968, cellulases have been commercially available although not in a very purified form. Nonetheless, the ready availability of these enzymes has been a major stimulus for protoplast research. The commercial cellulase preparations in most widespread use are derived from *Trichoderma viride*, whilst pectinase is produced from *Rhizopus*. Some of the more commonly used commercially available enzymes for protoplast isolation are listed in Table 4.1. Recently 'Onozuka' cellulase and a pectinase have become available in a more purified form; 'Onozuka' cellulase R10 and 'Macerozyme' R10.

The successful enzymatic isolation of protoplasts from plant tissue is dependent on several critical considerations. First, an osmotic stabilizer must be selected and be at a level which is sufficient to prevent the majority of protoplasts from bursting once the cell wall is weakened and yet not be

**Table 4.1.** Commercial enzymes used for the isolation of cells and protoplasts

| | | |
|---|---|---|
| *Cellulases* | | |
| 'Onozuka' | ex *Trichoderma* | All Japan Biochemicals Co., Nishinomiya, Japan |
| 'Meicelase' P | ex *Trichoderma* | Meiji Seika Kaisha Ltd., chuo—Ku Tokyo, Japan |
| 'Driselase' | ex *Basidiomycete* | Kyowu Hakko Kogyo Co. Ltd., Tokyo, Japan |
| 'Glusulase' | ex Snail gut juice | Endo Laboratories Inc., Garden City, New York, U.S.A. |
| 'Helicase' | ex Snail gut juice | L'Industrie Biologique Francaise, Gennevillies, France |
| *Hemicellulases* | | |
| 'Rhozyme' HP150 | | Rohm and Haas, Philadelphia, Pa., U.S.A. |
| Hemicellulase | ex *Rhizopus* | Sigma Chemical Co., St. Louis, Mo., U.S.A. |
| *Pectinases* | | |
| 'Macerozyme' | ex *Rhizopus* | All Japan Biochemical Co., Nishinomiya, Japan |
| 'Pectinol' R10 | ex *Aspergillus* | Rohm and Haas, Philadelphia, Pa., U.S.A. |
| Pectinase | ex *Aspergillus* | Sigma Chemical Co., St. Louis, Mo., U.S.A. |
| Pectinase | | Serva Feinbiochemicals, Heidelberg, W. Germany |

of too high a concentration so that irreversible plasmolytic damage is done to the protoplasts. This latter consideration is especially important when dealing with protoplasts isolated from the leaves of certain plant species. It is wise first to ascertain the correct level of plasmolyticum required for each tissue. A simple method is to measure the gain or loss in weight of a sample of tissue when it is incubated in solutions of various osmotic potential. The solution in which the tissue loses weight indicates that some of the cells have plasmolysed. Direct microscopic observation of the cells to determine the osmotic level at which the majority of cells become visibly plasmolysed is another method but it is not often possible with intact tissues. Generally, mannitol is regarded as the most suitable osmotic stabilizer as it is not readily metabolized or taken into the cytoplasm. Sorbitol, an isomer of mannitol, has the advantage of being more soluble. Sucrose has also been used (Power and Cocking, 1970), and at the concentrations (0·6–0·7 *M*) necessary for the isolation of leaf protoplasts, the tissue will often float. This fact can be put to good use in separating protoplasts from chloroplasts, cells and cell debris (see later). Sucrose is, however, metabolized and enzyme solutions, containing high concentrations of sucrose, can often be difficult to filter sterilize. Further, some workers claim sucrose actually to be detrimental to protoplast isolation (Fodil, Esnault and Trapy, 1971). Salt solutions, used extensively to plasmolyse cells in the early work on mechanical isolation, have not been frequently used as plasmolytica in enzyme isolation methods, although Kameya and Uchimiya (1972) found a mixture of potassium chloride and calcium chloride more suitable than mannitol for the isolation of protoplasts from carrot roots.

Secondly, it is necessary to formulate the correct combination of hydrolytic enzyme which will successfully degrade the cell walls of the tissue under investigation. Although much progress has recently been made in our understanding of the primary plant cell wall (Albersheim, 1973; Lamport, 1973) we are not yet in a position to add specific enzymes to degrade the various specific components and liberate viable plant protoplasts. In fact, all of the commercially available cellulase and pectinase preparations are mixtures of enzymes and this is probably an essential feature for successful wall breakdown. 'Onozuka' cellulase contains the cellulase complexes $C_1$ and $C_x$, the former attacking native and crystalline cellulose and the latter degrading amorphous cellulose. For a detailed consideration of the various components of the cellulase complex in relation to protoplast isolation, see Selby, 1973. These commercial cellulase preparations are also known to contain cellobiase, xylanase, glucanase pectinase, lipase, phospholipase, β-1,3-glucanase, chitinase and various nucleases, together with a considerable amount of low molecular weight material, some of which may be toxic to certain cell systems. The possible damaging effects to the protoplasts associated with enzyme incubation may be reduced if the cells are plasmolysed. This observation extends back to the work of Tribe (1955), who found that plasmolysed cells were less susceptible than turgid cells to the effects of cell-separating enzymes. As a routine procedure, therefore, it is beneficial to plasmolyse the cells before enzyme incubation.

It has proved advantageous in some cases to desalt the enzymes (Schenk and Hildebrandt, 1969; Keller, Harvey, Gamborg, Miller and Eveleigh, 1970). This often leads to an enzyme preparation giving more rapid release and a higher yield of protoplasts. This is especially true for the isolation of protoplasts from cells grown as callus or suspension cultures. The cell walls of these *in vitro* cultures have with some exceptions proved more resistant to breakdown than cells from leaf tissue. The newly available 'Onozuka' cellulase R10 and 'Macerozyme' R10 will presumably not need further purification.

The various commercially available cellulases and pectinases vary in their ability to produce protoplasts from different cell systems. This is a reflexion of the differing composition of the cell walls as well as the differing specificity of the enzymes. For instance, in certain tissues, particularly those of fruits where there is a high pectin content, protoplasts can be released by the action of pectinase alone (Gregory and Cocking, 1965). Keller *et al.* (1970), using a purified cellulase, were able to isolate protoplasts from cultured cells of only three out of six species tested. They found a fair degree of correlation between the composition of the cell wall and the ability of the enzymes to liberate protoplasts. When attempting to isolate protoplasts from a particular cell system or plant tissue, which has proved resistant to 'Onozuka' cellulase and 'Macerozyme', it is advisable to try several different cellulase and pectinase preparations both separately and in combination.

When stored dry at low temperatures cellulases and pectinases can be kept for long periods without loss of activity; in solution, however, activity appears to be less stable. It is advisable, therefore, for reproducible results to make up the enzyme solutions just prior to use.

A further consideration is that considerable care is required to obtain plant material in a suitable condition for protoplast isolation, and this applies equally for the isolation of cells. The plants must be of the correct age and have been grown under the correct environmental conditions. A protocol for protoplast isolation worked out for a given plant grown under one set of conditions may not be suitable if these conditions are changed. As yet we are not certain of all the environmental factors which influence the susceptibility of the plant cells to the hydrolytic enzymes and therefore the selection of suitable growth conditions must be somewhat empirical. These considerations also apply when it is required to isolate protoplasts from plant cells grown *in vitro*. Although the growth environment can be rigorously controlled, certain stages in the growth cycle of the culture will be more suitable for protoplast isolation than others.

When dealing with the isolation of protoplasts from plant tissues, the ease with which the hydrolytic enzymes can penetrate into the tissue to attack the walls of the internal cells is important. This is especially true for compact tissues, such as root or stem apices, and these need to be cut into small pieces before enzyme incubation. The leaf, with its large intercellular spaces, into which the enzymes can readily penetrate, particularly if the lower epidermis is removed and the tissue is vacuum infiltrated, is more suitable in this respect. Leaves from which the epidermis cannot be peeled, such as cereals, need to be cut into narrow shreds. The enzymic digestion of the epidermis with pectin glucosidase (Rohament P. Rohm GmbH, Darmstadt, W. Germany) may sometimes be useful (Schilde-Rentschler, 1973).

It is now possible to isolate protoplasts from a very wide range of plants and from virtually every tissue. In fact it is probably true that protoplasts are more readily obtained from some parts of the plant than are free cells. Roots (Power and Cocking, 1970), tubers (Lorenzini, 1973), petals (Potrykus, 1971), pollen mother cells (Bhojwani and Cocking, 1972), pollen (Power and Frearson, 1973a) and endosperm (Taiz and Jones, 1971) have all yielded protoplasts. But leaf tissue, because of the high yield ($2$–$3 \times 10^6$ protoplasts/gm leaf), the uniformity of the protoplasts and the ease with which preparations can be obtained free of cells and cell debris, has become the choice material for most workers. There are now several different approaches available for the production of protoplasts from leaf tissue. Free cells obtained by either mechanical or enzymatic methods already described can be converted into protoplasts by treatment with cellulase (Takebe, Otsuki and Aoki, 1968; Harada, 1973). Alternatively, intact tissue can be treated with a mixture of cellulase and pectinase to yield protoplasts directly from the leaf (Power and Cocking, 1970). It is this method which, under certain conditions, can lead to the formation of multinucleate protoplasts through spontaneous fusion (see later). Yet a further method treats the tissue with pectinase to loosen the cells and then, before free cells are formed, the pectinase is replaced with cellulase to yield protoplasts (Potrykus and Durand, 1972). Each technique has its

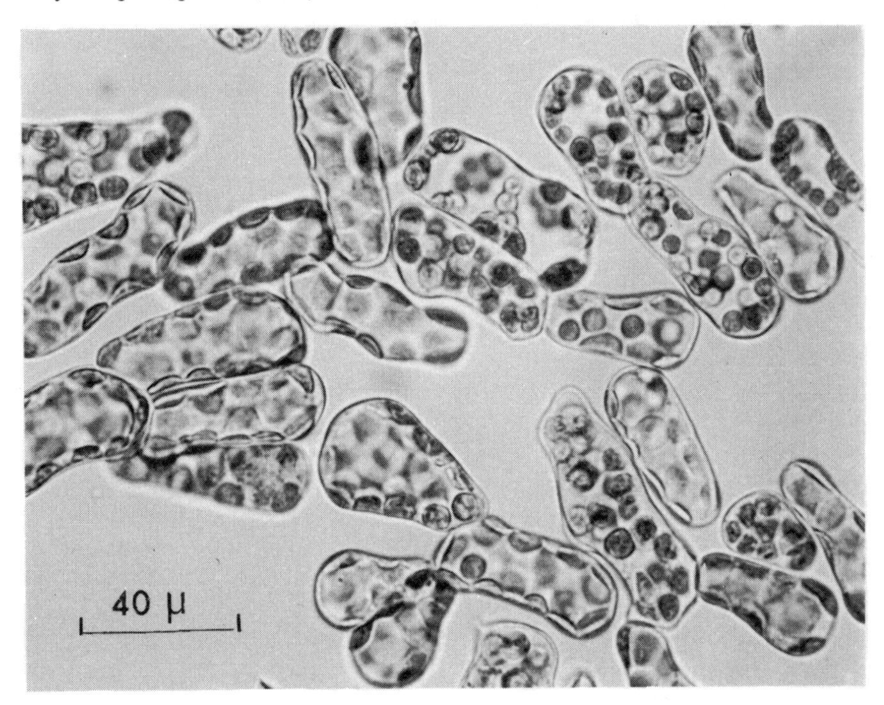


**Figure 4.1.** Mechanically isolated mesophyll cells of *Asparagus* (from M. R. Davey and I. A. Mackenzie, unpublished)

own particular advantage and they are therefore described here in some detail.

The liberation of tobacco leaf protoplasts directly from the leaf (Power and Frearson, 1973a) requires mature, fully expanded leaves from 50–60 day-old plants. These leaves are immersed for 30 sec in 70% ethanol and then transferred to 4% (v/v) sodium hypochlorite solution for 30 min. The hypochlorite is removed by five washes in sterile distilled water. The lower epidermis is removed (see earlier) and peeled leaf pieces are placed, exposed surface downwards, on the surface of a 0·7 *M* mannitol or sorbitol solution. After some 3 h this solution is replaced by a filter-sterilized enzyme solution (1·5% cellulase 'Onozuka' P1500 and 0·3% 'Macerozyme' in 0·7 *M* mannitol pH 5·8), approximately 2 g of leaf material to 20 ml of enzyme mixture, in a 14 cm plastic petri dish. After 16 h of static incubation at 26°C, protoplasts can be readily released by gently swirling the remaining leaf fragments. The protoplasts are sedimented by centrifugation at 100 *g* for 5 min and the enzyme solution removed by several washings in 0·7 *M* mannitol solution. A preparation of protoplasts, virtually free of debris can be obtained by centrifugation in 0·75 *M* sucrose at 100 *g* for 5 min. Debris and

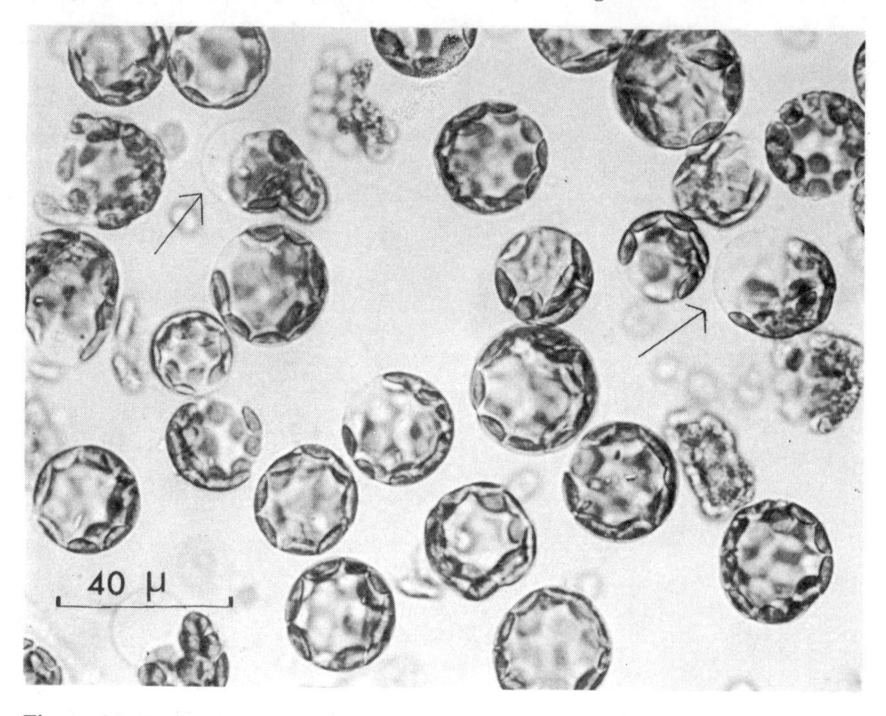


**Figure 4.2.** Isolated protoplasts derived from *Asparagus* cells by digestion of the cell wall with cellulase. Damaged protoplasts from which the vacuoles are emerging can also be seen (arrowed) (from M. R. Davey and I. A. Mackenzie, unpublished)

chloroplasts sediment, whilst protoplasts rise and collect at the surface. The yield of protoplasts is 1–2 × $10^6$ g fresh weight of leaf.

Isolation of protoplasts from *Petunia hybrida* leaves can be achieved by the following method (Potrykus and Durand, 1972). Fully expanded turgid leaves are sterilized in 10% v/v saturated sodium hypochlorite for 1 min and then washed with six changes of sterile water. The leaves are cut into thin cross-sections, washed in water three times and briefly vacuum infiltrated with a solution of 2% pectinase (Serva Heidelberg, W. Germany) in 0·04 *M* mannitol pH 5·4 and then statically incubated at 26°C for 25 min. This enzyme solution is replaced with 2% 'Onozuka' cellulase in 0·4 *M* mannitol pH 5·8 and incubated for a further 55 min. Gentle shaking of the tissue releases the protoplasts. The debris is removed by filtering the protoplast preparation through wire gauze and the protoplasts sedimented by centrifugation at 100 *g* for 2 min. The enzyme is removed by washing with 0·4 *M* mannitol.

Free, enzymatically isolated tobacco mesophyll cells can be converted into protoplasts (after Nagata and Takebe, 1970) by transferring the cells to 50 ml of filter-sterilized enzyme solution (4% 'Onozuka' cellulase 1500 and 0·7 *M* mannitol pH 5·2) in a conical flask. The cells are incubated at 36°C for 3 h with gentle shaking every $\frac{1}{2}$ h. The enzyme solution is centrifuged at 100 *g* for one minute and the sedimented protoplasts are resuspended in 0·7 *M* mannitol and 0·1 m*M* $CaCl_2$. This final step is repeated to remove residual enzyme and the protoplasts are ready for use.

Free, mechanically isolated *Asparagus* cells can be converted into protoplasts by the method of Mackenzie, Bui-Dang-Ha and Davey (1973). Cells from approximately 5 g of tissue are incubated in 120 ml of enzyme solution. (1% 'Macerozyme', 3% 'Onozuka' cellulase, 0·9 *M* mannitol pH 5·2) for 3–4 h. The isolated protoplasts are sedimented by centrifugation at 100 g for 2 min and washed with 0·7 *M* mannitol to remove residual enzyme (Figure 4.2).

## IV. CULTURE OF CELLS AND PROTOPLASTS

Although the motivation for the isolation of cells and protoplasts from plants has always been to obtain a cell system which could be used to answer biological problems, workers were at first satisfied if successful isolation could be achieved. But now that isolation has become largely a routine procedure the emphasis has moved to the successful manipulation of these cell preparations, particularly to achieve cell division in culture. The obvious problem in the culture of isolated protoplasts which is not encountered in the culture of complete cells, is the fragile nature of the system and the need of the protoplast to re-form a new cell wall. At the same time the culture must be maintained under conditions of relatively high osmotic potential. The isolated protoplasts may, however, differ in other, more subtle ways, from the intact cell. Recently protoplasts isolated from cultured crown gall cells of *Parthenocissus tricuspidata*, which normally grow without added growth hormones, have been found to have a hormone requirement for division. Once the protoplasts regenerated into cells and underwent division, this requirement ceased (Scowcroft, Davey and Power, 1973). After this wall is synthesized, regenerated cells appear to share the same requirements in culture as normal isolated cells.

After the failure of Haberlandt (1902) to induce cell division in isolated leaf cells, there was some controversy as to whether these cells were suitable or even capable of successful culture. It has now been established, however, that the act of isolating the cells combined with their incubation in a suitable nutrient medium, induces a considerable proportion of the cells to undergo a process of dedifferentiation leading to cell division. The readily isolated mesophyll cells of *Macleaya cordata* were the first system in which successful cell cultures were obtained. Shumucker (1930), reported divisions in mesophyll cells cultured in leaf sap and in 1959, Kohlenbach observed divisions in these cells when they were cultured in liquid White's medium with coconut milk and 2,4-dichlorophenoxyacetic acid (2,4-D). Ball and Joshi (1965) firmly established that isolated cells can divide by following the division of a single isolated mesophyll cell of peanut in liquid culture in the light, using time-lapse photomicrography. They noted, like Haberlandt, that after 3–5 days in culture, the leaf cell increased in size such that it no longer resembled a palisade cell and the chloroplasts increased in size but lost much of their chlorophyll. Systrophy, the accumulation of plastids around the nucleus, preceded mitosis and the new cell plate appeared to have no consistent orientation to the original axis of the cell. Over a period of 14 days, colonies of 20–30 cells developed. Kohlenbach (1966) observed that the isolated palisade mesophyll cells of *Macleaya cordata* could develop in one of several ways depending on the composition of the nutrient medium. In a simple salt medium containing glucose and leaf extract, produced by grinding young leaves and shoot apices with sand, followed by filtration, the cells expand but do not divide. When coconut milk and 2,4-D are added, the cells expand and subsequently divide. Continued division was dependent on the behaviour of the chloroplasts, which in turn was dictated by the nutrient medium. In a medium with simple salts, coconut milk, vitamins and 2,4-D the chloroplasts did not divide but were distributed among the new cells, thus the number per cell decreased. When only a few plastids were present in each cell, division ceased. In contrast, in a medium with a more complex salt composition and kinetin as well as 2,4-D, the chloroplasts fragmented and cell division continued and produced proliferating cell aggregates.

In a detailed study of the growth and nutritional requirements of isolated leaf cells grown in shake culture in the light, Joshi and Ball (1968) developed an entirely synthetic medium. They found that the major salts of Heller (1953) and the minor salts listed in Table 4.2, sustained the growth of peanut cells. Notably the mineral salts which Murashige and Skoog (1962) and White (1963), used extensively for callus culture, were found to be unsuitable. They also noted that only the palisade cells survived and divided, the spongy mesophyll cells dying after 1 or 2 days of culture. The addition of complex nutrients such as coconut milk, yeast extract, yeast hydrolysate, corn hydrolysate and enzymatic casein hydrolysate was detrimental and produced no growth. However, blood hydrolysate and lactoalbumin hydrolysate supported

**Table 4.2.** Nutrient medium for the culture of isolated cells of *Arachis hypogea* (Joshi and Ball, 1968)

| Salts | mg/l |
|---|---|
| *Major* | |
| KCl | 750 |
| $NaNO_3$ | 600 |
| $NaH_2PO_4.2H_2O$ | 141 |
| $CaCl_2.6H_2O$ | 112 |
| $MgSO_4.7H_2O$ | 250 |
| $NH_4Cl$ | 5·35 |
| *Minor* | |
| $H_3BO_3$ | 0·056 |
| $MnCl_2.4H_2O$ | 0·036 |
| $ZnCl_2$ | 0·15 |
| $CoCl_2$ | 0·02 |
| $CuCl_2.2H_2O$ | 0·054 |
| $NaMoO_4.2H_2O$ | 0·025 |
| $FeCl_3.6H_2O$ | 0·5 |
| Disodium salt of ethylene dinitrilotetraacetic acid | 0·8 |
| Amino acids: | |
| Acid-hydrolysed salt and vitamin-free casein hydrolysate | 400 |
| Hormones: | |
| 2,4-Dichlorophenoxyacetic acid | 1·0 |
| Kinetin | 0·1 |
| Carbon source: | |
| Sucrose | 20 g/l |

The nutrient medium was filter sterilized.

growth, but a better growth rate was achieved with salt and vitamin-free casein hydrolysate (400 mg/l). Glutamine could substitute for casein hydrolysate and when these were both present in the medium there was a substantial increase in growth. But the best growth was produced by the addition of a small amount of ammonium chloride (5·3 mg/l) in the presence of casein hydrolysate. As a carbon source, sucrose (at 20–30 g/l) appeared to be the most suitable. The presence of an auxin in the absence of a cytokinin proved inhibitory to growth and the converse was also true. However, when both 2,4-D and kinetin were present, growth was stimulated. Joshi and Noggle (1967) came to the conclusion that for sustained growth these peanut mesophyll cells required only mineral salts, an appropriate source of ammonia, an energy source and an auxin and cytokinin. The requirement for ammonium ions and the failure to use nitrate successfully was presumably related to the lack of an efficient nitrate reductase. The colonies of cells arising from these isolated mesophyll cells could be maintained in culture by transfer to fresh liquid medium at appropriate intervals. It was interesting that when the spheres of tissue were transfered to agar-solidified medium, the modified Heller's medium listed in Table 4.2 was no longer suitable. Instead the medium of Linsmaier and Skoog (1965) supported growth.

Rossini (1969), studying isolated cells from leaves of bindweed, *Calystegia sepium*, was also able to obtain growth on a synthetic medium (Table 4.3) and

**Table 4.3.** Culture media for growth of isolated leaf cells. After Rossini, 1969

| Salts | mg/l |
|---|---|
| *Major* | |
| $KNO_3$ | 950 |
| $NH_4NO_3$ | 720 |
| $MgSO_4.7H_2O$ | 185 |
| $CaCl_2$ | 166 |
| $KH_2PO_4$ | 68 |
| *Minor* | |
| $MnSO_4.4H_2O$ | 25 |
| $H_3BO_3$ | 10 |
| $ZnSO_4.4H_2O$ | 10 |
| $NaMoO_4.2H_2O$ | 0·25 |
| $CuSO_4.5H_2O$ | 0·025 |
| Iron solution: | |
| NaEDTA 7·5 g/l | |
| $FeSO_4.7H_2O$ 5·57 g/l | 2·5 ml/l |
| Vitamins: | |
| Myo-inositol | 100 |
| Nicotinic acid | 5 |
| Pyridoxin HCl | 0·5 |
| Thiamin HCl | 0·5 |
| Folic acid | 0·5 |
| Biotin | 0·05 |
| Amino acid: | |
| Glycine | 2·0 |
| Hormones: | |
| 6-Benzylaminopurine | 0·1 |
| 2,4-Dichlorophenoxyacetic acid | 1·0 |
| Carbon source: | |
| Sucrose | 10 g/l |

This nutrient medium can be sterilized by autoclaving

also recorded a requirement for the medium to contain both an auxin and a cytokinin for division to occur. More recently, Jullien (1970) re-examined the growth of leaf cells of peanut and found that, unlike Joshi and Ball he was able to induce division in both palisade and spongy mesophyll cells. He attributed this difference to the isolation procedure rather than to the nutrient medium, both of which were the same as used by Rossini, suggesting that removal of the cells by a scalpel is more likely to damage, both by mechanical shock and desiccation, the large delicate cells of the spongy mesophyll than the palisade cells.

Preparations of mesophyll cells produced enzymatically have also been shown to undergo division. Tobacco cells isolated in this way by Usui and Takebe (1969) divided in a simple salt solution containing mannitol at 0·7 *M*, kinetin (0·1 mg/l) and 2,4-D (1 mg/l) and also in White's medium (White, 1963) containing mannitol (0·6 *M*) and 3% sucrose and the same level of auxins and cytokinins. It was significant that the presence of high levels of mannitol in the nutrient medium did not prevent division. In fact, the cells recovered from plasmolysis after 2–3 days in culture. The division, however, did not progress much beyond the 6–8 cell stage and Usui and Takebe made a similar observation to Kohlenbach (1966) in that the chloroplasts appeared not to multiply.

Takebe, having firmly established that enzymatically isolated leaf cells could divide in the presence of high levels of mannitol, went on to culture isolated protoplasts (Nagata and Takebe, 1970). The initial response of the protoplasts was to expand in spite of this high osmotic potential. Subsequently this expansion has been observed by other workers using leaf protoplasts from widely differing plant species. Indeed, failure of the protoplast to expand is often an indication that the cell is dead.

After 2 days in culture some 60–80% of the protoplasts showed some fluorescent staining in the presence of 'Calcafluor White' ST (American Cyanamide Co., Wayne, New Jersey, U.S.A.), an optical brightener which binds to cell wall material. A day later the fluorescence covered the entire surface of most of the cells. Since freshly isolated protoplasts completely free of cell wall show no fluorescence it appears to be a good method of detecting wall synthesis. Protoplasts can be treated in either liquid or agar culture. The cells are incubated in 0·1% 'Calcafluor' in the appropriate osmotic stabilizer (0·6–0·7 *M* mannitol) for 5 min. The protoplasts are then washed to remove any excess dye and mounted on a slide in mannitol. The fluorescence is observed using a mercury vapour lamp with an excitation filter BG12 and suppression filter K510. 'Tinapol' B.O.P.T. (Geigy U.K. Ltd., Dye Stuff and Textile Chemicals Division, Simonsway, Manchester, U.K.) behaves in a similar fashion. For an account of techniques involving optical brighteners, see Paton and Jones (1971). Following wall synthesis cell division occurred and by the fourth day of culture 60–80% of the regenerated cells had divided.

The medium used was the same in salt composition as Takebe had used earlier for the culture of isolated leaf cells but, thiamin, myo-inositol and glycine were added. Nonetheless these regenerated cells behaved in the same way in culture as did the isolated leaf cells in that division continued up to the 6–8 cell stage and then ceased. Again there appeared to be no division of the chloroplast. The presence of the cytokinin was essential for division and although some division occurred in the absence of the auxin, it was far more vigorous when present. Fragmentation of the chloroplasts and continued division of the regenerated cells into colonies was subsequently observed by Takebe, Labib and Melchers (1971) when tobacco mesophyll protoplasts were cultured in Kohlenbach's medium II (Kohlenbach, 1966) which contained coconut milk but with the modification that glucose and leaf extract were omitted and naphthaleneacetic acid (NAA) (3 mg/l) replaced 2,4-D. Mannitol (0·8 *M*) was present as an osmotic stabilizer. By transferring these colonies of cells to an appropriate nutrient medium it was possible to induce the formation of small shoots which could give rise in turn to roots and finally, complete flowering plants. Takebe *et al.* (1971) had clearly demonstrated that isolated leaf protoplasts were capable of developing to give rise to a complete plant and were indeed totipotent. A technique which employs a somewhat simpler defined nutrient medium into which isolated protoplasts were 'plated' was subsequently developed by Nagata and Takebe (1971). By this method they were able to induce some 60 % of the cultured protoplasts to develop into colonies.

The concept of culturing plant cells plated directly onto solidified medium was pioneered by Bergmann (1960). He 'plated' isolated cells from suspension cultures of tobacco, *Nicotiana tabacum*, and bean, *Phaseolus vulgaris*, and observed that about 20 % of the cells divided repeatedly to form small colonies. Plating of cells in this way has the distinct advantage that the development of individual colonies from the single-cell stage can be followed and also the proportion of cells giving rise to colonies can be readily determined and expressed as a 'plating efficiency'. The use of the plating technique, therefore, opens up the possibility of a quantitative study of the effect of plating density and environmental conditions on the 'plating efficiency' of cells. The technique requires the cells or protoplasts to be suspended in a nutrient medium at twice the desired final concentration. An aliquot of these cells is gently mixed in a petri dish with an equal volume of nutrient medium containing 0·9–1·2 % liquid agar which has been kept molten by maintaining the temperature at about 45°C. Gentle swirling of the dish allows an even distribution of the cells. The dish is then sealed with parafilm and incubated under the desired conditions. Cells can be observed by inverting the dish and viewing through the agar. It is advisable to use the highest quality agar (Noble agar, Difco, Detroit, U.S.A. or Ionagar Oxoid, London, U.K.) for the clearest view of the cells. Nagata and Takebe (1971) observed that this 'plating efficiency' was markedly influenced by the density at which the protoplasts were cultured. Protoplasts plated at 5–7·5 × $10^3$/ml produced the highest 'plating efficiency', approximately 60–70%, whereas protoplasts plated at, or below, 1 × $10^3$/ml normally failed to produce colonies. A similar effect of 'plating' density was found by Jullien (1973) in a study of the growth requirements of plated intact mesophyll cells of *Asparagus*. Some 40% of the cells divided when 'plated' at high density 3·5 × $10^5$/ml but if the density was reduced to 5 × $10^4$/ml only 5% divided.

Light intensity also influenced the 'plating efficiency' of tobacco leaf protoplasts. Plates incubated at 700 lux having less than half as many colonies as plates exposed to 2300 lux (Nagata and Takebe, 1971). But light intensity at this level has not been found to be suitable for all leaf cell systems, indeed complete darkness is the most suitable condition for both *Asparagus* cells (Jullien, 1973) and protoplasts (Mackenzie, Bui-Dang-Ha and Davey, 1973).

Nagata and Takebe made the interesting observation that isolated palisade cells plated into nutrient agar medium generally had a lower 'plating efficiency', compared to protoplasts isolated from the same tissue. One possible explanation is the difficulty in detecting damaged and non-viable cells. In contrast, damaged protoplasts may either burst or collapse and the resulting debris disappear from the protoplast preparation. It is, therefore, of interest in studies of the effect of environmental conditions on 'plating efficiency' of cells and protoplasts to know the proportion of viable cells in the initial inoculation. Widholm (1972) used fluorescein diacetate as a test for viability of cultured cells. Viable cells incubated in fluorescein diacetate and then examined under blue light, will fluoresce whereas dead cells will not. This fluorescence appears to depend on membrane integrity (Rotman and Papermaster, 1966). The polar fluorescein diacetate molecule passes readily into the cell where it is cleaved by esterase activity but the free fluorescein molecule passes out of the cell more slowly, with the net result that fluorescein accumulates within the cell, giving rise to the fluorescence. This technique can also be applied as a test for protoplast viability. A stock solution of fluorescein diacetate (Koch-light Ltd., U.K.) in acetone containing 5 mg/ml is diluted to 0·01% with culture medium containing the appropriate osmotic stabilizer. The protoplasts are incubated for 5 min and then examined using a mercury vapour lamp with an excitation filter BG12 and suppression filter K510. The fluorescence takes a few minutes to develop. Debris, damaged protoplasts and free chloroplasts show no fluorescence whereas intact protoplasts fluoresce brightly (Figure 4.3). However, some protoplasts which optically appear healthy, give no fluorescence. The proportion of protoplasts which fluoresce corresponds closely to the proportion of protoplasts which expand during the first few days of culture, this expansion being another indication of viability (see earlier).

Although the isolation of protoplasts and cells directly from the plant has now been achieved from a considerable number of species covering a wide range of families, the number of cases in which successful culture has been achieved is only small. Currently protoplasts from several species of tobacco (Takebe *et al.*, 1971; Carlson *et al.*, 1972) including both diploid and haploid lines of *N. tabacum* (Nitsch and Ohyama, 1971), *Petunia* (Frearson, Power and Cocking, 1973; Durand, Potrykus and Donn, 1973), carrot (Kameya and Uchimiya, 1972) and *Asparagus* (Bui-Dang-Ha and Mackenzie, 1973) have developed and produced entire plants. The culture of isolated cells has

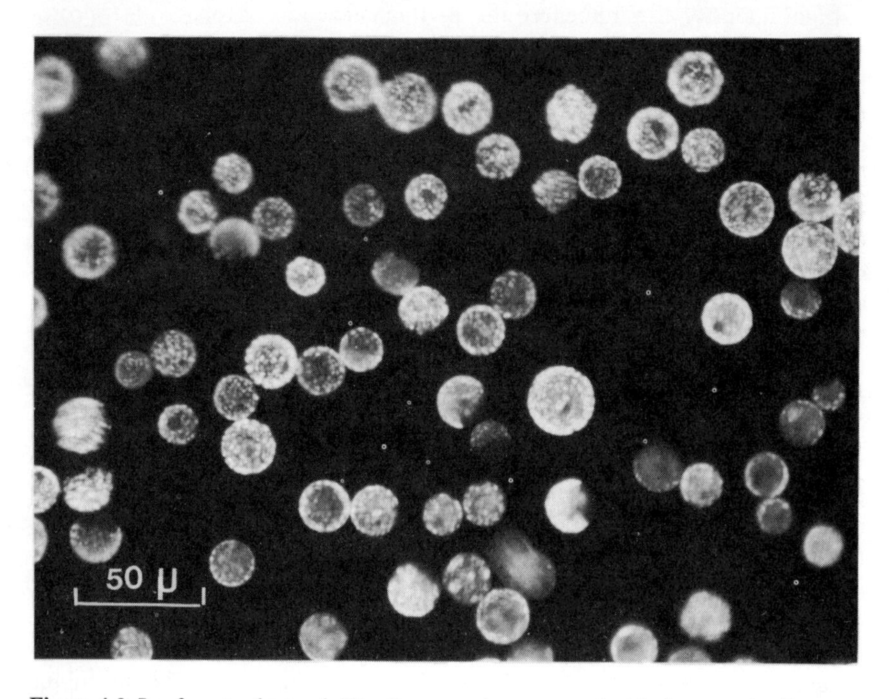


**Figure 4.3.** Leaf protoplasts of *Nicotiana otophora* treated with fluorescein diacetate soon after isolation and examined using blue light. The viable protoplasts fluoresce green, dead and damaged protoplasts do not fluoresce

received less attention although plants from isolated leaf cells of diploid and haploid lines of *N. tabacum* have been obtained (Jullien, 1971). This list of species in which isolated cells and protoplasts can be cultured so as to give rise to whole plants will, no doubt, lengthen. In some cases the isolated protoplasts can be cultured but it is not yet known how to produce plantlets from the resulting cell colonies. Legume leaf protoplasts, the subject of study in a number of laboratories, may be a case in point as reports of plant regeneration from legume callus are few.

Cereals, because of their economic importance, are of particular interest but unfortunately isolated cereal leaf protoplasts have proved rather difficult to culture. Leaf protoplasts of wheat, rye and barley will readily regenerate a new cell wall but cell division occurs only rarely (Evans, Keates and Cocking, 1972; Evans, Woodcock and Keates, 1973; Wenzel, 1973).

Another problem which has proved difficult to overcome has been bacterial contamination in some protoplast preparations from leaves. Although the surface sterilization appears to be adequate there is evidence that bacteria are present within the leaves. Unlike animal cell cultures antibacterial and antifungal agents are not normally included in protoplast culture media, since many of these antibiotics which allow the growth of animal cells are toxic to plant protoplasts.

## V. PROTOPLAST FUSION AND SOMATIC CELL HYBRIDIZATION

Unlike the animal cell, the somatic plant cell is surrounded by a semi-rigid cellulosic wall. Although this is a highly permeable structure and contact between adjacent cells does exist in the form of cytoplasmic connexions through the cell walls (the plasmodesmata) the cell wall does form a very effective barrier to cell fusion. Once this wall has been removed in the formation of an isolated protoplast, then the possibility of cell fusion exists. It became apparent from some of the initial studies on the phenomenon of protoplast fusion (Power, Cummins and Cocking, 1970) that this fusion could be of two distinct types. Under some conditions protoplasts observed directly after isolation were found to contain more than one nucleus, suggesting that fusion had occurred during the actual isolation procedure. The term 'spontaneous fusion' was used to describe this process. In contrast, freely isolated protoplasts did not appear to fuse unless they were treated with an inducing agent, this process being termed, therefore, 'induced fusion'.

### A. Spontaneous Fusion

A detailed study of spontaneous fusion has been carried out by Withers and Cocking (1972). They observed the gradual expansion of the plasmodesmata between adjacent cells as the cell wall was digested. Eventually this expansion was sufficient to allow the passage of cell organelles and, finally, a complete coalescence of the cells occurred. Thus spontaneous fusion takes place as a result of the symplastic nature of plant tissues; the protoplasts are never truly isolated units and genuine plasmalemma fusion does not appear to take place. Since spontaneous fusion occurs within the plant tissue, it can only give rise to the formation of homokaryons and is thus of no value for somatic hybridization. The exception would be if plasmodesmata occurred between the different cells in a chimeral tissue and spontaneous fusion could then result in a hybrid cell. Careful manipulation of the protoplast isolation procedure can lead to the formation of large quantities of multinucleate protoplasts. Naturally, it is essential to use the mixed enzyme isolation method. Indeed, when freely isolated cells are converted to protoplasts, only a very few multinucleate protoplasts are found.

Woodcock (1973) in a study of the factors which influence spontaneous fusion of mesophyll protoplasts of White Burley tobacco, could repeatedly obtain populations containing 30–40% multinucleate protoplasts of which some 20% were bi- or trinucleate, although some protoplasts containing as many as 15 or 16 nuclei were commonly seen. He found that mannitol or sorbitol produced higher levels of spontaneous fusion than sucrose at the equivalent osmotic potential. Increasingly high levels of plasmolyticum diminished the proportion of multinucleate protoplasts, as expected observation if integrity of the plasmodesmata were necessary for spontaneous fusion. The presence of potassium dextran sulphate in the enzyme incubation medium was essential for high levels of spontaneous fusion, this compound possibly acting here as a membrane-stabilizing agent. Although of little value for somatic hybridization, these multinucleate protoplasts provide an interesting system for the study of the growth and division of multinucleate cells. Indeed, synchronous mitosis has been observed in multinucleate protoplasts (Power, Frearson and Cocking, 1971; Motoyoshi, 1974) as well as nuclear fusion (Miller, Gamborg, Keller and Kao, 1971).

## B. Induced Fusion

In contrast to spontaneous fusion, induced fusion involves the fusion of freely isolated protoplasts. This opens up the possibility of the formation of both homo- and heterokaryons. Power, Cummins and Cocking (1970) found sodium salts, more especially sodium nitrate, to be suitable inducing agents. They were able to fuse isolated protoplasts from the meristematic region of the oat root with protoplasts from the vacuolated cells of maize roots.

One of the difficulties encountered in experiments on cell fusion, is the identification of the resulting hybrid cells. As a first step towards this objective Potrykus (1971) isolated coloured protoplasts from petals and was able to obtain, under the influence of sodium nitrate, heterokaryon protoplasts which could be identified by their coloured constituents. Giles (1972) induced fusion between green leaf protoplasts of crab grass, *Digitaria*, and colourless protoplasts from cultured soybean cells. In this case the fusion product proved capable of forming a cell wall. Common cell wall formation around a heterokaryon, consisting of a tobacco leaf protoplast and a protoplast from a cultured crown gall cell of *Parthenocissus tricuspidata*, was also obtained by Power and Frearson (1973b).

A differential staining technique based on carbol fuchsin was developed by Keller, Harvey, Kao, Miller and Gamborg (1973) for the purpose of identifying constituent nuclei of interspecific hybrids between *Vicia hajastana* and soybean, *Glycine max*. With this technique they were able to identify heterokaryons after the fusion body had rounded up and cytoplasmic mixing had occurred, something which is not always possible when green chloroplasts are used as markers. Keller *et al.* noted that heterokaryon formation was a rare event and frequencies of interspecific fusion of 0·1–1% were recorded. They tried a range of inducing agents which included Sendai virus, lysolecithin, poly-L-ornithine as well as sodium nitrate. All produced a similar level of fusion but, in some cases, treatment led to protoplast lysis, particularly treatment with lsyolecithin, and poly-L-ornithine.

A number of workers have investigated various compounds in order to find a method which will enhance the fusion rate as well as give insight into the mechanism of plasmalemma fusion. Recently a new technique for achieving the close aggregation of the protoplasts has been developed by Hartmann, Kao, Gamborg and Miller (1973). Using an antibody reaction protoplasts can be made to agglutinate but as yet there is no evidence that membrane fusion follows this close adhesion. Sendai virus was found by Withers (1973) to adhere to the plasmalemma and could induce some protoplast fusion but the effect was markedly dependent on the condition of the Sendai virus. Lysolecithin, glycerol monooleate and trioleate, which have induced fusion in animal cells, although at the same time causing considerable membrane damage leading to some loss of cell viability (Ahkong, Cramp, Fisher, Howell and Lucy, 1972) caused a high degree of membrane instability in the protoplasts. Concanavalin A caused aggregation of the protoplasts but did not promote fusion. Withers (1973) came to the conclusion that treatment with sodium nitrate still remained the most controlled and least damaging method for protoplast aggregation and fusion. Kameya and Takahashi (1972) from a study of the influence of various salts on protoplast fusion, concluded that sodium ions were responsible rather than the nitrate, confirming the earlier observation of Power *et al.* (1970).

From these studies on the effect of various fusion agents it became important to differentiate between protoplast fusion and aggregation. The aggregation and close adhesion of isolated protoplasts is an essential first step towards fusion. From studies of the electrophoretic mobilities of protoplasts (Grout and Coutts, 1974) it would appear that sodium nitrate leads to a marked reduction in the electronegativity of the protoplasts producing in turn a reduction in the mutual repulsive forces and allowing membrane adhesion to occur. Once the plasmalemmata are in close contact, breakdown and rearrangement of the membrane appears to take place leading to limited areas of cell fusion (Withers and Cocking, 1972; Withers, 1973). The rate at which the subsequent cell fusion and merging of the cytoplasm proceeds, appears to depend to a large extent on the physical structure of the cells. The cytoplasmic merging of fused mesophyll protoplasts which have only a very thin layer of cytoplasm surrounding a large central vacuole is a relatively slow process. If cell wall synthesis is actively taking place, complete cell fusion could be prevented (Power and Frearson, 1973b). Fusion between less vacuolated protoplasts, such as those from meristematic regions of the root, occurs much more rapidly.

It would presumably be of advantage in the fusion of the highly vacuolate leaf protoplasts to have rather more control over the process of wall synthesis, allowing cytoplasmic fusion to go to completion before wall synthesis is initiated. Protoplasts derived from some cultured cell systems appear to be less active in wall synthesis than mesophyll protoplasts and nuclear division has been observed without wall synthesis (Reinert and Hellmann, 1973). Scowcroft *et al.* (1973) have reported the actual proliferation of protoplasts from cultured crown gall cells of *Parthenocissus tricuspidata*. Small uninucleate non-vacuolate protoplasts appeared during culture but the mechanism by which they arose is not yet clear.

Recently, it has been demonstrated that animal cells subjected to alkaline conditions (pH 10·5) together with exposure to calcium ions will fuse to form multinucleate cells (Toister and Loyter, 1973). This procedure, with calcium ions being present during the exposure to the alkaline conditions, has now been applied to isolated leaf protoplasts and considerable protoplast fusion has been observed (Keller and Melchers, 1973). A notable feature of this technique is the rapidity with which the protoplasts merge together to form multinucleate fusion bodies (see Figure 4.4). If the fusion products subsequently prove to be stable and viable then this technique would appear to dispense with the requirement for less vacuolated protoplasts for rapid fusion and considerations of the effects of wall synthesis on the fusing protoplasts will not be of such importance.

The technique used in Nottingham (Power and Frearson, 1973b) to achieve protoplast aggregation and fusion, is to suspend the protoplasts from two different sources, normally in the ratio of 1:1, in a round-bottomed screw-cap tube. The protoplasts are sedimented at 100 *g* for 5 min and resuspended in 0·3 *M* sodium nitrate with the balance of the plasmolyticum made up with sorbitol or mannitol. The protoplasts are warmed to 35°C for 5 min, and, after gently resuspending the protoplasts, they are sedimented at 300 *g* for 5 min. This produces a tightly packed pellet of protoplasts. The tube is kept for a further 20 min at 35°C. The sodium nitrate solution is carefully removed and very gently replaced with culture medium containing 0·1% sodium nitrate and sucrose as plasmolyticum. During the next hour the protoplast aggregate normally rises to the surface of the medium. This aggregate is cultured under these conditions for one week. The large aggregates are then gently dispersed into smaller groups of cells by slowly drawing the protoplasts into a Pasteur pipette. These protoplasts can then be plated in nutrient medium as previously described.

## C. Selection

At present all the published accounts indicate that the frequency of complete cytoplasmic fusion between protoplasts from different species leading

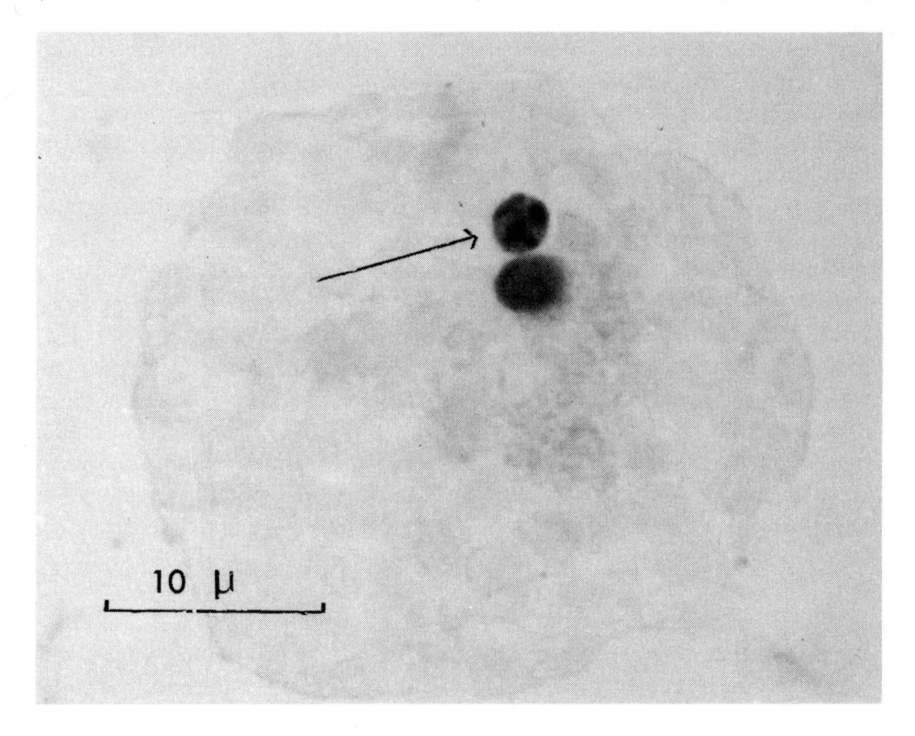


**Figure 4.4.** A heterokaryon resulting from the fusion of a *Nicotiana tabacum* leaf protoplast with a leaf protoplast of *Nicotiana otophora*. The *N. otophora* nucleus (arrowed) has large characteristic densely stained heterochromatic regions. The nuclei were stained with carbol fuchsin

to the formation of a viable heterokaryon is low. Furthermore, nuclear fusion must take place within this heterokaryon before a hybrid cell can be said to have formed. This factor will lower still further the frequency of full hybrid cell formation to a level where it can only be regarded as a rare event. However, this in itself is not a major drawback to somatic hybridization, since from a single leaf it is possible to isolate several million protoplasts and it is quite practical to perform experiments involving $1 \times 10^7$ protoplasts. Selection of the rare fusion product from amongst the non-fused and those protoplasts which have fused with like protoplasts, does then become a major problem. Carlson, Smith and Dearing (1972), using sodium nitrate to induce fusion, were able to obtain and select somatic hybrids between two different species of *Nicotiana*, *N. glauca* and *N. langsdorffii*. Their selection procedure was based on the observation that protoplasts isolated from leaves of *N. glauca* and *N. langsdorffii* did not undergo sustained divisions in a culture medium which would allow the successful culture of protoplasts isolated from the leaves of the hybrid produced by the sexual cross. When a mixture containing $1 \times 10^7$ protoplasts from each parent was treated with sodium nitrate and subsequently plated in the nutrient medium and cultured for 6 weeks, some 33 colonies were formed which were suspected to have arisen from hybrid cells because of their capability of growing in this medium. Further analysis of the growth characteristics of these colonies supported this contention and several of the colonies were manipulated to produce shoots. The leaf morphology, the density of hairs on the leaves and the ability of the plant spontaneously to form tumourous outgrowths were all characteristic of the sexual hybrid rather than the parent species. Further, the perioxidase isoenzyme patterns were a summation of those of the two parent species. A chromosome number of 42 was determined for the somatic hybrid which was the sum of the parent numbers (24 plus 18). From the accumulated evidence there can be little doubt that Carlson *et al.* (1972) had produced a hybrid plant through protoplast fusion and that somatic hybridization of these two sexually compatible species was a reality. Unfortunately their selection screen for the hybrid is not a generally applicable procedure, relying, as it does, on the prior knowledge of the cultural behaviour of the sexual hybrid.

What is required is a more general selection system for the isolation of hybrids formed as a result of fusion between protoplasts arising from plants which cannot be crossed sexually. Auxotrophic mutants would seem to offer an approach but Carlson *et al.* (1972) reported that initial attempts to utilize auxotrophic mutants of tobacco in a selection procedure have so far proved unsuccessful, possibly because of cross-feeding. Carlson *et al.* (1972) have also suggested a selection technique based on the use of complementing recessive nuclear albino mutations. The development of green colonies in culture would indicate the presence of hybrid cells. Indeed, genetic complementation of this type has been demonstrated by Giles (1973). He induced the fusion of green maize leaf protoplasts with protoplasts from an albino mutant and observed the development of chlorophyll in the plastids derived from the albino partner. Unfortunately it is not possible, at present, to obtain sustained cell division in cultures of maize leaf protoplasts.

A selection system utilizing mutant cell lines in a similar way to the H.A.T. system, so successfully exploited in animal cell fusion experiments (Littlefield, 1964) is not possible at present as such mutants in higher plants are not available. Some progress in this direction has recently been achieved with the isolation of a 5-bromodeoxyuridine-resistant cell line from callus of haploid tobacco (Maliga, Marton and Breznovits, 1973). Recently streptomycin-resistant cell lines of tobacco have also been recovered and furthermore it has been possible to regenerate from these cells fertile plants which carry the resistance trait (Maliga, Breznovits and Marton, 1973). It would be interesting to know if isolated leaf cells or protoplasts from these plants would grow in the presence of streptomycin.

Attempts to isolate mutant cell lines have in general used cultured cells but cells and protoplasts isolated directly from the plant would appear to be an attractive starting material. After treating isolated haploid tobacco leaf protoplasts with ethyl methyl sulphonate, Carlson (1973) was able to isolate a mutant cell line capable of growing in the presence of the amino acid analogue methionine sulphoximine and plants derived from this cell line were also resistant to methionine sulphoximine.

The use of haploids would appear to offer a distinct advantage for the recovery of mutant cell lines. However, many of the crop plants, the group in which somatic cell fusion is of particular interest, are polyploid in origin Furthermore, the range of species in which haploids derived from another culture have been achieved is, at the moment, limited. It might seem, therefore, that a general selection procedure, which is not based on the use of haploids would have the most widespread application in plant somatic hybridization. Moreover, it should be noted that most selection procedures for animal cell somatic hybrids do not depend on the use of haploid cell lines.

The capability developed in recent years, of culturing cells isolated directly from the plant, often on defined media and with a high plating efficiency, is now bringing about a renewed interest in this area of plant cell biology. The large numbers of cells and isolated protoplasts which can be obtained allows rare events such as mutations to be recovered. Furthermore, the isolated plant protoplast resembles, in some respects, the cultured animal cell and offers an alternative eukaryotic system. Increasingly, attempts are being made to exploit this plant protoplast system either by somatic cell fusion or transformation-type and transduction-type experiments with the aim of plant cell modification. No doubt, as it becomes possible to culture a wider variety of plant cell and protoplast systems and as further selection procedures are developed, the details of hybridization will become more established. When all these techniques are combined with the present ability to derive fertile plants from single cells and protoplasts, a new approach to plant genetics will become available.

## VI. REFERENCES

Ahkong, Q. F., F. Cramp, D. Fisher, J. I. Howell and J. A. Lucy (1972) 'Studies on chemically induced cell fusion', *J. Cell Sci.*, **10**, 769.

Albersheim, P. (1973) 'The primary cell wall', in 'Protoplastes et fusion de cellules somatique végétales', *Colloque Internationaux*, *C.N.R.S.*, **212**, 15.

Ball, E. and P. C. Joshi (1965) 'Divisions in isolated cells of palisade parenchyme of *Arachis hypogaea*', *Nature*, **207**, 213.

Bergmann, L. (1960) 'Growth and division of single cells of higher plants *in vitro*', *J. Gen. Physiol.*, **43**, 841.

Bhojwani, S. S. and E. C. Cocking (1972) 'The isolation of protoplasts from pollen tetrads', *Nature New Biol.*, **239**, 29.

Bui-Dang-Ha, D. and I. A. Mackenzie (1973) 'The division of protoplasts from *Asparagus officinalis* L. and their growth and differentiation', *Protoplasma*, **78**, 215.

Carlson, P. S. (1973) 'The uses of protoplasts for genetic research', *Proc. U.S. Natl. Acad. Sci.*, **70**, 598.

Carlson, P. S., H. H. Smith and R. D. Dearing (1972) 'Parasexual interspecific plant hybridisation', *Proc. U.S. Natl. Acad. Sci.*, **69**, 2292.

Chambers, R. and K. Hofler (1931) 'Microsurgical studies on the tonoplast of *Allium cepa*', *Protoplasma*, **12**, 338.

Chayen, J. (1952) 'Pectinase technique for isolating plant cells', *Nature*, **170**, 1070.

Cocking, E. C. (1960a) 'Some effects of chelating agents on roots of tomato seedlings', *Biochem. J.*, **76**, 51p.

Cocking, E. C. (1960b) 'A method for the isolation of plant protoplasts and vacuoles', *Nature*, **187**, 927.

Cocking, E. C. (1972) 'Plant cell protoplasts isolation and development', *Ann. Rev. Plant Physio.*, **23**, 29.

Coutts, R. H. A. (1973) in 'Isolation of protoplasts', Evans and Cocking in *Plant Tissue and Cell Culture* (Ed. H. E. Street), Blackwell's Botanical Monographs, Oxford.

Coutts, R. H. A., E. C. Cocking and B. Kassanis (1972) 'Infection of tobacco mesophyll protoplasts with tobacco mosaic virus', *J. Gen. Virol.*, **17**, 289.

Davey, M. R. and E. C. Cocking (1972) 'Uptake of bacteria by isolated higher plant protoplasts', *Nature*, **239**, 455.

Doy, C. H., P. M. Gresshoff and B. G. Rolfe (1973) 'Biological and molecular evidence for the transgenosis of genes from bacteria to plant cells', *Proc. U.S. Natl. Acad. Sci.*, **70**, 723.

Durand, J., I. Potrykus and G. Donn (1973) 'Plantes issues de protoplastes de *Pétunia*', *Z. Pflanzenplysiol.*, **69**, 26.

Edwards, G. E. and C. C. Black, Jr. (1971) 'Isolation of mesophyll cells and bundle sheath cells from *Digitaria sanguinalis* (L) Scop. leaves and a scanning microscopy study of the internal leaf cell morphology', *Plant Physiol.*, **47**, 149.

Evans, P. K., A. G. Keates and E. C. Cocking (1972) 'Isolation of protoplasts from cereal leaves', *Planta*, **104**, 178.

Evans, P., Woodcock and Keates (1973) 'Steps towards cell fusion: studies on cereal protoplasts and protoplasts of haploid tobacco' in 'Protoplasts et fusion de cellules somatique végétales', *Colloques Internationaux*, *C.N.R.S.*, **212,** 469.

Fodil, Y., R. Esnault and G. Trapy (1971) 'Remarques sur l'isolement de protoplastes végétaux; étude de l'influence de l'agent de plasmolyse', *C. R. Acad. Sc. Paris.*, **272**, 948.

Frearson, E. M., J. B. Power and E. C. Cocking (1973) 'The isolation, culture and regeneration of *Petunia* leaf protoplasts', *Dev. Biol.*, **33**, 130.

Giles, K. L. (1972) 'An interspecific aggregate cell capable of cell wall regeneration', *Plant and Cell Physiol*, **13**, 207.

Giles, K. L. (1973) 'Attempts to demonstrate genetic complementation by the technique of protoplast fusion', in 'Protoplastes et fusion de cellules somatique végétales', *Colloques Internationaux, C.N.R.S.*, **212**, 485.

Gnanam, A. and G. Kulandaivelu (1969) 'Photosynthetic studies with leaf cell suspensions from higher plants', *Plant Physiol*, **44**, 1451.

Gregory, D. W. and E. C. Cocking (1965) 'The large scale isolation of plant protoplasts', *J. Cell Biol.*, **24**, 143.

Grout, B. W. W. and R. H. A. Coutts (1974) 'Additives for enhancement of fusion and endocytosis in higher plant protoplasts: an electrophoretic study', *Plant Sci. Letters* **2**, 397.

Haberlandt, G. (1902) 'Kulturversuche mit isolierten pflanzenzellen', *Sitzber. Akad. Wiss. Wien, Math-Naturw. KI*, **III**, 69 (see Krikorian and Berquam, 1969).

Harada, H. (1973) 'A new method for obtaining protoplasts from mesophyll cells', *Z. Pflanzenphysiol*, **69**, 77.

Harada, H., K. Ohyama and J. Cheruel (1972) 'Effects of coumarin and other factors on the modification of form and growth of isolated mesophyll cells', *Z. Pflanzenphysiol*, **66**, 307.

Hartmann, J. X., K. N. Kao, O. L. Gamborg and R. A. Miller (1973) 'Immunological methods for the agglutination of protoplasts from cell suspension cultures of different genera', *Planta*, **112**, 45.

Heller, R. (1953) 'Recherches sur la nutrition minérale des tissus végétaux cultivés, *in vitro*', *Thèse, Paris et Anne. Sc. Nat. Bot. Biol. Veg. Paris*, **14**, 1.

Hoffmann, F. and D. Hess (1973) 'Die Aufnahme radioaktiv Markierter D.N.S. in isolierte protoplasten von *Petunia hybrida*', *Z. Pflanzenphysiol.*, **69**, 81.

Johnson, C. B., D. Grierson and H. Smith (1973) 'Expression of λ plac 5 DNA in cultured cells of a higher plant', *Nature New Biol.*, **244**, 105.

Joshi, P. C. and E. Ball (1968) 'Growth of isolated palisade cells of *Arachis hypogaea in vitro*', *Dev. Biol.*, **17**, 308.

Joshi, P. C. and G. R. Noggle (1967) 'Growth of isolated mesophyll cells of *Arachis hypogaea* in simple defined medium *in vitro*', *Science*, **158**, 1575.

Jullien, M. (1970) 'Sur l'aptitude à la diversion *in vitro* des cellules séparées du parenchyme foliaire *d'Arachis hypogaea*', *L. C. R. Acad. Sc. Paris*, **270**, 3051.

Jullien, M. (1971) 'Regeneration de plantes entières à partir de cellules séparées de feuilles de *Nicotiana tabacum* diploids et haploides', *C. R. Acad. Sci. Paris*, **273**, 1287.

Jullien, M. (1973) 'Division et croissance *in vitro* de cellules séparées du parenchyme foliaire d'*Asparagus officinalis* L', *Z. Pflanzenphysiol*, **69**, 129.

Jyung, W. H., S. H. Wittwer and M. J. Bukovac (1965) 'Ion uptake by cells enzymically isolated from green tobacco leaves', *Plant Physiol.*, **40**, 410.

Kameya, K. and H. Uchimiya (1972) 'Embryoids derived from isolated protoplasts of carrot', *Planta*, **103**, 356.

Kameya, T. and N. Takahashi (1972) 'The effects of inorganic salts on fusion of protoplasts from roots and leaves of *Brassica* species', *Japan, J. Genetics*, **47**, 215.

Keller, W. A., B. Harvey, O. L. Gamborg, R. A. Miller and D. E. Eveleigh (1970) 'Plant protoplasts for use in somatic cell hybridisation', *Nature*, **226**, 280.

Keller, W. A., B. L. Harvey, K. N. Kao, R. A. Miller and O. L. Gamborg (1973) 'Determination of the frequency of interspecific protoplast fusion by differential staining', in 'Protoplastes et fusion de cellules somatique végétales', *Colloque Internationaux, C.N.R.S.*, **212**, 455.

Keller, W. A. and G. Melchers (1973) 'The effect of high pH and calcium on tobacco leaf protoplast fusion', *Z. Naturforschung*, **28**, 737.

Kohlenbach, H. W. (1959) 'Streckungs–Und Teilungs Wachstum isolierter mesophyllzellen von *Macleaya cordata*', *Naturwissenschaften*, **46**, 116.

Kohlenbach, H. W. (1966) 'Die Entwicklungspotenzen explantierter und isolierter Dauerzellen', *Z. Pflanzenphysiol.*, **55**, 142.

Krikorian, A. D. and D. L. Berquam (1969) 'Plant cell and tissue cultures; The role of Haberlandt', *Bot. Rev.* **35**, 59.

Lamport, D. T. A. (1973) 'Is the primary cell wall a protein glycan network?' in 'Protoplastes et fusion de cellules somatiques végétales', *Colloques Internationaux, C.N.R.S.*, **212**, 27.

Lin, M. L. and E. J. Staba (1961) 'Peppermint and spearmint tissue cultures. I. Callus formation and submerged culture', *Lloydia*, **24**, 139.

Linsmaier, E. M. and F. Skoog (1965) 'Organic growth factor requirements of tobacco tissue cultures', *Physiol. Plant.*, **18**, 100.

Littlefield, J. W. (1964) 'The selection of hybrid mouse fibroplasts', in *Cold Spring Harbor Symp. Quant. Biol.*, **29**, 161.

Lorenzini, M. (1973) 'Obtention de protoplastes de tubercule de pomme de terre', *C. R. Acad. Sc. Paris*, **276**, 1839.

Mackenzie, I. A., D. Bui-Dang-Ha and M. R. Davey (1973) 'Some aspects of the isolation, the structure and growth of protoplasts from *Asparagus officinalis* L.', in 'Protoplastes et fusion de cellules somatique végétales', *Colloques Internationaux, C.N.R.S.*, **212**, 291.

Maliga, P., L. Marton and A. Sz-Breznovits (1973) '5-Bromodeoxyuridine-resistant cell lines from haploid tobacco', *Plant Sci. Letters*, **1**, 118.

Maliga, P. A., Sz-Breznovits and L. Marton (1973) 'Streptomycin-resistant plants from callus culture of haploid tobacco', *Nature New Biol.*, **244**, 29.

Miller, R. A., O. L. Gamborg, W. A. Keller and K. N. Kao (1971) 'Fusion and division of nuclei in multinucleated soybean protoplasts', *Can. J. Genet. Cytol.*, **13**, 347.

Motoyoshi, F. (1971) 'Protoplasts isolated from callus cells of maize endosperm', *Exp. Cell Res.* **68**, 452.

Murashige, T. and F. Skoog (1962) 'A revised medium for rapid growth and bioassays with tobacco tissue cultures', *Physiol. Plantarum*, **15.** 473.

Nagata, T. and I. Takebe (1970) 'Cell wall regeneration and cell division in isolated tobacco mesophyll protoplasts'. *Planta*, **92,** 301.

Nagata, T. and I. Takebe (1971) 'Plating of isolated tobacco mesophyll protoplasts on agar medium', *Planta*, **99**, 12.

Nitsch, J. P. and K. Ohyama (1971) 'Obtention de plantes à partir de protoplastes haploides cultures *in vitro*', *C. R. Acad. Sc. Paris*, **273**, 801.

Ohyama, K., O. L. Gamborg and R. A. Miller (1972) 'Uptake of exogenous DNA by plant protoplasts', *Can. J. Bot.*, **50**, 2077.

Otsuki, Y. and I. Takebe (1969) 'Isolation of intact mesophyll cells and their protoplasts from higher plants', *Plant and Cell Physiol.*, **10**, 292.

Paton, A. M. and S. M. Jones (1971) 'Techniques involving optical brightening agents', in *Methods in Microbiology* **5a,** 135 (Eds. J. R. Norris and D. W. Ribbons), Academic Press, p. 135.

Pilet, P. M. (1973) 'Transaminase activity in root protoplasts enzymatically and mechanically prepared', in 'Protoplastes et fusion de cellules somatique végétales', Colloques Internationaux, C.N.R.S., **212**, 99.

Potrykus, I. (1971) 'Intra- and interspecific fusion of protoplasts from petals of *Torenia bailloni* and *Torenia fournieri*', *Nature New Biol.*, **231**, 57.

Potrykus, I. (1973) 'Transplantation of chloroplasts into protoplasts of *Petunia*', *Z. Pflanzenphysiol.*, **70**, 364.
Potrykus, I. and J. Durand (1972) 'Callus formation from single protoplasts of *Petunia*', *Nature New Biol.*, **237**, 286.
Potrykus, I. and F. Hoffmann (1973) 'Transplantation of nuclei into protoplasts of higher plants', *Z. Pflanzenphysiol*, **69**, 287.
Power, J. B. and E. C. Cocking (1970) 'Isolation of leaf protoplasts: Macromolecule uptake and growth substance response', *J. Exp. Bot.*, **21**, 64.
Power, J. G., S. E. Cummings and E. C. Cocking (1970) 'Fusion of isolated plant protoplasts', *Nature*, **225**, 1016.
Power, J. B., E. M. Frearson and E. C. Cocking (1971) 'The preparation and culture of spontaneously fused tobacco leaf spongy mesophyll protoplasts', *Biochem. J.*, **123**, 29.
Power, J. B. and E. M. Frearson (1973a) in 'Isolation of protoplasts', P. K. Evans and E. C. Cocking in *Plant Tissue and Cell Culture* (Ed. H. E. Street), Blackwell's Botanical Monographs, Oxford.
Power, J. B. and E. M. Frearson (1973b) 'The inter- and intraspecific fusion of plant protoplasts: subsequent developments in culture with reference to crown gall callus and tobacco and petunia leaf systems', in 'Protoplastes et fusion de cellules somatique végétales', *Colloque Internationaux, C.N.R.S.*, **212**, 409.
Racusen, D. W. and S. Aronoff (1953) 'A homogeneous cell preparation from soybean leaves', *Science*, **118**, 302.
Reinert, J. and S. Hellmann (1973) 'Aspects of nuclear division and cell wall formation in protoplasts of different origin', in 'Protoplastes et fusion de cellules somatique végétales', *Colloques internationaux, C.N.R.S.*, **212**, 274.
Rossini, L. (1969) 'Une nouvelle méthode de culture *in vitro* de cellules parenchymateuses séparées des feuilles de *Calystegia sepium* L.', *C. R. Acad. Sc. Paris*, **268**, 683.
Rotman, B. and B. W. Papermaster (1966) 'Membrane properties of living mammalian cells as studied by enzymatic hydrolysis of fluorogenic esters', *Proc. U.S. Natl. Acad. Sci.*, **55**, 134.
Ruesink, A. W. (1971) 'The plasma membrane of *Avena* coleoptile protoplasts', *Plant Physiol.*, **47**, 192.
Ruesink, A. W. and K. V. Thimann (1965) 'Protoplasts from the *Avena* coleoptile', *Proc. U.S. Natl. Acad. Sci.*, **54**, 56.
Schenk, R. U. and A. C. Hildebrandt (1969) 'Production of protoplasts from plant cells in liquid culture using purified commercial cellulases', *Crop Sci.*, **9**, 629.
Schilde-Rentschler, L. (1973) 'Preparation of protoplasts for infection with *Agrobacterium tumefaciens*', in 'Protoplastes et fusion de cellules somatique végétales', *Colloques Internationaux, C.N.R.S.*, **212**, 479.
Schmucker, T. (1930) 'Isolierte Gewebe und Zellen Von Blutenpflanzen', *Planta*, **9**, 339.
Scowcroft, W. R., M. R. Davey and J. B. Power (1973) 'Crown gall protoplasts–isolation, culture and ultrastructure', *Plant Sci. Letters*, **1**, 451.
Selby, K. (1973) 'The components of cell wall degrading enzymes with particular reference to the cellulases', in 'Protoplastes et fusion de cellules somatiques végétales', *Colloques Internationaux, C.N.R.S.*, **212**, 33.
Street, H. E. (1973), *Plant Tissue and Cell Culture*, Blackwell's Botanical Monographs, Oxford.
Sunderland, N. (1973) 'Nuclear cytology', in *Plant Tissue and Cell Culture* (Ed. H. E. Street), Blackwell's Botanical Monographs, Oxford.

Taiz, L. and R. L. Jones (1971) 'Isolation of barley aleurone protoplasts', *Planta*, **101**, 95.

Takebe, I., Y. Otsuki and S. Aoki (1968) 'Isolation of tobacco mesophyll cells in intact and active state', *Plant and Cell Physiol.*, **9**, 115.

Takebe, I., G. Labib and G. Melchers (1971) 'Regeneration of whole plants from isolated mesophyll protoplasts of tobacco', *Naturwissenschaften*, **58**, 318.

Tempé, J. (1973) 'Protoplastes et fusion de cellules somatiques végétales', *Colloques Internationaux, C.N.R.S.*, **212.**

Toister, Z. and A. Loyter (1973) 'The mechanism of cell fusion. II. Formation of chicken erythrocyte polykaryons', *J. Biol. Chem.*, **248**, 422.

Törnävä, S. R. (1939) 'Expansion capacity of naked plant protoplasts', *Protoplasma*, **32**, 329.

Tribe, H. L., (1955) 'Studies in the physiology of parasitism XIX On the killing of plant cells by enzymes from *Botrytis cinerea* and *Bacterium aroideae*'. *Ann. Bot. N.S.* **19,** 351.

Usui, H. and I. Takebe (1969) 'Division and growth of single mesophyll cells isolated enzymatically from tobacco leaves', *Dev., Growth and Differentiation*, **II**, 143.

Wenzel, G. (1973) 'Isolation of leaf protoplasts from haploid plants of petunia, rape and rye', *Z. Pflanzenzuchtg*, **69**, 58.

Whatley, F. R. (1956) 'Cytochemical methods', in *Modern Methods of Plant Analysis*, Vol. I (Eds. K. Paech and M. V. Tracey). Springer-Verlag, Berlin.

White, P. R. (9163), *The Cultivation of Animal and Plant Cells*, 2nd ed., The Ronald Press, New York.

Widholm, J. M. (1972) 'The use of fluorescein diacetate and phenosafranine for determining viability of cultured plant cells', *Stain Technology*, **47**, 189.

Wilson, S. B., P. J. King and H. E. Street (1971) 'Studies on the growth in culture of plant cells. XII. A versatile system for the large scale batch or continuous culture of plant cell suspensions', *J. Exp. Bot.*, **22**, 177.

Withers, L. A. (1973) 'Plant protoplast fusion: methods and mechanisms', in 'Protoplastes et fusion de cellules somatique végétales', *Colloques Internationaux, C.N.R.S.*, **212**, 515.

Withers, L. A. and E. C. Cocking (1972) 'Fine-structural studies on spontaneous and induced fusion of higher plant protoplasts', *J. Cell Sci.*, **II**, 59.

Woodcock, J. (1973) 'The culture, behaviour and organogenesis of cells derived from higher plants', *M. Phil Thesis*, University of Nottingham.

Zaitlin, M. (1959) 'Isolation of tobacco leaf cells capable of supporting virus multiplication', *Nature*, **184**, 1002.

Zaitlin, M. and R. N. Beachy (1973) 'The use of protoplasts and separated cells in plant virus research', in *Advances in Virus Research* (in press).

CHAPTER 5

# New techniques in detection of antibodies to viral antigens and tumour-associated antigens

Chou-Chik Ting and Ronald B. Herberman
*Cellular and Tumour Immunology Section,*
*Laboratory of Cell Biology,*
*National Cancer Institute,*
*National Institutes of Health,*
*Bethesda, Maryland, U.S.A.*



## I. INTRODUCTION

New antigens are induced in cells transformed by oncogenic viruses. In RNA virus-transformed cells (Gross, Friend, Moloney, Rauscher, Rous sarcoma, Avian tumour viruses, etc.), the infectious viral particles are continuously synthesized and released. In DNA virus-transformed cells (SV40, polyoma, adenoviruses, etc.), the viral genomes are integrated into the host cellular genome but they do not synthesize infectious viral particles; thus, they are in a 'lysogenic' state. In either case, the virus itself and its products are foreign to the host cells; consequently, the new antigens induced by the viral transformation may be directly related to the virion, or they may be only coded for by the viral genome. The antigens which are directly related to the virion are: (1) Viral envelope antigen: this is associated with viral coat protein; (2) Group-specific antigen (gs antigen): this is associated with the viral core, and is specific to a group of viruses (Geering *et al.*, 1966; Armstrong, 1968; Roth and Dougherty, 1969). There are species-specific gs antigens and interspecies gs antigens (Huebner *et al.*, 1971). Both viral envelope antigens and gs antigens are located intracellularly, but they may be extracellular if infectious particles are released from the cells.

The antigens which are not directly related to the virion but appear to be coded for by the viral genome include:

1. T antigen: these antigens are present in DNA virus (SV40, polyoma, adenoviruses) infected or transformed cells (Black *et al.*, 1963; Huebner *et al.*, 1963; Rapp *et al.*, 1964; Takemoto and Habel, 1965; Takemoto *et al.*, 1966). The antigens are presumably coded for by the early genes of these viruses (Rapp *et al.*, 1965; Rowe *et al.*, 1965; Eckhart, 1969). Recently it has been reported that the T antigen is not a viral structure component, nor a precursor of a viral structural protein (Villano and Defendi, 1973).
2. Tumour-associated cell surface antigens (TASA): these are the cell surface antigens detected by various *in vitro* assays.
3. Tumour-associated transplantation antigens (TATA): these are the antigens defined by *in vivo* immunorejection experiments, and are presumably located on the cell surface, as are the histocompatibility transplantation antigens.

TATA is of prime importance in induction of tumour immunity. It is specific for a particular tumour or for tumours induced by a particular virus. This antigen is defined by *in vivo* experiments. The transplantation tests are usually time consuming and very often are difficult to perform. Therefore, a variety of *in vitro* assays have been developed to detect the tumour-associated cell surface antigens (TASA). In this chapter, we shall mainly discuss various serological assays for the detection of these antigens. Only those techniques which have been developed recently will be described in detail. Others, which are commonly used in many laboratories, will be discussed briefly. The details for these techniques can be found in the references given here. We shall also emphasize the methods for the detection of TASA, the relationship between TASA and TATA, and the implications for biological science and clinical medicine.

## II. TECHNIQUE

### A. Cytotoxicity Test

This assay is commonly used for detection of various cell surface antigens.

It is limited to certain cell types which are susceptible (lysable) to humoral antibodies, e.g. lymphoid leukaemia cells, and cannot be used with some cell types which resist lysis by humoral antibodies, e.g. many sarcoma tumour cells. There are several modifications of the techniques which will be discussed here:

*1. Dye-exclusion cytotoxicity test*

The viable cells with intact membrane will exclude the uptake of dye (e.g. trypan blue), but the dead cells with damaged cell membrane will be stained by the dye.

Procedure: This assay was originally developed by Gorer and O'Gorman (1956). Equal volumes of test cells, antiserum and complement (guinea pig or rabbit serum) are incubated for 45 min at 37°C, then the cells are stained by 0·2% trypan blue. The proportions of stained and unstained cells are determined microscopically and the per cent lysis calculated. Minor modifications have been made in different laboratories. The test procedure can be separated into two steps. First, test cells are incubated with antiserum for 30–45 min at 37°C, then complement is added and the mixture incubated for another 30–45 min at 37°C. The cells are stained by trypan blue to determine viability. Fass and Herberman have modified the test by addition of antiglobulin reagent to the test system. This would increase the cytotoxic titres of the antiserum (Fass and Herberman, 1969). The increase in cytotoxic activity has been attributed to increased complement binding and to the conversion of non-complement-fixing complex by antiglobulin (Borsos *et al.*, 1968). In this test, the antiglobulin reagent (0·1 ml) is added after test cells ($1 \times 10^5$) have been incubated with antiserum (0·1 ml) at 37°C for 30 min. They are allowed to incubate for another 30 min at 37°C. Then the complement (0·1 ml) is added and the mixture is incubated for an additional 30 min at 37°C. Then the cells can be stained by trypan blue for examination.

In these complement-dependent cytotoxicity tests, a good source of complement is essential. Many of the normal guinea pig or rabbit sera used as the source of complement are quite toxic to the heterologous mouse or human target cells (Terasaki *et al.*, 1961; Wakefield and Batchelor, 1966; Herberman, 1969). Several methods have been devised to reduce or eliminate this interfering toxicity. A common solution has been to use diluted sera, at a concentration which no longer gives significant lysis. The difficulty with this is that the amount of complement is also reduced, and this can result in lower antibody titres, especially in weak systems. Absorption of the natural cytotoxic antibodies in the sera is a more satisfactory method. Boyse *et al.* (1970) developed a method of absorption of complement sera in the presence of ethylenediaminetetraacetate (EDTA). This allows complete absorption without appreciable removal of complement activity. However, this method necessitates absorption of small aliquots of serum with large amounts of cells or tissues, a rather laborious process. Another common method has been to screen several batches of rabbit or guinea pig sera and select the least toxic serum. Again, because of the widespread distribution of natural cytotoxic antibodies, this can be a laborious task. The last general method for handling this problem is the addition to the sera of substances which can inhibit toxicity yet have no effect on complement activity. Cohen and Schlesinger (1970) found that agar reduces the toxicity of guinea pig sera. Herberman (1970) reported that small amounts of human IgM would inhibit the cytotoxic activity of normal rabbit serum for human lymphoid cells and for some mouse cells.

Application: This assay has been used to detect the tumour-associated cell surface antigens of leukaemias induced by Gross (Old *et al.*, 1965; Geering *et al.*, 1966), Friend, Moloney, Rauscher (Old *et al.*, 1964; Klein and Klein, 1964), and radiation leukaemia viruses (Old *et al.*, 1963; Ferrer and Kaplan, 1968); of TL positive leukaemias (Boyse *et al.*, 1965), and mammary leukaemia antigen (Stück, *et al.*, 1964), etc. This technique is relatively simple, requires very little equipment (the major equipment is a centrifuge and microscope). However, to determine the viability by microscopic examination is time consuming, subjective and may be misinterpreted if performed by inexperienced personnel.

*2. $^{51}$Cr cytotoxicity test*

Procedure: This assay is similar to the dye-exclusion cytotoxicity test, except that the test cells are prelabelled with $^{51}$Cr. $^{51}$Cr labels intracellular proteins and under the conditions of a short-term antibody assay, is only released upon lysis of the cell. The amount of cytotoxicity is determined by the release of radioactivity from the labelled cells. In labelling the cells, usually 2–3 $\times$ $10^7$ cells are incubated with 200–300 $\mu$Ci of $Na^{51}CrO_4$ in 1 ml volume for 30–45 min at 37°C. The cells are then washed twice with Eagle's minimal essential medium containing 10% foetal bovine serum (MEM–FBS). Then the labelled cells, usually 0·1 ml at 1 $\times$ $10^6$/ml, can be used for testing. After incubating the labelled cells with equal volumes of antiserum and complement at 37°C for 30–45 min, 2 ml of MEM-FBS are added to each tube, the cells sedimented at 500 *g* for 10 min, and the supernatant fluid poured into counting tubes to be counted in a gamma-scintillation counter. The amount of cytolysis is determined by comparison of the $^{51}$Cr released into the supernatant and the $^{51}$Cr released by freezing and thawing the cells three times.

Application: This technique has been used to detect the tumour-associated cell surface antigens of leukaemia induced by Gross virus (Herberman and Oren, 1971), Moloney leukaemia virus (Klein *et al.*, 1966), an adenovirus-induced tumour (Ankerst and Sjögren, 1969), and an SV40-induced tumour (Wright and Law, 1971). The advantage of this technique is that the amount of cytolysis can be determined by counting the radioactivity in a gamma-scintillation counter; therefore, it is objective and less time consuming. The sensitivity of the $^{51}Cr$ cytotoxicity test and dye-exclusion techniques is about the same.

*3. $^{3}H$-Uridine or $^{3}H$-thymidine cytotoxicity test*

Procedure: This technique was developed by Hashimoto and Sudo (1971) and is similar to the $^{51}Cr$ cytotoxicity test. The test cells (usually at $1 \times 10^6$/ml) are labelled with 0·5 μCi/ml of $^{3}H$-uridine or $^{3}H$-thymidine. Equal volumes (0·1 ml) of the cells at $2{\cdot}5 \times 10^6$/ml, antiserum and complement are incubated at 37°C for 45 min. The reaction is stopped by chilling, and the reaction mixture is filtered through a Millipore membrane filter (0·45 μ pore size, 24 mm diameter). The filters are washed with 5% trichloroacetic acid, dried and then transferred to a vial with toluene-based scintillation fluid. The radioactivity is determined by a beta liquid scintillation counter.

$$\text{Amount of cytolysis} = \left[100 - \frac{\text{antiserum, cpm} - \text{background}}{\text{serum control, cpm} - \text{background}}\right] \times 100$$

Application: This assay has been used to detect antibodies to mouse H-2 histocompatibility antigens. It can also be used for tumour-associated surface antigens. This technique can be used in experiments which require prolonged incubation, where the cells may remain viable for more than 48 h. While the $^{51}Cr$-labelled cells can usually only be used in short-term incubation, the spontaneous release of $^{51}Cr$ from many cells is very high at 10–12 hours after labelling. However, this test is not suitable for cells which will continue to grow in culture, because the released radioactivity can be reutilized.

*4. $^{125}$IUDR cytotoxicity test*

Procedure: This assay was developed by Cohen *et al.* for detection of HLA antigens (Cohen *et al.*, 1971), and was modified by LeMevel and Wells for tumour antigens (1973). Two thousand test cells in 0·2 ml medium are seeded into each well of a microtest II plate (Falcon Plastics). After incubation at 37°C for 24 h, the medium is aspirated and 25 μl serum are added to each well for 30 min, then 25 μl of complement are added and the plates are rocked for 6 h. The serum and complement are removed, and 0·5 μCi $^{125}$IUDR ($^{125}$I-iododeoxyuridine) added for 18 h. Then the plates are washed three times, dried and sprayed with plastic film. The wells are separated from the microtest plate with a band-saw and counted in a gamma-scintillation counter.

$$\text{Amount of cytolysis} = \left[1 - \frac{\text{antiserum, cpm} + \text{complement}}{\text{normal serum, cpm} + \text{complement}}\right] \times 100$$

Application: This assay has been used in detection of TASA of chemically induced tumours (LeMevel and Wells, 1973) and it can be used in virus-induced tumours. This technique is objective and relatively simple. The $^{125}$IUDR will not be reutilized after it has been released, but $^{125}$IUDR itself may be cytotoxic at 48 h after labelling.

*5. Colony inhibition test, cell inhibition test and microcytotoxicity test*

Procedure: The colony inhibition test was originally developed by Hellström and Sjögren (1965), using petri dishes. It was then modified by Hellström *et al.* (1969), and Takasugi and Klein (1970), using microtest plates. The original colony inhibition test is rarely used nowadays. We shall only describe the latter two modifications. With Takasugi's modification, the microtest plate I (Falcon Plastics, Los Angeles, California) was used and the assay was called the 'Microcytotoxicity Test'. The microtest plate I has 60 wells, each with a capacity slightly greater than 100 $\mu$l, and $5 \times 10^2$ to $5 \times 10^4$ cells can be seeded into each well. The microtest plate II (Falcon Plastics, Los Angeles, California) was used by Hellström *et al.*, and the assay was called the 'Cell Inhibition (CI) Test'. These plates have 96 wells, each with a capacity of about 250 $\mu$l and therefore, more cells can be seeded. The number of cells seeded is usually determined by the plating efficiency of each cell line. The microcytotoxicity test and cell inhibition test are mainly used in studying cell-mediated immunity *in vitro*, although they have also been used to measure antibodies. Bloom (1970) had adapted the microcytotoxicity test to be used in the detection of antibodies against chemically induced tumours. In this test, $2{\cdot}5 \times 10^3$ cells are seeded into each well and are allowed to attach for 16–72 h. The period required for the target cells' adherence to the well appears to be a property of the particular tumour. The culture medium is aspirated before the addition of test serum. Five $\mu$l of serum is added to each well and incubated at 37°C for 30 min in a humidified 5% $CO_2$ atmosphere. Then 5 $\mu$l of rabbit complement is added to each well and incubated for another 3 h. The wells are washed twice with minimal essential medium, the remaining cells are fixed for 30 min with methanol and stained with Giemsa. The fraction of cells surviving is calculated as the ratio of the number of cells in wells treated with experimental serum to the number of cells in wells treated with normal serum. The survival value is plotted against the $\log_2$ of the antiserum dilution, and the titre was designated as $\log_2$ of the dilution interpolated to 0·50 survival. Smith and Mora (1972) have used the microtest plate I to perform short-term incubation cytotoxicity tests. In this test, 20 $\mu$l of undiluted antiserum is mixed with $1 \times 10^5$ test cells in $10 \times 75$ mm test-tubes and incubated at 4°C for 1 h. The cells are washed once with Tris-buffered saline (NaCl 0·154 *M*, Tris-aminomethane 0·30 *M*, pH 7·2) containing 5% foetal bovine serum (TD–FBS). Then they are resuspended in 0·2 ml of TD-FBS. Two $\mu$l containing about 1000 cells are transferred to the wells of microtest plate I. Five $\mu$l of rabbit complement are then added, and the wells sealed with mineral oil to prevent evaporation and incubated at

22°C for 2 h. One μl of 0·1 % trypan blue is then injected beneath the oil in each well, and the degree of cytolysis is determined by examination under a phase-contrast microscope. This group also adopted the $^{51}$Cr assay in a similar manner.

Application: These assays are widely used in many laboratories. The original colony inhibition test has been used in detecting TASA of polyoma virus (Hellström and Sjögren, 1966), and adenovirus-induced tumours (Sjögren and Motet, 1969), a Moloney-virus induced mouse sarcoma (Hellström and Hellström, 1969) and Shope-virus-induced rabbit papillomas (Hellström *et al.*, 1969). The microcytotoxicity test and cell inhibition test have also been used with chemically induced tumours (Bloom, 1970) and SV40-induced tumours (Smith and Mora, 1972). These techniques are quite sensitive. Since no radioisotope is used, prolonged incubation can be performed without worrying about damage due to the toxicity of the radioisotope. However, they can only be used with those tumour cells which can grow in tissue culture as monolayers, and cannot be used with cells that grow in suspension or those which cannot be adapted to culture. Similar to the dye-exclusion cytotoxicity test, counting the cells is time consuming, subjective and may require experienced personnel.

## B. Haemadsorption, Haemadsorption Inhibition and Mixed Haemadsorption Test

Procedure:

### *1. Haemadsorption and Haemadsorption inhibition test*

The haemadsorption test was originally developed by Boyden (1951) for detection of tuberculoprotein, and was later applied to detection of viral antigen. The 2 % sheep erythrocyte (SRBC) suspension in phosphate-buffered saline (PBS, pH 7·2) is mixed with an equal volume of 1:20,000 tannic acid solution in saline, kept at 37°C for 10 min, then washed with PBS. Equal volumes of tanned SRBC and virus are mixed and allowed to incubate at room temperature for 10–15 min. The sensitized cells are washed twice and resuspended in PBS containing 1 % normal rabbit serum (PBS/NRS). Twofold serial dilutions of antiviral serum (0·0025 ml) in PBS/NRS are mixed with equal volumes of 0·8 % antigen(viral)-coated SRBC. The mixtures are allowed to settle at room temperature for at least 2 h. The titre is the highest dilution of antiserum which still gives complete agglutination of the erythrocytes. The haemadsorption inhibition assay is performed with the antiserum which has been preincubated with viral preparations for 10 min at 37°C before adding the sensitized SRBC. This modification permits the measurement of antigens in the preparations.

*2. Mixed haemadsorption test*

This assay was developed by Coombs *et al.* (1956) for detection of platelet antibodies. The principle was that if the platelets and erythrocytes were both sensitized by antibodies from the same species, both cell types would possess a common antigen in the form of the adsorbed antibody globulin and this would allow a corresponding antiglobulin serum to bring about a mixed erythrocyte–platelet agglutination. This test was found to be very sensitive and was applied to the detection of other cell surface antigens. There are several variations to this technique: The test can be done on agar (Barth *et al.*, 1967), coverslips (Gillespie, 1968), test-tubes (Häyry and Defendi, 1970) or microtitre plates (Hilgers *et al.*, 1971). The test system includes (*a*) test cells (*b*) indicator cells and (*c*) antiserum. For instance, in the experiments by Barth *et al.* (*a*) The test cells used are Moloney-virus-infected embryonic fibroblasts; (*b*) The indicator cells are sheep erythrocytes coated with amboceptor (mouse anti-sheep erythrocytic antiserum). This is prepared by incubation of equal volumes of the 2% suspension of SRBC with amboceptor for 1 h at room temperature, then washing twice with saline and further incubation with an equal volume of the antiglobulin reagent (rabbit anti-mouse globulin serum) for 1 h at room temperature; (*c*) The antiserum used was produced by the immunization of mice with syngeneic Moloney lymphoma cells. In the mixed haemadsorption test, the test cells are grown as a confluent monolayer, then overlaid with 0·3 cm thickness of 0·75% Difco agar in Eagle's medium with 5% calf serum. Sterilized filter-paper discs of 0·5 cm in diameter are moistened with 0·02 ml of heat-inactivated test serum in serial 10-fold dilutions and placed on the solidified agar surface. The flasks are stoppered and kept at room temperature for 48 h, then the agar is dislodged by rinsing with 10 ml saline, and poured off. Twelve ml of 0·2% suspension of indicator cells are added to each flask. They are incubated at room temperature for 1 h, then the supernatant fluid is carefully poured off and the flasks are allowed to dry, then stained with benzidine, and the diameters of the adsorption zones are determined. In other modifications of this test, the procedure is essentially the same except for the test cells which are grown in different culture vessels (tubes, plates or coverslips, etc.)

*3. Isotopic mixed haemadsorption test*

The tests described above require visual examination of the agglutination pf erythrocytes. Sundqvist and Fagraeus (1972) have modified this test by using $^{51}$Cr-labelled sheep erythrocytes as indicator cells. All reactions are performed in scintillation tubes on the bottom of which the test cells have been cultivated to a confluent monolayer. The procedure is similar to the mixed haemadsorption test described above. After the reaction is complete, the unattached indicator cells are removed, and the test-tubes counted in a gamma-scintillation counter.

#### 4. *Immune adherence test*

This is another modification of mixed haemadsorption test by Nishioka *et al.* (1969). In this test, the antiglobulin reagent is replaced by guinea pig complement. The test cells (mouse mammary tumour cells) are first incubated with antiserum at 37°C for 30 min, followed by incubation at 0°C for 60 min, the cells are washed and complement is added. Then human type O erythrocytes are added and allowed to incubate at 37°C for 15 min and at 24°C for 10 min. The haemadsorption pattern is determined under a microscope. The principle of this test is that when antigen (Ag), antibody (Ab) and complement (C′) are present, they form an Ag–Ab–C′ complex which causes the erythrocytes to adhere to the cell surface. (Human erythrocytes have a surface receptor for C′.)

Application: The haemadsorption or haemadsorption inhibition test has been used to detect the viral antigens of polyoma virus (Eddy *et al.*, 1958), murine leukaemia virus (Sibal *et al.*, 1968) and mammary tumour virus (Fink *et al.*, 1968; Sibal *et al.*, 1969). The inhibition test can also be used to determine the specificity of these reactions. The mixed haemadsorption test has been used to detect the viral antigens and tumour-associated cell surface antigens of Moloney lymphoma (Barth *et al.*, 1967), Gross leukaemia (Gillespie, 1968), mouse mammary tumour (Hilgers *et al.*, 1971) and SV40 tumours (Häyry and Defendi, 1970). The immune adherence test has been used to detect the tumour-associated cell surface antigen of mouse mammary tumours (Nishioka *et al.*, 1969). Although the mixed adsorption test appears to be a very sensitive assay, it may be subjective, requires great care in performance, and very often the test results may not be reproducible. The isotopic mixed haemadsorption test (Sundqvist and Fagraeus, 1972) was said to be more sensitive and reproducible, but it has only been applied to the detection of histocompatibility antigens. It may be useful for the detection of tumour antigens.

### C. Immunofluorescence Tests

This technique was originally developed by Coons and Kaplan (1950), and has been widely used in many immunological studies. An indirect technique, using viable target cells in suspension, was developed by Möller (1961) and has been used extensively in detection of various cell surface antigens. For detection of intracellular antigens, the assays are performed with fixed cells or tissue sections.

Procedure:

#### 1. *Direct immunofluorescence test*

It was developed by Coons and Kaplan (1950). In this test, the antiserum is directly conjugated with fluorescein (isothiocyanate). A drop of the conjugate is added to a quick-frozen, unfixed tissue section on a slide, which is covered with a small inverted dish containing a piece of moist filter-paper to prevent evaporation. The slides are incubated for 20–30 min, the conjugate is shaken off and the slides are then immersed in buffered saline in a Coplin jar. After a few seconds the jar is filled with fresh buffered saline, and the slide is gently agitated for 10 min, and then the saline is discarded. The slide is removed from the jar and wiped dry on the periphery of the section. A drop of glycerol (1 part of glycerol in 10 parts of buffered saline) is put over the section, and the slide is covered by a clean coverslip. It is then ready to be examined under ultraviolet light with a fluorescence microscope.

*2. Indirect test*

This test was developed by Möller (1961). It is a two-step procedure. Instead of antiserum itself, the antiglobulin reagent is conjugated with fluorescein. The method differs slightly with the nature of test sample.

*a. Solid tissue.* Tissues are cut into approximately 0·5 $cm^2$ in area and 3–5 mm in thickness, and are immediately frozen with carbon dioxide snow on a microtome object holder. The frozen tissue is cut at 4 $\mu$ in thickness with a cryostat at −20°C. The sections are placed on a chilled slide and rapidly thawed by placing a finger under the slide. The slides are immersed either in Ringer's solution or in 96% ethanol for 10 min. The latter procedure is used only when the sections are to be counter stained with rhodamine-conjugated bovine albumin. The unfixed, thawed tissue sections are covered with 0·05–0·1 ml of antiserum. They are incubated in a moist chamber for 15–30 min at room temperature, then washed three times with excess Ringer's solution. Then 0·05–0·1 ml of fluorescein-conjugated antiglobulin reagent is added. They are further incubated for 15–30 min at room temperature. After three more washings, the sections are mounted in a medium containing 9 parts of anhydrous glycerine and 1 part of Ringer's solution. The preparations are then ready to be examined with the fluorescence microscope.

*b. Living cell preparation.* Single-cell suspensions of tumour cells are washed once with Ringer's solution and brought to a final concentration of 5 to 20 $\times$ $10^6$ cells/ml. Tissue-culture cells, growing in monolayers, are usually brought into suspension with 0·1% to 0·3% trypsin, washed once with Eagle's medium containing 10% foetal bovine serum and three times with Tris-buffered saline at pH 7·4. Solid tissue can be made into single-cell suspension by pressing through a 60 mesh stainless-steel screen or by treating with 0·25% trypsin for 15 min to 1 h at 37°C.

The cells (1 to 5 $\times$ $10^6$) are pelleted in a test-tube by centrifugation, and the cells are then incubated with 0·05–0·1 ml of antiserum for 15 min to 1 h at room temperature or 37°C. The cells are washed two to three times with Ringer's solution or Tris-buffered saline. Then they are incubated with fluorescein-conjugated antiglobulin reagent for 20–30 min at room temperature or 37°C. After two to three more washings, the cells are suspended in Ringer's solution or in buffered glycerol (pH 7·0). Then the cells are ready to be examined in the fluorescence microscope. Usually 0·05 ml of cell suspension is placed on the slide and overlayed with a coverslip. Care is taken to avoid pressure of the coverslip against the slide. Some preparations are fixed by exposing the cell suspension with 95% ethanol for about 10 minutes at room temperature after the staining procedure is completed. This procedure preserves the cell well and will prevent some cells from swelling during inspection in ultraviolet light.

*c. Fixed cells.* Cell smears and cells grown as a monolayer on the coverslips can be fixed with acetone for 10–15 min. They can then be processed for the immunofluorescence test as described above, or can be stored at −20°C until testing.

On examination, diffuse fluorescence of the whole cell indicates death of the cells similar to the supravital staining by eosin or trypan blue dye. These dead cells are excluded from calculation. Only those cells showing rings or patches of bright-green fluorescence on the cell surfaces are counted as positive. Cells having only the faint non-specific fluorescence characteristics of unstained cells are classified as negative. The controls used in these experiments should include cells lacking the appropriate antigenic specificity, and replacement of the antiserum by normal serum. The test is defined as positive when 50% or more of the test cells show specific immunofluorescence. Alternatively, the results can be expressed by the 'fluorescence index' which is defined as the proportion of negative cells in a sample treated with control (normal) serum minus the proportion of unstained cells in the sample exposed to a given antiserum, divided by the former figure. The reaction is considered positive when the immunofluorescence index is greater than 0·2.

Application: The direct test of Coons and Kaplan has been used to detact rickettsial and mumps virus antigen (Coons *et al.*, 1950), and pneumococcal capsular polysaccharide antigen (Coons and Kaplan, 1950). It has also been used to detect the Epstein–Barr-virus(EBV)-associated antigens in Burkitt lymphoma (Klein *et al.*, 1969a; Pearson *et al.*, 1969; Nadkarni *et al.*, 1970; De Schryver *et al.*, 1972). The indirect test of Möller using viable cells has been widely used in virology and tumour immunology. It has been used to detect the cell surface antigen of tumour cells transformed by Friend–Moloney–Rauscher leukaemia viruses (Klein and Klein, 1964; Glynn *et al.*, 1968), Gross virus (Aoki *et al.*, 1966), mammary tumour virus (Hilgers *et al.*, 1971), SV40 (Tevethia *et al.*, 1968), polyoma virus (Irlin, 1967; Malmgren *et al.*, 1968), adenovirus (Vasconcelos-Costa, 1970). Shope papilloma virus (Ishimoto and Ito, 1969) and the EBV-associated antigens (Henle and Henle, 1966; Klein *et al.*, 1966a; Pearson *et al.*, 1969). The fixed-cell technique has been used to detect the tumour (T) antigen of tumour cells induced by SV40

(Rapp *et al.*, 1964), polyoma virus (Takemoto and Habel, 1965; Takemoto *et al.*, 1966) and adenovirus (Pope and Rowe, 1964).

The immunofluorescence test is a very sensitive assay. It can be used to detect viral antigen, and the antigens in tumour cells that are induced by viruses, either on the cell surface or within the cells. This test does not rely on whether the antibodies are cytotoxic to the cells or whether the cells are susceptible to the cytotoxic effect of the antibodies. However, examining the cells in a fluorescence microscope may be a tedious and time-consuming job, and it requires experienced personnel.

## D. Immunoelectron Microscopy

Ferritin can be coupled to immunoglobulins with bivalent chemical agents, such as metaxylylene diisocyanate (Singer, 1959), toluene-2,4-diisocyanate (Singer and Schick, 1961) or *p,p'*-difluoro-*m,m'*-dinitrodiphenyl sulphone (SriRam *et al.*, 1963), but the product is usually heterogenous and requires further purification, and these chemical manipulations may result in loss of antibody activity. Recently, Hämmerling *et al.* (1968) have developed a hybrid antibody composed of anti-immunoglobulin and anti-ferritin specificities, to detect cell surface antigens by electron microscopy.

Procedure: The hybrid antibody technique was based on Nisonoff's finding that antibody molecules of dual specificities can be produced by combining the univalent Fab fragments of pepsin-treated antibodies of different specificities (Nisonoff and Rivers, 1961).

### *1. Preparation of hybrid antibody*

Anti-ferritin and anti-IgG can be purified by a chromatographic procedure (DEAE–'Sephadex' A-50 in 0·08 in Tris HCl buffer, pH 8·0 at 5°C). The purified globulin is treated with pepsin to make $F(ab')_2$ dimer. The $F(ab')_2$ of anti-ferritin and anti-IgG are mixed with 2-mercaptoethylamine, which gives rise to Fab′ fragment. Then the mixture is reoxidized gently by stirring in an oxygen atmosphere. The Fab′ fragments recombine to form $F(ab')_2$ dimers. Some of these dimers possess both anti-ferritin and anti-immunoglobulin specificities.

### *2. Electron microscopy*

Viable tumour cells, in single-cell suspension, are washed twice with serum-free medium 199 and pelleted by centrifugation at 4°C. Usually $3–10 \times 10^6$ cells are used. They are incubated with 0·05 ml of undiluted antiserum at room temperature for 30 min, washed twice with serum-free medium 199, and then incubated with hybrid antibody which possesses both anti-IgG and anti-ferritin specificities. After two more washes, the cells are spun down and resuspended, with ferritin added at a concentration of 0·5 mg/ml. The cells are incubated on ice for 30 min, with periodic agitation. The cells are further washed twice, and the cell pellet fixed with 1 % glutaraldehyde and 1 % osmium tetroxide. The cells are dehydrated in alcohol and embedded in 'Epon'. The blocks are sectioned and can be stained with lead citrate to add contrast to the ferritin. As other visual markers, the ferritin can be replaced by southern bean mosaic virus (Aoki *et al.*, 1970) or tobacco mosaic virus (Aoki *et al.*, 1971). In these modifications, anti-southern bean mosaic virus antiserum or anti-tobacco mosaic virus antiserum is used to prepare the hybrid antibody.

Application: The hybrid antibody immunoelectron microscopy has been used to study viral envelope antigens and tumour-associated cell surface antigens of Gross-virus-induced leukaemia (Aoki *et al.*, 1970), Avian RNA tumour viruses, infected or transformed chicken cells (Gelderblom *et al.*, 1972), cells transformed by murine sarcoma virus (Aoki *et al.*, 1973a), and cells which release endogenous C-type viruses (Aoki and Todaro, 1973b). Though this technique is tedious and requires skilled personnel to perform, it appears to be very sensitive. More importantly, it provides a direct localization of various antigens. By using this technique, one can distinguish viral envelope antigens budding on the cell surface from the tumour-associated cell surface antigens which are free of viral envelope component (Aoki *et al.*, 1970; Aoki *et al.*, 1973a). Therefore, it can help to dissociate these two different antigens which may both be present on the surface of virus-producing cell lines.

## E. Radioimmune Precipitation Assay

This assay was originally described by Farr (1958), and was based on the observation that $^{131}$I-labelled bovine serum albumin was soluble in 50 % saturated ammonium sulphate but became insoluble after reaction with specific antibody in the zone of antigen excess. A two-step procedure was developed by Gerloff *et al.* (1962) by using an antiglobulin reagent. This test consists of primary reaction between labelled virus ($^{32}$P-labelled poliovirus) and specific immunoglobulin (antibody). The subsequent reaction of the IgG with the antiglobulin reagent causes precipitation of radioactivity. Recently, this radioimmune precipitation assay has been used in detection of the antigens of oncogenic viruses.

Procedure: Scolnick *et al.* (1972) have adapted the radioimmune precipitation assay in detection of mammalian type C viral protein (the group-specific gs antigen). The purified gs antigen was first labelled with $^{125}$I. (The labelling procedure will be discussed in Section II.F, page 173). To assay for antibody to gs antigen, a small amount of $^{125}$I-labelled gs antigen (0·1 ng with approximately 5000 cpm) is incubated with serial dilutions of specific antiserum in a total volume of 0·5ml at 37°C for 3 h and then at 4°C for 18 h. The reaction mixtures included 0·01 *M* potassium phosphate, pH 7·8, 0·01 *M* EDTA, pH 7·8 and 0·1 % ovalbumin. Anti-IgG reagent (0·01 ml was then added and the mixture allowed to incubate at 37°C for 1 h and at 4°C for 3 h. They were then centrifuged at 2500 rpm in a PR-6 centrifuge at 4°C for 15 min and the supernatant counted in a gamma-scintillation counter. To assay for gs antigen, the concentration of antiserum used was that which precipitated 50 % of the $^{125}$I-labelled gs antigen. This volume of antiserum was incubated with unlabelled test antigen at 37°C for 2 h, $^{125}$I-labelled gs antigen was then added, and incubated at 37°C for 1 h and at 4°C for 16–18 h. Next, the anti-IgG reagent was added and the reaction continued as in the assay for antibody.

Ihle *et al.* (1973) developed a radioimmune precipitation assay for RNA tumour virus (murine leukaemia virus). They used an AKR mouse embryo cell line which gave spontaneous virus synthesis. The virus was labelled with $^{3}$H by feeding the tissue-culture cells with $^{3}$H-leucine. The labelled virus was collected, layered on a 15–60 % linear sucrose gradient in 0·05 *M* Tris, 0·1 *M* NaCl, 1 m*M* EDTA (TNE), and purified by isopycnic banding in a Spinco SW 25·1 rotor at 25,000 rpm for 12 h. The purified virus was diluted with TNE to an appropriate concentration and aliquots were frozen and stored at −70°C. In the test, 0·2 ml test serum at serial dilutions in TNE was mixed with 0·05 ml of labelled virus (about 6000 cpm), and incubated at 37°C for 1 h. Then, 0·2 ml of antiglobulin reagent at appropriate dilution was added, and the incubation continued, at 37°C for 1 h and then at 4°C for 2 h. The precipitates were collected by centrifugation at 1200 *g* for 10 min. The supernatant was removed for determination of radioactivity, and the pellets were washed three times with TNE, resuspended in 0·4 ml TNE and prepared for counting. Precipitation was expressed as the percentage of counts in the precipitate relative to the combined counts in the precipitate and the first supernatant.

Application: As already mentioned in the procedure, the radioimmune precipitation assay has been used to detect group-specific antigen of mammalian C-type viruses (Scolnick *et al.*, 1972) and of Avian C-type virus (Stephenson *et al.*, 1973) and the viral envelope antigen of Gross virus (Ihle *et al.*, 1973). It can also be used to detect the antibodies reacting with these antigens. This assay is very sensitive; according to Scolnick *et al.* (1972), the antibody titres were 200 to 1000 times higher than the complement fixation or immunodiffusion test, and the antigen titres were 100 to 500 times greater. Stephenson *et al.* (1973) also found that this assay was 10- to 100-fold more sensitive than the complement fixation test. According to Ihle *et al.* (1973), the radioimmune precipitation assay was 500 times more sensitive than the virus neutralization test. This assay also appears to be very specific, if purified antigen and specific antiserum can be obtained for testing.

## F. Radioiodine-labelled Antibody Technique

Recently, several isotopic techniques have been developed for study of antigens on cell surfaces, using radioiodine-labelled antibody. In general, they can be classified into direct and indirect methods. The direct technique uses the labelled immunoglobulin of specific immune serum, and the indirect technique uses a labelled anti-IgG reagent. Both techniques require purification of the antibody (IgG) and iodination of these globulin molecules. Therefore, we shall first describe the methods used for these procedures.

### *1. Purification of antibody*

The more purified are the antibodies (IgG), the more specific the reactions become. Usually, only isolation of the immunoglobulin fraction, predominantly IgG, is performed. One procedure has been to separate the IgG from sera by diethylaminoethyl (DEAE) cellulose or DEAE–'Sephadex' A-50 chromatography, equilibrated with 0·1 *M* sodium phosphate buffer, pH 6·5, with the IgG eluted by the method of Baumstark *et al.* (1964). The separated immunoglobulin can be stored at −20°C to −70°C for prolonged periods of time.

### *2. Iodination of antibody*

There are two methods, using iodine monochloride and chloramine T, which are commonly used for iodination of protein. In addition, an enzymatic method using lactoperoxidase has recently been developed. In these procedures, the objective is to couple approximately one atom of iodine to each molecule of protein.* Labelling in excess of 1·5–2 atoms/molecule has been shown to cause denaturation of some proteins.

*a. Iodine monochloride* (*ICl method.* This method was first developed by MacFarlane (1958) and was modified by Helmkamp *et al.* (1967).

* The amount of labelling compound needed to iodinate proteins, with one mole of iodine per mole of protein, can be determined by the following formula:

1. $\text{mg iodinating compound used} = \text{M.Wt. iodinating compound} \times \dfrac{\text{mg protein}}{\text{M.Wt. protein}}$
2. Correct by multiplying by the percent labelling efficiency.

For instance, in the iodine monochloride method, M.Wt. of iodinating compound (ICl) is 163, M.Wt. of protein (IgG) is 150,000, and labelling efficiency is 30% (factor of 3·3). In order to label 10 mg of IgG:

$$\text{mg ICl used} = 163 \times \frac{10}{150{,}000}\,\text{mg} = 1 \times 10^{-2}\,\text{mg}$$

Correct with labelling efficiency

$$= 1 \times 10^{-2} \times 3{\cdot}3$$
$$= 3{\cdot}3 \times 10^{-2}\,\text{mg}.$$

The above calculation is based on coupling of one atom of iodine to each molecule of protein (1:1 ratio); if twice as much of iodinating compound is used, then the ratio will also be 2:1, and so on.

1. Reagents used for labelling. These include 0·064 *M* borate buffer, pH 8·0; 2 *M* NaCl and stock ICl solution (0·033 *M* iodine chloride in 1 *M* HCl); 10% solution of bovine serum albumin; sodium iodide ($^{125}I$) or ($^{131}I$) in 0·1 *M* NaOH; and purified protein. The 0·033 *M* ICl solution is prepared according to MacFarlane: dissolve 150 mg sodium iodide (NaI) in 8 ml of 6 *M* HCl, then add 108 mg of iodate ($NaIO_3.H_2O$) dissolved in 2 ml distilled water. The iodate solution should be forcibly injected into the iodide–hydrochloric acid to avoid precipitation of iodine. Dilute the mixture with distilled water to 40 ml, and shake in a glass cylinder with 5 ml carbon tetrachloride ($CCl_4$). The solution should have a clear yellow colour. If there is free iodine, the bottom $CCl_4$ layer will be pink, and the procedure should be repeated. Finally, the residual $CCl_4$ is removed by aerating with moist air for 1 h, and the volume is adjusted to 45 ml with distilled water.

2. Iodination procedure. To label 10 mg of protein, it is usually suspended in borate buffer to a final volume of 3 ml. This is mixed thoroughly with 0·1 ml of carrier-free $Na^{125}I$ in borate buffer, pH 8·0, with 10 mCi radioactivity. Then 0·6 ml of ICl solution at 1/100 dilution of 0·033 *M* stock solution in 2 *M* NaCl is added. After the mixture is allowed to react for 60 sec, the reaction is terminated by addition of 1 ml of 10% BSA. To separate the labelled protein from unbound iodine, the iodinated protein is dialysed immediately against 300 volumes of isotonic saline at 4°C for 24 h with three changes of dialysis fluid at 2 h intervals for the first two changes, and overnight for the last change. The labelling efficiency is about 30%. The final iodine–protein molar ratio is 1·0 to 1·5, and the specific activity 250 to 400 $\mu$c $^{125}I$ per mg of protein.

*b. Chloramine T method.* This method was developed by Greenwood and Hunter (1963; Hunter and Greenwood, 1964).

1. Reagents used. These include chloramine T, sodium phosphate buffer, pH 7·5, carrier-free $Na^{131}I$ or $Na^{125}I$ in 0·02 to 0·04 *M* NaOH, sodium metasulphate, potassium iodide and protein.

2. Iodination procedure. This method was used by Greenwood and Hunter (1963) to label human growth hormone. Five $\mu$g of growth hormone in 0·025 ml of 0·05 *M* sodium phosphate buffer is mixed with 2 mCi of $Na^{131}I$, followed by 100 $\mu$g of fresh chloramine T. Immediately after mixing the chloramine T, 0·1 ml of sodium metasulphate is added at 2·4 mg/ml to terminate the reaction. The residual iodide is diluted with 0·2 ml of carrier potassium iodide at 10 mg/ml. The $^{131}I$-labelled growth hormone is separated from the reaction mixture by gel filtration on a 'Sephadex' G-50 column, equilibrated with 0·07 *M* barbitone buffer, pH 8·6. The labelling efficiency is about 74%. Minor modifications of this labelling procedure are made in other laboratories. The amount of protein to be labelled, the amount of $^{131}I$ or $^{125}I$ and chloramine T can be varied, and the labelling time may range from a few seconds to 30 min. Various chromatography procedures ('Sephadex'

G-25, 'Dowex' column, 'Bio-Gel' 20, etc.) can be used to separate the labelled protein from reaction mixture. The labelling efficiency obtained from the chloramine T method is usually between 50 and 70%.

c. *Enzyme method* (*lactoperoxidase*). This was developed by Marchalonis (1969).

1. Reagents used. 0·05 *M* sodium phosphate buffer, pH 7·3, purified lactoperoxidase, carrier-free $^{125}I$ (iodide) solution, 8 m*M* $H_2O_2$ solution, 5 m*M* cysteine, or 5 m*M* 2-mercaptoethanol and purified IgG.

2. Iodination procedure. In the original protocol, 250 $\mu$g of protein, 1·25 $\mu$g of purified lactoperoxidase, and 1 $\mu$l of ($^{125}I$) iodide solution at 4 mCi/ml were mixed. (In some cases, 1 mCi of $^{125}I$ was used.) The volume of the total reaction mixture was 50 $\mu$l. The reaction is initiated by the addition of 1 $\mu$l of 8 m*M* $H_2O_2$ followed by vigorous mixing, and it is carried out at room temperature for 15–30 min. The reaction is stopped by the addition of 0·5 ml of 5 m*M* cysteine or 5 m*M* 2-mercaptoethanol. This method of iodination is quite gentle, and there is no evidence of denaturation when the labelled protein was analysed by electrophoresis and density gradient velocity ultracentrifugation.

### 3. *Test procedure*

As with the fluorescence assay, there is a direct technique and an indirect technique.

#### a. *Direct technique*

1. Paired radioiodine-labelled antibody technique (PRILAT). This technique was originally developed by Pressman *et al.* (1957) to determine the distribution and the localization of purified antitumour antibodies in a rat lymphosarcoma. They used the immunoglobulin of immune serum labelled with $^{131}I$, and the immunoglobulin of normal serum labelled with $^{133}I$. These two preparations were mixed and injected into the tumour-bearing host. The tumour specimens were examined for the two isotopes present. This technique has been modified as an *in virto* assay.

a. Experiments performed on tissue sections of cell smears. This method was developed by Tanigaki *et al.* (1967) who used frozen sections of human autopsy material or cell smears from tissue culture prepared on 25 × 18 mm coverslips. These test samples can be stored at −20°C. The antisera are produced in rabbits by immunization with human liver sediment or with various tissue-culture cell lines. The IgG fraction of antisera is labelled with $^{125}I$, and the IgG from normal serum is labelled with $^{131}I$. These are mixed at 1:1 ratio with respect to the amount of protein. The test specimens are fixed with acetone for 10 min before testing. A few drops of the labelled serum mixture applied to the specimen which was incubated in a moist chamber at room temperature for 60 min. After washing three times with phosphate-buffered saline the specimens are dried and the coverslips cut into two pieces with a diamond knife and inserted into test-tubes to be counted in a double-channel gamma-scintillation counter. The results are expressed as specific uptake quotient (SUQ). SUQ = $R/R_0$, where

$$R = \frac{^{125}\text{I cpm on test cells}}{^{131}\text{I cpm on test cells}} \text{ and } R_0 = \frac{^{125}\text{I cpm on control cells}}{^{131}\text{I cpm on control cells}}$$

The control cells used in the experiments are the cells which did not have the appropriate antigenic specificity. Any value of 1·0 or less indicates that there is no significant amount of immune IgG attached to the test cells. Any value greater than 1·0 indicates a specific uptake of $^{125}$I-labelled immune IgG. Evans and Yohn (1970) used 95% confidence limits to evaluate the results and the standard errors (SE) were calculated according to the method of Finney *et al.* (1964). Results can be considered significant when (SUQ$-$2SE $>$ 1·0).

b. Experiments performed on tissue-culture cells grown as monolayers on coverslips in Leighton tubes. This method was developed by Evans and Yohn (1970). The procedure of the assay is essentially similar except that the preparation of the test specimens differs.

c. Experiments performed with cell suspensions. This method was developed by Boone *et al.* (1972). Counted numbers of suspension cells are dispensed into 15 × 132 mm conical tipped centrifuge tubes, and pelleted by centrifugation for 5 min at (1500 rpm, 0°C, in the 1EC rotor 269), and resuspended in phosphate-buffered saline (PBS). A measured amount of paired labelled serum mixture is added to each tube to bring the final volume to 1 ml. The tubes are incubated at room temperature for 30 min, placed into ice-cold IEC cylindrical metal inserts in swinging buckets, and centrifuged at 4°C for 5 min at 1500 rpm. The cells are further washed six times with PBS, then transferred to plastic tubes to be counted in a double-channel gamma-scintillation counter. The results can be expressed as the absorption ratio (AR).

AR = $\mu$g of immune serum protein bound/$\mu$g of normal serum protein bound.

2. Single radioiodine-labelled antibody technique (RIB). This method was developed by Inoue and Klein (1970) and Hewetson *et al.* (1972). The purified IgG of immune serum was labelled with $^{125}$I by the chloramine T method. Twenty $\mu$l of cell suspension ($3 \times 10^5$ to $2 \times 10^6$ cells) and 0·1 ml of test serum are mixed in 11 × 50 mm siliconized glass tubes, and incubated at room temperature for 30 min. The tubes are then stirred with a mechanical mixer, 1·5 ml of minimal essential medium added and the tubes centrifuged. The cells are washed once more and the supernatant aspirated, leaving a small amount of fluid (0·1 ml) covering the cell pellet. Then, 10 $\mu$l of radioiodine-labelled antibody is added, and the mixture incubated at room temperature for 30 min. The tubes are then washed four times, and the cells are

counted in a gamma-scintillation counter. If the test serum has antibody with the same specificity as labelled antibody, then it will block the binding of labelled antibody to the appropriate test cells. The results are expressed as the blocking index:

$$\text{Blocking index} = \frac{\text{cpm in medium control} - \text{cpm in test sample}}{\text{cpm in medium control}}$$

*b. Indirect technique*

1. Isotopic antiglobulin technique (IAT). This method was originally developed by Harder and McKhann (1968), and was modified by Sparks *et al.* (1969). The isolated anti-IgG immunoglobulin reagent was labelled with $^{125}$I by the ICl method. 0·1 ml of test cells (1 to 5 $\times$ $10^5$ cells) suspended in Barbital-buffered saline (0·025 *M* sodium barbital, 0·7 *M* NaCl, pH 7·4) containing 10% foetal bovine serum (BBS/FBS) are mixed with 0·1 ml of each dilution of test sera (mouse) in 10 $\times$ 75 mm test-tubes, and another set of tubes containing cells are mixed with normal serum at the same dilutions. The mixtures are incubated at room temperature for 30 min, and the cells then washed five times with BBS/FBS. A second incubation with 2 $\mu$l of $^{125}$I-labelled antimouse immunoglobulin is performed at 4°C for 15 min. The cells are again washed five times with BBS/FBS and transferred to plastic tubes to be counted in a gamma-scintillation counter. Activity of the test sera is expressed as the counts per minute (cpm) of $^{125}$I-antiglobulin bound to the cells when a standard control serum is used in each experiment, or expressed as the absorption ratio (AR).

$$\text{AR} = \frac{^{125}\text{I counts obtained with antiserum}}{^{125}\text{I counts obtained with normal serum}}$$

A reaction is defined as positive when AR is equal to, or greater than, 2. Absorption experiments should always be performed to confirm the specificity of the reactions. In absorption experiments, usually 0·5 to 1·0 ml of a dilution of antiserum, close to the end-point, but which still gives an AR greater than 2, is incubated with varying numbers of cells at 37°C for 45–60 min, with frequent mixing. The cell suspensions are then centrifuged at 600 *g* for 15 min. The supernatant is carefully removed and centrifuged again at 1500 *g* for 15 min to remove any remaining cells. The supernatant is then tested for residual activity, against the appropriate target cells, as described earlier. The cpm obtained with the absorbed antiserum were compared to the cpm obtained with the same dilution of normal serum, absorbed in the same manner. The absorption is considered positive if the AR's obtained from the absorbed sera are reduced by more than 20%.

$$\%\ \text{AR reduction} = \frac{\text{AR before absorption} - \text{AR after absorption}}{\text{AR before absorption}} \times 100$$

To determine whether the antigen(s) present on different cell lines are identical, or to determine their relative quantity, the results of the absorption experiments can be plotted as the reduction of cpm in antiserum as a function of the number of cells used for absorption. Identical antigen should give identical absorption curves (Crisler *et al.*, 1966; Ting and Herberman, 1970, 1971). The relative quantity of the antigen is determined by the number of cells, corrected for surface area, to remove 50% of antiserum activity. Absorption of 50% antiserum activity is defined as the midpoint of the maximal and minimal cpm obtained with the absorbed antiserum. The minimal cpm that can be achieved should be in the range of the cpm of the absorbed normal serum. Absorption experiments performed with antipolyoma virus

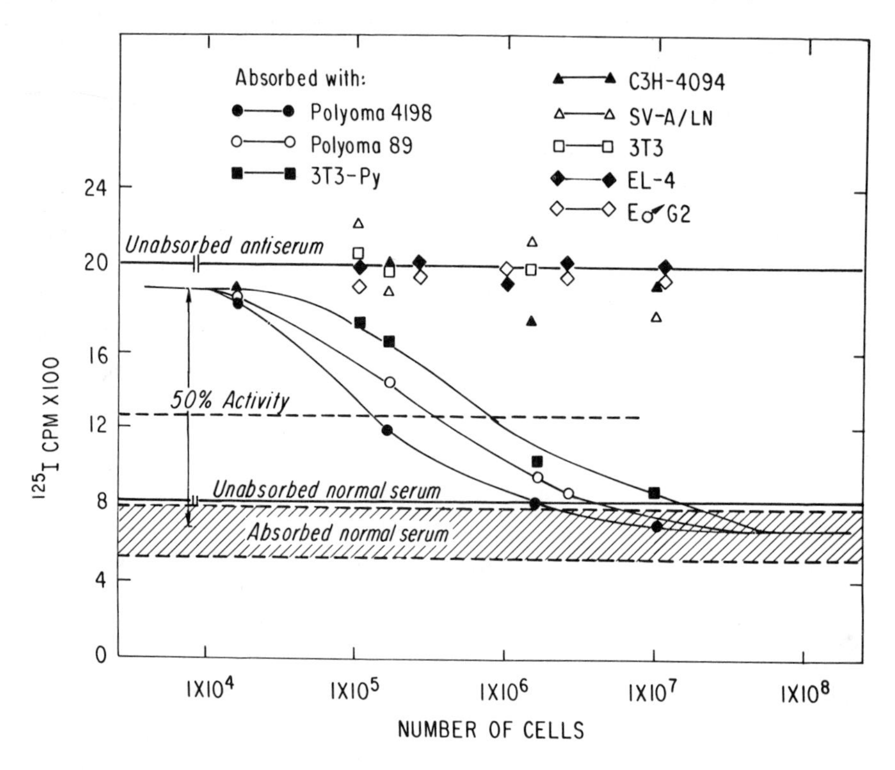


**Figure 5.1.** Antipolyoma antiserum (produced by immunization of C3H/HeN mice with syngeneic polyoma 4198) and normal mouse serum (C3H/HeN) at 1:70 dilution were absorbed with different numbers of polyoma cell lines (C3H/HeN polyoma 4198, C57/KaLw polyoma 89, Swiss 3T3-Py), and with non-polyoma cells (spontaneously transformed C3H/HeN 4094, SV40 transformed SV-A/LN, untransformed Swiss 3T3, dimethylbenzanthrene-induced lymphogenous leukosis EL-4, and Gross-virus-induced lymphoma E♂G2). Absorbed and unabsorbed sera were then tested for antibody activity with polyoma 4198 target cells. The range of the $^{125}$I cpm obtained with the absorbed normal serum is shown by the hatched area

antiserum are shown in Figure 5.1. In this experiment the radioactivity obtained with unabsorbed antiserum was 2000 cpm, and with unabsorbed normal serum was 800 cpm, giving a value of AR = 2·6. After absorption, the activity of the antiserum as indicated by the cpm was only removed by three polyoma tumour cells (C3H/HeN polyoma 4198, C57/KaLw polyoma 89, and the polyoma virus-transformed cell line 3T3-Py). Five non-polyoma cells gave no significant absorption. The slopes of the absorption curves obtained with the three polyoma cell lines were identical, thus indicating that they possess identical polyoma TASA. Since these cell lines were of similar size, the relative antigenic content of the different lines is determined by the number of cells needed to absorb 50% of the antiserum activity, corresponding to $^{125}$I cpm of 1250. The polyoma 4198 cells were found to have an antigen density about three times higher than polyoma 89 cells, and almost six times higher than that of 3T3-Py cells. Therefore, this technique permits a quantitative comparison of the surface antigenic content of tumours with the same origin.

2. Indirect paired-labelled antibody technique. This method is similar to the direct paired-labelled technique, and was developed by McCammon and Yohn (1971). Instead of labelling the IgG of the immune serum directly, the antiglobulin reagent, rabbit antimouse immunoglobulin, in the original study and immunoglobulin from normal serum of the same species which produced the antiglobulin reagent, rabbit IgG are labelled with $^{125}$I and $^{131}$I, respectively. In this test, the test cells are grown on coverslips as monolayers, and they are incubated with appropriate dilutions of antiserum at 37°C for 30 min. The coverslips are given three 5 min washes with phosphate-buffered saline. Then, a mixture of equal amounts of the $^{125}$I-labelled antiglobulin reagent and $^{131}$I-labelled normal IgG is placed on the coverslips, and is incubated further at 37°C for 30 min. The coverslips are again washed three times in phosphate-buffered saline and air dried. They are then crushed into counting vials, to be counted in a gamma-scintillation counter. The results are expressed as SUQ (specific uptake quotient) as in the direct paired label technique.

Applications: The radioiodine-labelled antibody techniques have been applied to detect a variety of virus-induced tumour-associated antigens. The direct paired radioiodine-labelled antibody technique (PRILAT) has been used to detect the T antigen of adenovirus 12 induced tumour (Evans and Yohn, 1970), gs antigen of Avian tumour virus (Weber and Yohn, 1972) and surface antigen of AKR-virus-induced rat lymphoma cells (Boone *et al.*, 1972). Since the amount of labelled antibody bound to the cells can be determined, the number of antigenic receptor sites can also be estimated (Boone *et al.*, 1972). The direct single-label technique has been used to detect the cell surface antigen associated with Epstein–Barr virus (Inoue and Klein, 1970; Hewetson *et al.*, 1972). The isotopic antiglobulin technique has been used to detect the tumour-associated cell surface antigen of polyoma tumours (Ting and Herberman, 1970), SV40 tumours (Ting and Herberman, 1971), leukaemias induced by Gross virus (Herberman *et al.*, 1973) and Friend virus (Ting *et al.*, unpublished observation). The indirect PRILAT has been used to detect the T antigen of adenovirus 12 induced tumours (McCammon and Yohn, 1971). In general, all these techniques appear to be very sensitive. For detection of adenovirus 12 induced tumour T antigens, the PRILAT was 100 to 200 times more sensitive than the complement fixation test, and ten times more sensitive than immunofluorescence (Evans and Yohn, 1970). The direct single-label technique was found to be four times more sensitive than the immunofluorescence test when it was used to detect the EBV-virus-associated cell surface antigen (Hewetson *et al.*, 1972). The isotopic antiglobulin technique could detect antibody at serum dilutions which gave negative results by other tests (Sparks *et al.*, 1969). Besides these techniques being very sensitive, they are also rapid and objective. In performing these tests, the most important factor is to obtain the purified IgG fraction of immune serum with high and specific activity. This will increase the specific binding and reduce non-specific binding of the labelled antibody, thereby increasing the sensitivity of the test. For both the isotopic antiglobulin technique and the indirect immunofluorescence test, an antiglobulin reagent of good quality is very important. The antiglobulin reagents should react strongly in immunodiffusion against immunoglobulin. They should either react broadly against the various immunoglobulin classes, or specifically with IgG or IgM. In either case, they should not be able to react with non-Ig serum proteins. Since most of the commercial reagents are not standardized (specific activity, specificity and other characteristics vary from one company to another, and also vary from one lot to another), they are often not satisfactory for detection of the antibodies reacting with tumour-associated antigens. Therefore, it is often worthwhile preparing one's own reagent.

The advantage of the direct technique is that the amount of antibody bound to the cells or test specimens can be more accurately estimated. However, it requires a fair amount of specific immune serum to be used for testing, and it also requires the labelling of each immune serum to be tested. Therefore, the direct assay is difficult to use for testing of mouse immune sera. The indirect test requires only a large supply of the antiglobulin reagent. (The rabbit, goat or sheep can be the source for producing the reagent.) It requires very small amounts of antiserum to perform the test, and the specificity of the reactions can always be confirmed by appropriate absorption experiments. The disadvantages of the indirect test are that two reaction steps are involved (first the immunoglobulin of antiserum binds to the cells, then the antiglobulin binds to the immunoglobulin), and the amount of antibody bound to the cells cannot accurately be calculated. Therefore, only the relative quantity of the antigenic density can be estimated between different cell lines. On the other hand, the indirect technique is more convenient and the same labelled anti-globulin reagent can be used to test a variety of antisera from the same species. Thus, it is appropriate to use the direct technique when a specific antigen is being tested, and if a specific antiserum in reasonable quantity can be obtained. On the other hand, when antigens of different specificities are being studied, or the supply of antisera is scarce, then it is more appropriate to use the indirect technique.

## III. APPRAISAL OF THE METHODS

In this chapter, we have discussed several major techniques which are commonly used for the detection of viral antigens and tumour-associated antigens induced by oncogenic viruses. There are numerous other more familiar techniques which are not discussed here, such as the complement fixation test and the immunodiffusion test. The techniques which have been discussed can be classified into six groups, namely cytotoxicity tests, haemadsorption and mixed haemadsorption tests, immunofluorescence tests, immunoelectron microscopy, radioimmune precipitation assays and radioiodine-labelled antibody techniques. Each type of assay has several variations. The cytotoxicity test is easy to perform, but it can only be used with the cells which are susceptible to the cytotoxic antibodies. This assay also cannot detect antibodies which do not bind complement. The TASA of SV40 tumours detected by this assay showed good correlation to the TATA which were defined by the *in vivo* immunorejection experiments (Wright and Law, 1971; Smith and Mora, 1972). The kinetics of the antibody response to the TASA of Gross-virus-induced leukaemia also varied with the growth of the tumours (Herberman and Oren, 1971).

The mixed haemadsorption test appears to be sensitive though it has been difficult to obtain reproducible results in experiments performed with tumour antigens. A surface antigen of SV40 tumours detected by this technique was shown not to be specific for SV40 tumours but was shared by non-SV40 cells after trypsin treatment. This antigen also did not correlate with the TATA (Häyry and Defendi, 1970).

The immunofluorescence test is very sensitive. It has been applied to detect the T antigen of the DNA virus-induced tumours, and the surface antigen of a variety of virus-induced tumours. The antibody response to the EBV-induced surface antigen detected by this technique appears to be correlated with the clinical conditions of the patients with Burkitt lymphoma or nasopharyngeal carcinoma (Klein *et al.*, 1969a; Einhorn *et al.*, 1970; Lin *et al.*, 1972). However, the surface antigen of SV40 tumours (S antigen) detected by this technique did not correlate with the TATA, and this antigen appeared to be a foetal antigen (Tevethia *et al.*, 1968b; Berman, 1972). The immunofluorescence test needs to be performed by experienced personnel.

Immunoelectron microscopy has the advantage of making the sites of the antigen–antibody reaction directly visible. It can distinguish easily between antibodies reacting with viral envelope antigen and those reacting with cell surface antigen, whereas with other assays, it may be necessary to perform absorption experiments to distinguish these two different antigens.

The radioimmune precipitation assay is very specific and sensitive. Its usefulness, however, is limited to soluble antigens or antigens of very small particle size, which remain in suspension. It cannot be used to detect the surface antigen of viable cells.

The radioiodine-labelled antibody techniques are rapid, sensitive, and objective. The observed reactions are specific if purified antiserum or antiglobulin reagent of high activity can be obtained for testing. The direct paired-labelled antibody technique is desirable when a specific immune serum can be obtained in large quantities for testing, and a particular type of antigen is being studied. When the amounts of immune serum are scarce or when several different antigens or different immune sera are being studied, then the indirect technique (isotopic antiglobulin technique or paired-labelled technique) is more desirable. Since both indirect methods require a normal serum for comparison, it is, therefore, important to have a good control sample. This is easy to obtain if the experiments are carried out in an animal system, especially when the inbred strains of animals are available. However, sometimes reactivity to certain antigens may be found in the sera of normal individuals. This is particularly difficult in human systems. In such a case, when a suitable normal control sample cannot be obtained, it may be better to perform the direct single-labelled antibody technique, if immune serum can be obtained in reasonable quantities. The specificity of the reaction can be checked by blocking experiments with other test sera and by testing with other cells with different antigenic specificities. Since these techniques are very sensitive, quantitative study of the antigenic density can be performed by determination of the amount of antibody bound to the cells in the direct technique (Boone *et al.*, 1972) or by appropriate absorption experiments in the indirect isotopic antiglobulin technique (Ting and Herberman, 1970, 1971).

Testing by the isotopic antiglobulin technique, the TASA of polyoma or SV40 tumours showed a very good correlation to the TATA of these tumours (Ting and Herberman, 1971; Ting *et al.*, 1972). It has been further shown that by immunizing the mice with syngeneic SV40 tumour (Ting *et al.*, unpublished observations), or Friend virus leukaemia (Ting *et al.*, unpublished observations), the antisera produced might be able to recognize at least two different antigenic specificities. One was TASA which was specific for SV40 tumour or Friend-virus-induced leukaemia, respectively; the other was a common antigen which was shared by other tumours and foetal cells. Only the TASA was found to be correlated with TATA; the common antigen

**Table 5.1.** Detection of viral antigens and tumour-associated antigens

| Assay | | Applications |
|---|---|---|
| (1) Cytotoxicity test | SA: | Leukaemias induced by Gross virus, FMR viruses, radiation leukaemia virus; TL leukaemia, ML antigen; and tumours induced by adenoviruses, polyoma virus and SV40 |
| (2) Haemadsorption and haemadsorption inhibition test | VEA: | Polyoma virus, murine leukaemia virus, mammary tumour virus |
| Mixed haemadsorption test | SA: | Moloney lymphoma, Gross leukaemia, mouse mammary tumour, SV40 tumour |
| (3) Immunofluorescence test | SA: | Leukaemias induced by Gross virus and FMR viruses; mouse mammary tumour; tumours induced by adeno-viruses, polyoma virus, SV40, and Shope papilloma virus; Burkitt's lymphoma and EBV-associated antigens |
| | TA: | Tumours induced by adenoviruses, SV40 and polyoma virus |
| (4) Immunoelectron microscopy | SA and VEA: | Gross leukaemia; tumours induced by Moloney sarcoma virus, and Avian tumour viruses; murine endogenous C-type viruses |
| (5) Radioimmune precipitation assay | VEA: | Murine leukaemia virus |
| | GSA: | Mammalian and Avian C-type viruses |
| (6) Paired radioiodine-labelled antibody technique | SA: | Gross leukaemia; Burkitt's lymphoma and EBV-associated antigens |
| | TA: | Adenoviruses-induced tumours |
| | VEA: | EBV-associated antigens |
| | GSA: | Avian tumour viruses induced tumours |
| Isotopic antiglobulin technique | SA: | Polyoma, SV40 tumour, Gross leukaemic, and FMR viruses induced leukaemia |

SA: Tumour-associated cell surface antigen.
VEA: Viral envelope antigen.
TA: T antigen.
GSA: Group-specific antigen.

was not; therefore, this common antigen resembled the S antigen of hamster SV40 tumours detected by the immunofluorescence test. Since the nature of cell surface antigens is complex, antigens of different specificities may be expressed on some tumour cell lines, and each antigenic specificity may provoke different antibody responses in different hosts. Therefore, the specificities of these serological reactions should be carefully studied by testing appropriate antisera with test cells which carry various specificities. If the antigens detected by these procedures show good correlation with TATA, the assay may be useful in monitoring the effects of immunoprophylaxis or immunotherapy of the tumours. Although some cell surface antigens do not correlate with TATA, if their expression is unique for tumour cells, they may be used for immunodiagnosis.

The application of various techniques is summarized in Table 5.1. We can only offer a brief glimpse of each technique here. Each one has its own specific usefulness, as well as having its limitations. Before selecting a particular technique or techniques for performing experiments, the following factors should be considered: the nature of the antigens to be studied, the availability of test samples or test sera, the skills of the personnel and the facilities available. Since no technique has been perfected, they can still be improved upon and modified according to the needs of the particular experiment.

## IV. ACKNOWLEDGEMENT

We deeply appreciate the help of Mrs. Sandy Ryan in the preparation of this manuscript.

## V. REFERENCES

Ankerst, J. and H. O. Sjögren (1969) 'Cross-reacting TSTAs in adeno 7 and 12 tumors demonstrated by $^{51}$Cr-cytotoxicity and isograft rejection test', *Int. J. Cancer*, **4**, 279.

Aoki, T., E. A. Boyse and L. J. Old (1966) 'Occurrence of natural antibody to the G(Gross) leukemia antigen in mice', *Cancer Res.*, **26**, 1415.

Aoki, T., E. A. Boyse, L. J. Old, E. deHarven, U. Hämmerling and H. A. Wood (1970) 'G(Gross) and H-2 cell-surface antigens: location on Gross leukemia cells by electron microscopy with visually labeled antibody', *Proc. Nat. Acad. Sci. U.S.A.*, **65**, 569.

Aoki, T., H. A. Wood, L. J. Old, E. A. Boyse, E. deHarven, M. P. Lardis and C. W. Stackpole (1971) 'Another visual marker of antibody for electron microscopy', *Virology*, **45**, 858.

Aoki, T., J. R. Stephenson and S. A. Aaronson (1973a) 'Demonstration of cell-surface antigen associated with murine sarcoma virus by immunoelectron microscopy', *Proc. Nat. Acad. Sci. U.S.A.*, **70**, 742.

Aoki, T. and G. J. Todaro (1973b) 'Antigenic properties of endogenous type-C viruses from spontaneously transformed clones of BALB/3T3', *Proc. Nat. Acad. Sci. U.S.A.*, **70**, 1598.

Armstrong, D. (1969) 'Multiple group-specific antigen components of avian tumor viruses detected with chicken and hamster sera', *Virology*, **3**, 133.

Barth, R. F., J. A. Espmark and A. Fagraeus (1967) 'Histocompatibility and tumor virus antigens identified on cells grown in tissue culture by means of the mixed hemadsorption reaction', *J. Immunol.*, **98**, 888.

Baumstark, J. S., R. J. Laffin and W. A. Bardawil (1964) 'A preparative method for the separation of 7S gamma globulin from human serum', *Arch. Biochem. Biophys.*, **108**, 514.

Berman, L. D. (1972) 'The SV40 S antigen: a carcinoembryonic-type antigen of the hamster?', *Int. J. Cancer*, **10**, 326.

Black, P. H., W. P. Rowe, H. C. Turner and R. J. Huebner (1963) 'A specific complement-fixing antigen present in SV40 tumor and transformed cells', *Proc. Nat. Acad. Sci. U.S.A.*, **50**, 1148.

Bloom, E. T. (1970) 'Quantitative detection of cyotoxic antibodies against tumor-specific antigens of murine sarcomas induced by 3-methylcholanthrene', *J. Nat. Cancer Inst.*, **45**, 443.

Boone, C. W., P. R. Brandchaft, D. N. Irving and R. Gilden (1972) 'Quantitative studies on the binding of syngeneic antibody to the surface antigens of AKR-virus-induced rat lymphoma cells', *Int. J. Cancer*, **9**, 685.

Borsos, T., H. R. Colten, J. S. Spalter, N. Rogentine and H. J. Rapp (1968) 'The C'la fixation and transfer test: examples of its applicability to the detection and enumeration of antigens and antibodies at cell surfaces', *J. Immunol.*, **101**, 392.

Boyden, S. V. (1951) 'The absorption of proteins on erthyrocytes treated with tannic acid and subsequent hemagglutination by antiprotein sera', *J. Exp. Med.*, **93**, 107.

Boyse, E. A., L. J. Old and E. Stockert (1965) 'The TL (thymus-leukemia) antigen: a review', in *Immunopathology, IVth International Symposium* (Eds. P. Grabar and P. A. Miescher), Schwabe and Co., Basel, p. 23.

Boyse, E. A., L. Hubbard, E. Stockert and M. E. Lamm (1970) 'Improved complementation in the cytotoxic test', *Transpl.*, **10**, 446.

Cohen, A. and M. Schlesinger (1970) 'Absorption of guinea pig serum with agar', *Transpl.*, **10**, 130.

Cohen, A. M., J. F. Burdick and A. S. Ketcham (1971) 'Cell mediated cytotoxicity: an assay using $^{125}$I-iododeoxyuridine-labeled target cells', *J. Immunol.*, **107**, 895.

Coombs, R. R. A., J. Marks and D. Bedford (1956) 'Specific mixed agglutination: mixed erythrocyte-platelet antiglobulin reaction for the detection of platelet antibodies', *Brit. J. Haemat.*, **2**, 84.

Coons, A. H. and M. H. Kaplan (1950) 'Localization of antigen in tissue cells. II. Improvements in a method for the detection of antigen by means of fluorescent antibody', *J. Exp. Med.*, **91**, 1.

Coons, A. H., J. C. Snyder, F. S. Cheever and E. S. Murray (1950) 'Localization of antigen in tissue cells. IV. Antigens of rickettsial and mumps virus', *J. Exp. Med.*, **91**, 31.

Crisler, C., H. J. Rapp, R. M. Weintraub and T. Borsos (1966) 'Forssman antigen content of guinea pig hepatomas induced by diethylnitrosamine: a quantitative approach to the search for tumor specific antibodies', *J. Nat. Cancer Inst.*, **36**, 529.

DeSchryver, A., G. Klein, G. Henle, W. Henle, H. M. Cameron, L. Santesson and P. Clifford (1972) 'EB-virus associated serology in malignant disease: antibody level to viral capsid antigens (VCA), membrane antigens (MA) and early antigens (EA) in patients with various neoplastic conditions', *Int. J. Cancer*, **9**, 353.

Eckhart, W. (1969) 'Complementation and transformation by temperature-sensitive mutants of polyoma virus', *Virology*, **38**, 120.

Eddy, B. E., W. P. Rowe, J. W. Hartley, S. E. Stewart and R. J. Huebner (1958) 'Hemagglutination with the S.E. polyoma virus', *Virology*, **6**, 290.

Einhorn, N., G. Klein and P. Clifford (1970) 'Increase in antibody titer against the EBV-associated membrane antigen complex in Burkitt's lymphoma and naso-pharyngeal carcinoma after local irradiation', *Cancer*, **26**, 1013.

Evans, M. J. and D. S. Yohn (1970) 'Application of the paired radioiodine-labeled antibody technique (PRILAT) to the detection of adenovirus 12 tumor (T) antigen', *J. Immunol.*, **104**, 1132.

Farr, R. S. (1958) 'A quantitative immunochemical measure of the primary interaction between I* BSA and antibody', *J. Inf. Dis.*, **103**, 239.

Fass L. and R. B. Herberman (1969) 'A cytotoxic antiglobulin technique for assay of antibodies to histocompatibility antigens', *J. Immunol.*, **102**, 140.

Ferrer, J. F. and H. S. Kaplan (1968) 'Antigenic characteristics of lymphomas induced by radiation leukemia virus (Rad LV) in mice and rats', *Cancer Res.*, **28**, 2522.

Fink, M. A., W. F. Feller and L. R. Sibal (1968) 'Methods for detection of antibody to the mammary tumor virus', *J. Nat. Cancer Inst.*, **41**, 1395.

Finney, D. J. (1964), *Statistical Method in Biological Assay*, 2nd ed., Hofner Publishing Co., New York, p. 21.

Geering, G., L. J. Old and E. A. Boyse (1966) 'Antigens of leukemias induced by naturally occurring murine leukemia virus: their relation to the antigens of Gross virus and other murine leukemia viruses', *J. Exp. Med.*, **124**, 753.

Gelderblom, H., H. Bauer and T. Graf (1972) 'Cell-surface antigens induced by avian RNA tumor viruses: detection by immunoferritin technique', *Virology*, **47**, 416.

Gerloff, R. K., B. H. Hoyer and L. C. McLaren (1962) 'Precipitation of radio-labeled poliovirus with specific antibody and antiglobulin', *J. Immunol.*, **89**, 559.

Gillespie, A. V. (1968) 'Detection of a tumor specific antigen (Gross) with the mixed antiglobulin reaction using erythrocytes from NZB/B1 mice', *Immunol.*, **15**, 855.

Glynn, J. P., J. L. McCoy and A. Fefer (1968) 'Cross-resistance to the transplantation of syngenic Friend, Moloney, and Rauscher virus-induced tumors', *Cancer Res.*, **28**, 434.

Gorer, P. A. and P. O'Gorman (1956) 'The cytotoxic activity of isoantibodies in mice', *Transplant Bull.*, **3**, 142.

Greenwood, F. C. and W. H. Hunter (1963) 'The preparation of $^{131}$I-labeled human growth hormone of high specific radioactivity', *Biochem. J.*, **89**, 114.

Hämmerling, U., T. Aoki, E. deHarven, E. A. Boyse and L. J. Old (1968) 'Use of hybrid antibody with anti-$\gamma$G and anti-ferritin specificities in locating cell surface antigens by electron microscopy', *J. Exp. Med.*, **128**, 1461.

Harder, F. H. and C. F. McKhann (1968) 'Demonstration of cellular antigens on sarcoma cells by an indirect $^{125}$I-labeled antibody technique', *J. Nat. Cancer Inst.*, **40**, 231.

Hashimoto, Y. and H. Sudo (1971) 'Evaluation of cell damage in immune reactions by release of radioactivity from $^{3}$H-uridine labeled cells', *GANN*, **62**, 139.

Häyry, P. and V. Defendi (1970) 'Surface antigen(s) of SV40-transformed tumor cells', *Virology*, **41**, 22.

Hellström, I. and H. O. Sjögren (1965) 'Demonstration of H-2 isoantigens and polyoma specific tumor antigens by measuring colony formation in vitro', *Exp. Cell Res.*, **40**, 212.

Hellström, I. and H. O. Sjögren (1966) 'Demonstration of common specific antigen(s) in mouse and hamster polyoma tumors', *Int. J. Cancer*, **1**, 481.

Hellström, I. and K. E. Hellström (1969) 'Studies on cellular immunity and its serum mediated inhibition in Moloney-virus-induced mouse sarcomas', *Int. J. Cancer*, **4**, 587.

Hellström, I., C. A. Evans and K. E. Hellström (1969) 'Cellular immunity and its serum-mediated inhibition in Shope-virus-induced rabbit papillomas', *Int. J. Cancer*, **4**, 601.

Helmkamp, R. W., M. A. Contraras and W. F. Bale (1967) '$^{131}$I-labelling of protein by the iodine monochloride method', *Int. J. Appl. Radiat.*, **18**, 737.

Henle, G. and W. Henle (1966) 'Immunofluorescence in cells derived from Burkitt's lymphoma', *J. Bact.*, **91**, 1248.

Herberman, R. B. (1970) 'Inhibition of natural cytotoxic rabbit antibody by human IgM: production of nontoxic rabbit serum for use as complement source', *J. Immunol.*, **104**, 805.

Herberman, R. B. and M. E. Oren (1971) 'Immune response to Gross virus-induced lymphoma. I. Kinetics of cytotoxic antibody response', *J. Nat. Cancer Inst.*, **46**, 391.

Herberman, R. B., T. Aoki and M. E. Nunn (1973) 'Solubilization of G(Gross) antigens on the surface of G leukemia cells', *J. Nat. Cancer Inst.*, **50**, 481.

Hewetson, J. F., B. Gothoskar and G. Klein (1972) 'Radioiodine labeled antibody test for the detection of membrane antigen associated with Epstein–Barr virus', *J. Nat. Cancer Inst.*, **48**, 87.

Hilgers, J., W. C. Williams, B. Myers and L. Dmochowski (1971) 'Detection of antigens of mouse mammary tumor (MTV) and murine leukemia virus (MuLV) in cells of culture derived from mammary tumors of mice of several strains', *Virology*, **45**, 470.

Huebmer, R. J., W. D. Rowe, H. C. Turner and W. Lane (1963) 'Specific adenovirus complement-fixation antigens in virus-free hamster and rat tumors', *Proc. Nat. Acad. Sci. U.S.A.*, **50**, 379.

Huebner, R. J., P. S. Sarma, G. J. Kelloff, R. V. Gilden, H. Meier, D. D. Myers and R. L. Peters (1971) 'Immunological tolerance to RNA tumor virus genome expressions: significance of tolerance and prenatal expressions in embryogenesis and tumorigenesis. *Ann. N.Y. Acad. Sci.*, **181**, 246.

Hunter, W. M. and F. C. Greenwood (1964) 'A radioimmunoelectrophoretic assay for human growth hormone', *Biochem. J.*, **91**, 43.

Ihle, J. N., M. Yurconic, Jr. and M. G. Hanna, Jr. (1973) 'Autogenous immunity to endogenous RNA tumor virus: radioimmune precipitation assay of mouse serum antibody levels', *J. Exp. Med.*, **138**, 194.

Inoue, M. and G. Klein (1970) 'Reactivity of radioiodinated serum antibody from Burkitt's lymphoma and nasopharyngeal carcinoma patients against culture lines derived from Burkitt's lymphoma', *Clin. Exp. Immunol.*, **7**, 39.

Irlin, I. S. (1967) 'Immunofluorescent demonstration of a specific surface antigen in cells infected or transformed by polyoma virus', *Virology*, **32**, 725.

Ishimoto, A. and Y. Ito (1969) 'Specific surface antigen in Shope papilloma cells', *Virology*, **39**, 595.

Klein, E. and G. Klein (1964) 'Antigenic properties of lymphomas induced by the Moloney agent', *J. Nat. Cancer Inst.*, **32**, 547.

Klein, G., P. Clifford, E. Klein and J. Stjernswärd (1966a) 'Search for tumor-specific immune reactions in Burkitt lymphoma patients by the membrane immunofluorescence reaction', *Proc. Nat. Acad. Sci. U.S.A.*, **55**, 1628.

Klein, G., E. Klein and G. Haughton (1966b) 'Variation of antigenic characteristics between different mouse lymphomas induced by the Moloney virus', *J. Nat. Cancer Inst.*, **36**, 607.

Klein, G., P. Clifford, G. Henle, W. Henle, G. Geering and L. J. Old (1966a) 'EBV-associated serological patterns in a Burkitt lymphoma patient during regression and recurrence', *Int. J. Cancer*, **4**, 416.

Klein, G., G. Pearson, G. Henle, W. Henle, G. Goldstein and P. Clifford (1969b) 'Relationship between Epstein–Barr viral and membrane immunofluorescence in Burkitt tumor cells. III. Comparison of blocking of direct membrane immunofluorescence and anti-EBV reactivities of different sera', *J. Exp. Med.*, **129**, 697.

LeMevel, B. P. and S. A. Wells, Jr. (1973) 'A microassay for the quantitation of cytotoxic antitumor antibody: use of $^{125}$I-iododeoxyuridine as a tumor cell label', *J. Nat. Cancer Inst.*, **50**, 803.

Lin, T. M., C. S. Yang, S. W. Ho, J. F. Chiou, C. H. Liu, S. M. Tu, K. P. Chen, Y. H. Ito, A. Kawamura and T. Hirayama (1972) 'Antibodies to Herpes-type virus in nasopharyngeal carcinoma and control groups', *Cancer*, **29**, 603.

MacFarlane, A. S. (1958) 'Efficient trace-labelling of proteins with iodine', *Nature* (*London*), **182**, 53.

Malmgren, R. A., K. K. Takemoto and P. G. Carney (1968) 'Immunofluorescent studies of mouse and hamster cell surface antigens induced by polyoma virus', *J. Nat. Cancer Inst.*, **40**, 263.

Marchalonis, J. J. (1969) 'An enzymic method for the trace iodination of immunoglobulins and other proteins', *Biochem. J.*, **113**, 299.

McCammon, J. R. and D. S. Yohn (1971) 'Application of an indirect paired radioiodine-labeled antibody technique to adenovirus 12 tumor serology', *J. Nat. Cancer Inst.*, **47**, 447.

Möller, G. (1961) 'Demonstration of mouse isoantigens at the cellular level by the fluorescent antibody technique', *J. Exp. Med.*, **114**, 415.

Nadkarni, J. S., J. J. Nadkarni, G. Klein, W. Henle, G. Henle and P. Clifford (1970) 'EB viral antigens in Burkitt tumor biopsies and early cultures', *Int. J. Cancer*, **6**, 10.

Nishioka, K., R. F. Irie, T. Kawana and S. Takeuchi (1969) 'Immunological studies on mouse mammary tumors. III. Surface antigens reacting with tumor specific antibodies in immune adherence', *Int. J. Cancer*, **4**, 139.

Nisonoff, A. and M. M. Rivers (1961) 'Recombination of a mixture of univalent antibody fragments of different specificity', *Arch. Biochem. Biophys.*, **93**, 460.

Old, L. J., E. A. Boyse and E. Stockert (1963) 'Antigenic properties of experimental leukemias. I. Serological studies *in vitro* with spontaneous and radiation-induced leukemias', *J. Nat. Cancer Inst.*, **31**, 977.

Old, L. J., E. A. Boyse and E. Stockert (1964) 'Mouse leukemia', *Nature* (*London*), **201**, 777.

Old, L. J., E. A. Boyse and E. Stockert (1965) 'The G(Gross) leukemia antigens', *Cancer Res.*, **25**, 813.

Pearson, G., G. Klein, G. Henle, W. Henle and P. Clifford (1969) 'Relation between Epstein–Barr viral and cell membrane immunofluorescence in Burkitt tumor cells', *J. Exp. Med.*, **129**, 707.

Pope, J. H. and W. P. Rowe (1964) 'Immunofluorescent studies of adenovirus 12 tumors and of cells transformed or infected by adenoviruses', *J. Exp. Med.*, **120**, 577.

Pressman, D., E. D. Day and M. Blau (1957) 'The use of paired labeling in the determination of tumor-localizing antibodies', *Cancer Res.*, **17**, 845.

Rapp, F., J. S. Butel and J. L. Melnick (1964) 'Virus-induced intranuclear antigen in cells transformed by papovavirus SV40', *Proc. Soc. Exptl. Biol. Med.*, **116**, 1131.

Rapp, F., J. S. Butel, L. A. Feldman, T. Kitahara and J. L. Melnick (1965) 'Differential effects of inhibitors on the steps leading to the formation of SV40 tumor and virus antigens', *J. Exp. Med.*, **121**, 935.

Roth, F. K. and R. M. Dougherty (1969) 'Multiple antigenic components of the group-specification antigen of the avian leukosis-sarcoma viruses', *Virology*, **38**, 278.

Rowe, W. P., S. G. Baum, W. E. Pugh and M. D. Hoggan (1965) 'Studies of adenovirus SV40 hybrid viruses. I. Assay system and further evidence for hybridization', *J. Exp. Med.*, **122**, 943.

Schreck, R. (1965) 'Differences in sensitivity of human blood cells to fresh rabbit serum', *Proc. Soc. Exptl. Biol. Med.*, **120**, 789.

Scolnick, E. M., W. P. Parks and D. M. Livingston (1972) 'Radioimmunoassay of mammalian type C viral proteins. I. Species specific reactions of murine and feline viruses', *J. Immunol.*, **109**, 570.

Sibal, L. R., M. A. Fink and D. D. Robertson (1968) 'Quantitative measurement of a murine leukemia virus (Rauscher) antigen', *Virology*, **35**, 498.

Sibal, L. R., W. F. Feller, M. A. Fink, B. E. Kohler, W. T. Hall and H. E. Bond (1969) 'Mammary tumor virus antigen: sensitive immunoassay', *Science*, **164**, 76.

Singer, S. J. (1959) 'Preparation of an electron-dense antibody conjugate', *Nature* (*London*), **183**, 1523.

Singer, S. J. and A. F. Schick (1961) 'The properties of specific stains for electron microscopy prepared by the conjugation of antibody molecules with ferritin', *J. Biophys. Biochem. Cytol.*, **9**, 519.

Sjögren, H. O. and D. Motet (1969) 'Separation of specific antibodies to TSTAs of adeno 12 and polyoma tumors', *Int. J. Cancer*, **4**, 14.

Smith, R. W. and P. T. Mora (1972) 'Cytotoxic microassays and studies of SV40 tumor-specific transplantation antigen', *Virology*, **50**, 233.

Sparks, F. C., C. C. Ting, W. G. Hammond and R. B. Herberman (1969) 'An isotopic antiglobulin technique for measuring antibodies to cell-surface antigens', *J. Immunol.*, **102**, 842.

SriRam, J., S. S. Tawde, G. B. Pierce, Jr. and A. R. Midgley, Jr. (1963) 'Preparation of antibody–ferritin conjugates for immuno-electron microscopy', *J. Cell Biol.*, **17**, 673.

Stephenson, J. R., R. E. Wilsnack and S. A. Aaronson (1973)' Radioimmunoassay for avian C-type virus group-specific antigen: detection in normal and transformed cells', *J. Virol.*, **11**, 893.

Stück, B., E. A. Boyse, L. J. Old and E. Carswell (1964) 'ML: a new antigen found in the leukemias and mammary tumors of the mouse', *Nature* (*London*), **203**, 1033.

Sundqvist, K. G. and A. Fagraeus (1972) 'A sensitive isotope modification of the mixed haemadsorption test applicable to the study of prozone effect', *Immunol.*, **22**, 371.

Takasugi, M. and E. Klein (1970) 'A microassay for cell-mediated immunity', *Transpl.*, **9**, 219.

Takemoto, K. K. and K. Habel (1965) 'Hamster ascitic fluids containing complement fixation antibodies against virus-induced tumor antigens', *Proc. Soc. Exptl. Biol. Med.*, **120**, 124.

Takemoto, K. K., R. A. Malmgren and K. Habel (1966) 'Heat-labile serum factor required for immunofluorescence of polyoma tumor antigen', *Science*, **153**, 1122.

Tanigaki, N., Y. Yagi and D. Pressman (1967) 'Application of the paired label radioantibody technique to tissue sections and cell smears', *J. Immunol.*, **98**, 274.

Terasaki, P. I., M. L. Esail, J. A. Cannon and W. P. Longmire, Jr. (1961) 'Destruction of lymphocytes in vitro by normal serum from common laboratory animals', *J. Immunol.*, **87**, 383.

Tevethia, S. S., L. A. Couvillion and F. Rapp (1968a) 'Development in hamsters of antibodies against surface antigens present in cells transformed by papovavirus SV40', *J. Immunol.*, **100**, 358.

Tevethia, S. S., G. Th. Diamandopoulos, F. Rapp and J. F. Enders (1968b) 'Lack of relationship between virus-specific surface and transplantation antigens in hamster cells transformed by Simian papovavirus SV40', *J. Immunol.*, **101**, 1192.

Ting, C. C. and R. B. Herberman (1970) 'Detection of humoral antibody response to polyoma tumor-specific cell-surface antigen', *J. Nat. Cancer Inst.*, **44**, 729.

Ting, C. C. and R. B. Herberman (1971) 'Detection of tumor-specific cell surface antigen of Simian virus 40-induced tumors by the isotopic antiglobulin technique', *Int. J. Cancer*, **7**, 499.

Ting, C. C., D. H. Lavrin, K. K. Takemoto, R. C. Ting and R. B. Herberman (1972) 'Expression of various tumor-specific antigens in polyoma virus-induced tumors', *Cancer Res.*, **32**, 1.

Ting, C. C., J. R. Ortaldo and R. B. Herberman (unpublished observation) 'Serological analysis of the antigenic specificities of SV40 transformed cells and their relationship to tumor associated transplantation antigen'. (Submitted for publication.)

Ting, C. C., J. R. Ortaldo and R. B. Herberman (unpublished observations).

Vasconcelos-Costa, J. (1970) 'Detection by immunofluorescence of surface antigens in cells from tumors induced in hamsters by adenovirus type 12', *J. Gen. Virol.*, **8**, 69.

Villano, B. C. D. and V. Defendi (1973) 'Characterization of the SV40 T antigen', *Virology*, **51**, 34.

Wakefield, J. D. and J. R. Batchelor (1966) 'The effect of natural antibody in guinea-pig serum on mouse lymphoma cells in vitro and in vivo', *Immunol.*, **11**, 441.

Weber, J. and D. S. Yohn (1972) 'Detection and assay of avian tumor virus group-specific antigen and antibody by the paired radioiodine-labeled antibody technique', *Virology*, **9**, 244.

Wright, P. W. and L. W. Law (1971) 'Quantitative in vitro measurement of Simian virus 40 tumor-specific antigens', *Proc. Nat. Acad. Sci. U.S.A.*, **68**, 973.

CHAPTER 6

# Dielectric spectroscopy as a tool for studying hydration

Alan Suggett
*Biophysics Division,*
*Unilever Research Colworth/Welwyn,*
*Sharnbrook, Bedford, England*

# I. INTRODUCTION

The concept of water as a sort of passive 'filler' in biological systems is now quite untenable. There are many indications of the controlling influence of the aqueous solvent on the conformations of biopolymers and of the mediation of water in the mechanisms of biological processes (see for example Tait and Franks, 1971). For example, the effects of various small solutes (electrolytes and non-electrolytes) on the stability of native forms of various proteins, nucleic acids (von Hippel and Schleich, 1969) and polysaccharides (Suggett, 1974) follows, with a few minor variations, a common sequence—the lyotropic or Hofmeister series. In view of the very different chemical structures of these macromolecules one must conclude that somehow the only common factor—the water—is implicated in the stability of the native structures. This stabilizing influence, it is argued (von Hippel and Schleich, 1969), is reduced by the modification of the 'structure' of the water by the added small solutes. One way in which water is thought to influence the native conformation of macromolecules is by means of the so-called 'hydrophobic interaction'. By this process, apolar chains in macromolecules are brought together by entropic forces thus minimizing their contact with the solvent. Only with water as solvent has such a phenomenon been detected.

A striking demonstration of the role of water in biological systems comes from the substitution of $D_2O$ for $H_2O$ in growth media—this leads to alterations in almost every biological trait of a cell (Ling, 1962). $D_2O$ is in fact a poison, and only certain algae, bacteria, yeasts, fungi and protozoa have so far been induced to grow in 99·8% $D_2O$ (Katz, 1965). There is therefore considerable interest at present in studying the states of water in biological systems; how, for example, solutes influence (*a*) the short-range, and (*b*) the long-range order in water. Although both of these undoubtedly have considerable relevance to the stability of biological systems, usually it is the short-range order which is referred to as *hydration*.

Of course, while a description of the 'structure' of liquid water itself is still a matter of conjecture (Eisenberg and Kauzmann, 1969; H. S. Frank, 1972) it is perhaps somewhat hazardous to attempt any sort of generalized definition of *hydration water* (or *bound water* as it is often referred to). Nevertheless out of the many possibilities, one convenient definition which is appropriate to the use of relaxation methods is to consider the water of hydration as those water molecules whose molecular motions are significantly restricted compared to those in liquid water at the same temperature and pressure. This must not be taken to imply that this water is in any case immobilized, as in many cases the reorientation time of a water molecule is lengthened merely from about $10^{-11}$ sec (liquid water, ambient temperatures), to $10^{-10}$ or $10^{-9}$ sec (cf the reorientation time in Ice I of about $10^{-5}$ sec).

Estimates of the degree of hydration of biological solutes vary widely depending upon the technique used. Dynamic methods (e.g. viscosity, sedimentation) have often produced results at extreme variance with those from equilibrium methods (e.g. measurement of thermodynamic properties). It also may be important to consider more generally the effect of the different time scales of the experiments; Figure 6.1, for example, shows the time intervals for which a number of experimental techniques have yielded information on ice and water. Generally speaking, the time scale of Figure 6.1 can be divided into techniques whose experimental observations relate to (*a*)

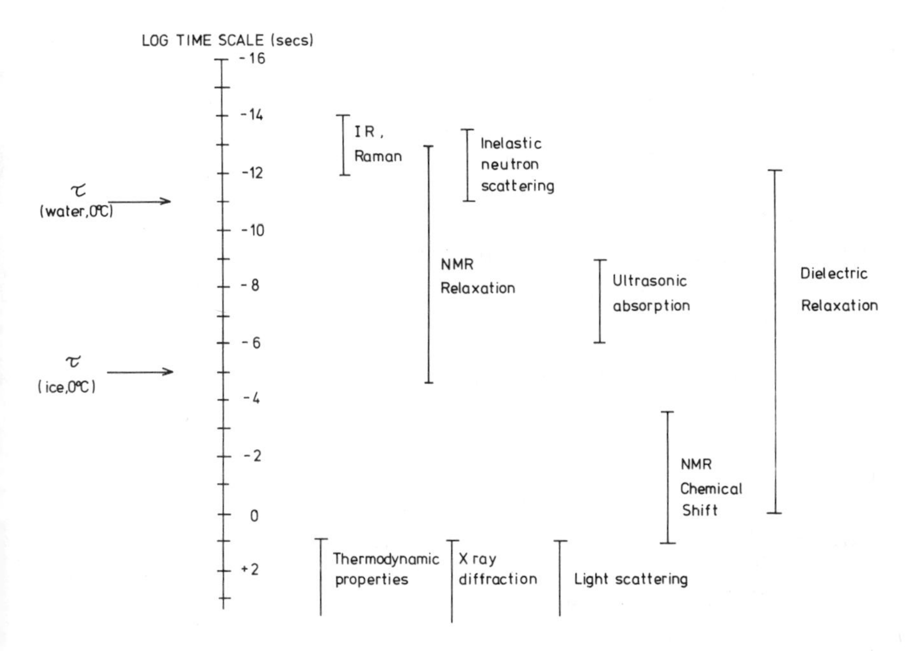


**Figure 6.1.** Time scale of various experimental techniques which have been used to study ice and water. The indicated $\tau$ values are the appropriate dielectric relaxation times (adapted from Eisenberg and Kauzmann, 1969)

intramolecular motions ($10^{-12}$–$10^{-14}$ sec), (*b*) molecular rotational and translational motions ($\sim 10^{-3}$–$10^{-12}$ sec) and (*c*) the 'structure' of the system averaged over all molecular reorientations ($> 10^{-3}$ sec). Therefore, using the Eisenberg and Kauzmann (1969) terminology, a Raman 'snapshot' of our system (exposure time $> 10^{-14}$ sec) may look quite different to one derived from say thermodynamic measurements (exposure time $> 10$ sec).

Despite the volumes of published data on this subject, the hydration structures of only a few biological solutes have been adequately described. Nuclear magnetic resonance methods have been applied very successfully to hydrated biopolymer fibres (Berendsen, 1965; Migchelsen, Berendsen and Rupprecht, 1968; Dehl and Hoeve, 1969; Chapman and McLauchlan, 1969), but hydration properties in aqueous solution are much less amenable to experimental study, and it may therefore be appropriate at this time to consider what new techniques or combination of techniques can be profitably applied to this area.

## II. POTENTIAL SITES FOR HYDRATION

### A. Hydrophobic Hydration

The introduction of non-bipolar (i.e. essentially hydrocarbon) residues into water produces unfavourable entropy changes thought to originate from changes in the arrangement of water molecules in the vicinity of the solute molecules (Frank and Evans, 1945). Studies of such processes clearly indicate that the diffusional and intramolecular modes of water are markedly affected (Franks, 1968). A related phenomenon is the so-called hydrophobic interaction between non-polar residues in macromolecules which is thought to be a consequence of the minimization of these unfavourable entropy changes (Kauzmann, 1959).

It has been suggested that the water molecules may be arranged in a clathrate-type structure around the hydrophobic group (Glew, Mak and Rath, 1968) in a manner similar to that in the crystalline gas hydrates (Feil and Jeffrey, 1961). Hydrophobic hydration has been shown to induce long-range ordering of the solvent (Franks and Smith, 1968); it is in general not orientation dependent, and is thermally very labile. The suggestion of a clathrate-type orientation of water molecules around hydrophobic groups was initially based upon analogy and inference from thermodynamic measurements. Recently, however, this orientation has been demonstrated by the NMR measurements of Hertz and Rädle (1973), and also suggested by the molecular dynamics calculations of Stillinger (1973).

### B. Hydrophilic Hydration (by direct hydrogen bonding)

The introduction of solutes such as peptides, amides or monosaccharides into water is not accompanied by the large changes in certain thermodynamic properties (e.g. heat capacity) which characterize solution of hydrophobically hydrated solutes. This indicates little long-range ordering of the solvent structure—a conclusion which is supported by the concentration independence of the apparent molal volumes of monosaccharide solutions at high dilutions (Franks, Reid and Ravenhill, 1972). Nevertheless specific solute–solvent interactions appear to be involved, and from both thermodynamic inference (Taylor and Rowlinson, 1955) and Raman observation (Walrafen, 1966) are indicated to be of a hydrogen-bonded nature. In view of this, the interactions must be highly orientation dependent.

In this context Warner (1965) pointed out that in many organic molecules the spacing of ether, ester, hydroxy and carbonyl oxygens is about 4·8 Å which is very close to the next-nearest oxygen spacing in a hypothetical ice lattice at 25°C (4·75 Å). This suggests that, provided there exists in liquid water some resemblance to an 'ice-like' structure (and X-ray diffraction methods (Danford and Levy, 1962) indicate a high concentration of water molecules 4·9 Å apart) then molecules like biotin, 1,4-quinone, $\beta$-D-glucose and triglyceride (Figure 6.2) could be cooperatively hydrogen bonded into the water structure without undue modification to the existing hydrogen-bonded system. In Figure 6.2 it can be seen that for sugar molecules, only the equatorial hydroxyls can fit into this particular hydration structure, and it is well known that mutarotation equilibria usually favour the anomer with the equatorial hydroxyl.

There are a great many polar groups in biological solutes which in principle can hydrogen bond to water molecules acting either as donor or acceptor molecules, for example:

$$\text{—OH, —NH}_2\text{, >NH, } \equiv\text{N, —CO}_2\text{H, >C=O, >O}$$

## C. Hydration of Ionic Groups

Ionic groups have the power to bind water in a highly orientation-dependent manner, the specific orientation depending upon the sign of the ionic charge. This has been elegantly demonstrated by Pottel and coworkers (Giese, Kaatze and Pottel, 1970) using dielectric relaxation, and Hertz and Rädle (1973) using various NMR probes. For example the structure of the hydration of strongly hydrated ions such as $Mg^{2+}$ is thought to be 6 water molecules with the oxygens in a distorted octahedron, the orientation of the electric dipoles of the water molecules being along the Mg—O directions (i.e. 'oxygen pointing in'). $F^-$ is however much less hydrated with the water molecules oriented with F—H—O atoms colinear ('proton pointing in'). It is probable that the above orientations of the hydration water molecules are typical of cations and anions respectively, although the numbers of water molecules involved may vary considerably.

In most biological solutes it is likely that the overall hydration structure is a mixture of all three types described above. The water molecules concerned may display a distribution of molecular properties (e.g. rotational mobility), thus further emphasizing the difficulty in studying macromolecular hydration properties. Non-ionic carbohydrates on the other hand, should be rather more amenable to study; the attachment of a hydroxyl to a carbon atom

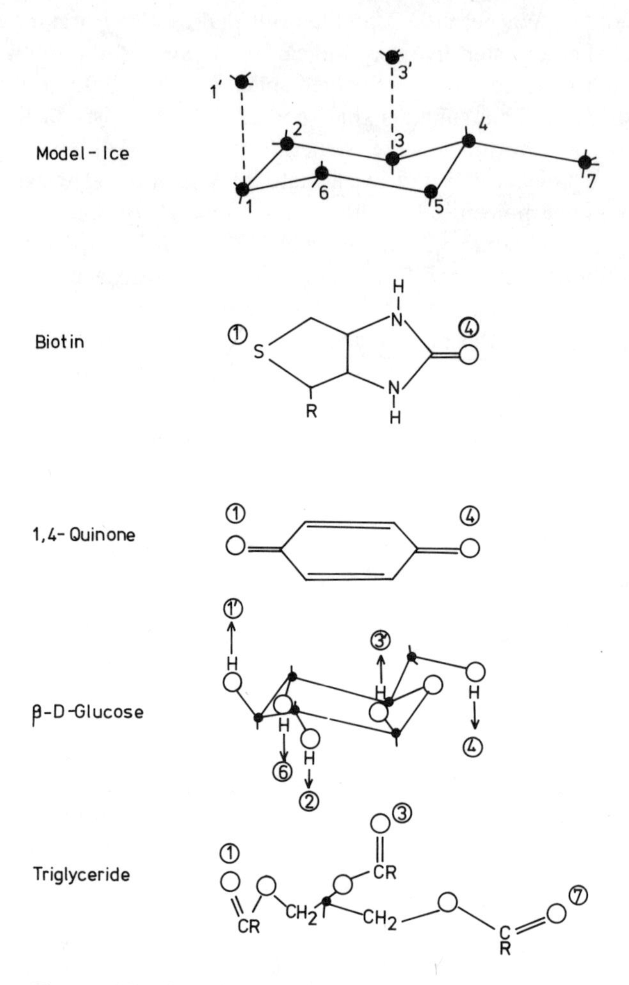


**Figure 6.2.** Correlation between oxygen spacings in organic molecules and in the ice lattice. Numbers in the formulae indicate interaction points with the correspondingly numbered oxygens in the ice lattice (adapted from Franks, 1968)

removes completely its hydrophobic character (Franks, Reid and Suggett, 1973), and the dominant mode of interaction with the water is by hydrogen bonding from the sugar hydroxyls. Despite this, physicochemical studies on carbohydrate/water systems have been relatively infrequent.

## III. ESTIMATION OF HYDRATION

It is not one of the objectives of this chapter to attempt a survey of existing methods for determining hydration. It is nevertheless pertinent to consider very briefly what are the limitations of various methods and whether dielectric methods can in any way complement or replace 'conventional' approaches.

### A. Heterogeneous Systems

There is a large number of documented methods for the estimation of the quantity of water bound by an insoluble substance. A common approach is, for example, to subject the specimen to a chemical dehydration (over $P_2O_5$) or a freezing procedure, and estimate the water which remains in an unchanged form. However the literature reveals that the agreement between the results obtained on the same material by such methods is often very poor, and in any case one is frequently interested in the nature as well as the extent of hydration. A more detailed understanding of the hydration structure requires the use of a spectroscopic probe, and, so far, NMR methods have probably been the most useful. They have been applied, for example, to studies of water on collagen (Berendsen, 1965; Dehl and Hoeve, 1969), on DNA (Migchelsen *et al.*, 1968), and in muscle and brain (Bratton *et al.*, 1965; Hazlewood *et al.*, 1969; Cope, 1969), hair (Clifford and Sheard, 1966), bacterial spores (Maeda *et al.*, 1968) and plant tissues (Fedotov *et al.*, 1969). The application of NMR techniques to frozen protein, polypeptide and nucleic acid solutions by Kuntz, Brassfield and coworkers (Kuntz, Brassfield, Law and Purcell, 1969; Kuntz, 1971; Kuntz and Brassfield, 1971) has demonstrated the existence of a relatively narrow proton magnetic resonance at sub-zero temperatures which has been interpreted as being due to the water bound to the macromolecule. Whether or not this 'unfreezable' water is equatable with the water of hydration in aqueous solution is of course another question!

### B. Homogeneous Aqueous Solution

The difficulties and inconsistencies in estimating hydration in aqueous solution are if anything even greater than for the previously considered case of heterogeneous systems. This is particularly apparent in our lack of knowledge of the *nature* of the hydration of biological solutes. The range of techniques commonly used for such determinations have been discussed in previous reviews (Edsall, 1953; Tanford, 1961; Ling, 1972), and will not be considered further. Many estimates have however been obtained by using 'sledge-hammer' methods, such as viscometry, dilatometry, sedimentation etc. Such techniques measure properties of the system as a whole, and interpolation of the amount of hydration from these bulk properties depends critically upon the validity of the assumptions which must be made. A more acceptable approach is to use a suitable spectroscopic probe to focus on the water in the system—or better still, to focus on just the water of hydration. We therefore select the most direct method with the minimum number of critical assumptions. Nuclear magnetic relaxation methods have again provided useful information about hydration properties in solution. In principle (Figure 6.1) NMR can provide information at the picosecond level (and does so for liquid water), but this approach in solution cannot be exploited as fully as one would wish because the real time scale of NMR measurements is determined by the frequency of the applied radiation, which may be much smaller than the rate of exchange of water molecules between the bulk and hydration states. What is obtained therefore is an average property of the total water in the system; thus a small fraction of water relatively strongly bound may be indistinguishable from a larger fraction less strongly bound, etc. This time-scale problem can be avoided by choosing a spectroscopic technique whose time scale of measurement encompasses the times associated with the motion of water molecules in its liquid ($\sim 10^{-11}$ sec) and solid ($10^{-5}$ sec) states. Figure 6.1 indicates that dielectric relaxation methods offer such a facility.

Hydration studies using dielectric methods have in general used two distinct approaches:

1. 'the absence approach' is which the microwave dielectric properties of the system are measured. Any water which does not contribute to the bulk water relaxation is assumed to be hydration water;

2. direct observation of the relaxation of the hydration water. This, although obviously a more satisfactory method in principle, requires a broad time range and high resolution. The novel dielectric methods known as Time Domain Spectroscopy (TDS), which will be discussed later in some detail, are a very convenient tool for such investigations.

## IV. DIELECTRIC RELAXATION

Relaxation methods in general can be thought of as monitoring the rate at which equilibrium is reattained after the application of some sort of perturbation. Thus, for example, the temperature-jump method, which has been widely exploited in the field of enzyme kinetics, may be classed as a relaxation method. The temperature perturbation in this case causes a shift in a particular equilibrium under investigation, and the rate at which equilibrium is re-established may be followed by some convenient optical property of the system.

For the simple equilibrium

$$A \underset{k_2}{\overset{k_1}{\rightleftharpoons}} B$$

the concentration change in A or B after a temperature jump is an exponential function of time (proportional to exp $(t/\tau)$), where $\tau$, the time constant of the exponential is called the relaxation time. For the simple equilibrium above,

$$\frac{1}{\tau} = k_1 + k_2$$

So the relaxation time can be considered very approximately as the reciprocal of a rate constant.

For relaxation methods in general the perturbation can take many forms—for example pressure, magnetic fields, electric fields—and may be applied either as a step function or as a periodic function. With the step-function approach (transient methods) the 'whole relaxation' may be observed in one experiment, while the periodic-function approach (steady-state methods) can be more sensitive but requires many experiments at different applied frequencies in order to obtain the relaxation profile. Dielectric relaxation employs an electric field of one sort or another as the perturbing force, and in the following section some of the basic principles are outlined. It should be noted, however, that unlike the T-jump analogy, dielectric relaxation is not normally used to follow the rate of a process which converts one species to another, but rather the *rate of reorientation* of individual species. Thus very simply, dielectric relaxation measures the rotational mobility of molecules—what this can tell us about the nature and environment of such molecules will be discussed later.

## A. Basic Principles

For a system of polar molecules in a liquid or gas, the molecular motions in the absence of any applied field will be purely random, and although the individual molecules have permanent dipole moments, the net average dipole moment of the system is zero. After the application of an electric field, however, there is a tendency for the molecules to orient themselves in the direction of the field, and the average orientation is a balance between this field-induced effect and the random thermal motions. The 'polarization', $P$, is a measure of the magnitude of this average orientation and is expressed as the dipole moment per unit volume. The 'electric displacement', $D$, correlates the field intensity, $E$, with the polarization, $P$:

$$D = \quad \varepsilon E + P \tag{6.1}$$

$P$ and $E$ are also related by

$$\epsilon_0 = 1 + \frac{P}{\varepsilon E} \tag{6.2}$$

where $\epsilon_0$ = relative permittivity (often referred to as the dielectric constant), and is the permittivity of free space.

Polarization in our system can arise by three mechanisms—electronic, atomic, and orientational polarization. The first two do not concern us directly in dielectric relaxation phenomena, as they respond in less than $10^{-14}$ sec to the application of an electric field. Now if a field is suddenly applied to the dielectric sample, the field-induced orientation of the molecule is not an instantaneous process—it takes a time characteristic of the rotational freedom of the particular molecular species (this is in turn determined by such factors as the size and shape of the molecule, and its interactions with the rest of the system). Thus the polarization (and hence the permittivity) will take a finite time to reach its maximum value (Figure 6.3). For the simplest of polar

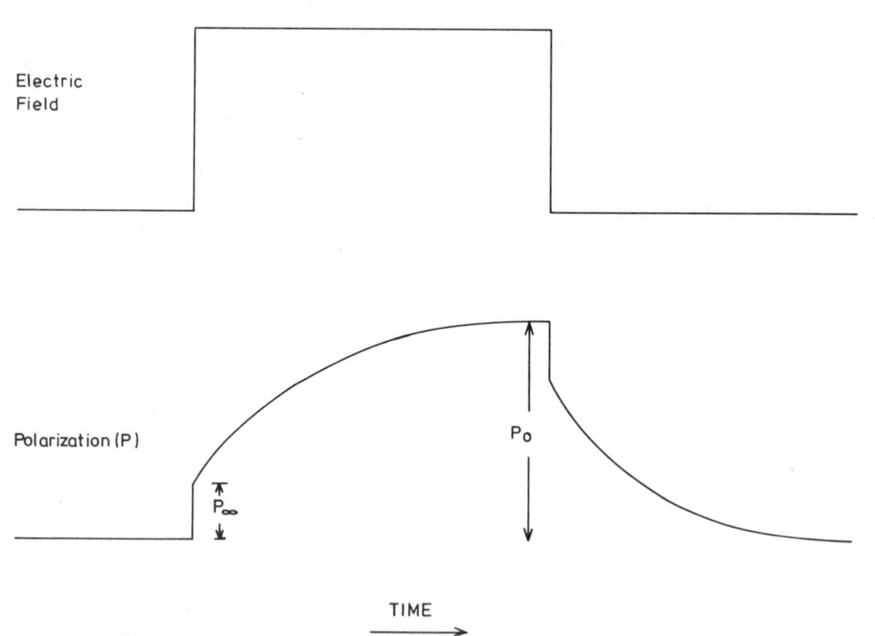


**Figure 6.3.** The polarization response to a pulse of electric field intensity

systems (a Debye-type dielectric) the decay of the polarization on switching off the field, is characterized by a single time constant, $\tau$ (the relaxation time). For times longer than about $10^{-12}$ sec, the expression

$$\epsilon(t) = \epsilon_0 - (\epsilon_0 - \epsilon_\infty) \exp(-t/\tau) \tag{6.3}$$

holds approximately. Thus at long times ($t \gg \tau$), $\epsilon(t)$ is equal to the static permittivity $\epsilon_0$, and at very short times ($t \ll \tau$), $\epsilon(t)$ is equal to the permittivity due only to electronic and atomic polarization, $\epsilon_\infty$. Debye (1929) interpreted this relaxation time in terms of a rigid (spherical) polar molecule rotating in a viscous medium, and using Stokes' law he obtained

$$\tau = \frac{3V\eta}{kT} \tag{6.4}$$

where $V$ is the molecular volume, and $\eta$ is the 'effective viscosity' (which may be significantly different from the macroscopic viscosity).

For more complex dielectric systems the transient permittivity, $\epsilon(t)$, may not be determined by a unique relaxation time. For example if a number of different dipolar species are present in the system then the reorientation of each of these species may be governed by a separate relaxation time. The observed transient permittivity may then be of the form.

$$\epsilon(t) = \epsilon_0 - (\epsilon_0' - \epsilon_\infty') \exp(-t/\tau') - (\epsilon_0'' - \epsilon_\infty'') \exp(-t/\tau'') \tag{6.5}$$

The individual $\tau$ values may be resolvable provided they are sufficiently different in magnitude (by at least a factor of 5). Other common variations from Equation (6.3) can occur when a single dipolar species exhibits some distribution of relaxation times (Cole and Cole, 1941; Davidson and Cole, 1951).

At this point it is appropriate to consider an alternative, although equivalent, approach to the phenomenon of dielectric relaxation—that of the *frequency-dependent*, rather than the *time-dependent*, permittivity. This is a very necessary consideration as classically the frequency domain methods† and the frequency domain presentation of the results are by far the more common. (Although relaxation methods can in principle employ either a step (Figure 6.3) or a periodic perturbation, the step-function approach, largely for technical reasons, has not really been successful until the last few years.)

For the periodic variables it is convenient to use complex number notation. For example the applied electric field can then be expressed as

$$E^* = E_0 \exp(j\omega t) \tag{6.6}$$

† Experimental methods which select preferentially the frequency (periodic-function) or time (step-function) variable are usually referred to as frequency domain or time domain methods respectively.

where $\omega$ = angular frequency, $j = \sqrt{-1}$, and the symbol * indicates a complex quantity. The (complex) electric displacement is similarly given by

$$D^* = D_0 \exp(j\omega t - \delta) \tag{6.7}$$

where $\delta$ = the phase difference between $D$ and $E$.

In an alternating field, the permittivity also becomes 'complex' and the complex relative permittivity is conveniently represented in the absence of appreciable conductivity by

$$\epsilon^* = \epsilon' - j\epsilon'' \tag{6.8}$$

where $\epsilon'$ is called the permittivity and $\epsilon''$ the loss factor. (In the past, $\epsilon'$ and $\epsilon''$ have been referred to as the real and imaginary parts of the dielectric constant, but as $\epsilon''$, for example, is neither imaginary nor constant it is preferable to use the nomenclature given above.) The frequency domain equivalents of Equations (6.3) and (6.5) are

$$\epsilon^* = \epsilon_\infty + \frac{\epsilon_0 - \epsilon_\infty}{1 + j\omega\tau} \tag{6.9}$$

and

$$\epsilon^* = \epsilon_{\infty 1} + \frac{(\epsilon_{01} + \epsilon_{\infty 1})}{1 + j\omega\tau_1} + \epsilon_{\infty 2} + \frac{(\epsilon_{02} - \epsilon_{\infty 2})}{1 + j\omega\tau_2} + \ldots \tag{6.10}$$

From Equations (6.8) and (6.9) we obtain

$$\epsilon' = \epsilon_\infty + \frac{\epsilon_0 - \epsilon_\infty}{1 + \omega^2\tau^2} \tag{6.11}$$

and

$$\epsilon'' = \frac{(\epsilon_0 - \epsilon_\infty)}{1 + \omega^2\tau^2} \tag{6.12}$$

Graphical representation of the frequency dependence of $\epsilon^*$ is often given in 2 ways:

(*a*) Plots of $\epsilon'$, $\epsilon''$ against log (frequency) as in Figure 6.4a. For a Debye dielectric (i.e. a dielectric following Equation 6.8) the $\epsilon'$ and $\epsilon''$ curves are symmetrical in log (frequency) about a critical angular frequency, $\omega_c$ which is equal to the reciprocal of the relaxation time,

(*b*) Plots of $\epsilon''$ against $\epsilon'$ (i.e. a complex plane diagram). For Debye dielectrics such plots are semicircular as in Figure 6.4b. For a system having two discrete relaxation times, the corresponding plots are shown in Figure 6.5.

The relaxation time can be treated as an inverse rate constant for the

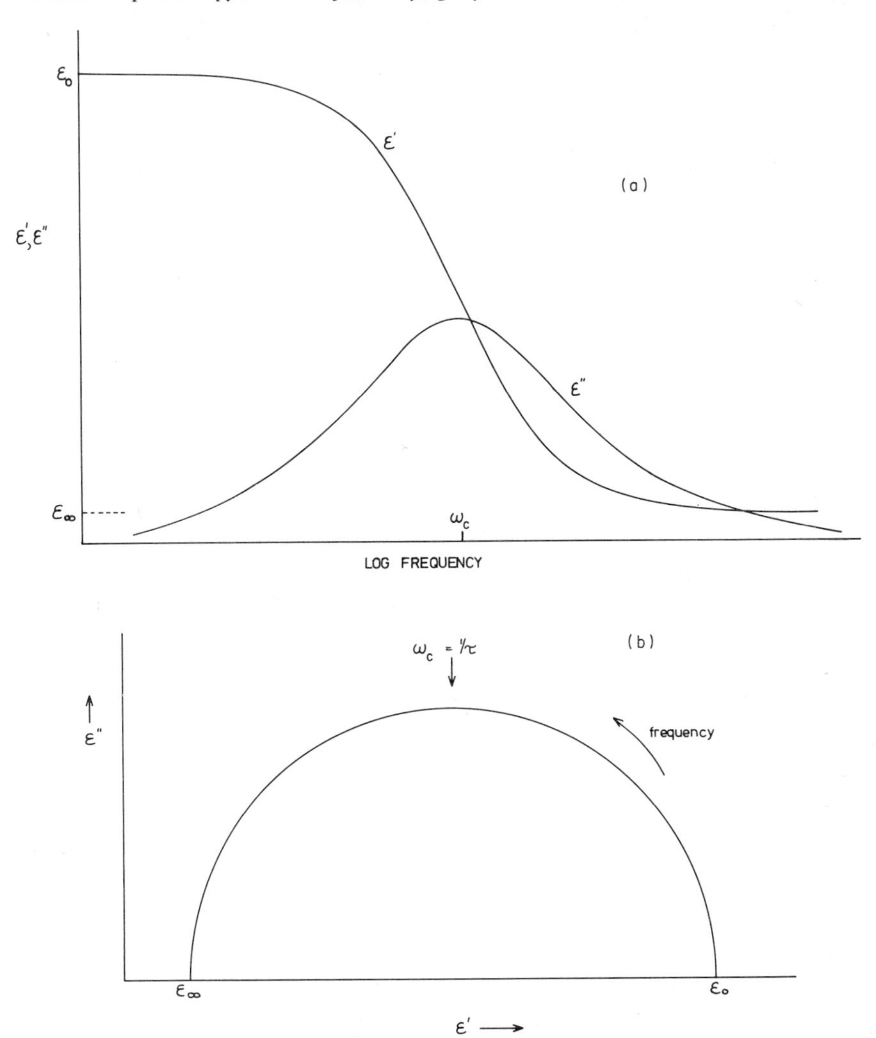


**Figure 6.4.** Graphical representation of a Debye-type dielectric relaxation process. (a) $\epsilon'$ and $\epsilon''$ as a function of frequency, (b) complex plane diagram

particular reorientation, and application of Eyring's theory of rate processes (Eyring, 1941; Kauzmann, 1942) gives

$$\tau = \frac{h}{kT} \exp \frac{(\Delta G^{\ddagger})}{(RT)} \tag{6.13}$$

where $h$ = Planck's constant and $\Delta G^{\ddagger}$ is the molar free energy of activation. Thus from the temperature dependence of $\tau$, the enthalpic and entropic barriers to the reorientation can be determined.

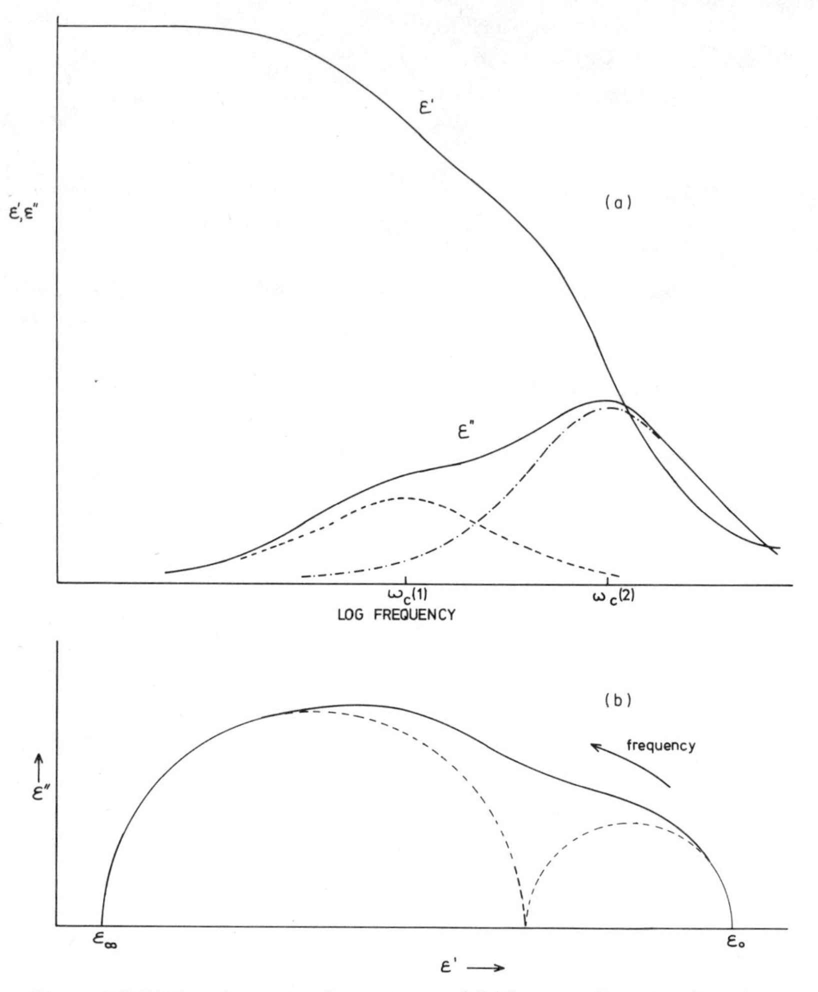


**Figure 6.5.** Dielectric spectra for systems exhibiting two discrete relaxation times. (a) Frequency spectrum; ——— observed $\epsilon'$, $\epsilon''$; – – – –, – · – · –, contributions to $\epsilon''$ from the individual processes. (b) Complex plane diagram; ——— observed $\epsilon'$, $\epsilon''$ data; – – – –, contributions from the individual processes. In this example the individual relaxation times are separated by a factor of 13, and the relaxation amplitudes differ by a factor of 2

The frequency dependence of $\epsilon'$ in a relaxation region is known as *Dielectric Dispersion*, and that of $\epsilon''$ as *Dielectric Absorption* ($\epsilon''$ is an energy absorption mechanism analogous to the absorption of infrared radiation, but at longer wavelengths). Either or both of these collectively can be known as *Dielectric Relaxation*, and the application of broad-band techniques and resolution of individual processes (as in Figure 6.5), *Dielectric Spectroscopy*.

The above treatment of the basic ideas of dielectric relaxation is necessarily only a very superficial one, and for more detailed information readers are referred to standard works (Böttcher, 1952; Hill, Vaughan, Price, and Davies, 1969). However, for the reader who perhaps is not too concerned with the details, it may suffice to reiterate that dielectric relaxation processes are usually associated with molecular rotation and that simple relaxation processes can be defined in terms of two main parameters:

1. *The relaxation time* ($\tau$) which is a measure of the rotational mobility of the relaxing molecular species (very loosely can be considered as the time taken for the molecule to rotate through an angle of about 70°). This parameter can tell us something about the size and shape of the molecule, and its interaction with the rest of the system.
2. *The relaxation amplitude* ($\epsilon_0 - \epsilon_\infty$, or $\Delta\epsilon$) which is determined primarily by the concentration of the relaxing species and its dipole moment.

## B. Application to Aqueous Systems

In aqueous solutions there will generally be more than a single dipolar species present and therefore, from the earlier discussion, it is likely that more than a single relaxation process will be observable. Individual relaxation times can be resolved and assigned to the solute, the bulk solvent and, in some cases, to the water of hydration. The relaxation time of the solute molecule depends upon its molecular dimensions and its interactions with the rest of the system—for an indication of the order of magnitude of individual relaxation times see Table 6.1. A markedly asymmetric solute may

**Table 6.1.** Relaxation times in aqueous solution

| Relaxing species | Conditions | $\tau$ (sec) |
|---|---|---|
| Water | 25°C | $8 \cdot 3 \times 10^{-12}$ (a) |
| Urea | 8 *M*, 25°C | $2 \cdot 6 \times 10^{-11}$ (b) |
| Glucose | 2 *M*, 5°C | $6 \cdot 9 \times 10^{-11}$ (c) |
| Glycylglycine | 0·25 *M*, 25°C | $1 \cdot 3 \times 10^{-10}$ (d) |
| Metmyoglobin | 1–3%, 15°C | $3 \cdot 2 \times 10^{-8}$ (e) |
| Poly-L-glutamate (M. Wt. ≃ 90,000) | 15°C | $1 \times 10^{-5}$ (f) |

(a) Grant, Buchanan and Cook (1957).
(b) D. Adolph and R. Pottel, unpublished work.
(c) Tait *et al.* (1972).
(d) Aaron, Grant and Young (1966).
(e) Lumry and Yue (1965).
(f) Takashima (1963).

exhibit relaxation times associated with rotation around different axes, and a very flexible solute may exhibit a pronounced distribution of relaxation times.

The major concern of this chapter is however with the states of the *solvent* in these systems. From the parameters associated with the bulk (or 'free') water relaxation viz. $\tau$, $(\epsilon_0 - \epsilon_\infty)$, $\Delta H^\ddagger$ and $\Delta S^\ddagger$, an indication of the long-range order in the solvent and the amount of water in each state, can be obtained. The equivalent parameters for the relaxation of the hydration water indicate the extent of the short-range ordering of the water—the amount 'bound', the degree of restriction of the motion of the water molecules relative to that in the pure liquid and the energetics of the interactions with the rest of the system. For water on solid materials, resolution of the relaxation data may give information about the amount and nature of the states of absorbed water.

One important restriction on the use of dielectric relaxation methods in aqueous systems is that they are not in general applicable to systems of high conductivity. The complex permittivity (Equation 6.8) in the presence of conductivity is described by

$$\epsilon^* = \epsilon' - j\epsilon'' - \frac{j\sigma}{\omega\varepsilon} \qquad (6.14)$$

where $\sigma$ is the specific electric conductance, $\omega$ the angular frequency and $\epsilon$ the permittivity of free space. The final term in Equation (6.14) has the effect of decreasing the sensitivity in measuring $\epsilon'$ and $\epsilon''$ as the conductance increases and, for a constant $\sigma$, as the frequency decreases.

## V. TECHNIQUES OF DIELECTRIC RELAXATION

### A. Frequency Domain Methods

In frequency domain methods a sinusoidal field of given frequency is applied to the suitably enclosed dielectric sample and the permittivity ($\epsilon'$) and loss factor ($\epsilon''$) determined at that frequency. By changing the frequency in a point-by-point manner, one can eventually construct dielectric spectra of the sort indicated in Figures 6.4 and 6.5. A range of different frequency domain methods will normally be required, as individual instruments will rarely cover more than about one decade of frequency. These methods also tend to be very time consuming and require considerable experimental expertise and experience. Some indication of the available techniques is given in Figure 6.6 and for experimental detail the reader is referred to standard works (e.g. Vaughan, 1969). At frequencies above about $10^8$ Hz it is particularly difficult to obtain the complex permittivity at enough frequency points to be able to

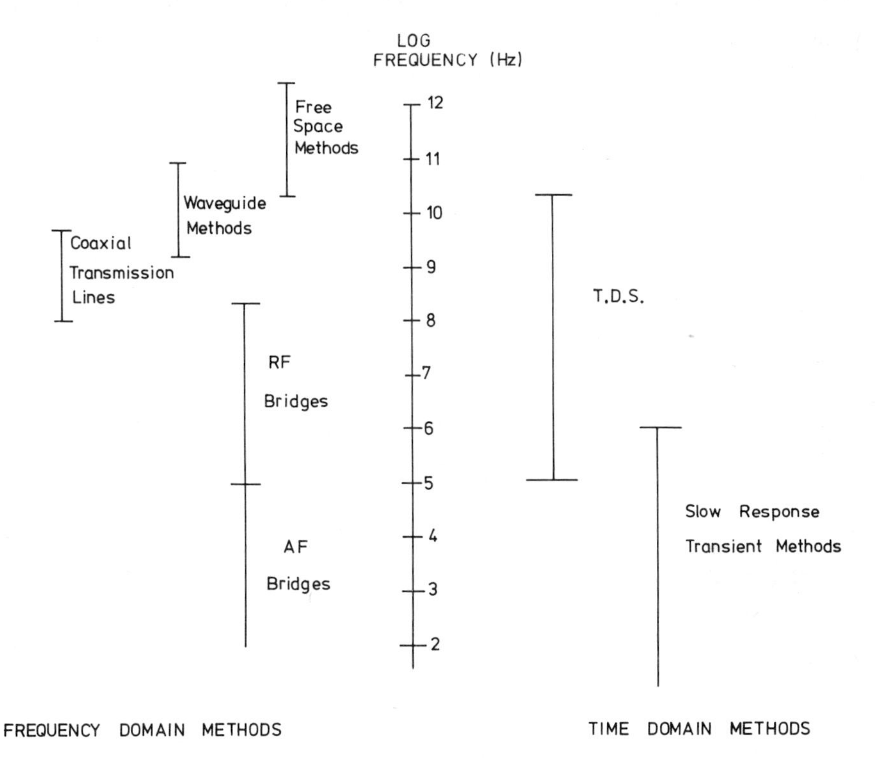


**Figure 6.6.** Frequency spectrum of dielectric techniques

attempt any sort of separation of relaxation processes. Time domain methods on the other hand are not so limited, and they are relatively rapid and inexpensive, and a single instrument can cover the equivalent of five decades in frequency.

## VI. TIME DOMAIN SPECTROSCOPY (TDS)

Before discussing the theoretical and experimental nature of TDS in some detail, it may aid some readers to have a more simplified account of what one is measuring in a TDS experiment and how it is carried out. In a similar way to that displayed in Figure 6.3, an electric field is suddenly applied to a dielectric sample in the form of a rectangular step pulse. This pulse is reflected from the sample surface in a way that is determined by the dielectric properties of the sample; for example, the size of the pulse will be decreased and its shape distorted and a decaying tail on the reflected pulse may often be visible (like the decay of the polarization in Figure 6.3). In a single TDS experiment, from the recorded profiles of the incident and reflected pulses, one aims to calculate all the required dielectric relaxation parameters and to interpret these in terms of the molecular nature of the sample.

The major breakthrough in time domain dielectric methods came in 1969 when Fellner-Feldegg (1969) had the foresight to recognize in a qualitative technique known as Time Domain Reflectometry (TDR) certain possibilities for quantitative measurement of dielectric properties. His time range was about 50 psec to 100 nsec (i.e. approximately 10 MHz to 5 GHz in frequency terms). Although the accuracy of the initial measurement (Fellner-Feldegg, 1969) was not very encouraging, it has been greatly improved as a result of the development and extension of this original idea by other workers (e.g. Suggett *et al.*, 1970; Nicolson and Ross, 1970; Loeb *et al.*, 1971; van Gemert, 1971, 1973; Suggett, 1972, 1973). There is now a number of related time domain methods which we call collectively Time Domain Spectroscopy. These all involve propagating a fast-rise voltage pulse in a coaxial line system and monitoring the changes in the characteristics of the pulse after reflexion from or transmission through a section of coaxial line filled with sample. In view of its very recent development, the basic principles underlying the methods will be briefly outlined.

## A. Basic Principles

Figure 6.7 shows a diagrammatic representation of a coaxial line partially filled with a dielectric sample. In this line the propagation of an electromagnetic signal will be governed by two parameters, the characteristic impedance and the propagation constant. If the air-filled line has an impedance $Z_0$ then the impedance of the dielectric-filled region is given by

$$Z = \frac{Z_0}{\sqrt{\epsilon^*}} \tag{6.15}$$

At the air–dielectric interface the sudden change in the impedance causes a partial reflexion and partial transmission of the incident signal. The voltage reflexion and transmission coefficients are given respectively by

$$\rho_{12} = \frac{Z - Z_0}{Z + Z_0} \tag{6.16}$$

$$\tau_{12} = \frac{2Z}{Z + Z_0} \tag{6.17}$$

The propagation constant $\gamma$ governs the propagation of the signal in the medium of complex permittivity $\epsilon^*$;

$$\gamma = \frac{j\omega}{c} \sqrt{\epsilon^*} \tag{6.18}$$

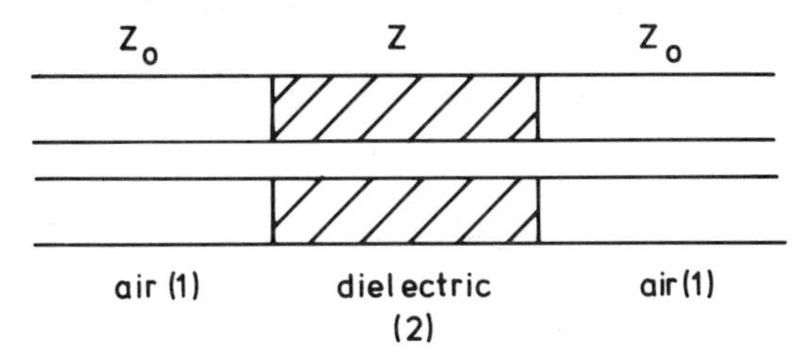


**Figure 6.7.** A section of coaxial transmission line partially filled with a dielectric sample

Thus the transmission coefficient between the faces of the dielectric sample is given by exp $(-\gamma d)$ where $d$ = thickness of sample. Including the interfacial transmission coefficients, the observed coefficient of transmission is thus given by

$$T = \tau_{12}\,\tau_{21}\exp(-\gamma d) \tag{6.19}$$

From the above equations we can now obtain the relationships between the complex permittivity of the sample and observed reflexion and transmission properties. For the reflexion case one obtains the simple relation

$$\epsilon^*(\omega) = \left[\frac{1-\rho(\omega)}{1+\rho(\omega)}\right]^2 \tag{6.20}$$

and for transmission the following equation holds:

$$T = \frac{4\sqrt{\epsilon^*}}{(1+\sqrt{\epsilon^*})^2}\exp\left(\frac{-j\omega d\sqrt{\epsilon^*}}{c}\right) \tag{6.21}$$

A number of different TDS methods are currently in use, and the basic differences between three of these methods are illustrated in Figure 6.8. In methods A and B we monitor the result of a direct reflexion from a single interface or a direct transmission through a single length of sample, whereas in method C the total scattered intensity is monitored (the result of a number of multiple reflexions and transmissions at the interfaces). Let us now briefly examine each of these methods.

## B. Direct Reflexion TDS

Figure 6.9 indicates the major instrumental components in a TDS system which are in this case arranged for direct reflexion measurements. The tunnel diode pulse generator, matched 'tee', remote sampler and sample cell are linked by highest quality coaxial transmission line. The tunnel diode provides a train of pulses, the 'tee' (or power divider) splits an incident pulse into two identical pulses of half the original amplitude, and the sampler is a device for

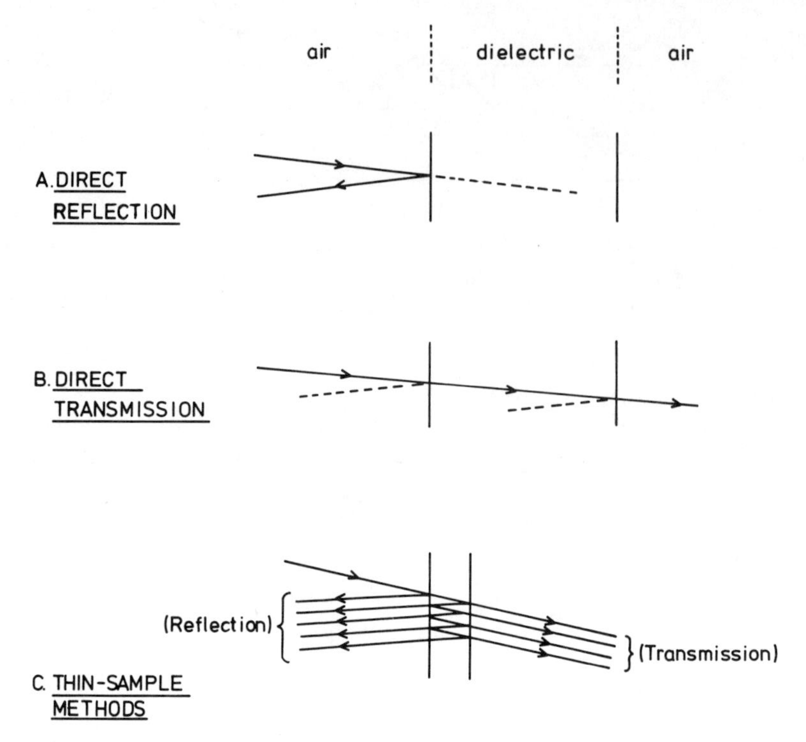


**Figure 6.8.** Diagrammatic representation of three different TDS methods

monitoring the pulse as it passes through. The tunnel diode pulse is orders of magnitude too fast to be followed by conventional oscilloscopes; the sampling oscilloscope system overcomes this by using a stroboscopic-like method to 'sample' successive pulses at slightly different points, and thus to build up a picture of the pulse on the oscilloscope screen in the form of a series of dots. The specimen to be examined is enclosed inside a length of coaxial line of the same dimensions as for the rest of the experimental system. The 50Ω termination on the coaxial line has the property of absorbing all signals impinging on it from other parts of the system without any significant reflexion. All the equipment discussed above is available commercially. Both Hewlett Packard and Tektronix manufacture sampling oscilloscope systems and pulse generators. So far the Hewlett Packard time domain reflectometer (TDR) test sets have been the most popular. Precision 7 mm (APC-7) coaxial line is available from Amphenol Ltd., and the power divider from General Radio Co. Ltd. As all TDS measurements involve the comparison of at least 2 waveforms, it is essential to be able to refer each of these to a common point in time. This may be achieved by some sort of time marker system, and in Figure 6.9 one possible way of providing such a time marker is indicated. One half of the

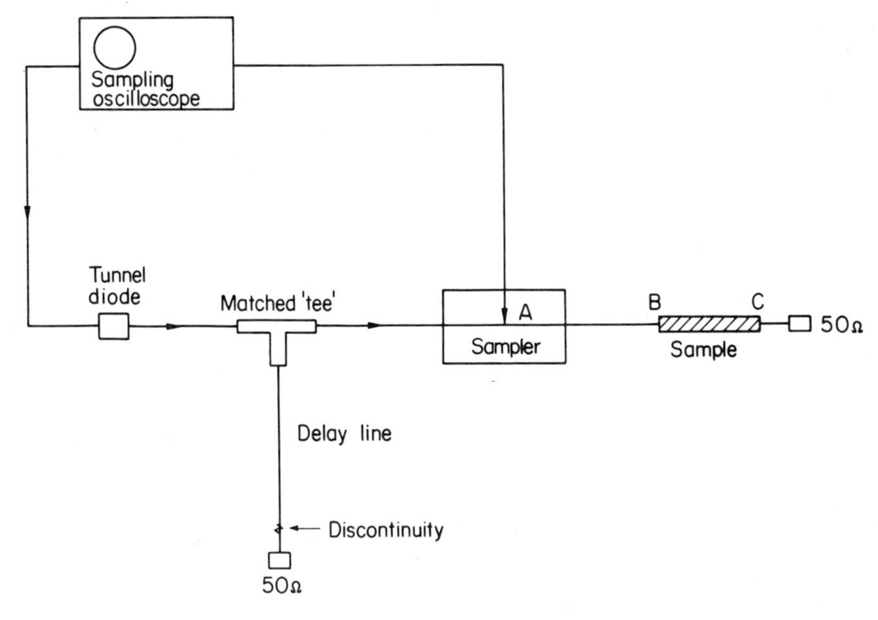


**Figure 6.9.** Experimental arrangement (direct reflexion)

tunnel diode pulse passes down a delay line (a length of high quality 50Ω coaxial cable) and is reflected in the form of a spike from a discontinuity (a piece of metal foil for example) deliberately introduced into a connector. This spike serves as a satisfactory time marker for many purposes. Figure 6.10 shows a somewhat idealized waveform corresponding to the experimental arrangement of Figure 6.9, and includes the echo pulses due to multiple reflexions between the air–dielectric and dielectric–air interfaces (B and C in Figure 6.9). The rise time of the experimental pulse is currently about 30 psec, and the extent of the available 'time window' is determined by the appearance of the first echo-reflected pulse. In actual measurements the incident pulse is not measured as it passes the sampling point; it has been found instead that a pulse reflected from a short circuit placed at a position equivalent to the air–dielectric interface (B in Figure 6.9) is a more faithful representation of the pulse incident *on the dielectric sample*. The short circuit is therefore used as a standard reference pulse for which, $\rho$, the reflexion coefficient, $= -1$.

In the section concerned with the basic principles of dielectric relaxation we have already seen how the polarization or permittivity would respond to the sudden application (or discontinuation) of an electric field. However, although the decay of the reflexion coefficient (Figure 6.10) is related to the decay of the permittivity (Figure 6.3), the conversion from one to the other is

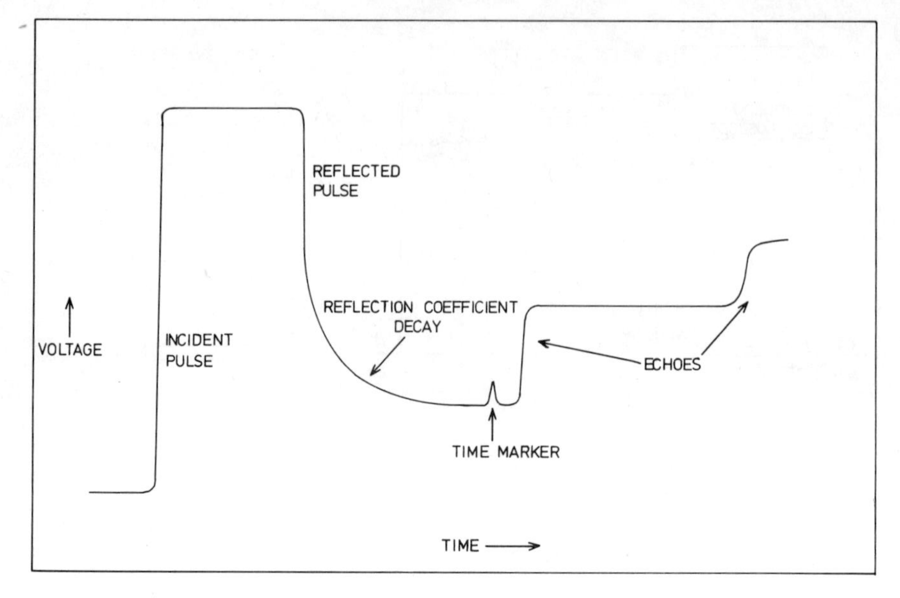


**Figure 6.10.** Idealized waveform from the arrangement of Figure 6.9

not usually so straightforward as Equation (6.20) seems to indicate. For example the substitution of a time-dependent reflexion coefficient, $\rho(t)$, into Equation (6.20) to get a time-dependent permittivity, $\epsilon(t)$, is strictly not mathematically valid as Equation (6.20) is derived for complex, i.e. frequency-dependent, quantities. This substitution is only valid when $\rho(\omega)$ and $\epsilon(\omega)$ are real numbers, i.e. outside any relaxation region. Nevertheless, it appears that when the relaxation time, $\tau$, is much greater than the rise time of the pulse, then for Debye dielectrics the use of Equations (6.3) and (6.20) leads to quite acceptable relaxation time values.

The more strictly correct method of analysing direct reflexion TDS data involves the Fourier transformation of each waveform into the frequency domain. So if $F'(\omega)$, $F''(\omega)$ are the Fourier transforms of the pulses reflected from sample and short circuit respectively, then

$$\rho(\omega) = \frac{F'(\omega)}{F''(\omega)} \tag{6.22}$$

and this $\rho(\omega)$ may be substituted into Equation (6.20) to obtain $\epsilon^*$. A suitable method for a discrete Fourier transform is given by Suggett *et al.* (1970), who applied this Fourier transform TDS approach to the relaxation of a number of aliphatic alcohols.

With direct reflexion TDS the major problem in the production of precise dielectric data for aqueous systems at frequencies above a few GHz ($10^9$ Hz)—a frequency region important for the characterization of the bulk water in most systems—is that the time marker method as described earlier is not really accurate enough. The reasons for this are considered in the Appendix and also by Loeb *et al.* (1971) and by Suggett (1973). How this problem can be resolved is discussed in the next section.

## C. Transmission TDS Methods

The experimental arrangement for transmission measurements is very similar to the reflexion case (Figure 6.9) except that the sample cell must be positioned between the pulse generator and sampler, and a rather different form of time marker must be incorporated. Figure 6.11 shows the corresponding idealized waveform again including echoes and a time marker spike. The experimentally determined parameter is the transmission coefficient which is obtained by referencing the Fourier transform of the pulse after transmission through a length of sample to that after transmission through the same length of air-filled line. Equation (6.21), which relates the observed transmission coefficient to the complex permittivity, cannot however be solved analytically for $\epsilon^*$. This problem is overcome by measuring the trans-

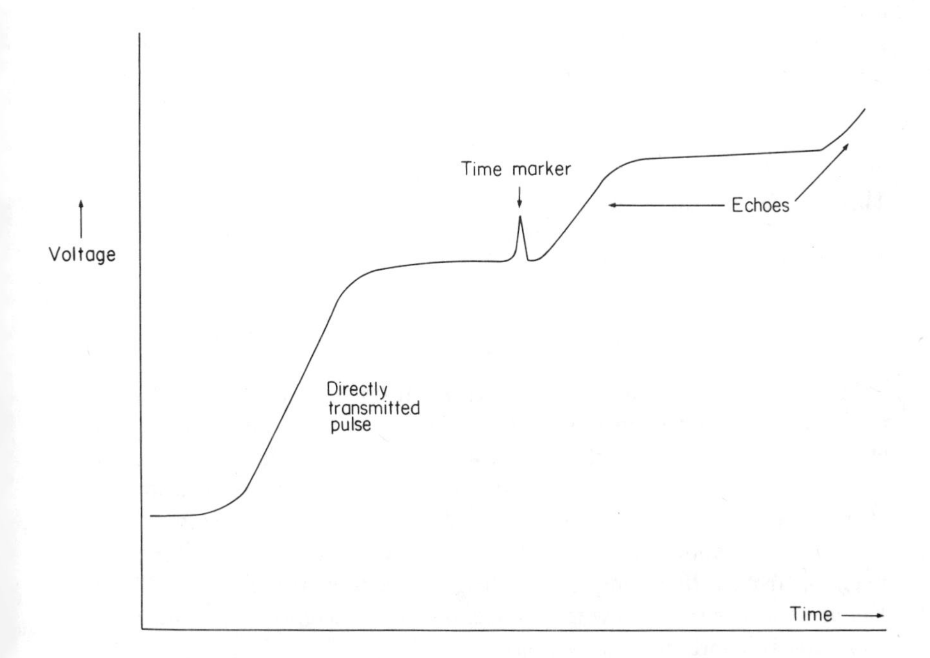


**Figure 6.11.** Idealized waveform from direct transmission TDS

mission properties through one length ($d_1$) of sample relative to those through a second length ($d_2$), in which case

$$\frac{T_1}{T_2} = \exp\left(j\omega\sqrt{\epsilon^*}(d_1 - d_2)/c\right) \tag{6.23}$$

Before showing how we can use Equation (6.23), it is worth recalling that the Fourier transformations of the pulses are complex quantities having real and imaginary parts. They can be expressed as $A + jB$ ($j = \sqrt{-1}$) or more conveniently as $R \exp(j\theta)$ ($R$ = amplitude, $\theta$ = phase; see also Appendix). So if we take the ratio of two transforms $T_1/T_2$ then these will be recorded in the form $R_1/R_2 \exp(j(\theta_1 - \theta_2))$.

Let us call this $S.\exp(j\phi)$. Putting $T_1/T_2 = S\exp(j\phi)$ in Equation (6.23) we obtain

$$\epsilon^* = \frac{c^2}{\omega^2 (d_1 - d_2)^2}\left(\phi^2 - (\ln S)^2 + 2j\phi \ln S\right) \tag{6.24}$$

Thus if we have $S$ and $\phi$ recorded as a function of frequency, we can through Equation (6.24) obtain $\epsilon^* = (\epsilon' - j\epsilon'')$ as a function of frequency. This is what we now call the *Transmission Coefficient Ratios* method.

Although the upper frequency limit for transmission methods is usually lower than for the direct reflexion method, they have one significant advantage in that the determined $\epsilon^*$ (via Equation 6.24) is much less sensitive to timing errors. Consequently by a judicious combination of the experimental data from reflexion and transmission coefficient ratio measurements on the same system, it is possible (Loeb *et al.*, 1971) to obtain results of relatively high precision in aqueous systems and at frequencies up to about 10–15 GHz. This is demonstrated most convincingly by Figure 6.12 which shows the dielectric relaxation of water itself at 278°K as determined by TDS and frequency domain methods.

## D. Multiple Reflexion TDS Methods

In both the direct reflexion and direct transmission cases, the presence of the echo pulses determines the extent of the available 'time window'. However if a cell is made very thin then the echoes can be totally enclosed within the rise time of the pulse (Fellner-Feldegg, 1972) and the relaxation of the sample monitored over a time interval determined only by the length of the applied pulse (about 5 $\mu$sec at present). An idealized form of a pulse after reflexion from a thin sample of a dielectric is shown in Figure 6.13. For a Debye dielectric (i.e. one with a single, discrete relaxation time) the decaying 'tail' closely approximates an exponential, $\propto \exp(-t/\tau)$.

In 'Thin-sample TDS', although, as usual, the most accurate conversion of

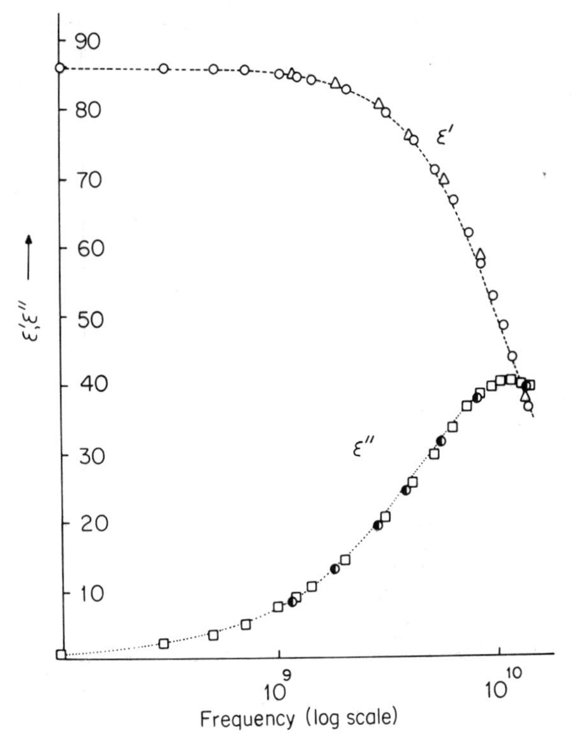


**Figure 6.12.** The complex permittivity of water at 278° K O, □, TDS data; – – – –, · · · ·, frequency domain data interpolated from Hasted (1972); △, ◑, frequency domain data from Pottel (private communication to Hasted (1972)

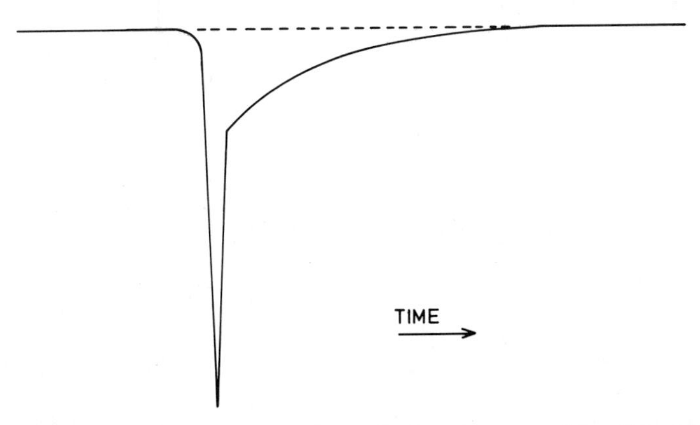


**Figure 6.13.** Waveform for reflexion from a thin dielectric sample

the experimental data is via Fourier transformation, a simple time domain conversion may in many cases be carried out, permitting a very rapid estimation of relaxation times (Fellner-Feldegg, 1972; Suggett, 1973). In the author's laboratory, for example, relaxation parameters for solutes ranging in size from monosaccharides and amino acids to globular proteins have been measured in this way and shown to be in good agreement with those obtained by the full Fourier transformation procedure. Cells for this method can be constructed using windows of PTFE or glass-filled PTFE in the manner described by Fellner-Feldegg (1972), and can vary in thickness from less than 0·1 mm (necessary for the study of very short relaxation times—for example, those of amino acids) to 5 mm (for the study of longer relaxation times—for example, those of globular proteins).

More recently it has been shown (de Loor *et al.*, 1973; Suggett, Clark and Quickenden, (1974) that the relationship between the totally reflected pulse from a dielectric sample, including all the echoes (see Figure 6.8C), can be solved numerically without the need to restrict the cell thickness. These rigorous methods, which differ depending upon whether the sample cell is terminated with a short circuit (de Loor *et al.*, 1973) or a 50Ω termination (Suggett, Clark and Quickenden, 1974), are much more sensitive than the thin-sample method. They have been shown to be capable of producing accurate dielectric information over the frequency range 50 kHz–10 GHz, and one of these (Suggett, Clark and Quickenden) has been applied in particular to aqueous systems including solutions of certain amino acids and globular proteins.

Overall, these multiple reflexion approaches have much to commend them. In many cases a simple time domain analysis of the data is possible thus obviating the need for a computer, smaller amounts of material (0·1–2 ml) are usually required and dielectric information can be generated over an exceedingly wide range of time or frequency ($>$ 5 decades). These methods are thus admirably suited to studies of, for example, aqueous solutions of large molecules, for example globular proteins, covering as they do the relaxation regions of both protein and solvent (in whatever state).

## VII. HYDRATION STUDIES BY DIELECTRIC RELAXATION

It might well be argued that some of the earlier work in this field could now be carried out more satisfactorily using time domain methods. This does not, however, invalidate the results obtained by classical methods and therefore the following section includes results of hydration studies on biochemical systems using a range of dielectric approaches.

## A. Hydration of Proteins

The dielectric relaxation characteristics of protein solutions were extensively studied in the 100 kHz–10 MHz region many years ago by Oncley and coworkers (Oncley, 1938, 1943; Oncley *et al.*, 1938, 1940), and later by Marcy and Wyman (1942), Shaw *et al.* (1944), Takashima (1962, 1964), Goebel and Vogel (1964), Lumry and Yue (1965), Grant and coworkers (Grant, 1966; Grant *et al.*, 1968, 1971) and many others. The measured relaxation behaviour in this frequency range and its relation to the solute molecular size and asymmetry, although a subject of great interest, is outside

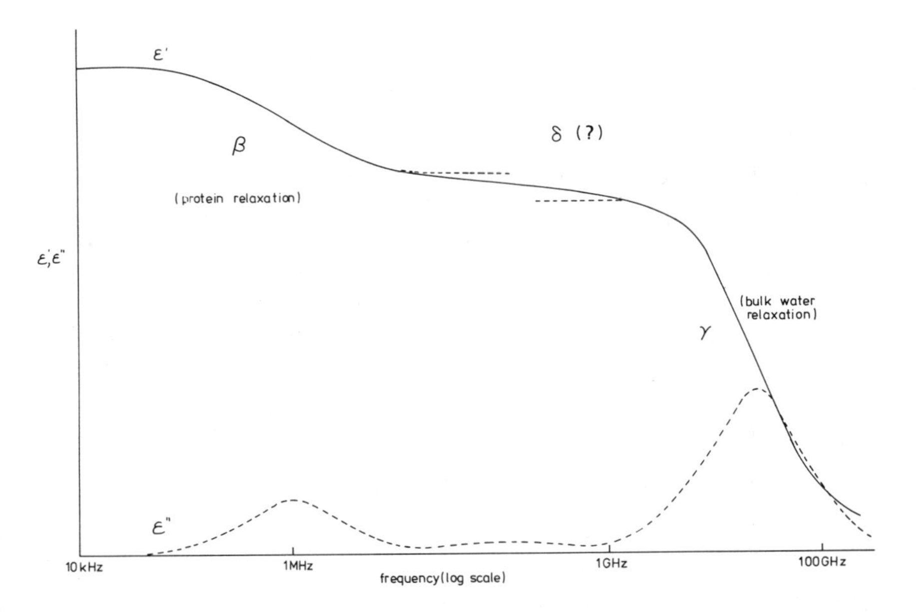


**Figure 6.14.** Typical dielectric relaxation spectrum for a protein solution

the terms of reference of this chapter and readers are referred to the references cited above.

Buchanan, Haggis, Hasted and Robinson (1952) examined the dielectric properties of a number of protein solutions at microwave frequencies between 3 and 24 GHz, i.e. in the frequency region where the bulk water is relaxing (see Figure 6.14). Following the procedure of Haggis *et al.* (1952) used previously for aqueous solutions of small molecules, they estimated the amount of water not contributing to the observed bulk water relaxation and equated this with the amount of water 'irrotationally bound' to the protein. At microwave frequencies the protein with its shell of bound water is considered as a spheroidal cavity of low permittivity in the water, and the 'static' permittivity ($\epsilon_0$) of such a mixture may be written (Fricke, 1924)

$$\epsilon_0 = \epsilon_{\text{water}} - \beta\rho\,(\epsilon_{\text{water}} - \epsilon_{\text{solute}}) \tag{6.25}$$

where $\epsilon_{\text{water}}$ is the static permittivity of water at the same temperature, $\rho$ is here the volume fraction of particles and $\beta$ is a parameter dependent upon the axial ratio of the spheroids, taking values of 1·5 for spheres, 1·6 for prolate spheroids of axial ratio 4:1, 1·67 for very long needles, 1·9 for oblate spheroids of axial ratio 1:4, etc. The difference between the extrapolated static permittivity ($\epsilon_0$) for the bulk water relaxation and that predicted by Equation (6.25) is then used to estimate the amount of water 'bound' to the protein. This procedure, or some modification of it, has been used extensively for the estimation of protein hydration, usually yielding values in the range 0·2–0·45 g water/g protein.

An extra dispersion, the $\delta$ dispersion, was first observed at frequencies between the protein relaxation region ($\beta$) and the bulk water relaxation region ($\gamma$) by Schwan (1957) and later by Grant (1966), and attributed to the relaxation of a layer of bound water (see Figure 6.14). However, Schwan (1965) also pointed out that the polar side chains of the globular protein might conceivably relax at frequencies much higher than those for the relaxation of the whole molecule, and thus make a contribution to the $\delta$ process. The later measurements of Pennock and Schwan (1969) in fact suggested that the relaxation of polar side chains occurred between 10 and 100 MHz and that of the bound water between 100 and 1000 MHz. They quote a relaxation time of about $2 \times 10^{-10}$ sec at 25°C and a $\Delta H^{\ddagger}$ of 7·3 kcal/mole for the bound water on haemoglobin. Finally, it is believed (Grant *et al.*, 1971) that the size of the $\delta$ dispersion is consistent with the relaxation of a hydration layer of up to 0·45 g water/g protein.

It is clear that the microwave estimates of protein hydration depend critically upon the assumption of a suitable dielectric mixture model (e.g. Equation 6.25) and a molecular shape factor. Furthermore the possibility exists (*vide infra*) that hydration water might still contribute to a significant extent to the measured permittivity at frequencies of a few GHz, and therefore estimates of hydration based upon the assumption that the permittivity is determined only by the amount of bulk water may be somewhat in error.* The alternative approach would be to attempt an estimation from the sub-

* This brings the 'missing signal' dielectric approach almost into the range of the sort of criticism levelled at hydration estimates via bulk solution measurements of expansibility, compressibility, viscosity, etc. Each of these methods incorporates its own particular set of assumptions and it is not easy to decide which of these, if any, is the most quantitatively reliable. In defence of the dielectric approach it is true to say that at least the observations are a property of the water *alone*, and the magnitudes of the possible errors are usually calculable.

sidiary $\delta$ process, but this, so far, has been difficult because of its small amplitude compared to those of the $\beta$ and $\gamma$ processes.

### B. Dielectric Properties of Hydration Water

The exact nature of the electrical processes occurring in the relaxation of hydration water is even less certain than of those in liquid water (Hasted, 1972) or in ice (Eisenberg and Kauzmann, 1969). However, the characterization of relaxation processes due to hydration water might be expected to lead to an alternative and perhaps more acceptable way of estimating hydration. The two parameters of most direct relevance to a dielectric interpretation of the nature of any hydration water are the relaxation time (or mean relaxation time) and the activation enthalpy.

Direct observation of the dielectric relaxation properties of hydration water has in general indicated that, in terms of the rotational mobility of such water molecules, the behaviour is more like liquid water than ice. Thus the observed relaxation time usually is less than two orders of magnitude longer than in liquid water, but more than four orders of magnitude shorter than in ice. This is exemplified by studies of the water adsorbed on solid starch (Abadie *et al.*, 1953) from which it appears that the mean relaxation time of the water varies from about $10^{-9}$ sec at 15% water content to about $2 \times 10^{-10}$ sec at 20% water content. The data appear to indicate a number of specific sites for the binding of water whose relaxation time approaches that of liquid water as the water content is increased.

In aqueous solution there has been a number of estimations of the relaxation time of hydration water in addition to those discussed above under 'Protein Hydration'. For example the work of Koide (1969) indicates that the relaxation times for water bound to dioxane, ethylene glycol and polyethylene glycol are between $2 \times 10^{-10}$ sec and that of liquid water at 25°C (viz about $8 \times 10^{-12}$ sec). Shepherd and Grant (1968a, b) analysed their dielectric data for solutions of some amino acids (proline, hydroxyproline, $\epsilon$-amino-n-caproic acid) in terms of three component relaxations. Although their restricted frequency range did not permit the sort of resolution that is now possible with TDS, the analysis revealed the presence of a relaxation due to a 'modified water' component together with relaxations assigned to the reorientations of the solute and bulk solvent. Figure 6.15 shows as an example their data for a 2 *M* $\epsilon$-amino-n-caproic acid solution.

### C. Applications of Time Domain Spectroscopy

There have been three recent applications of TDS methods to studies of the hydration of materials of biological interest. Firstly, a high-frequency relaxation process having a single relaxation time has been observed (Grigera

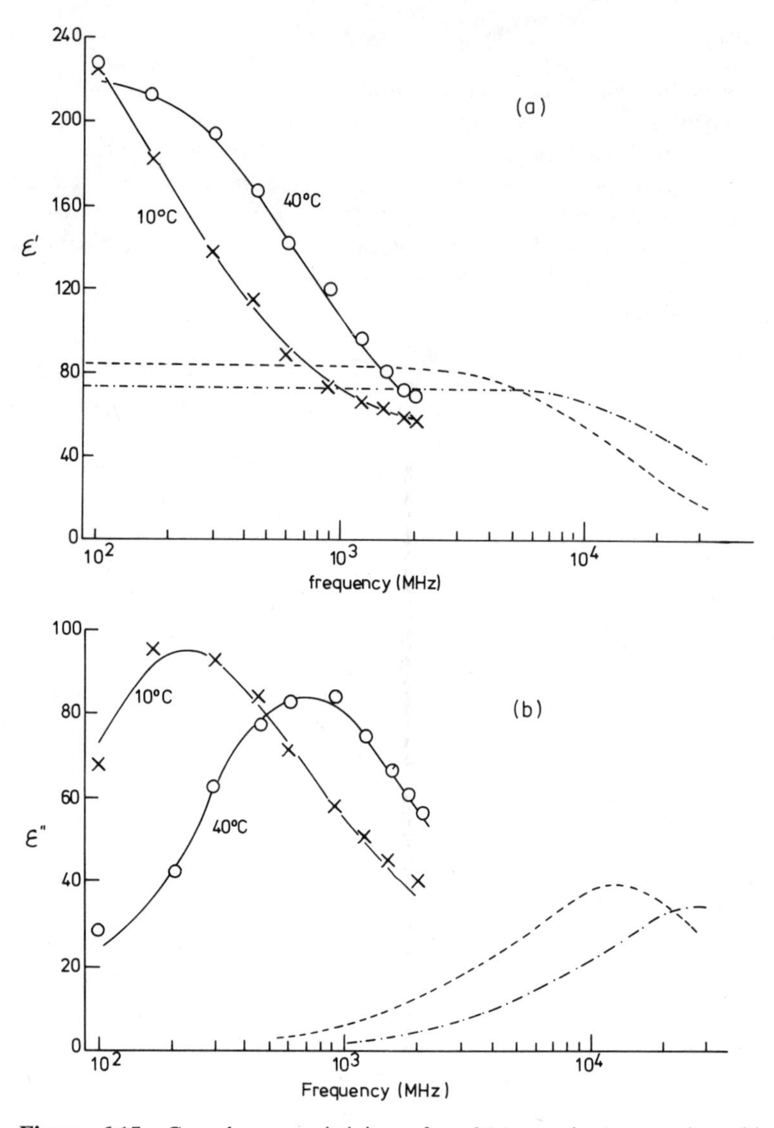


**Figure 6.15.** Complex permittivity of a $2M$ $\epsilon$-amino-n-caproic acid solution. (a) $\epsilon'$ versus frequency; (b) $\epsilon''$ versus frequency; ×, experimental data points at 10°C; ○, experimental data points at 40°C; ———, fitted curves based on three component relaxation model; – – – –, – · – · – · – ·, dielectric properties of liquid water at 10°C, 40°C respectively (adapted from Shepherd and Grant, 1968a and b)

Hallenga, and Berendsen, 1973), by direct reflexion TDS measurements on the collagen/water system. The position of this relaxation process depends somewhat on water content, but for a water content of 40% the observed relaxation time is $\sim 1{\cdot}5 \times 10^{-10}$ sec. Microwave measurements have indicated that there is a possibility of another higher frequency relaxation, and further characterization of the relative amplitudes and time constants of the water relaxations might be expected to consolidate the already relatively extensive knowledge about the hydration of collagen.

Direct reflexion TDS has also been applied (Harvey and Hoekstra, 1972) to a study of the state of surface-bound water on lysozyme (containing 0–0·6 g water/g protein). Two distinct relaxations were observed in this case which have been assigned to the first two layers of adsorbed water. The relaxation of the inner layer is centred around 0·3 GHz (i.e. $\tau \sim 5 \times 10^{-10}$ sec) at 25°C for a sample containing about 33% water, while the second relaxation occurs near 10 GHz ($\tau \sim 1{\cdot}5 \times 10^{-11}$ sec). However, there is some uncertainty about the precision of this TDS data, especially at the higher frequencies, in view of the rather less acceptable method of data analysis which was used in this work.

Hydration studies in aqueous solution are of course complicated by the presence of an excess of bulk water, the relaxation of which dominates the dielectric spectrum. This is particularly evident for small solute molecules whose own relaxation spectrum considerably overlaps that of the water. The problem of resolving unequivocally such a complex spectrum into contributions from the solute and solvent (bulk and modified forms?) is not inconsiderable, but there are grounds (Suggett, 1973) for believing that the nature of TDS makes this process somewhat easier than with classical frequency domain dielectric methods. Recent TDS measurements on solutions of monosaccharides (Tait *et al.*, 1972) have however provided, in combination with $^{17}$O NMR relaxation, hitherto unobtainable information about the nature and extent of hydration of sugars. The $^{17}$O NMR in this work provided the time-averaged properties of the water (over a temperature range 0–80°C), and TDS the dielectric spectrum (time dependence or frequency dependence) which was resolvable in terms of the different types of species present. Figure 6.16 shows the frequency dependence of the complex permittivity for a 2·8 molal ($\sim 2$ *M*) glucose solution at 5°C compared to that of pure water. Attempts were made to fit this data to a number of models. It appeared that the most simple model—a resolution into two Debye-type (i.e. discrete relaxation time) processes, one for the solute and one for the solvent—did not fit the data acceptably. A three-process model was preferred (although the longest time process was small in amplitude compared to the other two). The dependence of the resolved relaxation times upon solute concentration is shown in Figure 6.17 in terms of the resulting solution viscosity. Thus each relaxation time has a smooth positive dependence upon solution viscosity as is expected (see Section IV.A). Some of the relevant parameters for a 2·8 molal glucose solution are given in Table 6.2. Supporting evidence for the three-process model comes from (*a*) temperature-dependence studies which indicate (Table 6.2) that there is a correlation between the determined relaxation times and the corresponding enthalpy of activation, $\Delta H^{\ddagger}$—the process with the longest $\tau$ has the largest $\Delta H^{\ddagger}$, etc.,—and (*b*) corresponding analyses for other monosaccharide and disaccharide solutions.

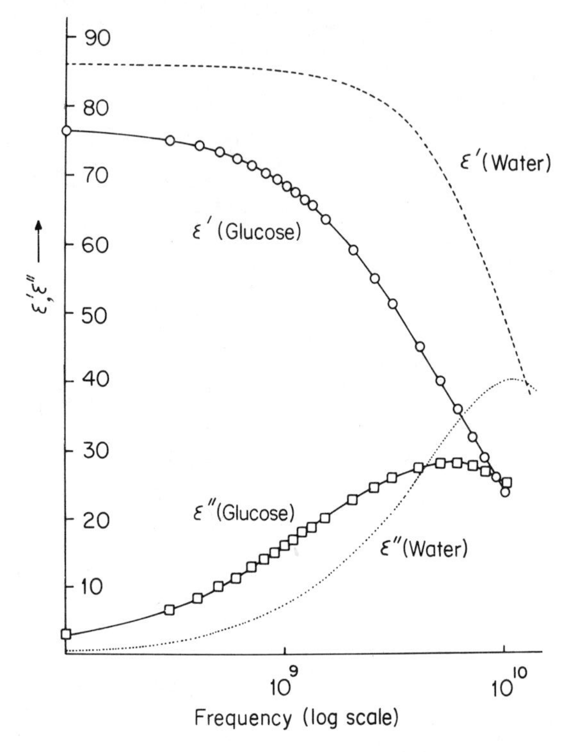


**Figure 6.16.** Dielectric relaxation of a 2·8 molal glucose solution at 278° K. ○, □, TDS data; ----, ····, dielectric relaxation of water at the same temperature

The origins of the separated relaxation processes were deduced from a consideration of, for example:

(*a*) how the relaxation amplitudes varied with solute concentration—with increasing solute concentration the contribution of the bulk water should decrease while that of the solute and any hydration water should increase;

(*b*) how the relaxation times changed when a disaccharide is substituted for a monosaccharide—solute relaxation time should be lengthened;

(*c*) whether the amplitude of the process assigned to the solute is consistent with its estimated dipole moment;

(*d*) whether the total amplitude of the water process(es) is consistent with that calculated from permittivity mixture models, of the type shown in Equation 6.25.

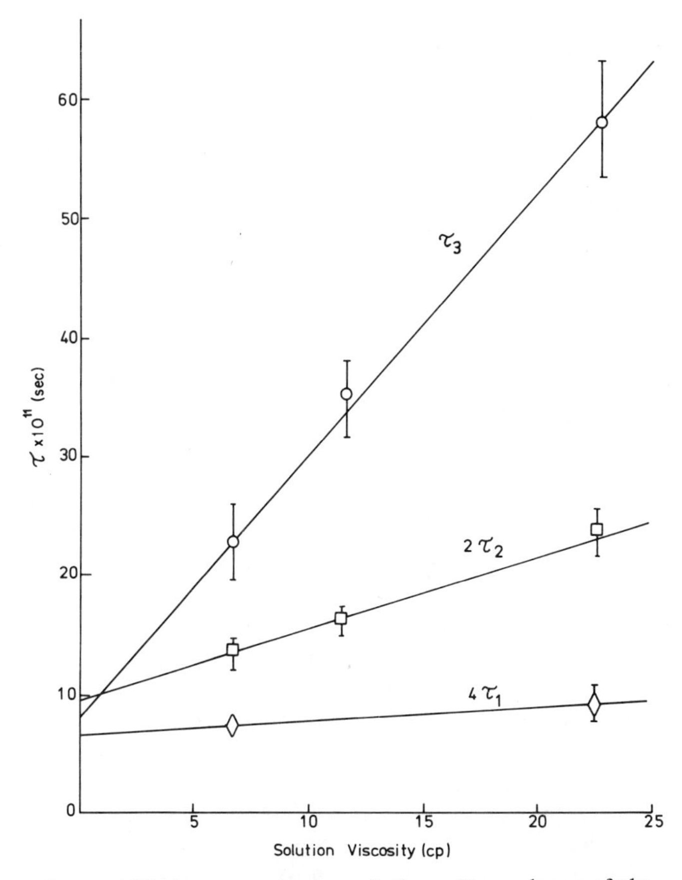


**Figure 6.17.** Aqueous glucose solutions. Dependence of the relaxation times upon solution viscosity. ○, $\tau_3$; □, $2\tau_2$; ◇, $4\tau_2$

From such considerations it was deduced (Franks, Reid and Suggett, 1973) that Process I was due to the reorientation of the bulk solvent molecules, and that Process II was due to the reorientation of the *hydrated* solute. The minor Process III was tentatively assigned to an exchange mechanism involving the hydration water. A consideration of the relaxation amplitudes leads to the conclusions that, on a time average, the hydration of glucose is about 6 mol

**Table 6.2.** Kinetics of the observed dielectric relaxation processes for 2·8 molal aqueous glucose at 5°C (from Tait *et al.*, 1972)

| | $10^{11}\tau$ (sec) | $\Delta H^{\ddagger}$(kcal/mole) | $\Delta S^{\ddagger}$(e.u.) |
|---|---|---|---|
| Process I | 1·85 | (4·5) | (7) |
| Process II | 6·9 | 5·5 ± 1 | 8 ± 3 |
| Process III | 25 | 9 ± 2 | 18 ± 7 |

$H_2O$/mol sugar, whereas that of ribose, for example, is only about 2·5 mol $H_2O$/mol sugar. Such work has clearly demonstrated that the interaction between water and chemically similar 'hydrophilic' solutes can be significantly different, depending critically upon the stereochemistry of the solute molecules. It is suggested further, that the reasons for certain stereochemical arrangements of hydroxyl groups being hydrated to a greater or lesser extent, can be understood in terms of their relative compatibilities with the long-range order in water.

## D. Complementary Nature of Dielectric and Nuclear Magnetic Relaxation for Hydration Studies

So far we have considered the value of dielectric spectroscopy as a tool for studying hydration phenomena, particularly from the point of view that it has the power to quantify the hydration with the minimum of assumptions. This is perhaps as far as one would normally go in a discourse on the application of dielectric methods. However, it may be that using a combination of techniques we can add an extra dimension to our studies. Very briefly, therefore, we shall consider something of the relative advantages and disadvantages of the dielectric and nuclear magnetic relaxation methods for studying hydration phenomena, and whether the two techniques can be used in a complementary manner for maximal effect.

The great advantage of nuclear magnetic relaxation which we would like to exploit is its facility, by suitable choice of nucleus, to examine solute properties in the 'absence' of solvent and vice versa. The motion of the solute may be examined by $^{1}H$ (in $D_2O$ solution), $^{13}C$, $^{14}N$, etc., while solvent behaviour may be studied, for example, by dissolving the solute in $H_2{}^{17}O$. For aqueous solution studies, however, the great drawback normally is that a NMR relaxation time ($T_1$ or $T_2$) is obtained which is an average over all the water molecules in the system. Thus it is the general rule that a small amount of relatively tightly bound hydration water would be indistinguishable from a larger amount of less tightly bound water. Thus, for example, whereas Tait *et al.* (1972) observed changes in $T_2$ for $H_2{}^{17}O$ in monosaccharide solutions which seemed to indicate that the hydration of ribose is much less extensive than that of glucose, galactose or mannose, they were not able to quantify these differences from NMR results alone. Dielectric relaxation on the other hand has the power to resolve the different states of water, but the overall relaxation spectrum consists of contributions from all the molecular species in the system, and, especially for small solutes, there is inevitably some difficulty in resolving unequivocally the relaxation spectrum into its components.

There are ways in which the information from the two relaxation techniques can be combined, and these can be demonstrated with reference to our running example of the monosaccharide solutions: $^{13}C$ and $^{1}H$ NMR studies of monosaccharide solutions in $D_2O$ lead to the calculation of a *rotational correlation time* for the solute motion. For the case of Brownian motion of spherical or near-spherical molecules (expected to be followed reasonably well by monosaccharide molecules in water) the NMR rotational correlation time will be one-third of the corresponding dielectric relaxation time. So from the NMR experiment one can obtain an estimate, within limits of say 20%, for the dielectric relaxation time for the solute rotation. This imposes a severe constraint on the fitting of the dielectric relaxation spectrum. Likewise the $^{17}O$ NMR studies can impose another constraint: from any model for the fitting of the dielectric data it is possible to calculate a mean nuclear magnetic relaxation time for the water expected on the basis of that model. This can be compared to the experimentally observed $T_2$, etc.

So, in principle, these two constraints should be sufficient to decide between the possible alternatives for analysing the dielectric data, and a much more confident prediction of the extent of hydration can be made from the final dielectric relaxation amplitudes. As a final check, the dielectric relaxation times of the water in its bulk and hydration states can be combined (in a similar manner to that described by Tait *et al.* (1972) and Franks *et al.* (1973)) with the NMR $T_1$ or $T_2$ for the motion of the $H_2{}^{17}O$ to yield yet another estimate of the extent of hydration. The combined NMR/TDS approach has very recently been exploited (Suggett, Clark and Quickenden, to be published) in just this manner.

## E. Concluding Remarks

Dielectric relaxation, because of its time scale of measurement (Figure 6.1) is inherently a particularly sensitive method for characterizing the different states of water in a system, while being also one of the oldest spectroscopic techniques (Drude, 1897). Despite many years of development, the experimental measurement and interpretation of dielectric spectra have constantly proved to be a difficult and tedious affair requiring a great deal of experience and technical expertise. Dielectric methods, as a result, have in general fallen well behind other spectroscopic techniques in the scientific popularity stakes. However, while TDS measurements still require some skill and experience to spot the 'artefacts', the advantages of these novel methods over the classical alternatives in terms of speed of measurement and band width are so enormous, that dielectric relaxation, from the point of view of the chemist or biochemist, can now be considered re-born.

We have earlier attached some significance to the importance in hydration studies of focussing the measurements on the hydration itself, and have shown how wide-band dielectric spectra can be used to characterize the state(s) of the hydration water. Table 6.3 shows a selection of dielectric

**Table 6.3.** Dielectric characteristics of bound water in various systems

| System | Conditions | $\tau$ (sec) | $\Delta H^{\ddagger}$ (kcal/mole) | Reference |
|---|---|---|---|---|
| Horse haemoglobin/$H_2O$ | 22% solution, 25°C | $2 \times 10^{-10}$ | 7·3 | a |
| Collagen/$H_2O$ | Powder, 40% water | $1{\cdot}5 \times 10^{-10}$ | | b |
| Polyethylene glycol/$H_2O$ | Various concentrations, 25°C | $10^{-10}$–$10^{-11}$ | | c |
| Lysozyme/$H_2O$ | Freeze-dried powder, 33% water, 25°C | $5 \times 10^{-10}$ | (−0·4) | d |
| ε-Aminocaproic acid/$H_2O$ | 2 *M* solution, 20°C | $10^{-10}$ | 7 | e |
| D-Glucose/$H_2O$ | 2 *M* solution, 5°C | $2 \times 10^{-10}$ | 9 ± 2 | f |

(a) Pennock and Schwan (1969).
(b) Grigera, Hallenga and Berendsen (1973).
(c) Koide (1969).
(d) Harvey and Hoekstra (1972).
(e) Shepherd and Grant (1968a).
(f) Tait *et al.* (1972).

parameters obtained for modified water components in a number of systems, indicating some similarities in the status of the 'bound' water. At this stage however, the overriding impression is of how little we really know about the physical properties of such water (particularly in aqueous solution), and future progress in this area depends in part upon being able to characterize the water bound to different types of chemical groupings in various stereochemical arrangements.

Finally the point must be made that, while this chapter has been largely concerned with one technique area, no single technique by itself is likely to unravel the mysteries of the hydration of biological molecules in solution. Major advances will in all probability come only from the complementary use of a number of different techniques.

# VIII. APPENDIX

## Additional Problems in TDS Measurement Associated with Aqueous Systems

### *a. Amplitude errors*

For the direct reflexion TDS method the relationship (Equation 6.20) between reflexion coefficient and complex permittivity

$$\epsilon^*(\omega) = \left[\frac{1 - \rho(\omega)}{1 + \rho(\omega)}\right]^2 \tag{6.20}$$

is such that for a given amplitude uncertainty in $\rho$ the resulting uncertainty in $\epsilon$ increases as $\epsilon$ itself increases. This is shown for the static permittivity $\epsilon_0$ (Equation 6.3, Figure 6.4) in Table 6.4. It can be seen that, to retain a reasonable sensitivity in measurement of $\epsilon$ (say 0·2%) at a static permittivity of about 80 (normal for aqueous solutions), the TDS pulse amplitude needs to be measured very accurately (better than 0·03%). This cannot be attained using simple chart recorders, and this factor is one reason for the use of signal averagers or on-line minicomputers with TDS system. Signal averagers are in some ways more useful in that they can accumulate many hundreds or even thousands of TDS scans in order to improve the signal to noise ratio, but of course other computing facilities are then required for the Fourier transformation. With direct on-line facilities the averaging (usually a reduced number of scans) and the Fourier transformations can be performed on the same instrument.

**Table 6.4.** The effect of amplitude errors on the measured 'static' permittivity

| | Percentage error in $\epsilon_0$ | |
|---|---|---|
| $\epsilon_0$ (true) | (a) for 1% error in $\rho_0$ | (b) for 0·1% error in $\rho_0$ |
| 2 | 0·7 | 0·07 |
| 4 | 1·5 | 0·15 |
| 9 | 2·6 | 0·26 |
| 25 | 4·8 | 0·48 |
| 49 | 6·8 | 0·68 |
| 81 | 8·9 | 0·89 |

### *b. Timing errors*

The reflexion coefficient, as computed from the ratio of the Fourier transforms of the sample reflexion and the short-circuit reflexion, is a complex quantity. It is customary, and most useful, to express this in terms of amplitude ($R$) and phase ($\theta$),

$$\rho = Re^{j\theta} \tag{6.26}$$

The relationship between the complex permittivity and complex reflexion coefficient can then be expressed (Loeb *et al.*, 1971) as

$$\epsilon' = \frac{(1 - R^2) - 4(R \sin \theta)^2}{(1 + R^2 + 2R \cos \theta)^2} \tag{6.27}$$

$$\epsilon'' = \frac{4(1 - R^2)R \sin \theta}{(1 + R^2 + 2R \cos \theta)^2} \tag{6.28}$$

The problems in measuring amplitude ($R$) have already been discussed. We are now concerned with phase ($\theta$) errors. In TDS phase is synonymous with 'timing', and an error in time referencing (i.e. the 'setting up' of the two reflected pulses in exactly the same time scale) produces an error in phase only. This phase error is linearly dependent upon the measurement frequency, in fact $\Delta\theta = 0{\cdot}36\Delta t.f$ where $\Delta\theta$ is the phase error in degrees of arc, $\Delta t$ is the timing error in psec ($10^{-12}$ sec) and $f$ is the frequency of measurement in GHz.

Using the type of time marker discussed in the main part of the chapter, a time reference accurate to $\pm$ 2 psec can be obtained. In terms of the equivalent phase error, this is 0·72° per GHz of frequency. The problem for measurement of aqueous solutions can be illustrated by the fact that the magnitude of the experimental phase for pure water (278° K) at 10 GHz is about 5° only, which is actually *less* than the possible uncertainty on the basis of a 2 psec timing error. This explains why for aqueous solutions *at frequencies above a few GHz*, a much more precise time-referencing method is required. The most effective method is simply to have one or two measurements of $\epsilon^*$ at fixed frequencies, either by a classical frequency domain technique or by another suitable TDS method (Transmission Coefficient Ratios). The phase error ($\Delta\theta$) at these frequencies can be calculated from Equations (6.27) and (6.28), and as $\Delta\theta$ is proportional to frequency the correction over the whole frequency range can be obtained.

## IX. REFERENCES

Aaron, M. W., E. H. Grant and S. E. Young (1966) 'Dielectric properties of some amino-acids, peptides, and proteins at decimetre wavelengths', in *Molecular Relaxation Processes*, Chem. Soc./Academic Press, pp. 77–82.

Abadie, P., R. Charbonniere, A. Gidel, P. Girard and A. Guilbot (1953) 'L'eau dans la cristallisation du maltose et du glucose et états de l'eau de sorption de l'amidon', *J. Chim. Phys.*, **50**, C46–52.

Berendsen, H. J. C. (1965) 'Hydration of fibrous macromolecules', *Ann. N.Y. Acad. Sci.*, **125**, 365–79.

Böttcher, C. J. F. (1952) *Theory of Electric Polarization*, Elsevier.
Bratton, C. B., A. L. Hopkins and J. W. Weinburg (1965) 'NMR studies of living muscle', *Science*, **147,** 738–9.
Buchanan, T. J., G. H. Haggis, J. B. Hasted and B. G. Robinson (1952) 'The dielectric estimation of protein hydration', *Proc. Roy. Soc.*, **A213**, 379–91.
Chapman, G. E. and K. A. McLauchlan (1969) 'Hydration structure of collagen', *Proc. Roy. Soc.*, **B173**, 223.
Clifford, J. and B. Sheard (1966) 'Nuclear magnetic investigation of the state of water in human hair', *Biopolymers*, **4**, 1057.
Cole, K. S. and R. H. Cole (1941) 'Dispersion and absorption in dielectrics. I. A.C. characteristics', *J. Chem. Phys.*, **9**, 341–51.
Cope, F. W. (1969) 'Nuclear magnetic resonance evidence for structured water in muscle and brain', *Biophys. J.*, **9**, 303.
Danford, M. D. and H. A. Levy (1962) 'The structure of water at room temperature', *J. Amer. Chem. Soc.*, **84**, 3965–6.
Davidson, D. W. and R. H. Cole (1951) 'Dielectric relaxation in glycerol', *J. Chem. Phys.*, **18**, 1417.
Debye, P. J. W. (1929), *Polar Molecules*, Dover publications, New York.
Dehl, F. E. and C. A. Hoeve (1969) 'Broadline NMR study of $H_2O$ and $D_2O$ in collagen fibres', *J. Chem. Phys.*, **50**, 3245.
de Loor, G. P., M. J. C. van Gemert and H. Gravesteyn (1973) 'Measurement of permittivity with time domain reflectometry. Extension to lower relaxation frequencies', *Chem. Phys. Letters*, **18**, 295–9.
Drude, P. (1897) 'Frequency dependence of the electric permittivity', *Z. Phys. Chem.*, **23**, 267.
Edsall, J. T. (1953) 'The size, shape and hydration of protein molecules', in *The Proteins* Vol. IB (Eds. H. Neurath and K. Bailey), pp. 550–726.
Eisenberg, D. and W. Kauzmann (1969) *The Structure and Properties of Water*, Oxford Univ. Press.
Eyring, H. (1941) *Theory of Rate Processes*, McGraw-Hill, New York.
Fedotov, V. D., F. G. Miftakhutdinova and Sh. F. Murtazin (1969) 'Spin-echo NMR studies of protein relaxation in plant tissues', *Biophysics*, **14**, 918.
Feil, D. and G. A. Jeffrey (1961) 'The polyhedral clathrate hydrates', *J. Chem. Phys.*, **35**, 1863–73.
Fellner-Feldegg, H. (1969) 'The measurement of dielectrics in the time-domain', *J. Phys. Chem.*, **73**, 616–23.
Fellner-Feldegg, H. (1972) 'Thin sample method for the measurement of permeability, permittivity and conductivity in the frequency and time-domain', *J. Phys. Chem.*, **76**, 2116–22.
Frank, H. S. (1972) in *Water—A Comprehensive Treatise* Vol. I (Ed. F. Franks), Plenum Press, New York, pp. 515–43.
Frank, H. S. and M. W. Evans (1945) 'Structure and thermodynamics in aqueous electrolytes, *J. Chem. Phys.*, **13**, 507.
Franks, F. (1968) 'Effects of solutes on the hydrogen-bonding in water', in *Hydrogen-bonded Solvent Systems* (Eds. A. K. Covington and P. Jones), Taylor and Francis, London, pp. 31–48.
Franks, F. and H. T. Smith (1968) 'Volumetric properties of alcohols in dilute aqueous solutions', *Trans. Faraday Soc.*, **64**, 2962–72.
Franks, F., D. S. Reid and J. R. Ravenhill (1972) 'Thermodynamic studies of dilute aqueous solutions of cyclic ethers and simple carbohydrates', *J. Solution Chem.*, **1**, 3–16.

Franks, F., D. S. Reid and A. Suggett (1973) 'Conformation and hydration of sugars and related compounds in dilute aqueous solution', *J. Solution Chem.*, **2**, 99–118.

Fricke, H. (1924) 'Electrical conductivity of suspensions of homogeneous spheroids', *Phys. Rev.*, **24**, 575.

Giese, K., U. Kaatze and R. Pottel (1970) 'Permittivity and dielectric and proton magnetic relaxation of aqueous solutions of alkali halides', *J. Phys. Chem.*, **74**, 3718–25.

Glew, D. N., H. D. Mak and N. S. Rath (1968) 'Water-shell stabilisation by interstitial non-electrolytes', in *Hydrogen-bonded Solvent Systems* (Eds. A. K. Covington and P. Jones), Taylor and Francis, London, pp. 195–210.

Goebel, W. and H. Vogel (1964) 'Dielectric properties of haemoglobin and their origin', *Z. Naturforsch.*, **19b**, 292–7.

Grant, E. H. (1966) 'Dielectric dispersion in bovine serum albumin', *J. Mol. Biol.*, **19**, 133–9.

Grant, E. H., T. J. Buchanan and H. F. Cook (1957) 'Dielectric behaviour of water at microwave frequencies', *J. Chem. Phys.*, **26**, 156–61.

Grant, E. H., S. E. Keefe and S. Takashima (1968) 'Dielectric behaviour of aqueous solutions of bovine serum albumin', *J. Phys. Chem.*, **72**, 4373–80.

Grant, E. H., G. P. South, S. Takashima and H. Ichimura (1971) 'Dielectric dispersion in aqueous solutions of oxyhaemoglobin and carboxyhaemoglobin', *Biochem. J.*, **122**, 691–9.

Grigera, R., K. Hallenga and H. J. C. Berendsen 'The dielectric properties of hydrated collagen', to be published.

Haggis, G. H., J. B. Hasted and T. J. Buchanan (1952) 'The dielectric properties of water in solutions', *J. Chem. Phys.*, **20**, 1452–65.

Harvey, S. C. and P. Hoekstra (1972) 'Dielectric relaxation spectra of water absorbed on lysozyme', *J. Phys. Chem.*, **76**, 2987–94.

Hasted, J. B. (1972) 'Dielectric properties', in *Water—A Comprehensive Treatise* Vol. I (Ed. F. Franks), pp. 255–310.

Hazlewood, C. F., B. L. Nichols and N. F. Chamberlain (1969) 'Evidence for the existence of two phases of ordered water in skeletal muscle', *Nature*, **222**, 747.

Hertz, H. G. and C. Rädle (1973) 'The orientation of water molecules in the hydration sphere of $F^-$ and in the hydrophobic hydration sphere', *Ber. Bunsenges. Phys. Chem.*, **77**, 521–31.

Hill, N. E., W. E. Vaughan, A. H. Price and M. Davies (1969), *Dielectric Properties and Molecular Behaviour*, van Nostrand.

Katz, J. J. (1965) 'Chemical and biological studies with deuterium', (Thirty-ninth annual Priestley Lecture), Pennsylvania State University.

Kauzmann, W. (1942) 'Dielectric relaxation as a chemical rate process', *Rev. Mod. Phys.*, **14**, 12–44.

Kauzmann, W. (1959) 'Some factors in the interpretation of protein denaturation', *Advan. Protein Chem.*, **14**, 1.

Koide, G. T. (1969) 'The dielectric properties of structured water', *Ph.D. Thesis*, University of Rochester, New York.

Kuntz, I. D. (1971) 'Hydration of macromolecules. III. Hydration of polypeptides', *J. Amer. Chem. Soc.*, **93**, 514–16.

Kuntz, I. D., T. S. Brassfield, G. D. Law and G. V. Purcell (1969) 'Hydration of macromolecules', *Science*, **163**, 1329.

Kuntz, I. D. and T. S. Brassfield (1971) 'Hydration of macromolecules. II. Effects of urea on protein denaturation', *Arch. Biochem. Biophys.*, **142**, 660–4.

Ling, G. N. (1962), *A Physical Theory of the Living State*, Blaisdell, New York, p. xxviii.

Ling, G. N. (1972) 'Hydration of macromolecules', in *Water and Aqueous Solutions* (Ed. R. A. Horne), Wiley (Interscience) New York p. 670.

Loeb, H. W., G. M. Young, P. A. Quickenden and A. Suggett (1971) 'New methods for measurement of complex permittivity up to 13 GHz and their application to the study of dielectric relaxation of polar liquids', *Ber. Bunsenges. Phys. Chem.*, **75**, 1155–65.

Lumry, R. and R. H. Yue (1965) 'Dielectric dispersion of protein solutions containing small zwitterions', *J. Phys. Chem.*, **69**, 1162–74.

Maeda, Y., T. Fujita, Y. Suguira and S. Koga (1968) 'Physical properties of water in spores of *Bacillus megaterium*', *J. Gen. Appl. Microbiol.*, **14**, 217.

Marcy, H. O. and J. Wyman, Jr. (1942) 'Dielectric studies on muscle haemoglobin', *J. Amer. Chem. Soc.*, **64**, 638–45.

Migchelsen, C., H. J. C. Berendsen and A. Rupprecht (1968) 'Hydration of DNA', *J. Mol. Biol.*, **37**, 235.

Nicolson, A. M. and G. F. Ross (1970) 'Measurement of the intrinsic properties of materials by time domain techniques', *IEEE*, **IM—90**, 377–82.

Oncley, J. L. (1938) 'Dielectric properties of protein solutions', *J. Amer. Chem. Soc.*, **60**, 1115–25.

Oncley, J. L. (1943) 'Dielectric properties of proteins', in *Proteins, Amino Acids and Peptides* (Eds. E. J. Cohn and J. T. Edsall), Reinhold, New York, p. 543.

Oncley, J. L., J. D. Ferry and J. Shack (1938) 'Dielectric properties of protein solutions', *Cold Spring Harbor Symp. Quant. Biol.*, **6**, 21.

Oncley, J. L., J. D. Ferry and J. Shack (1940) 'Dielectric properties of protein solutions', *Ann. N.Y. Acad. Sci.*, **40**, 371.

Pennock, B. E. and H. P. Schwan (1969) 'Further observations on the electrical properties of haemoglobin-bound water', *J. Phys. Chem.*, **73**, 2600–10.

Schwan, H. P. (1957) 'Electrical properties of tissues and cell suspension', *Advan. Biol. Med. Phys.*, **5**, 147–56.

Schwan, H. P. (1965) 'Electrical-properties of bound water', *Ann. N.Y. Acad. Sci.*, **125**, 344–57.

Shaw, T. M., E. F. Jansen and H. Lineweaver (1944) 'Dielectric properties of $\beta$-lactoglobulin in aqueous glycine solutions', *J. Chem. Phys.*, **12**, 439–48.

Shepherd, J. C. W. and E. H. Grant (1968a) 'Dielectric properties of amino acid solutions. I. Dielectric dispersion in aqueous $\epsilon$-aminocaproic acid solutions', *Proc. Roy. Soc.*, **A307**, 335–44.

Shepherd, J. C. W. and E. H. Grant (1968b) 'Dielectric properties of amino acid solutions. II. Dielectric dispersion in aqueous proline and hydroxyproline solutions', *Proc. Roy. Soc.*, **A307**, 345–57.

Stillinger, F. H. (1973) 'Structure in aqueous solutions of apolar solutes from the standpoint of scaled-particle theory', *J. Solution Chem.*, **2**, 141–58.

Suggett, A. (1972) 'Time domain methods', in *Dielectric and Related Molecular Processes*, (Ed. M. Davies), Chem. Soc. London, pp. 100–20.

Suggett, A. (1973) 'Time domain spectroscopic measurements', in *High Frequency Dielectric Measurement*, (Eds. J. Chamberlain and G. W. Chantry), IPC Science and Technology Press Ltd., pp. 96–103.

Suggett, A. (1974) 'Role of solute–solvent interactions in sols and gels of macromolecules. II. Polysaccharides', in *Water; A Comprehensive Treatise* Vol. 4, (Ed. F. Franks), Plenum Press, N.Y.

Suggett, A., P. A. Mackness, M. J. Tait, H. W. Loeb and G. M. Young (1970) 'Dielectric relaxation studies by time domain spectroscopy, *Nature*, **228**, 456–7.

Suggett, A., A. H. Clark and P. Quickenden (1974) 'Total reflection time domain spectroscopy. Application to dielectric relaxation studies of polar liquids and solids and aqueous solutions in the time range $10^{-10}$ to $10^{-4}$ sec, *J. Chem. Soc. Faraday II*, **70,** 1847–62.

Tait, M. J. and F. Franks (1971) 'Water in biological systems', *Nature*, **230**, 91–6.

Tait, M. J., A. Suggett, F. Franks, P. A. Quickenden and S. Ablett (1972) 'Hydration of monosaccharides: A study by dielectric and nuclear magnetic relaxation', *J. Solution Chem.*, **1**, 131–51.

Takashima, S. (1962) 'Dielectric dispersion of protein solutions in viscous solvents', *J. Polymer Sci.*, **56**, 257–65.

Takashima, S. (1963) 'Dielectric dispersion of polyglutamic acid solutions', *Biopolymers*, **1**, 171–82.

Takashima, S. (1964) 'Dielectric dispersion of albumins; studies of denaturation by dielectric measurement, *Biochim. Biophys. Acta*, **79**, 531–8.

Tanford, C. (1961) *Physical Chemistry of Macromolecules*, John Wiley, N.Y.

Taylor, J. B. and J. S. Rowlinson (1955) 'The thermodynamic properties of aqueous solutions of glucose, *Trans. Faraday Soc.*, **51**, 1183–9.

van Gemert, M. J. C. (1971) 'Dielectric measurements with time domain reflectometry', *J. Phys. Chem.*, **75**, 1323.

van Gemert, M. J. C. (1973) 'Time domain reflectometry',*P h.D. Thesis*, Leiden Univ., Netherlands.

Vaughan, W. E. (1969) 'Experimental methods', in *Dielectric Properties and Molecular Behaviour*, van Nostrand, p. 108.

von Hippel, P. H. and T. Schleich (1969)' The effects of neutral salts on the structure and conformational stability of macromolecules in solution', in *Structure and Stability of Biological Macromolecules* (Eds. S. N. Timasheff and G. D. Fasman), Marcel Dekker, New York, ff. 417–574.

Walrafen, G. A. (1966) 'Raman spectral studies of the effects of urea and sucrose on water structure', *J. Chem. Phys.*, **44**, 3726–7.

Warner, D. T. (1965) 'Proposed water–protein interaction and its application to the structure of the tobacco mosaic virus particle', *Ann. N.Y. Acad. Sci.*, **125**, 605–30.

CHAPTER 7

# Nanosecond pulsefluorimetry

Philippe Wahl
*Centre de Biophysique Moléculaire,*
*C.N.R.S., 45045 ORLEANS CEDEX,*
*France*

# I. INTRODUCTION

The various phenomena connected with fluorescence emission are concerned with the interaction of the emitting chromophores with other molecules or residues situated in the vicinity. These phenomena have been extensively studied and fully described (see for example Förster, 1951; Parker, 1968; Birks, 1970; Mataga and Kubota, 1970). Potentially, analysis of the emission of fluorescent chromophores situated in biological systems should be capable of providing interesting and useful information about such systems and their environment. These chromophores may be natural parts of the system or artificially introduced as fluorescent probes. For example, the natural fluorescent chromophores of proteins are the tyrosine and tryptophan residues and several reviews on the fluorescence of proteins have been published (Koniev, 1967; Chen, Edelhoch and Steiner, 1969; Longworth, 1971; Weinryb and Steiner, 1971). Nucleic acids may also be studied by fluorescence techniques and accounts of the nucleic acid fluorescence may be found in the reviews of Eisinger and Lamola (1971). A further, useful way of studying proteins and membranes is to label them with covalently linked fluorescent compounds or with probes adsorbed on their surface. This aspect of the subject has been reviewed by Weber (1953), Edelman and McClure (1968), Stryer (1968), Dandliker and Portmann (1971), Chance *et al.* (1971) and by Penzer (1974).

In order to obtain the fluorescence emission of a solution, one must use an exciting light of appropriate wavelength, and in many cases, a continuous exciting light source is used. In this way, such steady state properties as quantum yield, excitation spectra, emission spectra and static polarization can be studied (Parker, 1968). By using nanosecond pulse excitation one can measure transient properties of fluorescence emission from which further structural information can be obtained. This chapter is concerned with such measurements and their relevance to investigations of molecular structure.

In recent years, pulsefluorimetry methods have been greatly improved (Birks and Munro, 1967; Ware, 1971). Among these (cf. reviews: Ware, 1971; Yguerabide, 1972; Isenberg, 1975) the photocounting method appeared to be specially versatile and accurate. In this technique, excitation is induced by a very short flash of the order of one nanosecond. A detecting system can be constructed which is able to follow the fluorescence decay until the intensity is reduced to 1 % and even to 0·1 % of its value at the maximum intensity.

In simple cases, the fluorescence decays according to a simple exponential law and a pulse fluorometer then gives no more information than a phase fluorometer working with a single modulation frequency and measuring the fluorescence time constant. But there are a number of more complicated cases where the determination of the decay in a large time range becomes very useful.

Let us take some examples which will be treated with more detail in the following parts of this chapter.

When a solution contains different kinds of fluorescent compounds the emission spectra of which overlap each other, the fluorescence decay is generally the sum of several exponential functions. Each kind of compound has its own time constant and the contribution of the corresponding exponential term is proportional to the molar concentration. By measuring the decay at several wavelengths, one may solve the spectra of each of the constituent chromophores. This principle has been applied to the study of proteins containing one tryptophan residue only (Wahl and Auchet, 1972 b; Brochon, Wahl and Auchet, 1974). It has been possible to decompose the broad fluorescence spectrum into a component characteristic of tyrosine emission, and one or two components characteristic of the single tryptophan residue. This approach could be extended to proteins containing several tryptophans, and could sometimes advantageously replace previous methods (Weber, 1961; Teale, 1961; Vladimirov and Zimina, 1965).

The time constants of fluorescence of a chromophore residing in a protein depend on the interactions which occur between this chromophore and other chemical groups which are located close to it. These interactions have an effect upon the non-radiative constant of deactivation. A number of studies on model compounds have been made in order to correlate the fluorescence properties with the interactions of the different chemical groups present in proteins (Longworth, 1971; Steiner and Weinryb, 1971). Thus it has been shown that the peptide bond quenches the fluorescence of the tryptophan and tyrosine residues in aqueous or alcoholic solutions (Cowgill, 1963, 1970).

However this effect appears to be negligible in aprotic solvents. For example, the quantum yield of solutions in dimethyl sulphoxide of cycloglycyltryptophan, acetyltryptophan and scatole are very similar (Edelhoch, Bernstein and Wilchek, 1968).

On the one hand, the fluorescence decays of the solutions of acetyltryptophan and scatole are single exponentials with time constants of 7·3 and 7·5 nsec and the decays are independent of the emission wavelengths. By contrast, the decay of cycloglycyltryptophan is complex and depends on the emission wavelength (Donzell, Gauduchon and Wahl, 1974). These results may be explained by the presence of two conformers. For the folded form, the indole ring is in close contact with the diketopiperazine ring. In the open form, the indole ring is surrounded by solvent molecules. The equilibrium constant for these interactions has been measured in the ground state by NMR (Kopple and Ohnishi, 1969; Donzell, 1971). These data may be incorporated in the fluorescence decay analysis which allows determination of the rate constants of exchange and the equilibrium constant relative to the excited conformers. Furthermore, it is possible to determine the fluorescence spectra of the two forms. It would be impossible, with the use of continuous excitation only, to obtain such information which can contribute to a better interpretation of the fluorescence of proteins.

Fluorescence is generally polarized. This polarization is related to the rotational Brownian motion of the emitting chromophores (Perrin, 1929; Weber, 1953). The degree of the static polarization depends on the ratio of the fluorescence time to the Brownian correlation time. The Brownian rotation of a chromophore covalently linked to a protein molecule may be complex. For the sake of convenience, one may consider three categories of motion. First, a local motion around the chemical bond of the side chain which links the chromophore to the peptide backbone; secondly, a motion of deformation of the peptide backbone; and thirdly, the rotation of the whole molecule. Each of these motions is characterized by at least one correlation time.

In principle, the study of polarization as a function of $T/\eta$ (ratio of the absolute temperature to the solvent viscosity), should allow the determination of the correlation times and this was the method proposed by Weber (1952; 1953) for the study of protein molecules. However, this method is of limited use and very often leads to the determination of an apparent correlation time with no real significance (Brochon and Wahl, 1972). This is due to several causes; in particular the useful range of $T/\eta$ is generally too small, because of the solvent properties and of the relatively narrow temperature range of stability of the protein being studied. In addition, interactions between the fluorescent chromophores and the macromolecules can change with variation of the temperature or of the nature of the solvent (Wahl and Weber, 1967; Brochon and Wahl, 1972). These drawbacks are absent in the method based on flash excitation (Wahl 1966; Stryer, 1968; Yguerabide, 1972; Rigler and Ehrenberg, 1973; Wahl, 1975). In this case, correlation times are obtained as time constants of the anisotropy decay curve. The measurement is performed in a given solvent and at a defined temperature. The method is applicable even if the chromophore is simply adsorbed on the macromolecule.

Anisotropy decay measurements may be performed in order to follow the kinetics of energy transfer between like chromophores. This method has been applied to the study of the complexes of ethidium bromide (EB) and DNA (Genest and Wahl, 1973, 1974). A model of the complex has been adopted in which the dye is intercalated between the DNA base pairs. This intercalation induces a deformation of the DNA molecule characterized by an angle $\delta$, which corresponds to a winding ($\delta > 0$) or an unwinding ($\delta < 0$) of the helix. Energy migration is then simulated for different values of $\delta$. Comparison with experimental curves determines the value of $\delta$. It was found that $\delta = -16°$. This result is in good agreement with the value deduced by Bauer and Vinograd (1968) from the sedimentation measurements of the complex formed by EB with a circular DNA molecule. By using static polarization, Paoletti and Le Pecq (1971) were able to show a winding of the DNA helix, but these measurements are probably not sensitive enough to give a reliable value of $\delta$.

Fluorescence decay measurements should also be very useful for the study of transfers between unlike donors and acceptors (Haugland, Yguerabide and Stryer, 1969; Grinwald, Haas and Steinberg, 1972). Interesting information concerning the spatial distribution of the chromophores should be obtained by these studies.

In the following part of this chapter, we shall examine in detail how the properties of molecules in their excited states influence the fluorescence decay. Instances of applications to biological systems will be given when they are available. Then the remaining part of the paper will be devoted to descriptions of measurement techniques.

## II. ASPECTS OF FLUORESCENCE KINETICS

### A. Activation and Deactivation of Molecules in Solution

#### *1. General principles*

At ordinary temperatures, molecules in solution are usually in their ground state $S_0$ which is a *singlet*. When we illuminate a solution with visible or UV light, the solute molecules may absorb photons and consequently may be brought into one of their singlet excited states $S_1$, $S_2$, etc. (Figure 7.1). The process of absorption is very fast, rate constants being of the order of $10^{14}$ to $10^{15}$ quanta per second. The excited molecules lose their energy by various processes which are characterized by the order of magnitude of their rate constants. The *internal conversion* which leads from an upper state $S_n$ to the state $S_1$ has still a large rate constant of the order of $10^{11}$–$10^{12}$/sec. But the internal conversion from $S_1$ to $S_0$ is usually much slower, due to the greater energy gap between these two states. Therefore the slower deactivation by fluorescence may occur. The corresponding rate constant varies from $10^7$–$10^9$/sec. In addition, there is a possibility for the molecule to go to a *triplet state* by *intersystem crossing*, and from there to be deactivated by a still

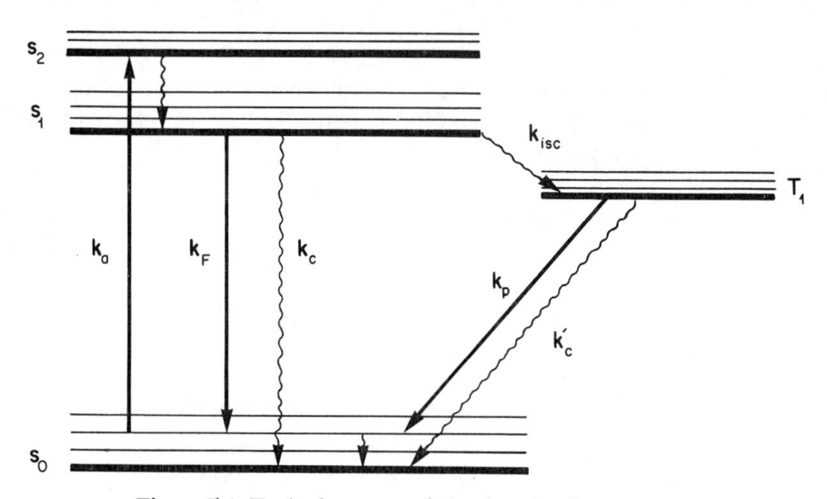


**Figure 7.1.** Excited states and the deactivation process

slower process (phosphorescence, internal conversion, etc.). The fluorescence emission then is directly linked to the rate of depopulation of the $S_1$ state.

Let us designate the concentration of the excited molecule at a given time $t$ by $[C^*]$. The rate at which $[C^*]$ decreases is

$$\frac{d[C^*]}{dt} = -(k_F + k_c + k_{isc})\,[C^*] \qquad (7.1)$$

The rate constants entering in this equation are:

$k_F$ the radiative rate constant corresponding to the fluorescence emission.
$k_c$ the rate constant of internal conversion $S_1 \rightarrow S_0$.
$k_{isc}$ the rate constant of intersystem crossing.

If the excitation is provided by an infinitely short flash producing $[C_0^*]$ molecules at zero time, Equation (7.1) resolves to the expression:

$$[C^*(t)] = [C_0^*]e^{-t/\tau} \qquad (7.2)$$

where the decay time $\tau$ is given by the relation:

$$\tau = \frac{1}{k_F + k_c + k_{isc}}$$

The fluorescence flux in photons per second per volume unit is

$$F(t) = k_F[C^*] = \frac{[C_0^*]}{\tau_0} e^{-t/\tau} \tag{7.3}$$

$\tau_0 = 1/k_F$ is the fluorescence decay time that one should obtain in the absence of non-radiative processes. It is often called the natural life time of fluorescence.

The quantum yield $\eta$ is defined as the ratio of the number of photons emitted to the number of photons absorbed; therefore it is expressed by:

$$\eta = \frac{1}{[C_0^*]} \int_0^\infty F(t)\mathrm{d}t = \frac{\tau}{\tau_0} \tag{7.4}$$

which may be written:

$$\eta = \frac{k_F}{k_F + k_c + k_{isc}} \tag{7.5}$$

The concentration of excited molecules is given by (Förster, 1951):

$$[C_0^*] = \frac{E(1 - 10^{-\epsilon[C]l})}{V} \tag{7.6}$$

where $E$ is the number of exciting photons, $V$ the volume of the solution, $\epsilon$ the molar absorbance, $[C]$ the molar concentration of solute molecules in the ground state and $l$ the length of the exciting beam in the solution.

If we take (7.4) and (7.6) into account, Equation (7.3) may be rewritten:

$$F(t) = \frac{E(1 - 10^{-\epsilon[C]l})}{V} \eta \frac{e^{-t/\tau}}{\tau} \tag{7.7}$$

and in dilute solution:

$$F(t) \approx \epsilon\,[C]\,\eta\,\frac{e^{-t/\tau}}{\tau} \tag{7.8}$$

Finally, from relation (7.7), one may also write:

$$F(t) \approx I\frac{e^{-t/\tau}}{\tau}$$

where $I$ is the fluorescence intensity in photons per second obtained with continuous excitation.

*2. Relation between natural lifetime and spectral properties*

Starting from *Einstein's coefficient of absorption* and of *spontaneous emission*, one can establish a relation between the rate of radiative deactivation and the absorption and emission spectra. Strickler and Berg (1962), have derived a formula applicable to the broad spectra observed in solution. In their derivation it is assumed that the nuclear configuration is the same in both states $S_0$ and $S_1$. The lifetime is given by the following relation:

$$\frac{1}{\tau_0} = 2{\cdot}88 \times 10^{-9}\, n^2 < \nu_F^{-3} >^{-1} \int \epsilon(\tilde{\nu}) \frac{d\tilde{\nu}}{\tilde{\nu}} \tag{7.9}$$

where $n$ is the refractive index of the solvent, $\epsilon$ the molar absorbance of the solute, $\tilde{\nu}$ the wave number; the interval of integration covers the absorption band corresponding to the transition $S_0 \rightarrow S_1$: $\langle \tilde{\nu}_F^{-3} \rangle$ is defined by the following relation:

$$\langle \nu_F^{-3} \rangle = \frac{\int F(\tilde{\nu})/\tilde{\nu}^3 \, d\tilde{\nu}}{\int F(\tilde{\nu}) d\tilde{\nu}} \tag{7.10}$$

$F(\tilde{\nu})$ is the fluorescence spectrum.

For most of the investigated compounds, the value of $\tau_0$ obtained by the formula (7.10) agrees well with the value obtained by the formula (7.4) in which $\eta$ and $\tau$ are experimentally determined. But that is not true when the configurations of the excited state and the ground state differ appreciably. An apparent discrepancy between the formulae (7.4) and (7.10) may also occur if there is a strong overlapping of the band corresponding to the electronic transitions $S_0 \rightarrow S_1$ and $S_0 \rightarrow S_2$. This is the case for indole and tryptophan and their derivatives. Hidden transitions may be revealed by this means and evidence for an $n$-$\pi^*$ transition has been found (Eisinger and Lamola, 1971).

*3. Mixture of non-interacting fluorescent species*

We consider a solution containing several fluorescent species which do not interact with one another, but which are characterized by several decay times. Such a situation often occurs in biological systems; for example, when a fluorescent label like the dansyl chromophore is chemically bound to a protein (Wahl and Lami, 1967; Brochon and Wahl, 1972), or adsorbed (Wahl, Kasai and Changeux, 1971). It is also the case with the intrinsic fluorescence of proteins (De Lauder and Wahl, 1970, Wahl and Auchet, 1972b).

Extension of equation (7.7) yields:

$$F(t) = E \sum_k (1 - 10^{-\epsilon_k [C_k] l})\, \eta_k \frac{e^{-t/\tau_k}}{\tau_k} \tag{7.11}$$

In dilute solution, that equation becomes:

$$F(t) \approx \epsilon_k\, [C_k]\, \eta_k \frac{e^{-t/\tau_k}}{\tau_k} \tag{7.12}$$

Analysis of the fluorescence decay may provide the time constants $\tau_k$ and the concentration $[C_k]$ of the fluorescent species. This principle has been applied to the study of the amino group titration of tryptophan (De Lauder and Wahl, 1970).

Expression (7.11) entails the relation:

$$F(t) \approx \sum_k I_k \frac{e^{-t/\tau_k}}{\tau_k} \tag{7.13}$$

where $I_k$ is proportional to the fractional fluorescence intensity of the component $k$, obtained with a continuous excitation. If we assume that each component $k$ has a characteristic fluorescence spectrum, then the shape and the contribution of this spectrum to the whole emission may be determined. This is realized in measuring and analysing the whole decay as a function of the wavelength of emission. A discussion of this method and its application to the spectral resolution of serum albumin fluorescence is given elsewhere (Wahl and Auchet, 1972b). The same principle could be applied to the resolution of the excitation spectra.

## B. Molecular Interactions

The interaction of an excited molecule with its surroundings may influence the fluorescence decay. We will first examine the solvent relaxation effect; then some specific interactions will be considered: fluorescence quenching formation of excimers and exciplexes, protonation and deprotonation in the excited state and intramolecular complexes. Finally we will survey the energy transfer mechanism.

### 1. *Solvent relaxation*

Polar chromophores very often have a stronger *dipolar moment* in the excited state than in the ground state. As a result, the equilibrium configuration of polar solvent molecules around the solute molecule is different in both states. In a rigid medium (a glass), the solvent molecules cannot reorientate themselves and remain, during the fluorescence emission, in their ground-state configuration. In a fluid medium, like water at ordinary temperature, the reorientation is fast and is complete before the fluorescence emission. The fluorescence spectrum is red shifted with respect to the spectrum observed in rigid solvents or in apolar solvents. In the intermediate case, where the solvent relaxation time is of the same order of magnitude as the fluorescence decay time, the spectrum shifts during the decay (Bakshiev *et al.*, 1966). At the same time, the fluorescence decay depends on the wavelength of emission. On the blue side of the spectrum, the decay rate is increased, while on the red side of the spectrum the intensity may first increase, flatten off to a maximum and finally decay. This effect has been observed in viscous solutions of

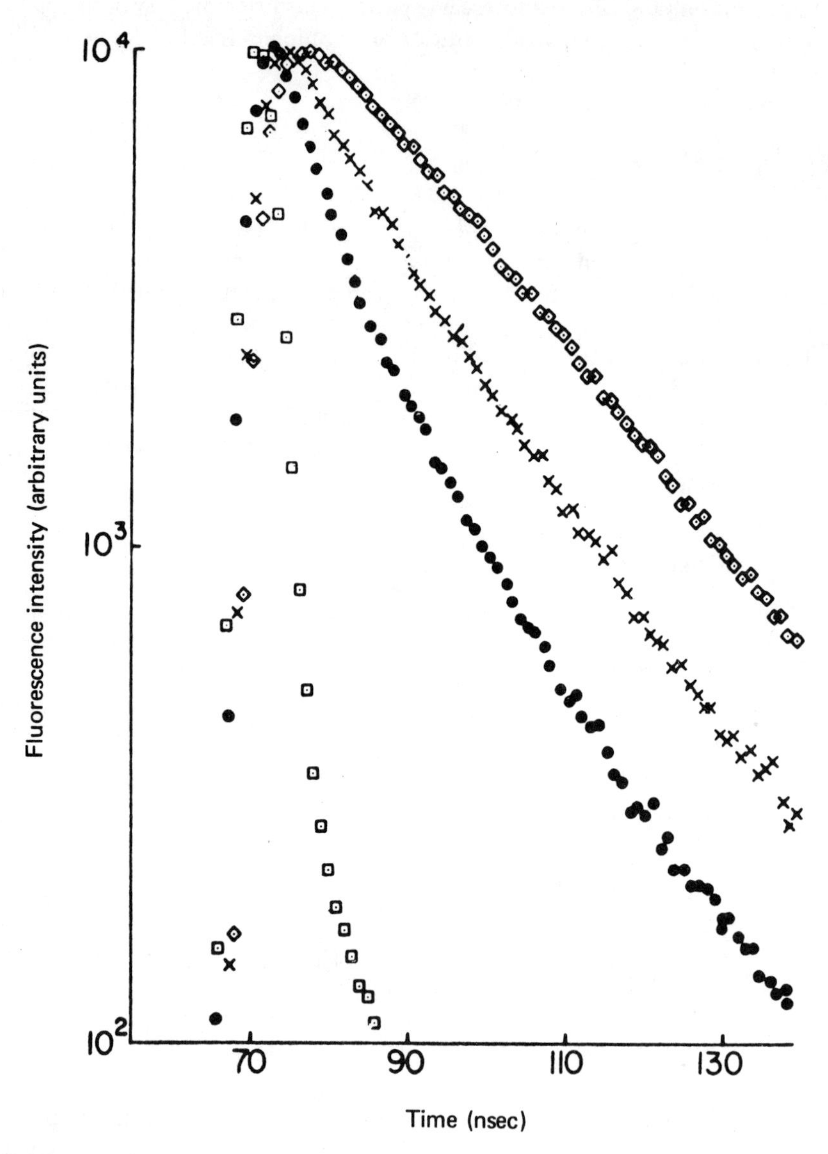


**Figure 7.2.** Wavelength dependence of fluorescence decay of 2-*p*-toluidinylnaphthalene-6-sulphonate in glycerol at 4°. Emission at 410 nm (●) 435 nm (×) and 490 nm (◇). After Brand and Gohlke, 1971

chromophores by Ware *et al.* (1968) (see also Ware, 1975) and Brand and Gohlke (1971) (Figure 7.2). These last authors proposed a method, based on this property, of probing the fluidity of the site of chromophores bound to protein molecules.

*2. Fluorescence quenching*

A compound is a quencher of fluorescence if by a specific interaction it decreases the quantum yield of a fluorescent compound. If the interaction occurs in the ground state of the fluorescent molecule, one speaks of static quenching; if the interaction occurs in the excited state one speaks of dynamic quenching.

In the *static quenching*, the fluorescent molecule A and the quencher molecule Q are in equilibrium with the non-fluorescent complex AQ according to the equation:

$$A + Q \rightleftharpoons AQ \tag{7.14}$$

The ratio of the quantum yield $\eta'$, in the presence of the quencher, to the quantum yield $\eta$ without quencher is:

$$\frac{\eta'}{\eta} = \frac{1}{1 + K[Q]} \tag{7.15}$$

where $K$ is the association constant for the equilibrium (7.14) and [Q] is the quencher concentration. On the other hand, the decay time corresponds to the emission of the free molecule A and does not change with [Q]. An example of such a quenching process has been found for the titration of the tyrosine hydroxyl in the basic pH range (Fayet and Wahl, 1971).

Measurements of fluorescence lifetime and determination of the decay can lead to interesting information in quenching studies of biological systems. For instance, such measurements show that proflavine intercalates in DNA molecules at at least two kinds of site. In the first kind of site, the fluorescence is quenched while in the second, the fluorescence is not quenched (Weil, 1965). In the case of serotonin, decay measurements show that the fluorescence is entirely quenched by intercalation into DNA (Helene, *et al.*, 1971).

In *dynamic quenching*, the formation of the complex occurs in the excited state according to the kinetic equation:

$$A^* + Q \rightarrow AQ^* \tag{7.16}$$

The reaction is in competition with the other deactivating processes. The time constant is given by the following expression:

$$\frac{1}{\tau'} = k_F + k_c + k_{isc} + k_Q [Q] \tag{7.17}$$

where $k_Q$ is the rate constant of the reaction (7.16). The quantum yield and the decay time ratio are given by the expression (7.18) which is called the Stern–Volmer law.

$$\frac{\eta'}{\eta} = \frac{\tau'}{\tau} = \frac{1}{1 + K'[\mathrm{Q}]}\,; \quad K' = k_{\mathrm{Q}}\tau \tag{7.18}$$

For the quenching to be efficient, $K'$ must be sufficiently large and therefore $k_{\mathrm{Q}}$ must not be too small compared to $1/\tau$. The upper limit of $k_{\mathrm{Q}}$ is the diffusion-controlled rate which is given approximately by the relation:

$$k_{\mathrm{Q}} \approx \frac{8RT}{3000\,\eta}\,\gamma \tag{9.19}$$

where $\gamma$ is a numerical coefficient close to one, $R$ the gas constant, $T$ the absolute temperature, $\eta$ the solvent viscosity (Birks, 1970). This simplified formula implies several approximations (Weller, 1961; Noyes, 1961). In particular, it is assumed that the conditions of steady-state kinetics are fulfilled. This assumption is not always true (Noyes, 1961; Ware and Novros, 1966).

### 3. *Exciplexes and excimers*

An excited solute molecule may form a fluorescent complex with another non-excited solute molecule. That complex is called an excimer or an exciplex according as the two interacting molecules are of the same species or of different species. The kinetic scheme of the deactivation process must include the conversion reactions of the species. For example, in the exciplex case one has:

| | rate constant | |
|---|---|---|
| $\mathrm{A^* + B \rightarrow A^*B}$ | $k$ | (7.20) |
| $\mathrm{A^*B \rightarrow A^* + B}$ | $k'$ | |

The time courses of the fluorescence intensities, obtained with an infinitely short excitation, result from the solution of two simultaneous differential equations: one finds (Birks, Dyson and Munro, 1963):

$$F_{\mathrm{A}} \simeq \exp(-t/\tau_{\alpha}) + c\exp(-t/\tau_{\beta}) \tag{7.21}$$

$$F_{\mathrm{AB}} \simeq \exp(-t/\tau_{\alpha}) - \exp(-t/\tau_{\beta}) \tag{7.22}$$

where

$$\frac{1}{\tau_{\alpha,\beta}} = \frac{1}{2}\,[X + Y \mp \sqrt{(X - Y)^2 + 4kk'\,[\mathrm{B}]}$$

$$X = \frac{1}{\tau_1} + k[\mathrm{B}] \tag{7.22 bis}$$

$$Y = \frac{1}{\tau_2} + k'$$

$\tau_1$ is the fluorescence decay time of A in the absence of B, and $\tau_2$ the decay time at infinite concentration of AB*; $c$ is a coefficient expressed in terms of the other parameters. Expression (7.22) predicts that the fluorescence $F_{AB}$ is null at the excitation time, increases to a maximum and then decays. This variation corresponds to the fact that the AB* concentration equals 0 at the excitation time and that it is formed from the species A* (Equation (7.20)). The validity of the expressions (7.21) and (7.22) has been verified in the case of pyrene (Birks, Dyson and Munro, 1963), which had previously been found to form excimers at high concentration (Förster and Kasper, 1954). The fluorescence kinetics of the pyrene—N,N-dimethylaniline system, in which an exciplex is formed, have been studied by Ware and Richter (1968). The spectrum shift of indole fluorescence, which is induced by a small added quantity of alcohol, has been explained by an exciplex formation (Walker, Bedmar and Lumry, 1967). Fluorescence decay measurements at different emission wavelengths are in agreement with that hypothesis and the different rate constants have been determined (De Lauder and Wahl, 1971b; Wahl and Auchet, 1972a), (Figure 7.3).

### 4. *Acid–base reactions*

The interactions of a proton with a base are usually diffusion-controlled reactions (Eigen *et al.*, 1961). For this reason the ionization reaction may interfere with the emission of aromatic acids or bases. In its ground state, an acid obeys the following equilibrium equation:

$$AH \rightleftharpoons A^- + H^+ \tag{7.23}$$

In the excited state, we have the two reactions:

$$\begin{aligned} AH^* &\rightarrow A^{-*} + H^+ \\ A^{-*} + H^+ &\rightarrow AH^* \end{aligned} \tag{7.24}$$

The kinetic scheme of deactivation is analogous to the exciplex kinetics scheme. Equation (7.23) however shows that the two species AH and $A^-$ may be present together so that the two excited species can be obtained by direct excitation. If the ionizable group is adjacent to an aromatic nucleus, the fluorescence spectra of the two species are distinct. Compounds of that kind have been studied by Förster (1951) and Weller (1961). Dehydroluciferine, a competitive inhibitor of luciferase, contains a phenol group which may be ionized giving a phenolate form. By measuring the decay time at different wavelengths, De Luca *et al.* (1971) have reconstituted the time course of the fluorescence spectrum corresponding to that reaction. In the pH range corresponding to the carboxyl titration, the fluorescence spectrum of tyrosine is, for all practical purposes, not shifted. However, the reactions of protonation and deprotonation in the excited state have an appreciable effect

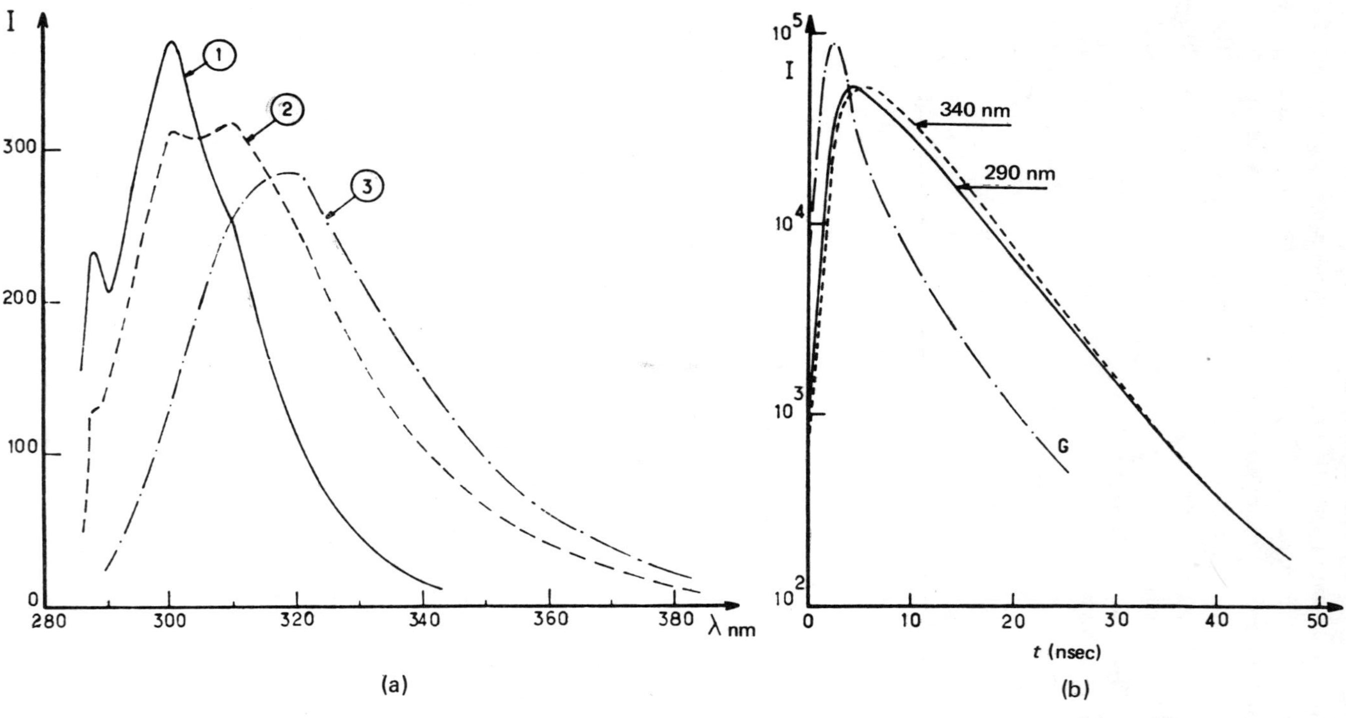


**Figure 7.3.** (a) Fluorescence spectra of indole dissolved ① in pure cyclohexane; ② in a mixture of cyclohexane and 0·2 *M* methanol; the spectrum ③ corresponds to the exciplex and is obtained by difference between ① and ② normalized on the 290 nm maximum. (Method of Walker *et al.*, 1967). (b) Fluorescence decay of indole, in the mixed solvent cyclohexane–methanol, is represented at the emission wavelengths of 290 nm (where the non-exciplex species is dominating) and of 340 nm (where the exciplex species is dominating). Note that this last emission is delayed compared to the first one. (After Wahl and Auchet, 1972a)

on the fluorescence decay. The rate constants of these reactions can be determined (Fayet and Wahl, 1971) (Figure 7.4).

*5. Intramolecular interactions*

In many biological compounds, there are intramolecular interactions which are conformation dependent. We shall restrict our discussion here to the case of only two conformations which we designate by $A_1$ and $A_2$. As for the interpretation of fluorescence properties, we have also to consider the conformers in their excited states $A_1^*$ and $A_2^*$. The reactions which are involved are given by the scheme:

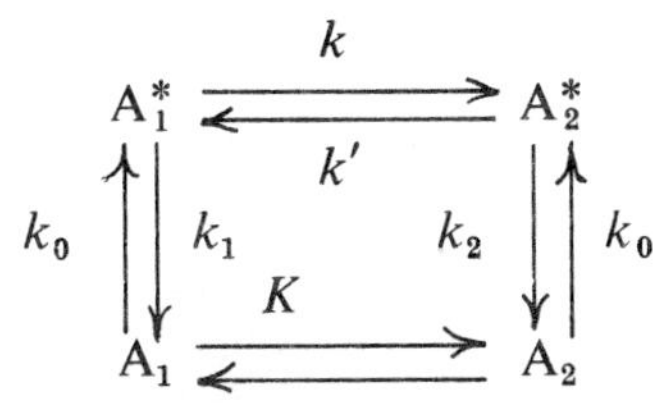


The meanings of the constants are as follows. $k_0$ is the rate of photon absorption and is proportional to the optical density of the solution. For the sake of simplicity, $k_0$ is assumed to be the same in the two conformers. $k_1$ and $k_2$ are the sums of the rate constants of deactivation by radiative and non-radiative processes. $\tau_1 = 1/k_1$ and $\tau_2 = 1/k_2$ are the time constants which would be measured if the exchange reaction did not occur in the excited state. $K$ is the equilibrium constant in the ground state, $k$ an $k'$ the rates of conformation changes.

This scheme is similar to that relating to the exciplexes and the acid–base reactions (Birks, 1970). Two examples of its application will be given here. First we report the study of Weber (1966) and Spencer and Weber (1972) concerning flavine adenine dinucleotide (FAD).

In this compound, the isoalloxazine ring interacts with the adenine moiety resulting in a quenching of the isoalloxazine fluorescence. The fluorescence quantum yield of FAD is ten times smaller than the quantum yield of the flavine mononucleotide (FMN). This last compound is quenched by Adenosine monophosphate (AMP), leading to a non-fluorescent complex. It can be then concluded that the FAD fluorescence arises only from a conformer in which the isoalloxazine is not interacting with adenine. This is confirmed by the analysis of fluorescence decay which is found to be a single exponential (Wahl, Auchet, Visser, Müller, 1974).

Let $A_2$ be the fluorescent form of FAD. It is reasonable to assume that the rate of deactivation of this form is equal to that of FMN. This leads to:

$$k_2 = 1/\tau_{FMN}$$

The decay times of FAD and FMN measured with a phase fluorometer are found to be 4·7 nsec and 2·49 nsec respectively.

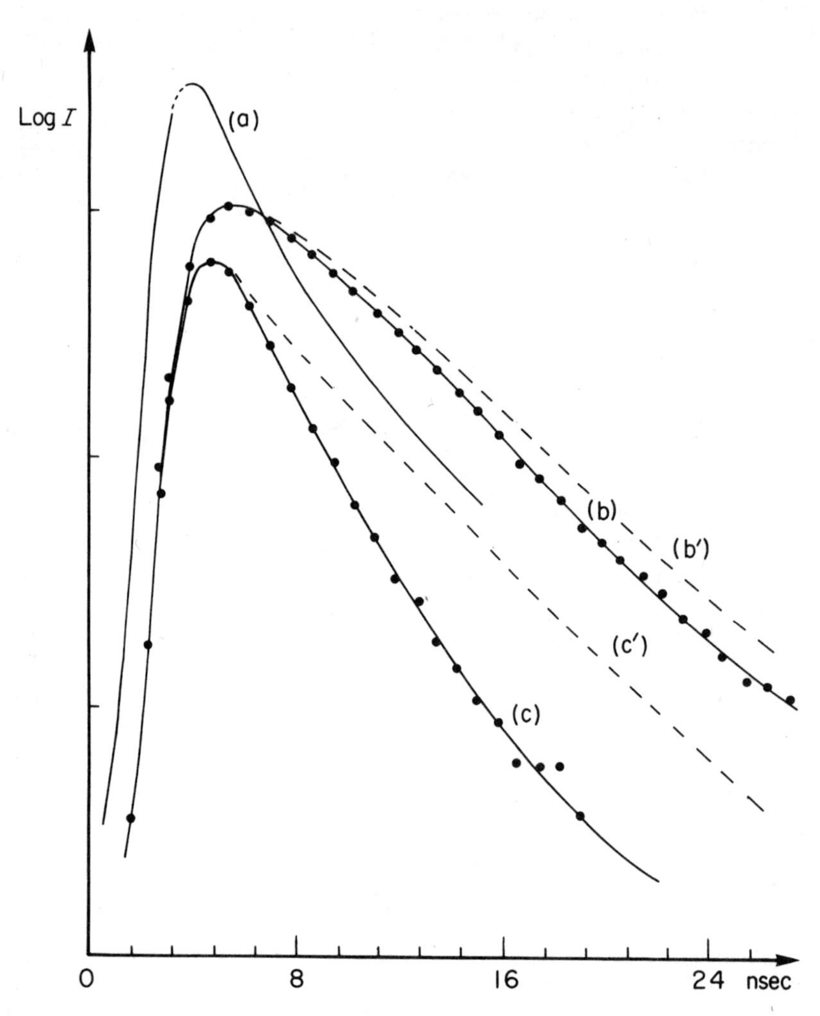


**Figure 7.4.** Fluorescence decay of tyrosine in the pH region of the carboxyl titration. The points represent experimental determinations. The dotted lines are the curve computed by assuming that the protonated and deprotonated molecules are only linked by the ground-state equilibrium (Equation 7.23). The solid lines are obtained by assuming that the reactions in Equation 7.24 occur in the excited state with the rate constants $5{\cdot}8 \times 10^8$ sec$^{-1}$ for the dissociation and $3{\cdot}6 \times 10^{10}$ mole $^{-1}$/sec for the association. Curve (a) is the response function, (b) tyrosine fluorescence at pH 2·6, (c) at pH 1·5. (After Fayet and Wahl, 1971)

Since $\tau_{FAD}$ is smaller than $\tau_{FMN}$, there must be a dynamic quenching of the isoalloxazine fluorescence by the adenine moiety. In other words, the reaction $A_2^* \longrightarrow A_1^*$ occurs with a rate constant given by the relation:

$$k' = \frac{1}{\tau_{FAD}} - \frac{1}{\tau_{FMN}} = 1{\cdot}88 \times 10^8 \text{ sec}^{-1}$$

In addition, the ratio of the time constants for FAD and FMN is greater than the ratio of the quantum yields. This implies a static quenching. In other words, an interaction between the two aromatic rings already exists in the FAD ground state. The association constant in the ground state is given by:

$$K = \frac{Q_{FMN}}{Q_{FAD}} \times \frac{\tau_{FMN}}{\tau_{FAD}} - 1$$

Numerical evaluation of this formula gives $K = 5{\cdot}2$.

In the case just described, the determination of time constants by a phase fluorometer working at a single modulation frequency allowed the complete analysis of the reaction scheme. This will not be possible with the following example where the determination of the actual decay becomes very useful. This second example concerns the study of two aromatic diketopiperazines: cycloglycyltryptophan and cycloalanyltryptophan (Donzell *et al.*, 1974).

These compounds have been examined in dimethylsulphoxide (DMSO) and water solutions. Their fluorescence decays may be fitted by the sum of two exponentials. For instance, the time constants of c-gly.trp in DMSO are $\tau_\alpha = 1{\cdot}9$ nsec and $\tau_\beta = 7$ nsec, whereas scatole and acetyltryptophanamide (Actrp.$NH_2$) decay as single exponential functions with time constants of 7·5 nsec and 7·3 nsec respectively.

It has been shown that c-gly.trp has two conformers in DMSO solutions (Kopple and Onishi, 1969; Donzell, 1971). In the folded form, $A_1$, the indole ring is in close contact with the diketopiperazine ring, whereas in the open form, $A_2$, the indole ring is surrounded by the solvent. The equilibrium constant of association is $K = 1{\cdot}9$ (Donzell, 1971).

The presence of two exponential terms in the decay may be explained if one assumes that each of the conformers has its own rate constant of deactivation, $k_1$ and $k_2$ respectively. The decay of the whole fluorescence is obtained by application of the general kinetics scheme given above. We can write

$$F(t) = F_1(t) + F_2(t) = \frac{1}{1/\tau_\alpha - 1/\tau_\beta}\left[\left(M - \frac{1}{\tau_\beta}\right)\exp\left(-\frac{t}{\tau_\alpha}\right) + \left(\frac{1}{\tau_\alpha} - M\right)\exp\left(-\frac{1}{\tau_\beta}\right)\right]$$

with

$$M = \frac{k_1 K + k_2}{1 + K}$$

$\tau_\alpha$ and $\tau_\beta$ are related to the kinetics constant by the formula (7.22 bis) where $k[B]$ is replaced by $k$.

The ratio of the quantum yield $Q$ of the cyclopeptide to the quantum yield $Q_A$ of the acetyltryptophanamide may be written:

$$\frac{Q}{Q_A} = \frac{1}{\tau_A} [\tau_\alpha + \tau_\beta - M\tau_\alpha \tau_\beta]$$

where $\tau_A$ is the lifetime of acetyltryptophanamide. This relation has been obtained with the reasonable assumption that the rate constant of radiation and the rate constant of deactivation relative to the open form are identical to the corresponding rate constants of acetyltryptophanamide. Then one can put:

$$k_2 = \frac{1}{\tau_A}$$

$k_1$ may be obtained from the expression of $M$. Finally, $k$ and $k'$ are determined by the relation (7.22 bis). The results of the calculation are: $\tau_1 = 1/k_1 = 4{\cdot}2$ nsec; $k = 2{\cdot}7 \times 10^8$ sec$^{-1}$ and $k' = 2{\cdot}3 \times 10^7$ sec$^{-1}$. The equilibrium constant of association in the excited state may be calculated by the relation:

$$K_e = \frac{k'}{k} = 0{\cdot}085$$

$K_e$ is appreciably smaller than $K$. This result means that the folded form is less stable in the excited state than it is in the ground state.

The values of $k$ and $k'$ throw some light on the height of the potential barrier which hinders the rotation around the $C_\alpha$—$C_\beta$ bonds of the cyclopeptides.

In addition, the decay law of the cyclopeptides varies as a function of the emission wavelength. The two time constants remain the same, but the coefficients change. This is explained by a shift of 9 nm between the emission spectra of the two conformers. The analysis of the data allows the reconstitution of these spectra.

Similar results have been obtained for solutions of c-gly.trp in water. In this last case, the folded form is much more stable than in DMSO.

Such experiments should lead to a better understanding of the protein fluorescence emission.

## 6. *Energy transfer*

An excited molecule, called a donor, may transfer its energy to another molecule, called an acceptor, by a radiationless transfer mechanism. Several cases must be distinguished according to the multiplicity of the excited states involved and the strength of the interaction (Förster, 1965). Here, we are going to consider the singlet–singlet radiation transfer and the case of very weak interaction. The rate of such an energy transfer has been calculated by Förster (1948, 1959) and is given by the following relation:

$$k_T = \frac{9000\ (ln\ 10)\ k_F}{128\ \pi^5\ n^4\ N} \times \frac{\kappa^2}{R^6} J \qquad (7.25)$$

where $n$ is the refractive index of the solvent, $N$ the Avogadro number, $J$ the overlap integral of the donor emission spectrum and the acceptor absorption spectrum. $J$ is defined by the following relation:

$$J = \int \frac{\epsilon(\tilde{\nu})F(\tilde{\nu})}{\tilde{\nu}^4}\, \mathrm{d}\tilde{\nu}$$

where $\epsilon(\tilde{\nu})$ and $F(\tilde{\nu})$ are respectively the molar absorptivity and the fluorescence intensity at the wave number $\tilde{\nu}$. $R$ is the distance between the donor and acceptor molecules, $\kappa$ an orientation factor which depends on the relative orientation of these two molecules. The critical distance of transfer $R_0$ is defined as the distance at which the probability of transfer is equal to the probability of emission. Its value is easily obtained from the expression (7.25).

If the acceptor is in a fixed position relatively to the donor, $k_T$ has a defined value and the donor emission has an exponential decay with a time constant given by the following expression:

$$\frac{1}{\tau} = k_F + k_c + k_{isc} + k_T \qquad (7.26)$$

An example of this type of situation has been studied by Haugland, Yguerabide and Stryer (1969). The donor was a carboxyl group and the acceptor a methylindole moiety, both of which were maintained by covalent links in a structure made rigid by a steroid frame.

In a rigid, homogeneous and isotropic solution, the configurations of the acceptor molecules around the donor molecules are variable. There is, therefore, a broad distribution of rate transfer values. The fluorescence decay of the donor is no longer an exponential and is given by (Birks, 1970):

$$i_D = k_F\,[D^*] \left(\exp\left(\frac{-t}{\tau}\right)\right) \exp - \left[2q\left(\frac{t}{\tau}\right)^{\frac{1}{2}}\right] \qquad (7.27)$$

with

$$q = \frac{2\pi^{3/2}\, R_0^3\, N[A]}{3000}$$

where [A] is the acceptor concentration, [D*] the excited donor concentration, $N$ the Avogadro number and $R_0$ the critical distance of transfer.

In a fluid solution, the Brownian motion induces molecular mixing and the energy transfer obeys Stern–Volmer kinetics (cf. Equation 7.18). In intermediate cases, with moderate viscosity, the theoretical treatment becomes very complicated. Several theories have been proposed and detailed discussions on these problems may be found in the reviews of Voltz (1968), Birks (1970), Mataga and Kubota (1970).

## C. Decay of Anisotropy

### 1. *Experimental definition of emission anisotropy*

Viscous solutions of simple chromophores emit a fluorescence which is polarized. Polarization of fluorescence is also present in solutions of chromophores linked to macromolecules, even in solvents of low viscosity. The fluorescence polarization may be characterized by the emission anisotropy (Jablonski, 1960); with a polarized exciting beam, the emission anisotropy is given by the following expression:

$$r = \frac{I_{\parallel} - I_{\perp}}{I_{\parallel} + 2I_{\perp}} \tag{7.28}$$

where $I_{\parallel}$ and $I_{\perp}$ are the principal components of the polarized fluorescence (Perrin, 1929) vibrating in the directions respectively parallel and perpendicular to the exciting vibration (Figure 7.5). With an exciting beam of natural light, the emission anisotropy is:

$$r_{\mathrm{n}} = \frac{I_{\parallel} - I_{\perp}}{2I_{\parallel} + I_{\perp}} \tag{7.29}$$

where $I_{\parallel}$ and $I_{\perp}$ are the components vibrating in the directions parallel and perpendicular to the vibration plane of excitation.

$I_{\parallel}$ and $I_{\perp}$ are conveniently determined by measuring the fluorescence intensities at right angles to the exciting beam, through a rotating polarizer alternatively transmitting the two directions of vibration (Figure 7.5). It is usual to measure the two polarized components on the same photomultiplier, in order to avoid complications arising from differences in sensitivity and time response, if two photomultipliers are used. The fluctuations of flash intensity may be eliminated by measuring the two components many times and alternatively (Wahl, 1969; Auchet and Wahl, 1972).

### 2. *Relation between polarization and molecular properties*

The basic property which causes the fluorescence polarization is the anisotropic character of the absorption and the emission of photons by an aromatic chromophore. This property is characterized by the transition moment associated with the corresponding electronic transition and the fluorescence polarization is a function of the orientation distribution of the transition moments at the time of emission.

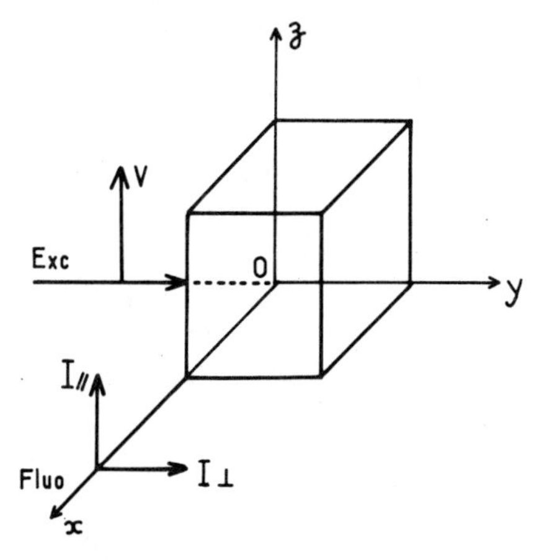


**Figure 7.5.** Experimental definition of the polarized components of the fluorescence

Let us assume, for example, excitation by linearly polarized light; the whole fluorescence intensity is proportional to

$$S = I_{\parallel} + 2I_{\perp}$$
$$\text{or } S = 2I_{\parallel} + I_{\perp}$$

in the case of natural light excitation.
After an excitation by an infinitely short flash, $S$ decays as the whole fluorescence emission decays. If there is any factor which produces a change in the orientation distribution of the transition moments during the decay, then the anisotropy of emission will also vary with time. Such factors are rotational Brownian motion and excitation migration between like chromophores. Let us consider the theory of these two phenomena.

### 3. *Brownian depolarization*

The transition moments of excited molecules constantly change their directions because of the thermal motion occurring in liquids. As a result, the initial angular distribution of these moments, set up at the time of excitation, is perturbed and the fluorescence anisotropy decreases with time. The decreasing rate of change may be computed in applying the theory of Brownian motion, provided that the fluorescent solute molecules are sufficiently large compared to the solvent molecules and assuming that the solvent is continuous. The theory of Brownian depolarization has been worked out for spherical and ellipsoidal molecules by Perrin (1929, 1936). For spherical molecules, the anisotropy decays as a single exponential:

$$r = r_0 e^{-t/\theta} \tag{7.30}$$

where $\theta$, the correlation time is given by:

$$\theta = \frac{\eta V}{kT} \tag{7.31}$$

with $\eta$ the viscosity of the solvent, $V$ the molecular volume, $k$ the Boltzman constant, $T$ the absolute temperature. In the case of ellipsoidal molecules, the absorption transition and the emission transition are defined by *tensors of second order* (Perrin, 1936). The general formulae obtained are applicable to linear moments as well as circular moments. The more restricted case of symmetric ellipsoids bearing linear moments has been considered by Memming (1961) while Tao (1969) assumed that the transition moments of absorption and emission are parallel. The emission anisotropy decay is a sum of five exponentials (Perrin, 1936; Tao, 1969; Belford *et al.*, 1972; Chuang and Eisenthal, 1972; Ehrenberg and Rigler, 1972). For a symmetric ellipsoid, the number of exponentials is reduced to three and one may write:

$$r(t) = A_1\,\mathrm{e}^{-t/\theta_1} + A_2\,\mathrm{e}^{-t/\theta_2} + A_3\,\mathrm{e}^{-t/\theta_3} \tag{7.32}$$

The three correlation times $\theta_1$, $\theta_2$, $\theta_3$ may be written as follows:

$$\theta_i = \frac{\eta V}{kT} f_i(\rho) \tag{7.33}$$

where $f_i(\rho)$ are functions of the axial ratio $\rho$ of the ellipsoid which may be determined from Perrin's work (1936). These functions have been tabulated by Tao (1969). $A_1$, $A_2$, $A_3$ may be expressed as functions of the three director cosines, $\gamma_1$, $\gamma_2$, $\gamma_3$, of the absorption transition moment and of the corresponding quantities $\delta_1$, $\delta_2$, $\delta_3$ of the emission moment. Here the reference axes are the principal axes of the ellipsoid. Thus, (Wahl, Meyer and Parrod, 1970)

$$\begin{aligned}
A_1 &= \frac{3}{5}\left[\frac{(\gamma_1^2 - \gamma_2^2) + (\delta_1^2 - \delta_2^2)}{2} + 2\gamma_1\gamma_2\delta_1\delta_2\right] \\
A_2 &= \frac{6}{5}\,[\gamma_2\gamma_3\delta_2\delta_3 + \gamma_1\gamma_3\delta_1\delta_3] \\
A_3 &= \frac{1}{10}\,[3\gamma_3^2 - 1]\,[3\delta_3^2 - 1]
\end{aligned} \tag{7.34}$$

If the solution contains a mixture of molecules bearing chromophores randomly orientated with respect to the molecular axes, $A_1$, $A_2$, $A_3$ have to be averaged and the following values are obtained:

$$A_1 = A_2 = 2A_3 = \frac{2r_0}{5} \tag{7.35}$$

It may be reasonably assumed that such a situation occurs with many proteins labelled with a synthetic chromophore such as the dansyl (dimethylnaphthalene sulphonyl) group (Weber, 1952, 1953).

Chromophores linked to a macromolecule frequently have a certain local freedom of rotation around the point of attachment to the macromolecule. For example, rotation may occur around a valence bond. A simple model corresponding to that situation consists of a small spherical chromophore rotating freely around an axis bound to a spherical macromolecule.

The decay of anisotropy is given by the following formula (Gottlieb and Wahl, 1963; Wallach, 1967):

$$r(t) = e^{-t/\theta} (A_1 e^{-2t/3\theta_i} + A_2 e^{-t/6\theta_i} + A_3) \qquad (7.36)$$

where $\theta$ is the relaxation time of the macromolecule and $\theta_i$ the relaxation time of the chromophore.

If the macromolecule is very large, its motion is negligible, and $r(t)$ is reduced to:

$$r(t) = A_1 e^{-2t/3\theta_i} + A_2 e^{-t/6\theta_i} + A_3 \qquad (7.37)$$

An infinite time, $r(t)$ reaches a constant value $A_3$. Generalizing this result, one expects the emission anisotropy to reach a finite value, when the fluorescent chromophore is attached to a rigid structure.

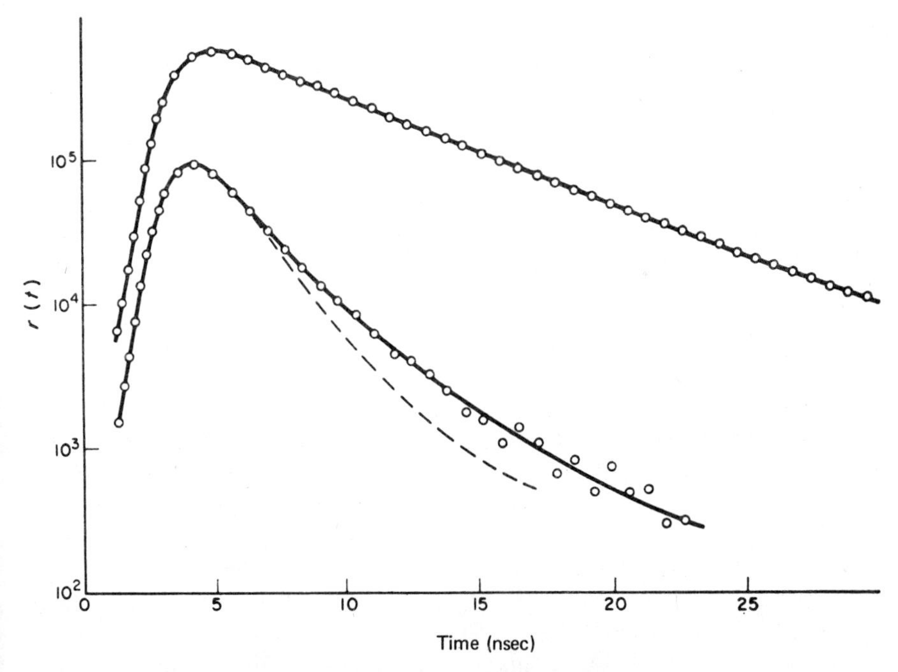


**Figure 7.6.** Decay of polarized fluorescence of a styrylphenylanthracene residue included in a polystyrene polymer. The upper curve ($s(t)$) and the lower curve ($d(t)$) are obtained respectively by adding and subtracting the $I_{\parallel}(t)$ and $I_{\perp}(t)$ decays, and are defined by: — $s(t) = I_{\parallel}(t) + 2I_{\perp}(t)$; — $d(t) = I_{\parallel}(t) - I_{\perp}(t)$. (From Wahl, Meyer and Parrod, 1970)

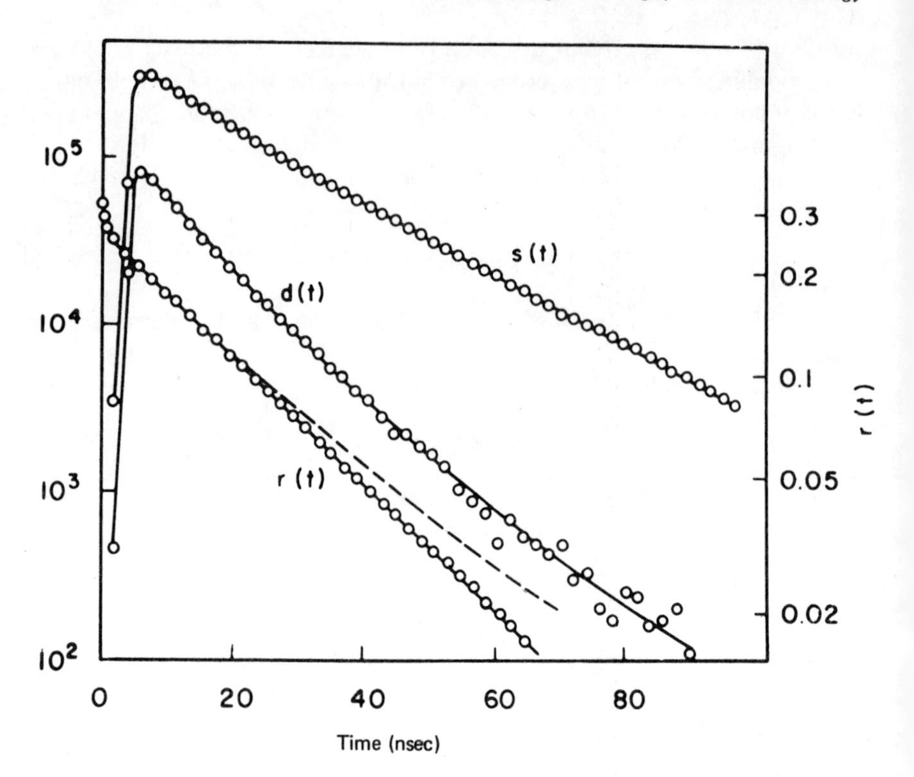


**Figure 7.7.** Fluorescence decay curves of the dansylated $\beta$-lactoglobulin-A dimer. $r(t)$ is the decay of emission anisotropy: the solid line is calculated for an ellipsoid of axial ratio 2. The dotted line for an axial ratio 3. (After Wahl and Timasheff, 1969)

By following the decay of anisotropy, it is thus possible to reveal the internal flexibility of macromolecules, provided that the characteristic correlation times are in the nanosecond range.

With polystyrene, we have an example of a flexible molecule. With this compound labelled with a fluorescent monomer, complete depolarization occurs solely by local motion (Figure 7.6). This motion involves a segment made of a few monomer residues situated in the neighbourhood of the fluorescent residue. An equivalent ellipsoid may be determined, the size of which corresponds to a molecular chain segment of about thirty monomers (Wahl, Meyer and Parrod, 1967, 1970).

Anisotropy decay determinations performed on labelled globular proteins show that these molecules are rigid (Figure 7.7) (Wahl, 1966; Wahl and Timasheff, 1969; Tao, 1969; Yguerabide, Epstein and Stryer, 1970). The correlation times measured are generally greater than the value expected from the known dimensions of the molecule. This discrepancy may be explained by including in the molecular volume a layer whose thickness corresponds to one water molecule.

An interesting example of the application of this technique was to the study of the immunoglobulin IgG molecule, since IgG is known to consist of three globular fragments (two Fab and one Fc) linked together by a region preferentially attacked by proteolytic enzymes. The anisotropy decay of dansylated IgG shows the presence of an internal motion, but the relaxation involved is too small to be attributed to the free rotation of the Fab or Fc fragments (Wahl, 1969; Brochon and Wahl, 1972). On the other hand, the anisotropy decay of the anti-(dansyl-lysine) antibody–dansyl-lysine complex can be interpreted on the basis of partial free rotation of the Fab fragments (Yguerabide *et al.*, 1970). We think however that further measurements are needed in order to establish that point definitely (Brochon and Wahl, 1972).

Ethidium bromide dye intercalated in the DNA helix shows a restricted motion which has been interpreted as the result of a local flexibility of the DNA molecule (Wahl, Paoletti and Le Pecq, 1970). Anisotropy decay measurements have been performed on the membrane of the electric organ of *electrophorus electricus*. The dansyl moiety covalently linked to the protein phase of the membrane shows a very restricted motion, as may be seen by the rapid decrease of the anisotropy followed by a plateau (Figure 7.8). Furthermore, the anisotropy of the dye attached to the membrane remains constant through the time range of the measurement, showing that the membrane structure is rigid (Wahl, Kasai and Changeux, 1971). The technique of anisotropy decay allows the study of molecular motions with correlation times ranging from fractions of nanoseconds to several hundred nanoseconds. On the other hand, with the spin label technique, the correlation times measurable are in the range of several picoseconds to a few nanoseconds. In this respect, the two techniques are complementary.

### 4. *Energy migration*

An excited molecule can transfer its energy to a molecule of the same kind situated in its proximity; this process may be repeated many times and is termed energy migration. Energy migration does not lead to any change in the quantum yield or in fluorescence decay. Fluorescence depolarization does however occur and is known as concentration depolarization. Most of the theoretical work on concentration depolarization concerns rigid isotropic solutions. A rigorous treatment of this problem would involve very complicated calculations and different approximate methods of calculation have been proposed. A critical review of these theories has been given by Knox (1968). Simulation by the *Monte Carlo method* has recently been used for the calculation of the average polarization (Paoletti and Le Pecq, 1971) and the decay of anisotropy (Genest and Wahl, 1973, 1974). These methods seem to be very promising.

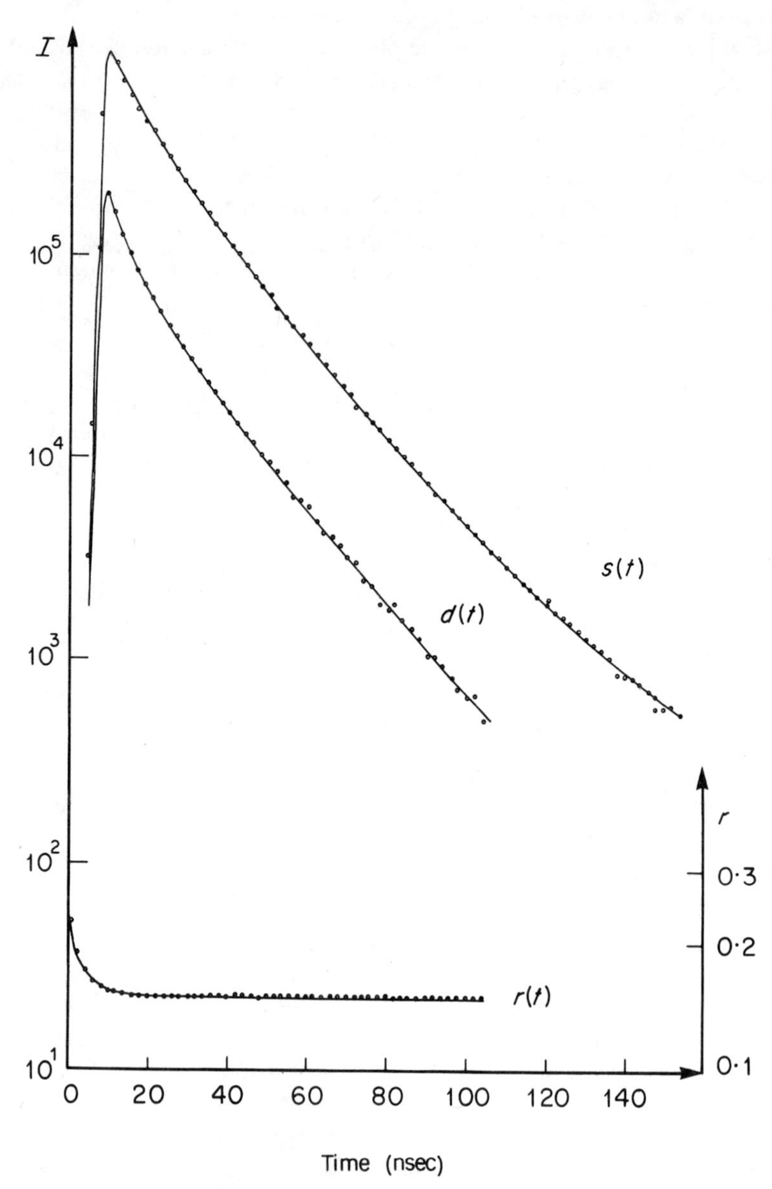


**Figure 7.8.** Fluorescence decay of dansyl residues attached to native membrane fragments of the electric organ of *electrophorus electricus*. (From Wahl, Kasai and Changeux, 1971.) $s(t)$ and $d(t)$ have the same definition as in Figure 7.6; $r(t)$ is the decay of anisotropy

The energy migration may occur with high efficiency between chromophores linked to the same macromolecule. The anisotropy decay will be the result of the superposition of the Brownian motion and the energy migration phenomenon. An extrapolation to zero concentration of labelling dye allows the determination of the correlation time of the macromolecule (Fayet and Wahl, 1969). The anisotropy decay of the complex formed between ethidium bromide and DNA has been measured: this decay is accelerated as the ratio $D/P$ (number of dyes per nucleotide) is increased. The result has been compared to formulae computed on the basis of various theoretical approximations. *Jablonski's approximation* of the active sphere does not satisfactorily describe the experimental anisotropy decay. On the other hand, Förster's approximation in which the energy exchange occurs only between the primary excited molecule and its nearest neighbour, is able to reproduce the experimental data obtained at low $D/P$ (Genest and Wahl, 1972, 1973).

A much better fit to the experimental data has been obtained by computing the anisotropy decay with a Monte Carlo method. The curves obtained depend markedly on the assumed deformation angle induced by the ethidium bromide in the DNA molecule (winding or unwinding of the DNA molecule). It is therefore possible to determine that angle by comparing the calculated anisotropy decay with the experimental decay (Genest and Wahl, 1973, 1974). This method may be generalized to other systems in which identical chromophores are disposed in a rigid matrix and could be useful in elucidating the structure of these systems.

## III. MEASUREMENTS OF FLUORESCENCE DECAY

### A. Different Methods of Measurements

Many apparatuses for measuring fluorescence decay times have been described in the literature. The different types may be classed under two main headings:

*a.* The *Phase and Modulation Fluorometers*, in which excitation is produced by a light source modulated with a periodic device at high frequency. The phase and/or modulation of the fluorescence emission is compared with the phase and/or modulation of the exciting light.

*b.* The *Pulse Fluorometers*, in which excitation is provided by a flash of very short duration, yielding a fluorescence pulse. This pulse is then recorded with a fast electronic device, and analysed by various methods. I shall not enter into a description of the numerous machines which have been built on these principles. An excellent account is to be found in the reviews by Birks and Munro (1967) and by Ware (1971). Let us however notice that an improved phase and modulation fluorometer working with two high frequencies has recently been built by Spencer and Weber (1969), and has been used by them for studying biological systems. A relatively simple pulse fluorometer is commercially available and has been described at length (Meserve, 1971). I shall turn to a more detailed description of the single photoelectron counting method, the principle of which has been utilized in different apparatuses: (Bollinger and Thomas, 1961; Koechlin, 1961; Pfeffer, 1965; Pfeffer *et al.*, 1963; Tao, 1969; Wahl, 1969; Ware, 1971; Breuze, 1971; Yguerabide, 1972; Isenberg 1975).

## B. The Single Photoelectron Counting Method

### 1. *Historical survey*

The origin of the method is found in the delayed coincidence method used to measure the lifetime of the excited states of nuclei (Bell, 1965). The method was then adapted for the measurement of lifetimes of atoms excited by a pulsed electron beam (Heron *et al.* 1956; Bennett *et al.*, 1965). In this work, the principle of the detection of photon arrival time on a photomultiplier was adopted. Independently, Bollinger and Thomas (1961) and Koechlin (1961) devised an apparatus for time measurements of the pulse of light emitted in scintillation. In these apparatuses, multichannel delayed coincidence is introduced, which implies the use of a time to amplitude converter (TAC), and of a multichannel analyser (MCA). The apparatus of Bollinger and Thomas had relatively poor resolution and a large range of time analysis, called window. Koechlin's study was directly linked to the building of a new generation of fast photomultiplier, the Radiotechnique 56 AVP (Pietri, 1961). The apparatus of Koechlin achieved very good time resolution but had a narrow window of analysis. Pfeffer (1965) and Pfeffer *et al.* (1963) used an excitation light flash and studied the fluorescence of various scintillator solutions. This apparatus embodied good time resolution with a large window and, in this form, became a true pulse fluorometer. The same type of apparatus was used in the first measurements on biological material (Wahl and Lami, 1967) by the photocounting method, and in the first measurement of anisotropy decay of macromolecules (Wahl, 1966; Wahl, Meyer and Parrod, 1967).

### 2. *Performance of the technique*

The electronics associated with the method are essentially those used in fast time measurements in nuclear physics and the technique has benefited from all the improvements introduced in that area. In its present form, the method offers a whole combination of advantages not found in most other methods of lifetime measurement, viz:

*a.* the determination of fluorescence decays is possible without an *a priori* hypothesis as to the decay law;

*b.* maximum sensitivity, since single photons are counted;

*c.* practically all wavelengths of excitation are available;

*d.* good time resolution;

*e.* large range of window available, commonly varying from 50 nsec to tens of $\mu$sec;

*f.* digital data, ready for the treatment by computers, are provided.

### 3. *Principle of the method*

The block diagram of a simple device is shown in Figure 7.9. The exciting light is produced by the discharge of a capacitor across two electrodes set in gas. The flash is repeated at an average frequency of about 10 kHz. The light from the flash is filtered through a monochromator or a filter, which determines the range of wavelength of excitation, before falling on the cell containing the fluorescent solution. The fluorescence is viewed at 90° by a fast high-gain photomultiplier (PM) in front of which has been put a second monochromator or filter. The intensity of each pulse of light reaching the cathode of that PM must be low enough so that no more than one photoelectron should be ejected at each burst of the flash. We shall discuss this condition below in more detail. In front of the photomultiplier, a diaphragm may be necessary to reduce the light flux. A photoelectron ejected at the cathode produces a pulse at the anode of the PM; that pulse is sent to the stop input of a time to amplitude converter (TAC) at time $t_1$. The start pulse is provided by an electric pulse synchronous with the light flash, and determines the time of reference $t_0$. At its output, the TAC delivers a pulse whose amplitude is proportional to $t_1 - t_0$. The multichannel analyser (MCA) measures that amplitude and stores one count in the channel the address of which is proportional to that amplitude. Therefore, to a given channel of the MCA corresponds a given value of $t_1 - t_0$. After many excitation cycles, the number of counts accumulated in a channel is proportional to the probability of the emission of a photon at a given time after the excitation, and the curve representing the content of each channel versus its address reconstitutes the shape of the fluorescence pulse. The data may be read out with various devices, such as printer, plotter, punch tape recorder, etc., or they may be fed directly into a computer to be analysed.

It is necessary to establish the correspondence between the channel address and the time difference. This is done by inserting various known delays (coaxial cables of known length) in the stop signal branch, and measuring the corresponding displacement of the channel address of a given point of a curve (usually the maximum of a sharp peak). The slope of the straight line representing the time delay versus the channel address, gives the calibration factor in nanoseconds for each channel.

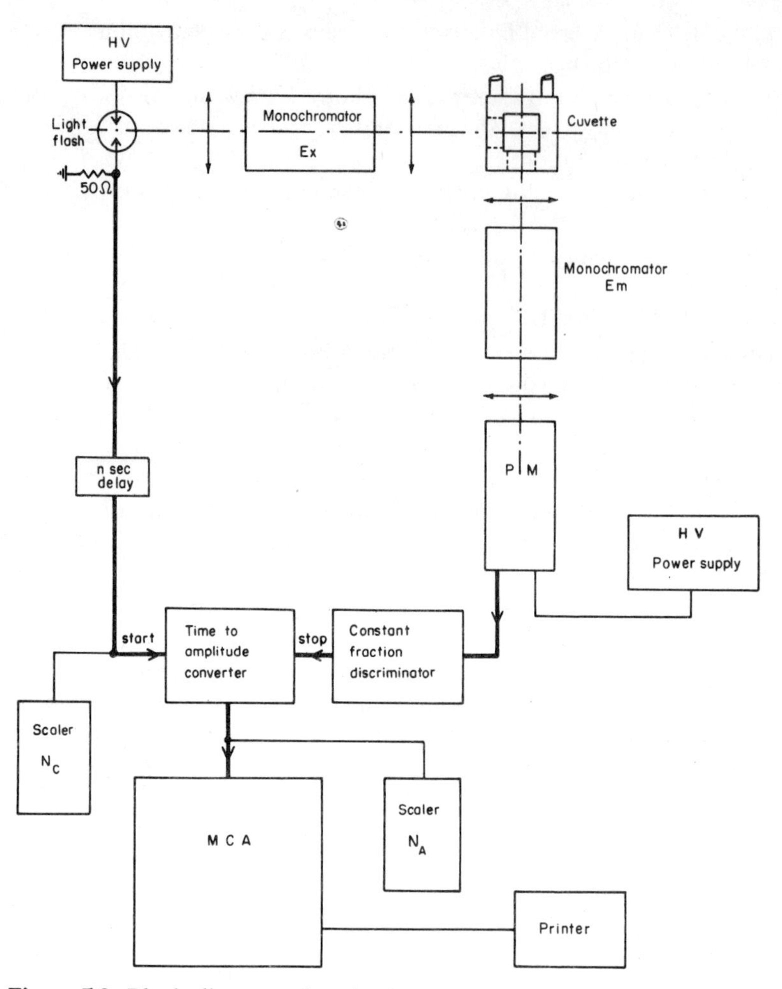


**Figure 7.9.** Block diagram of a simple photocounting pulse fluorometer (Auchet and Wahl, 1972). MCA, multichannel analyser

It is useful to have two scalers, one counting the total number of flash bursts (number of cycles $N_c$), and the other counting the number of correlated anode pulses $N_A$. One can control the single photocounting conditions by means of the value of $N_A/N_c$. To that basic scheme may be added several devices, about which we shall talk later.

### 4. *Measure of the decay of anisotropy*

In these measurements, the decays of the polarized components of the fluorescence $I_{\parallel}(t)$ and $I_{\perp}(t)$ must be determined under the same conditions (Wahl, 1969). This is realized by measuring these components repeatedly and alternately and summing the results so that fluctuations of the flash intensity are averaged out. Figure 7.10 gives a schema of an automatic apparatus used in our laboratory (Auchet and Wahl, 1972). The memory of an 800 MCA is split into two halves, one accumulating $I_{\parallel}(t)$, the other $I_{\perp}(t)$. The apparatus performs many successive cycles of measurement automatically. Each cycle includes:

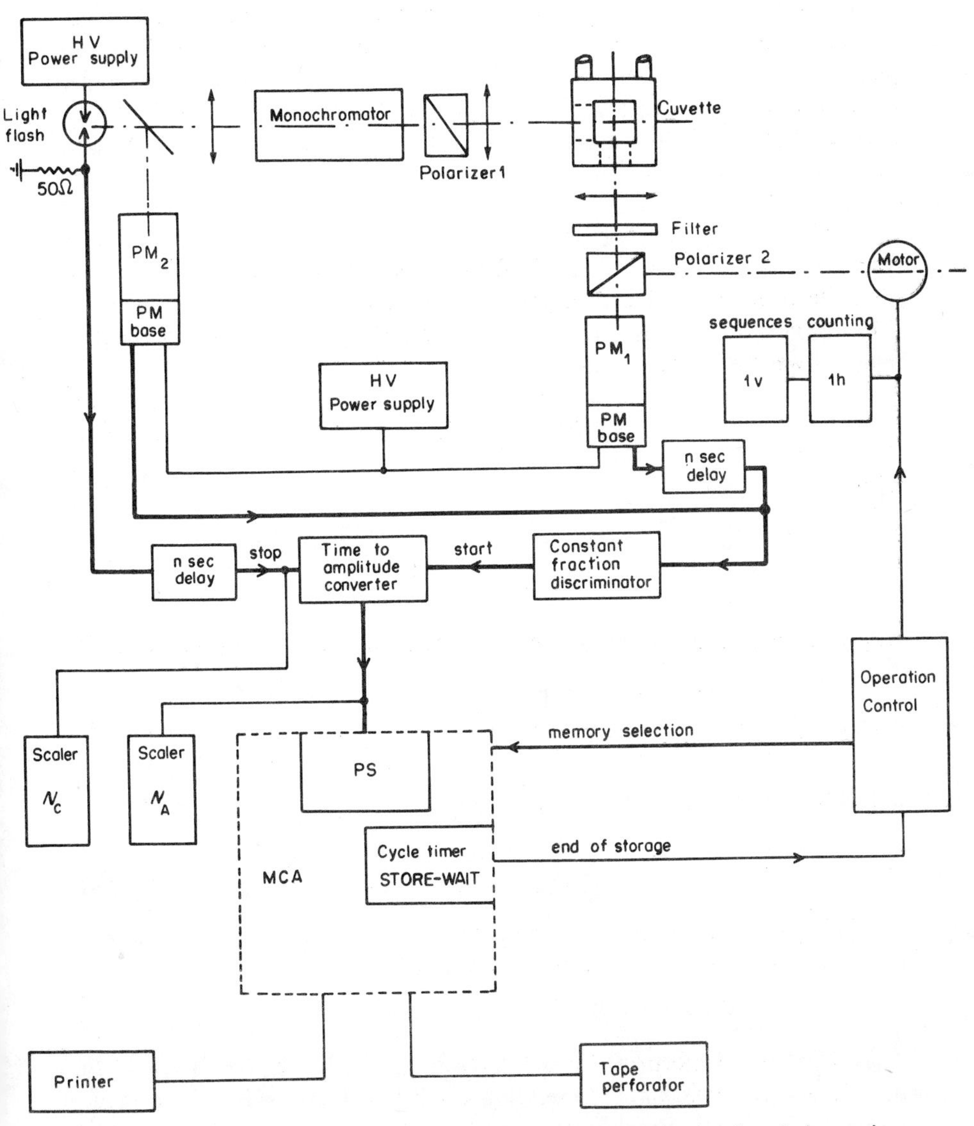


**Figure 7.10.** An apparatus for the determination of the decay of emission anisotropy. PS, peak stabilizer MCA, multichannel analyser. (Auchet and Wahl, 1972)

*a.* a storage time while one of the components is measured, and
*b.* a waiting time used to prepare for the measurement of the other component during which the polarizer is rotated and the storage grouping changed The number of measurements of each component is registered on counters.

### 5. *Photocounting*

We assume that, during an excitation cycle (one burst of the flash lamp), an average number of $Z$ photons impinge on the photocathode. Some of those photons arrive in the time interval $t_{i-\frac{1}{2}}$ to $t_{i+\frac{1}{2}}$, corresponding to the channel $i$. Let us assume that their average number is $z_i$. These photons will eject an average number of photoelectrons, $x$, from the PM photocathode. Among them, $x_i$ occurs at the time $t_i$ with:

$$x_i = \beta z_i \tag{7.38}$$

where $\beta$ is the quantum efficiency of the photocathode. More precisely, the photoelectron emission is a random process which obeys a Poisson distribution (Mandel and Wolf, 1965). The probability of emission of $k$ photoelectrons between $t_{i-\frac{1}{2}}$ and $t_{i+\frac{1}{2}}$ is:

$$p_k(i) = \frac{(x_i)^k}{k!} \exp(-x_i) \tag{7.39}$$

with

$$\sum_{k=0}^{\infty} p_k(i) = 1 \tag{7.40}$$

Applying the formula for zero, one and more than one gives (Koechlin, 1961):

$$\begin{aligned} p_0(i) &= \exp(-x_i) \\ p_1(i) &= x_i \exp(-x_i) \\ p_{k>1}(i) &= 1 - p_0(i) - p_1(i) = 1 - (1 + x_i) \exp(-x_i) \end{aligned} \tag{7.41}$$

Let us assume that $x_i \ll 1$. From (7.41):

$$\begin{aligned} p_1(i) &\approx x_i \\ p_{k>1}(i) &\approx x_i^2 \end{aligned} \tag{7.42}$$

and $p_{k>1}(i)$ is then negligible compared to $p_1(i)$. After a great number of cycles $N_c$, the number $N_i$ of anode pulses at a time $t_i$ is:

$$N_i = N_c(p_1(i) + p_{k>1}(i)) \approx N_c x_i \approx N_c \beta z_i \tag{7.43}$$

$N_i$ is therefore proportional to the intensity of the fluorescence pulse at the time $t_i$, in the single photon counting condition. Nevertheless $N_i$ does not represent the count stored in the multichannel analyser. This is because the TAC only detects the first photon emitted in a given cycle. Coates (1968) has shown that the true count $N_i$ is related to the experimental count $M_i$ stored in the analyser by the formula:

$$N_i = \frac{M_i}{1 - \frac{1}{N_c} \sum_{j=1}^{i-1} M_i} \tag{7.44}$$

One has necessarily:

$$\sum_{j=1}^{j=i} M_i \leqslant N_A \tag{7.45}$$

Therefore, if $N_A \ll N_c$, then $N_i = M_i$, and no correction is needed. $N_A/N_c \ll I$, (say $N_A/N_c = 1\%$), is the condition where no more than one photoelectron is emitted per cycle. Under this condition, the emission of a single photon at a time $t_i$ is fulfilled *a fortiori*. In this case, the curve stored in the multichannel is the true fluorescence curve. Even if $N_A/N_c$ is not very small but, for all channels $i$, $x_i$ is very small, we may still determine the true shape of the fluorescent pulse by applying the formula (7.44).

Miehe *et al.* (1970), Davis and King (1970) have considered the case where all cycles corresponding to the emission of more than one photoelectron are excluded from counting. The probability of detecting a pulse at the time $t_i$ is now (Miehe *et al.*, 1970):

$$P_1(i) = p_1(\mathrm{i}) \prod_{j \neq i}^{\infty} p_0(j) = x_i \exp(-x_i) \times \exp - \left( \sum_{j \neq i}^{\infty} x_i \right)$$

and finally

$$P_1(i) = x_i \exp(-x)$$

The number $M_i$ of count is therefore

$$M_i = N_c \beta z_i \exp - \beta z$$

This number is proportional to the fluorescence intensity $z_i$ without restriction of counting rate. The total number of useful counts $N_A$ is:

$$N_A = \sum_{i=1}^{m} M_i = N_c x \exp - x$$

The maximum rate of useful counts is obtained for:

$$\frac{N_A}{N_c} = \mathrm{e}^{-1} = 0{\cdot}37 \qquad \text{when } x = 1$$

The exclusion of cycles corresponding to the emission of several photoelectrons may be obtained in several ways by the use of an 'inhibit function'

(Davis and King, 1970), or better by selection of amplitude (Miehe *et al.*, 1970). This method is only possible if one uses a recent photomultiplier like the 8850 RCA, which has a narrow amplitude distribution of the anode pulse corresponding to a single photoelectron. The selection of amplitude is made with a single-channel analyser on a slow channel parallel to the fast channel, (Figure 7.11).

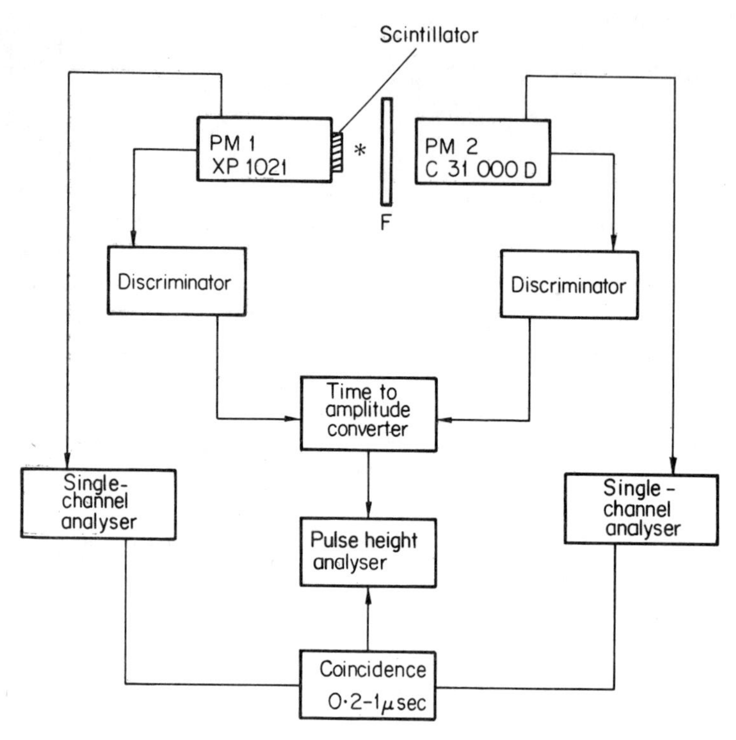


**Figure 7.11.** Block diagram of a photocounting apparatus with a slow channel used for amplitude selection. (After Miehe, Ambard, Zampach and Coche, 1970)

### 6. *Response of the apparatus*

*a. Introduction.* The function that we want to determine is the fluorescence decay $F(t)$ which is defined as the time course of fluorescence after excitation by an infinitely short flash ($\delta$ excitation); because of the imperfection of the apparatus, we measure a function $f(t)$ which is related to $F(t)$ by a convolution integral (Birks and Munro, 1967):

$$f(t) = \int_{-\infty}^{t} g(s)\, F(t-s)\, \mathrm{d}s \qquad (7.46)$$

$g(t)$ is the response function of the apparatus. It is conventionally written:

$$f(t) = g(t) * F(t)$$

We will consider here for simplicity that all the functions are normalized to 1; $g(t)$ itself must also be considered as the following product of convolution:

$$g(t) = E(\lambda_E, t) * H(\lambda_e, t) * K(t) \tag{7.47}$$

where

$E(\lambda_E, t)$ is the excitation function; in other words, $E$ is the time distribution of the flash intensity at the wavelength of excitation $\lambda_E$;

$H(\lambda_e, t)$ is the response function of the photomultiplier at the wavelength of emission;

$K(t)$ is the response function of the electronic circuits.

There are two important points concerning $g(t)$ from the point of view of decay measurements:

(i) The precision of the determination of $F(t)$ depends on the shape of $g(t)$, its width, slope, etc. This will dictate the time resolution of the apparatus.

(ii) It is necessary to know $g(t)$ in order to extract $F(t)$, from the measurement of $f(t)$.

*b. Simple parameters to characterize* g(t). The shape of $g(t)$ is rather complex: it rises abruptly then passes through a maximum and decreases with a decreasing slope, ending with a flat tail. The width of $g(t)$ may be characterized by the mean square deviation (or variance, or radius of gyration) $\sigma$ defined by the relation:

$$\sigma_g^2 = \int_{-\infty}^{+\infty} (t - \mu_1)^2 g(t)\, dt \tag{7.48}$$

where $\mu_1$ is the abscissa of the centroid, (or centre of gravity); $\mu_1$ is also the reduced moment of order one of the curve $g(t)$. From the relation of Bay—see below—(Bay *et al.*, 1955; Weaver and Bell, 1960), it follows that:

$$\sigma_g^2 = \sigma_E^2 + \sigma_H^2 + \sigma_K^2 \tag{7.49}$$

where $\sigma_E$, $\sigma_H$, $\sigma_K$ are the radius of gyration of the curves $E$, $H$, $K$. Another convenient quantity to characterize the width of $g(t)$ is the full width at half maximum (FWHM).

For a gaussian response, the relation exists:

$$\text{FWHM} = 2{\cdot}36\,\sigma. \tag{7.50}$$

One may also introduce the rise time (time elapsed between the points having intensities equal to 10% and 90% of the maximum), and a decay time to characterize the tail.

7. *Description of the different parts of the apparatus*

*a. The flash lamp* (Figure 7.12). The simpler way to operate the flash is in the relaxation mode. A high voltage supply charges the capacitance of the electrodes through a resistor R (several times ten MΩ). When the voltage reaches the breakthrough potential $V_D$, the flash fires. $V_D$ depends on the nature of the gas, on its pressure and on the gap between the electrodes (Paschen's law). If the power supply delivers a constant voltage $V_0 \gg V_D$, the frequency of the flash is:

$$f = \frac{V_0}{V_D RC}$$

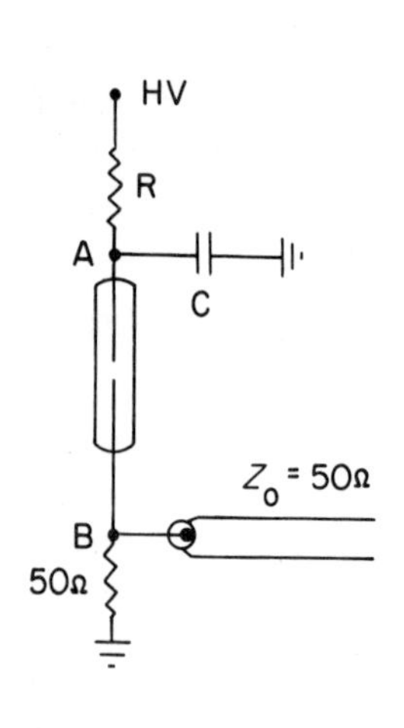


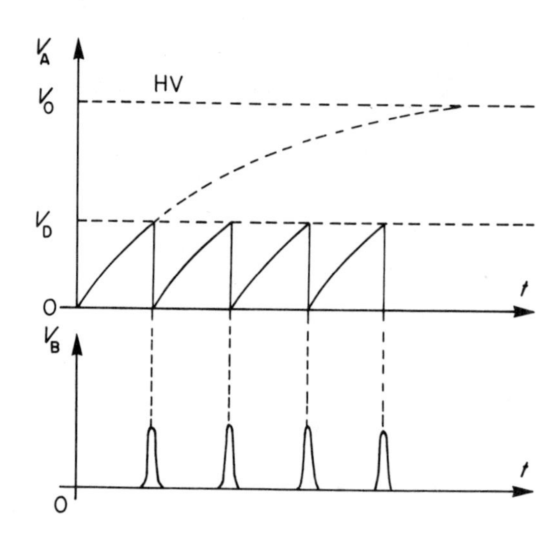


**Figure 7.12.** Principle of the flash operated in relaxation mode (after Pfeffer, 1965): (a) scheme of the lamp; HV: high voltage power supply, C: stray capacity, R: charge resistor, (b) Time course of potentials at points A ($V_A$) and B ($V_B$). $V_D$ is the breakthrough potential. The electric pulse ($V_B$) is sent to one of the 'time to amplitude converter' inputs (see Figure 7.9 and Figure 7.10)

where $R$ is the charge resistance and $C$ the stray capacitance of the electrodes (usually 10 to 30 pF). $f$ may be varied by changing the value of $R$ and $V_0$. The gap between the electrodes ranges from 1 mm to some tenths of a millimetre. The pressure may vary from a fraction of an atmosphere to 20 atmospheres. The spark obtained in free air is very convenient for excitation wavelengths ranging between 300 nm and 400 nm. Its spectrum shows several strong lines, as one may see in Figure (7.13). Furthermore, the tail of the flash is particularly short. The electrodes are made of tungsten.

For excitation in the visible, we use high pressure (10 bars or more) nitrogen:

the structure of the lamp is then an envelope of brass with a glass or quartz window. In the ultraviolet region, good results are obtained with hydrogen or deuterium at high pressure. The electrodes are silver wetted with mercury (Auchet and Wahl, 1972). Descriptions of lamp flashes may be found in the literature (Malmberg, 1957; Pfeffer *et al.*, 1963; D'Alessio *et al.*, 1964; Yguerabide, 1965; Hundley *et al.*, 1967; Ware 1971). Commercial apparatus is also available (Radiotechnique; Ortec see p. 277).

The true shape of the excitation function is not easy to obtain since one has to take into account the response of the measuring apparatus. A special device with an image converter has been built in order to examine very short flashes. A rise time of 0·2 nsec and a FWHM smaller than 0·5 nsec have been

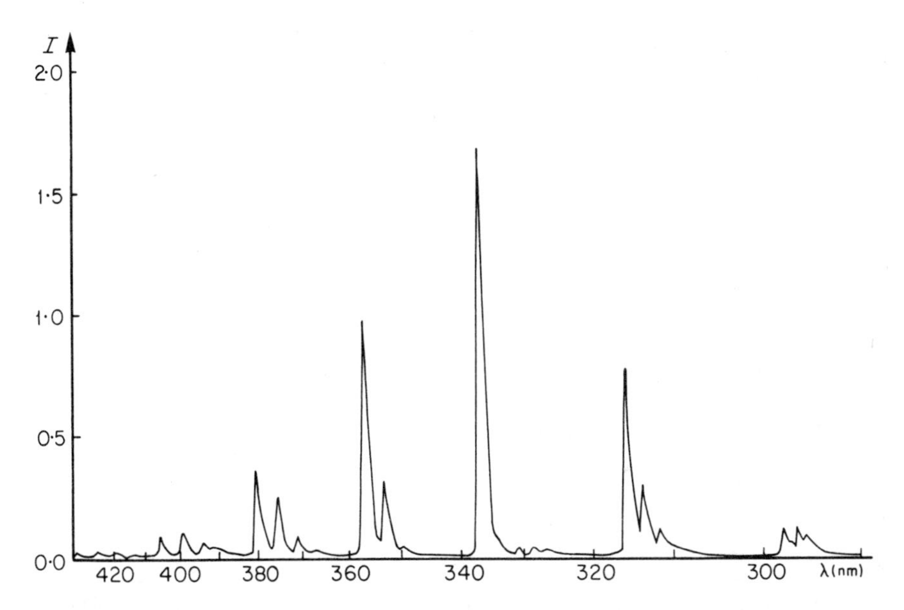


**Figure 7.13.** Emission spectrum of an airflash. The spectrum is not corrected and is measured with a spectrofluorometer Jobin Yvon. In this measurement, the flash replaces the cell compartment. (Auchet and Wahl, 1972)

measured (D'Alessio *et al.*, 1964; Yguerabide, 1965). It is also possible to to use a fast diode (rise time 0·2 nsec to 0·4 nsec) with a wide-band oscilloscope (for example a sampling oscilloscope) to study the excitation function.

A study of the characteristics of the flash under various conditions has been done by Yguerabide (1965). It has been observed (Wahl, Auchet and Donzell, 1974; Lewis *et al.*, 1973) that the excitation function depends on the wavelength of excitation. That result is of some importance with regard to the analysis of the experimental data. We shall come back to this point later.

*b. The photomultiplier.* A photon impinging on the photocathode of a PM ejects a photoelectron; the photoelectron is accelerated in the space between the cathode and the first dynode by the electrostatic field; its impact produces several secondary electrons. This process of multiplication is repeated in each interdynode stage. Finally, on the anode, one obtains an electric pulse called the single electron response (SER), the shape of which is characteristic of the photomultiplier structure. The electric charge of the SER expressed in electron charges is equal to the gain of the photomultiplier. An example of SER is given in Figure 7.14. Until recently, the amplitude fluctuations of the SER were very large. It was not possible to distinguish between an anode pulse corresponding to one, two or three photoelectrons. Recently, the use of a first dynode coated with gallium phosphate activated with caesium

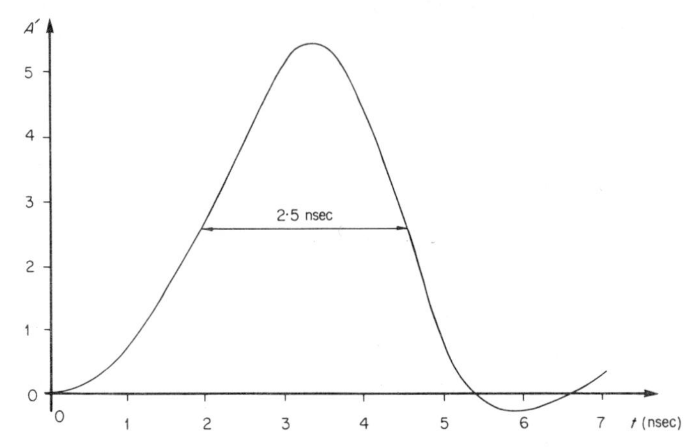


**Figure 7.14.** Single photoelectron response of the PM—Radio-technique 56 AVP. (After Pfeffer, 1965)

(GAP) allows the amplitude of the SER to be distinguished from the amplitude of anode pulses corresponding to 1, 2, 3, 4, etc. photoelectrons.

Time resolution. In ordinary pulse fluorometry, the response function of a photomultiplier is identical to its SER. In the photocounting method, the response function which we have called $H(t)$ (see Equation 7.47, p. 267), results from the fluctuations of the transit time (time elapsed between the ejection of a photoelectron and the arrival of the corresponding anode pulse). Theoretical studies (Gatti and Svelto, 1959), have shown that $H(t)$ and its mean square deviation $\sigma_H$ depend essentially on the fluctuations of the flight time of the photoelectrons in the space between the photocathode and the first dynode. The average flight time is longer for a photoelectron leaving the rim of the cathode than for a photoelectron leaving its centre; consequently $\sigma_H$ increases with the fraction of illuminated surface of the photocathode. The photoelectrons are ejected with various initial velocities. The greater the energy $h\nu$ of the impinging photons, the broader is that velocity distribution (Wahl, Auchet and Donzell, 1974). The influence of this initial velocity on the total time of flight of the photoelectron decreases when the electric field of acceleration in the proximity of the photocathode increases. In the space between the photocathode and the first dynode, one puts various electrodes (focusing and accelerating electrodes) in order to minimize the various causes of the time of flight dispersion and this determines the input optics of the photomultiplier. All these considerations will determine the choice of a photomultiplier for fluorescence decay measurements by the photocounting method. It must, in addition, have a high gain, between $10^7$ and $10^8$, and a small photocathode noise.

The Radiotechnique photomultipliers of the 56 AVP and 56 DVP (or DUVP) series, used to display the best characteristics from these points of view. Various measurements of the resolution function FWHM range from 1 nsec to 1·6 nsec (Pfeffer, 1965; Vallat, 1969; Breuze and Savine, 1969; Lami *et al.*, 1971). Improved performances have been obtained with photomultipliers of the RCA 8850 series (with GAP first dynode), which exhibit a FWHM of the order of 0·50 nsec (Present and Scarl, 1970; Miehe and Knispel, 1971) and with the Radiotechnique 2106 PM (FWHM = 0·65 nsec) (de la Barre, 1972).

It has been found that the response function of the 56 DUVP depends considerably on the wavelength of the incident light. On the contrary the type 2106 has a practically constant response over a large wavelength range (Wahl, Auchet and Donzell, 1974). That is explained by the fact that the input optics have been improved and, in particular, the electric field in the proximity of the cathode is much higher than it is with the 56 DUVP. The behaviour of RCA 8850 is between the two Radiotechnique tubes (Wahl and Auchet, unpublished observation).

*c. Fast electronics.* The time to amplitude converter TAC (see Figure 7.9) requires pulses of standard shape and amplitude at its inputs, in order to give the optimum time resolution. These pulses are generated from the photomultiplier pulse with a discriminator circuit. The discriminator triggers its output pulse when the input pulse reaches a given level. With pulses having a broad distribution of amplitude, like the SER pulses, a discriminator with a fixed triggering level (leading edge discriminator) defines the time with a large uncertainty. A much better timing definition is obtained (0·1 nsec) if the discriminator has its level adjusted to a constant fraction of the amplitude (constant fraction timing discriminator) (Gedcke and McDonald, 1968).

The resolution may be improved with the use of a single-channel analyser on a lateral slow channel which prohibits all the pulses outside a narrow range of amplitude.

A circuit especially well suited for the single photocounting technique and achieving a very good time resolution has been described by Odru, Prieur-Drevon and Vacher (1970), and tested by Breuze (1971).

*d. Electronic drift.* When performing long experiments, lasting several hours, one may observe a drift in the electronics provoking a shift of the address of the curve in the MCA; such a drift often comes from the instability of the TAC and may be corrected with a second photomultiplier viewing the flash directly (Wahl, 1969) see Figure 7.10). The impulses of this second photomultiplier are sent to the same input of the TAC as the measuring photomultiplier, but they are separated by means of delaying cables. The response of the second photomultiplier is stabilized in the memory of the MCA with a peak stabilizer (Wahl, 1969).

## C. Analysis of the Experimental Curves

### *1. Introduction*

If $g(t)$ is very short relatively to $F(t)$, then $f(t)$ is practically identical to $F(t)$. For an exponential decay of time constant $\tau$, $\log f(t)$ is a straight line and its slope is equal to $-1/\tau$. For a sum of $n$ exponentials, the decay law depends on $2n$ parameters ($n$ time constants and $n$ coefficients). Several methods have been devised to extract these parameters from an experimental decay and a bibliography may be found in Dyson and Isenberg (1971). But in many cases, concerning the fluorescence in solution, the response function $g(t)$ has an appreciable influence on the fluorescence response $f(t)$. The problem then is to extract $F(t)$ from the relation (Equation 7.46, 266). As a first step, one has to determine $g(t)$. Knowing $g(t)$ and $f(t)$, the problem is to find a numerical procedure in order to determine $F(t)$. Let us examine the first point.

### *2. Determination of the apparatus response function*

The response function which we will write $g(t, \lambda_E, \lambda_e)$, in order to emphasize its wavelength dependence, is given by the Equation (7.47). It is generally and implicitly assumed that the contribution of the photomultiplier $H(\lambda_e, t)$, entering in that expression, is independent of the wavelength of emission. In that case, $g(t, \lambda_E, \lambda_e)$ is simply the response of the apparatus to the exciting flash, and is obtained by performing a measurement when the fluorescent solution is replaced by a turbid solution or a reflecting plate. As we have seen, $H(\lambda_e, t)$ may depend on the incident light wavelength and this would introduce appreciable error in using that procedure. In that case, $g(t)$ can be computed by a numerical method which is an extension of a graphical method devised by Koechlin (Koechlin, 1961; Koechlin and Raviart, 1964). We assume that we have at our disposal a simple standard compound which displays a single exponential decay $F(t)$ with a time constant $\tau$, independent of the wavelength of emission and of excitation. A simple derivation of the expression (7.46) yields:

$$g(t, \lambda_E, \lambda_e) = f(t, \lambda_E, \lambda_e) - \tau \frac{\partial f(t, \lambda_E, \lambda_e)}{\partial t}, \tag{7.51}$$

where the wavelength dependence originates only from the apparatus characteristics. If we know $\tau$, a measurement of $f(t, \lambda_E, \lambda_e)$ under a given condition allows $g(t, \lambda_E, \lambda_e)$ to be computed numerically using equation (7.51). $\tau$ may be determined by measuring the response to the fluorescence of the standard compound under such conditions that the apparatus response is given by the flash response; that is to say where $\lambda_e$ and $\lambda_E$ are close to each other (Wahl, Auchet and Donzell, 1974) and by using a fast PM like the 2106.

### 3. *Numerical analysis of the data*

*a. Numerical inversion of the convolution integral.* In order to perform a direct numerical inversion of the convolution integral, a method has been proposed which would allow numerical determination of $F(t)$ without any hypothesis as to its mathematical form (Phillips, 1962). Use of Fourier analysis for the same purpose has also been proposed (Munro and Ramsay, 1968). After a study of experimental inaccuracies, Knight and Selinger (1971) concluded that direct deconvolution is not satisfactory in the case of pulse fluorometry. The phase plan method may be used when the decay is a single exponential (Demas and Adamson, 1971).

*b. Synthetic method.* We assume that we know the mathematical decay law which depends on several parameters. The problem is then to determine these parameters. For instance $F(t)$ is an exponential law and we want to determine $\tau$. First, we compute numerically the convolution integral (7.46) for several values of $\tau$; the computed curves are plotted and compared with the experimental data. The comparison is made on semilogarithmic graphs, the shapes of which are independent of a multiplicative factor. The computed curve for the true value of $\tau$ must fit the experimental curve. If no good fit is obtained, one can try a sum of two exponentials, then of three, and so on. For a sum of $n$ exponentials, there are $2n$ parameters to be fixed ($n$ time constants plus $n$ coefficients. The number of trials to be performed increases rapidly with $n$ and the method becomes very time consuming. The numerical calculation of the convolutions and sums of exponentials may be done with a recurrence formula. This result allows us to obtain such convolutions with a relatively small calculator having only a few storage memories (see Appendix 1). With a large computer, the best synthetized curve may be obtained by an automatic process based on the least mean squares method (Grinwald and Steinberg, 1974). The synthetic method also works with any decay law and even if the decay is given by a numerical table.

*c. Method of moments.* This method is applicable where $F(t)$ is a sum of exponentials. The reduced moments of order $k$ of the experimental fluorescence curve $f(t)$ is defined as:

$$M_k = \frac{\int_{-\infty}^{+\infty} t^k f(t)\mathrm{d}t}{\int_{-\infty}^{+\infty} f(t)\mathrm{d}t} \tag{7.52}$$

The corresponding moment of the response function for the excitation is:

$$\mu_k = \frac{\int_{-\infty}^{+\infty} t^k g(t)\mathrm{d}t}{\int_{-\infty}^{+\infty} g(t)\mathrm{d}t} \tag{7.53}$$

Let us assume that the fluorescence decay is a sum of exponentials which may be written as follows:

$$F(t) = \sum_{1}^{n} \frac{a_i}{\tau_i} \exp\left(-\frac{t}{\tau_i}\right) \tag{7.54}$$

with $\sum_{1}^{n} a_i = 1$. In this form one has: $\int_{-\infty}^{+\infty} F(t)\mathrm{d}t = 1$. $F(t)$ depends on $2n-1$ parameters ($n$ for $\tau_i$, and $n-1$ independent values of $a_i$). Let us define the following quantities:

$$\begin{aligned} S_1 &= \sum_{1}^{n} a_i \tau_i \\ S_2 &= \sum_{1}^{n} a_i \tau_i^2 \\ &\vdots \\ S_p &= \sum_{1}^{n} a_i \tau_i^p \\ &\vdots \end{aligned} \tag{7.55}$$

In applying the general formula given by Bay *et al.* (1955) for the moment of a convolution product, one may write (Wahl and Lami, 1967; Isenberg and Dyson, 1969):

$$M_k = \sum_{r=0}^{k} \frac{k!}{r!(k-r)!} \mu_{k-r} S_k \tag{7.56}$$

From that general formula, it can be deduced that:

$$\begin{aligned} S_1 &= M_1 - \mu_1 \\ S_2 &= \tfrac{1}{2}(M_2 - \mu_2 - 2\mu_1 S_1) \\ S_3 &= \tfrac{1}{6}(M_3 - \mu_3 - 3\mu_2 S_1 - 6\mu_1 S_2) \end{aligned}$$

If one calculates the moments until the order $p = 2n-1$, it is possible in theory to resolve the system of Equations (7.55) and find the parameters defining $F(t)$. The average time constants of increasing order are given by:

$$\begin{aligned}\tau' &= S_1\\ \tau'' &= S_2/S_1\\ \vdots\ &\quad \vdots\\ \tau^{(p)} &= S_p/S_{p-1}\end{aligned}$$

which, for positive coefficients $a_i$ are linked to one another by the following relation:

$$\tau' < \tau'' < \ldots < \tau^{(p)} < \ldots .$$

The integrals involved in the moment calculations have to be taken over an infinite time domain. In the actual calculations, the time interval is necessarily limited, and a cut-off error is introduced. Isenberg and Dyson (1969) use a correction involving an *iterative calculus* and carry out the procedure using an electronic computer. In the calculation of moments of high order, the tail of the decay curves is heavily weighted (Dyson and Isenberg, 1971). In addition to the cut-off error, other errors arising from the statistical fluctuations and from the noise become important. Good results may be obtained for the resolution of a sum of two exponentials. For the resolution of a sum of three exponentials, it is necessary to determine the moments of the fourth and fifth order which are generally of low accuracy causing the method to be inefficient. The determination of the moments however may be of great help, even if they do not suffice to resolve the experimental curve entirely. For example, one may couple these calculations with the analysis of the initial part of the experimental curve. The parameters which define the tangent at the origin of the $F(t)$ curve provide the two following quantities (Brochon and Wahl, 1972):

$$S_{-1} = \sum_1^n \frac{a_i}{\tau_i}$$

$$S_{-2} = \sum_1^n \frac{a_i}{\tau_i^2}$$

Other information may also be introduced into the calculation, such as known time constants or coefficients (Wahl and Auchet, 1972b).

*d. Method of modulation functions.* This method has been recently proposed by Valeur and Moirez (1973).

The modulation functions can be chosen in such manner that all the parts of an experimental curve are taken into account with an equal weight. In contrast, analysis by the method of moments tends to emphasize the final part of experimental decay curves. For this reason, the method of modulation functions succeeds in resolving multiexponential decays, where the method of moments very often fails.

This method has been successfully used for the analysis of fluorescence decay in several cases (Brochon, Wahl and Auchet, 1974; Donzell *et al.*, 1974).

## IV. APPENDIX I—SIMPLE ALGORITHM FOR COMPUTING CONVOLUTION INTEGRALS

The convolution integral of an exponential or sum of exponentials may be computed by a recurrence formula. Let $f(t)$ be the fluorescence intensity given by

$$f(t) = \int_{t_0}^{t} g(t)\,\mathrm{e}^{-\frac{(t-x)}{\tau}}\,\mathrm{d}x$$

where $g(t)$ is the response function of the apparatus. One may separate this integral into two parts:

$$f(t) = \int_{t_0}^{t-h} g(t)\,\mathrm{e}^{-\frac{(t-x)}{\tau}}\,\mathrm{d}x + \int_{t-h}^{t} g(t)\,\mathrm{e}^{-\frac{(t-x)}{\tau}}\,\mathrm{d}x \tag{7.57}$$

where $h$ is a given time increment. Equation (7.57) may be written as:

$$f(t) = f(t-h)\,\mathrm{e}^{-h/\tau} + \Delta f(h)$$

where $\Delta f(h)$ is the second integral of (7.57). If $h/\tau$ is small, $\mathrm{e}^{-\frac{(t-x)}{\tau}}$ may be replaced by its average value in the interval $t-h$, $t$ which is $\mathrm{e}^{-h/2\tau}$, and we obtain:

$$f(t) \approx kf(t\text{–}h) + k^{\frac{1}{2}} \int_{t-h}^{t} g(x)\mathrm{d}x \tag{7.58}$$

with $k = \mathrm{e}^{-h/\tau}$. Assuming that $h$ is the time interval between two channels of the analyser, the count stored in the channel $n$ may be considered as:

$$N_n = \int_{(n-\frac{1}{2})h}^{(n+\frac{1}{2})h} f(t)\mathrm{d}t \approx \frac{f[(n+\frac{1}{2})h] + f[(n-\frac{1}{2})h]}{2} \tag{7.59}$$

and taking into account the expression (7.58), we find:

$$N_n = kN_{n-1} + k^{\frac{1}{2}} \times \frac{G_n + G_{n-1}}{2} \tag{7.60}$$

with:

$$G_n = \int_{(n-\frac{1}{2})h}^{(n+\frac{1}{2})h} g(x)\mathrm{d}x$$

which is the count obtained for the response of the apparatus to the excitation function.

Expression (7.60) allows computation of the convolution integral step by step. It is easily programmed for a small desk electronic calculator having only a few memory registers. For instance, with a calculator of the size of the Olivetti Programma 101, one may compute the convolution of three exponentials simultaneously or the convolution of a sum of two exponentials. Approximations of higher order may also be introduced in expressions (7.58) and (7.59), but in that case a larger memory space is needed.

## V. APPENDIX 2—LIST OF COMMERCIAL INSTRUMENT MAKERS

ORTEC INC., 100 Midland Road, Oak Ridge, Tennessee 37830, U.S.A.
RADIOTECHNIQUE-COMPELEC, 130 Avenue Ledru Rollin.—75, Paris 11ème, France.

## VI. GLOSSARY

p. 237–238: *Singlet, triplet*

Electronic states are characterized by their multiplicity; the multiplicity corresponds to the number of degenerate substates in the absence of a magnetic field. In organic molecules the ground state is usually a singlet (no degeneracy). If an electron is excited without change of spin, the resultant excited state is a singlet. If the spin of the electron is reversed, the excited electronic state of the molecule is a triplet.

p. 238: *Internal conversion*

Internal conversion is a radiationless transition between two electronic states of the same multiplicity (two singlets or triplets) or between two vibrational states of the same electronic state.

p. 238: *Intersystem crossing*

This is a radiationless transition between states of different multiplicities, for example between singlet and triplet.

p. 240: *Einstein coefficients of absorption and spontaneous emission*

These coefficients correspond to probabilities of photon absorption and spontaneous emission relative to the transition between two vibronic states (states defined by their electronic and vibrational levels). By thermodynamics arguments, Einstein has established a relation of proportionality between these two coefficients.

p. 241: *Dipolar moment*

This vectorial quantity characterizes the anisotropy of charge distribution in a molecule. It is defined by the following expression

$$\vec{\mu} = \Sigma \vec{r_i} e_i$$

where $\vec{r_i}$ is a vector which defines the position of the charge $e_i$ in the molecule.

p. 243: *Dynamic quenching*

Dynamic quenching occurs when molecules of a quencher interact with fluorescent molecules in their excited state and not in their ground state. Therefore dynamic quenching is controlled by mutual diffusion of these two kinds of molecules.

p. 254: *Tensors of second order*

They are defined by a generalization of vectors. In a three-dimensional space, these tensors have nine components, which, by a change of the co-ordinate axes transform according to the following rule:

$$t_{ij} = \sum_{i'=1}^{3} \sum_{j'=1}^{3} \alpha_{ii'} \alpha_{jj'} t_{i'j'}$$

$t_{ij}$ and $t_{i'j'}$ are the tensor components in the new and the old axes respectively; $\alpha_{ii'}$ is the cosine of the angle between the new axis $i$ and the old axis $i'$ (see also p. 383).

p. 257. *Methods of Monte Carlo*

They are mathematical methods used for resolving problems by statistical sampling. They are useful when analytical methods are too complicated or do not exist. These methods involve a great number of calculations which necessitate the use of a powerful computer.

p. 259. *Jablonski approximation for computing energy transfers*

To simplify the computation of energy transfer and in particular the fluorescence depolarization by transfers, Jablonski introduces an 'active sphere' which surrounds the molecule directly excited by the exciting beam. This initially excited molecule can only transfer its energy to the neighbouring molecules situated within the active sphere and not to the molecules situated outside the active sphere.

p. 275. *Iterative calculus*

A calculus which proceeds by successive approximations obtained when repeating (iterating) a given procedure. After the $n$th approximation, one evaluates an estimate of the error. The following approximation is calculated by taking this estimate into account. The process is valuable if the error estimate diminishes at each iteration, in other words, if there is convergence.

## VII. ACKNOWLEDGEMENTS

The author's research work referred to in this article has been supported by the Centre National de la Recherche Scientifique and the Délégation Générale à la Recherche Scientifique et Technique, Comité de Biologie Moléculaire.

It is a pleasure to me, to express my gratitude to Professor Ch. Sadron, Director of the Centre de Biophysique Moléculaire, for the steady encouragement, during the progress of our research work.

## VIII. REFERENCES

Auchet, J. C. and Ph. Wahl (1972), work to be published.

Bakshiev, N. G., Yu. T. Mazurenko and I. V. Piterskaya (1966) 'Luminescence decay in different portions of the luminescence spectrum of molecules in viscous solutions', *J. Opt. Spectr.* (English translation), **21**, 307–9.

Bauer, W. and J. Vinograd (1968) 'The interaction of closed circular DNA with intercalative dyes. I. The superhelix density of S.V 40 DNA in the presence of dye', *J. Mol. Biol.*, **33**, 141–71.

Bay, Z., V. P. Henri and H. Kanner (1955) 'Statistical theory of delayed-coincidence experiments', *Phys. Rev.*, **100**, 1197–208.

Belford, G. G., R. L. Belford and G. Weber (1972) 'Dynamics of fluorescence polarization in macromolecules', *Proc. Natl. Acad. Sci. U.S.*, **69**, 1392.

Bell, R. E. (1965) 'Coincidence techniques and the measurement of short mean lives', in *Alpha, Beta and Gamma-ray Spectroscopy*, Vol. 2 (Ed. Siegbahn), North-Holland Publishing Company, Amsterdam, p. 905.

Bennett, W. R., P. J. Kindlmann and G. N. Mercer (1965) 'Measurement of excited state relaxation rates', *Appl. Opt. Suppl.*, **2**, 34–57.

Birks, J. B. (1970) *Photophysics of Aromatic Molecules*, Wiley–Interscience Publisher, London.

Birks, J. B., D. J. Dyson and I. H. Munro (1963) 'Lifetime studies of pyrene solutions', *Proc. Roy. Soc.*, **A291**, 244.

Birks, J. B. and I. H. Munro (1967) 'The fluorescence lifetimes of aromatic molecules', *Progr. Reaction Kinetics*, **4**, 239.

Bollinger, L. M. and G. E. Thomas (1961) 'Measurement of the time dependence of scintillation intensity by a delayed-coincidence method', *Rev. Sci. Instr.*, **32**, 1044–50.

Brand, L. and J. R. Gohlke (1971) 'Nanosecond time resolved fluorescence spectra of a protein–dye complex', *J. Biol. Chem.*, **246**, 2317–19.

Breuze, G. (1971) 'Dispositif de mesure des temps de fluorescence de certaines molécules organiques dans le domaine de la nanoseconde', *Rapport du Commissariat à l'Energie Atomique*, R. 4151.

Breuze, G. and P. Savine (1969) 'Fluctuations temporelles de la réponse des tubes photomultiplicateurs, Dario 56 AVP, XP 1021, XP 1210', *Rapport du Commissariat à l'Energie Atomique*, R. 3802.

Brochon, J. C. and Ph. Wahl (1972) 'Mesures des déclins de l'anisotropie de fluorescence de la $\gamma$-globuline et de ses fragments Fab, Fc et $F(ab)_2$ marqués avec du l-sulfonyl 5-dimethylaminonaphtalene', *Eur. J. Biochem.*, **25**, 20–32.

Brochon, J. C., Ph. Wahl and J. C. Auchet (1974) 'Fluorescence time resolved spectroscopy and fluorescence anisotropy decay of the staphylococcus aureus endonuclease', *Eur. J. Biochem.*, **41**, 577–83.

Chance, B., C. P. Lee and J. K. Blaste (1971), *Probes of Structure and Function of Macromolecules and Membranes*, Academic Press, New York.

Chen, R. F., H. Edelhoch and R. F. Steiner (1969) 'Fluorescence of proteins', in *Physical Principles and Techniques of Protein Chemistry*, Part A, (Ed. S. Y. Leach), Academic Press, New York, p. 171.

Chuang, T. J. and K. B. Eisenthal (1972) 'Theory of fluorescence depolarization by anisotropic rotational diffusion', *J. Chem. Phys.*, **57**, 5094.

Coates, P. B. (1968) 'The correction of photon "pile up" in the measurements of lifetimes', *J. Sci. Instr.*, Ser. **E1**, 878.

Cowgill, B. W. (1963) 'Fluorescence and the structure of proteins. I. Effects of substituents on the fluorescence of indole and phenol compounds', *Arch. Biochem. Biophys.*, **100**, 36–44.

Cowgill, B. W. (1970) 'On the mechanism of peptides quenching', *Biochim. Biophys. Acta*, **200**, 18–25.

D'Alessio, J. T., P. K. Ludwig and M. Burton (1964) 'Ultraviolet lamp for the generation of intense, constant pulses in the subnanosecond region', *Rev. Sci. Instr.*, **35**, 1015.

Dandliker, W. D. and A. J. Portmann (1971) 'Fluorescent protein conjugates', in *Excited States of Proteins and Nucleic Acids* (Eds. R. F. Steiner and I. Weinryb), Macmillan, pp. 199–262.

Davis, C. C. and T. A. King (1970) 'Correction methods for photon "pile up" in lifetime determination by single photon counting', *Proc. Phys. Soc. J. Phys.*, **A3**, 101.

De la Barre, F. *et al.* (1972) 'Influence of transit time differences on photomultiplier time resolution', *Nuclear Instrum. Meth.*, **102**, 77–88.

De Lauder, W. B. and Ph. Wahl (1970) 'pH dependence of the fluorescence decay of tryptophan', *Biochemistry*, **9**, 2750–4.

De Lauder, W. B. and Ph. Wahl (1971a) 'Fluorescence studies on human serum albumin', *Biochem. Biophys. Res. Comm.*, **41**, 398–403.

De Lauder, W. B. and Ph. Wahl (1971b) 'Effect of solvent upon the fluorescent decay of indole', *Biochim. Biophys. Acta*, **243**, 153–63.

De Luca, M., L. Brand, T. Acebula, H. H. Seliger and A. F. Makula (1971) 'Nanosecond time resolved proton transfer studies with dehydroluciferin and its complex with luciferase', *J. Biol. Chem.*, **246**, 6702–4.

Demas, J. N. and A. W. Adamson (1971) 'Evaluation of photoluminescence lifetimes', *J. Phys. Chem.*, **75**, 2463–6.

Donzell, B. (1971) 'Untersuchung der bevorzugten konformationen einiger tryptophan haltiger cyclo-dipeptide in Lösung', *Thesis*, Zurich.

Donzell, B., P. Gauduchon and Ph. Wahl (1974) 'Study of the conformation in the excited state of two tryptophanyl diketopiperazines', *J. Am. Chem. Soc.*, **96**, 801–8.

Dyson, R. and I. Isenberg (1971) 'Analysis of experimental curves by a method of moments with special attention to sedimentation equilibrium and fluorescence decay', *Biochemistry*, **10**, 3233.

Edelhoch, H., and R. S. Bernstein and M. Wilchek (1968) 'The fluorescence of tyrosyl and tryptophanyl diketopiperazines', *J. Biol. Chem.*, **243**, 5985–92.

Edelman, G. M. and W. O. McClure (1968) 'Fluorescent probes and the conformation of proteins', *Accounts Chem. Res.*, **1**, 65–70.

Ehrenberg, M. and R. Rigler (1972) 'Polarized fluorescence and rotational Brownian motion', *Chem. Phys. Letters*, **14**, 539.

Eigen, M., W. Kruse, G. Maase and L. de Maeyer (1961) 'Rate constants of protolytic reactions in aqueous solution', *Progr. Reaction Kinetics*, **2**, 285.

Fayet, M. and Ph. Wahl (1969) 'Etude du déclin de la fluorescence polarisée de la $\gamma$-globuline de lapin conjuguée avec l'isothiocyanate de fluoresceine', *Biochim. Biophys. Acta*, **181**, 373–80.

Fayet, M. and Ph. Wahl (1971) 'Variation avec le pH du rendement quantique et du déclin de la fluorescence de la tyrosine', *Biochim. Biophys. Acta*, **229**, 102–12.

Förster, Th. (1948) 'Zwischenmolekulare Energiewanderung und Fluoreszenz', *Ann. Physik*, **2**, 55–75.

Förster, Th. (1951) *Fluoreszenz Organischer Verbindungen*, Vandenhoeck and Ruprecht, Göttingen.

Förster, Th. (1959) 'Transfer of electronic excitation', *Disc. Faraday Soc.*, **27**, 7–17.

Förster, Th. (1965) in *Delocalized Excitation and Excitation Transfer in Modern Quantum Chemistry* (Ed. O. Sinanoglu), Academic Press, p. 93.

Förster, Th. and K. Kasper (1954) *Z. Phys. Chem.* (*N.F.*), **1**, 275.

Gatti, E. and V. Svelto (1959) 'Theory of time resolution in scintillation counters', *Nucl. Instr. Meth.*, **4**, 189–201.

Gedcke, D. A. and W. J. McDonald (1968) 'Design of the constant fraction of pulse height trigger for optimum time resolution', *Nucl. Instr. Meth.*, **58**, 253.

Genest, D. and Ph. Wahl (1972) 'Etude des transferts d'énergie dans le complexe DNA–bromure d'éthidium au moyen du déclin de l'anisotropie de fluorescence', *Biochim. Biophys. Acta*, **259**, 175–88.

Genest, D. and Ph. Wahl (1973) 'Energy transfer study in the DNA–ethidium bromide complex by means of anisotropy decay', in *Dynamical Aspects of Conformation Changes in Biological Macromolecules* (Ed. Ch. Sadron), Reidel-Dordrecht-The Netherlands, p. 350.

Genest, D. and Ph. Wahl (1974) 'The fluorescence anisotropy decay due to energy transfers occurring in the ethidium bromide DNA complex. Determination of the deformation angle of the DNA helix', *Biophys. Chem* **1**, 266–78.

Gottlieb, Yu. Y. A. and Ph. Wahl (1963) 'Etude théorique de la polarisation de fluorescence des macromolécules portant un groupe émetteur mobile autour d'un axe de rotation', *J. Chim. Phys.*, **60**, 849–56.

Grinwald, A. and E. Haas and I. Z. Steinberg (1972) 'Evaluation of the distribution of distances between energy donors and acceptors by fluorescence decay', *Proc. Natl. Acad. Sci. U.S.*, **69**, 2273–7.

Haughland, R., J. Yguerabide and L. Stryer (1969) 'Dependence of the kinetics of singlet–singlet energy transfer on spectral overlap', *Proc. Natl. Acad. Sci. U.S.*, **63**, 23–30.

Helene, V., J. L. Dimicoli and F. Brun (1971) 'Binding of tryptamine and serotin to nucleic acids. Fluorescence and proton magnetic resonance studies', *Biochemistry*, **10**, 3802–9.

Herons, R. W., P. McWhirter and E. H. Rhoderick (1956) 'Measurements of lifetimes of excited states of helium atoms', *Proc. Roy. Soc.*, **A234**, 565.

Hundley, L., T. Coburn, E. Garwin and L. Stryer (1967) 'Nanosecond fluorimetry', *Rev. Sci. Instr.*, **38**, 488.

Isenberg, I. and R. Dyson (1969) 'The analysis of fluorescence decay by a method of moments', *Biophys. J.*, **9**, 1337–50.

Jablonski, A. (1960) 'On the notion of emission anisotropy', *Bull. Acad. Pol. Sci. Ser. Sci. Math. Astr. Phys.*, **8**, 259–64.

Knight, A. W. and B. A. Selinger (1971) 'The deconvolution of fluorescence decay curves. A non-method for real data', *Spectrochim. Acta*, **27A**, 1223.

Knox, R. S. (1968) 'Theory of polarization quenching by excitation transfer', *Physica*, **39**, 361–86.

Koechlin, Y. (1961) 'Determination de la forme des impulsions lumineuses très brèves. Application aux phémomènes de scintillation', *Thèse*, Paris.

Koechlin, Y. and A. Raviart (1964) 'Analyse par échantillonnage sur photons individuels des liquides fluorescents dans le domaine de la nanoseconde', *Nucl. Instr. Meth.*, **29**, 45.

Koniev, S. V. (1967) *Fluorescence and Phosphorescence of Proteins and Nucleic Acids*, Plenum Press, New York.

Kopple, K. D. and M. Ohnishi (1969) 'Conformations of cyclic peptides. II. Side chain conformation and ring shape in cyclic dipeptides', *J. Am. Chem. Soc.*, **91**, 962.

Lami, H., J. T. D. D'Alessio, J. Zampach, R. Radicella, and J. M. Kesque (1971) 'Measurement of the instrument response function in the single photoelectron technique', *Nucl. Instr. Meth.*, **92**, 333.

Longworth, J. W. (1971) 'Luminescence of polypeptides and proteins', in *Excited States of Proteins and Nucleic Acids* (Eds. R. F. Steiner and I. Weinryb), MacMillan, London.

Malmberg, J. H. (1957) 'Millimicrosecond duration light source', *Rev. Sci. Instr.*, **28**, 1027.

Mandel, L. and E. Wolf (1965) 'Coherence properties of optical fields', *Rev. Mod. Phys.*, **37**, 231–87.

Mataga, N. and T. Kubota (1970) *Molecular Interactions and Electronic Spectra*, Marcel Dekker, New York.

Memming, R. (1961) 'Theorie de Fluoreszenzpolarisation für nicht kugelsymmetrische molekule', *Z. Phys. Chem. (NF)*, **28**, 168–89.

Meserve, E. T. (1971) 'Direct measurement of fluorescence lifetimes', in *Excited States of Proteins and Nucleic Acids* (Eds. R. F. Steiner and I. Weinryb), MacMillan, London.

Miehe, J. A., G. Ambard, J. Zampach and A. Coche (1970) 'Statistic and timing of the first photoelectron with a high gain dynode photomultiplier', *I.E.E.E. Trans. Nucl. Sci. NS*, **17**.

Miehe, J. A. and G. Knispel (1971) 'Influence des fluctuations de la vitesse initiale des photoélectrons sur la réponse percussionnelle d'un photomultiplicateur fonctionnant en régime de photoélectron unique', *C.R. Acad. Sci.*, **273**, 285–8.

Munro, I. H. and I. A. Ramsay (1968) 'Instrumental response time corrections in fluorescence decay measurements', *J. Sci. Instr.*, Ser. 2, **1**, 147.

Noyes, R. M. (1961) 'Effects of diffusion rates on chemical kinetics', *Progr. Reaction Kinetics*, **1**, 129.

Odru, R., M. Prieur-Drevon and J. Vacher (1970) 'Amélioration de la mesure du temps de décroissance des impulsions de lumière brève', *C.E.A. Conférence*, 1572.

Paoletti, J. and J. B. Le Pecq (1971) 'Resonance energy transfer between ethidium bromide molecules bound to nucleic acids', *J. Mol. Biol.*, **59**, 43–62.

Parker, C. A. (1968), *Photoluminescence of Solutions*, Elsevier, Amsterdam.

Penzer, G. R. (1974) 'Fluorescence', in *Amino Acids, Peptides and Proteins*, Vol. 5, Chemical Society, London, p. 203.

Perrin, F. (1929) 'La fluorescence des solutions', *Ann. Chim. Phys.*, **12**, 170–275.

Perrin, F. (1936) 'Mouvement Brownien d'un ellipsoïde (II). Rotation libre et dépolarisation de fluorescence', *J. Phys. Radium*, **7**, 1–11.

Pfeffer, G. (1965) 'Etude des propriétés statistiques de photomultiplicateurs rapides. Application à des dispositifs permettant la détermination des spectres d'émission et des constantes de temps des scintillateurs organiques', *Thèse*, Strasbourg.

Pfeffer, G., H. Lami, G. Laustriat and A. Coche (1963) 'Application de la technique du photoélectron unique à la détermination des constantes de temps des scintillateurs', Congrès d'Electronique nucléaire. Publication de l'O.C.D.E., Paris.

Phillips, D. L. (1962) 'Technique for the numerical solution of certain integral equations of the first kind', *J. Assoc. Comp. Mach.*, **9**, 84.

Pietri, G. (1961) 'Les photomultiplicateurs, instruments de physique expérimentale', *Acta Electronica*, **1**, 7.

Present, C. and R. Scarl (1970) 'Single photon time resolution of photomultipliers with gallium phosphite first dynode', *Rev. Sci. Instr.*, **41**, 771.

Rigler, R. and M. Ehrenberg (1973) 'Molecular interactions and structure as analysed by fluorescence relaxation spectroscopy', *Quart. Rev. Biophys.*, **6**, 139–99.

Spencer, R. D. and G. Weber (1969) 'Measurements of subnanosecond fluorescence lifetimes with a cross-correlation phase fluorometer', *Annals New York Acad. Sci.*, **158**, 361–76.

Spencer, R. D. and G. Weber (1972) in *Structure and Function of Oxidation Reduction Enzymes* (Eds. Åkeson and Ehrenberg), Pergamon, pp. 393–9.

Steiner, R. F. and I. Weinryb (1971) *Excited States of Proteins and Nucleic Acids*, MacMillan, London.

Strickler, S. J. and R. A. Berg (1962) 'Relationship between absorption intensity and fluorescence lifetime of molecules', *J. Chem. Phys.*, **37**, 814.

Stryer, L. (1968) 'Fluorescence spectroscopy of proteins', *Science*, **162**, 526–33.

Tao, T. (1969) 'Time dependent fluorescence depolarization and brownian rotational diffusion coefficients of macromolecules', *Biopolymers*, **8**, 609–32.

Teale, F. W. J. (1961) *Biochem. J.*, **80**, 14P.

Valeur, B. and J. Moirez (1973) 'Analyse des courbes de décroissance multiexponentielles par la méthode des fonctions modulatrices. Applications à la fluorescence', *J. Chim. Phys.*, **70**, 500–6.

Vallat, D. (1969) 'Mesures des fluctuations du temps de transit des photomultiplicateurs. L'onde électrique', No. **508–509**, 3–7.

Vladimirov, Yu. A. and G. M. Zimina (1965) 'Luminescence of certain proteins and tryptophan during monochromatic excitation in solution', *Biokhimiya*, **30**, 1105–14.

Voltz, R. (1968) 'Electronic energy transfer in irradiated aromatic materials', *Radiation Res. Rev.*, **1**, 301, 360.

Wahl, Ph. (1965) 'Sur l'étude des solutions de macromolécules par la décroissance de la fluorescence polarisée', *C.R. Acad. Sci.*, **260**, 6891.

Wahl, Ph. (1966) 'Détermination du temps de relaxation brownienne de la serum albumine en solution par la mesure de la décroissance de la fluorescence polarisée', *C.R. Acad. Sci.*, **263**, 1525–8.

Wahl, Ph. (1969) 'Mesure de la décroissance de la fluorescence polarisée de la gamma globuline D.N.S.', *Biochim. Biophys. Acta*, **175**, 55.

Wahl, Ph., G. Meyer and J. Parrod (1967) 'Etude de la décroissance de la fluorescence polarisée. Détermination du temps de relaxation d'un polymère vinylique en solution', *C.R. Acad. Sci.*, **C264**, 1641.

Wahl, Ph. and H. Lami (1967) 'Etude du déclin de la fluorescence du lysozyme l-dimethylaminonaphtalène-5 sulfonyl', *Biochim. Biophys. Acta*, **133**, 233–42.
Wahl, Ph. and G. Weber (1967) 'Fluorescence depolarization of rabbit gamma globulin conjugates', *J. Mol. Biol.*, **30**, 371–82.
Wahl, Ph. and S. N. Timasheff (1969) 'Polarized fluorescence decay curves for β lactoglobulin A in various states of association', *Biochemistry*, **8**, 2945–9.
Wahl, Ph., G. Meyer and J. Parrod (1970) 'Etude de la dépolarisation brownienne pendant le déclin de fluorescence de polymères vinyliques', *Europ. Polym. J.*, **6**, 585–608.
Wahl, Ph., J. Paoletti and J. B. Le Pecq (1970) 'Decay of fluorescence emission anisotropy of the ethidium bromide DNA complex. Evidence for an internal motion in DNA', *Proc. Natl. Acad. Sci. U.S.*, **65**, 417–21.
Wahl, Ph., M. Kasaï and J. P. Changeux (1971) 'A study on the motion of proteins in excitable membrane fragments by nanosecond fluorescence polarization spectroscopy', *Europ. J. Biochem.*, **18**, 332–41.
Wahl, Ph. and J. C. Auchet (1972a) 'Cinétique de l'interaction de l'indole a l'état excité avec le méthanol', *C.R. Acad. Sci.*, **B274**, 1334–7.
Wahl, Ph. and J. C. Auchet (1972b) 'Résolution des spectres de fluorescence au moyen des déclins. Application à l'étude de la serum albumine humaine', to be published in *Biochim. Biophys. Acta.*, **285**, 99–117.
Wahl, Ph., J. C. Auchet and B. Donzell (1974) 'The wavelength dependence of the response of a pulse fluorometer, when using the single photoelectron counting method', *Rev. Sci. Instr.*, **45**, 28–32.
Walker, M. S., T. W. Bedmar and R. Lumry (1967) 'Exciplex studies. II. Indole and indole derivatives', *J. Chem. Phys.*, **47**, 1020–8.
Wallach, D. (1967) 'Effect of internal rotation on angular correlation functions', *J. Chem. Phys.*, **47**, 5258–68.
Ware, W. R. (1971) 'Transient luminescence measurements', in *Creation and Detection of the Excited States* (Ed. A. A. Lamola), M. Dekker, New York, p. 213.
Ware, W. R. and J. S. Novros (1966) 'Kinetics of diffusion controlled reactions', *J. Phys. Chem.*, **70**, 3246–53.
Ware, W. R., P. Chow and S. K. Lee (1968) 'Time resolved nanosecond emission spectroscopy: spectral shifts due to solvent–solute relaxation', *Chem. Phys.*
Ware, W. R. and H. P. Richter (1968) 'Fluorescence quenching via charge transfer', *J. Chem. Phys.*, **48**, 1595–601.
Weaver, R. S. and R. E. Bell (1960) 'Method of evaluating delayed coincidence experiments', *Nucl. Instr. Meth.*, **9**, 149.
Weber, G. (1952) 'Polarization of the fluorescence of macromolecules', *Biochem. J.*, **51**, 145–67.
Weber, G. (1953) in *Advances in Protein Chemistry*, Vol. 8 (Eds. M. L. Anson, N. Bailey and J. T. Edsall), Academic Press, New York, pp. 415–59.
Weber, G. (1961) 'Enumeration of components in complex systems by fluorescence spectrophotometry', *Nature*, **190**, 27.
Weber, G. (1966) *Intramolecular Complexes of Flavins in Flavins and Flavoproteins* (Ed. Slater), Elsevier, Amsterdam, pp. 16–21.
Weil, G. (1965) 'Rendement quantique, temps de vie de fluorescence et hétérogénéité des sites de fixation de la proflavine sur le DNA, *Biopolymers*, **3**, 567, 572.
Weinryb, I. and R. F. Steiner (1971) 'The luminescence of aromatic aminoacids', in *Excited States of Proteins and Nucleic Acids*, MacMillan, London, p. 277.
Weller, &. &. (1961) 'Fast reactions of excited molecules', *Progr. Reaction Kinetics*, **1**, 187.

Yguerabide, J. (1965) 'Generation and detection of subnanosecond light pulse', *Rev. Sci. Instr.*, **36**, 1734.
Yguerabide, J. (1972) 'Nanosecond fluorescence spectroscopy of macromolecules', in *Methods in Enzymology*, Vol. XXVI, part C (Eds. C. H. W. Hirs and S. N. Timasheff), New York, Academic Press, p. 498.
Yguerabide, J., H. F. Epstein and L. Stryer (1970) 'Segmental flexibility in an antibody molecule', *J. Mol. Biol.*, **51**, 573–90.

**Additional References**

Eisinger, J. and A. A. Lamola (1971) 'The Excited States of Nucleic Acids' in *Excited States of Proteins and Nucleic Acids* (Eds. R. F. Steiner and I. Weinryb), 207, MacMillan, London.
Grinwald, A. and I. Z. Steinberg (1974) 'On the analysis of Fluorescence Decay Kinetics by the Method of Least Squares', *Anal. Biochem.*, **59**, 583–98.
Isenberg, I. (1975) 'Time Decay Fluorometry by Photon Counting' in *Concepts in Biochemical Fluorescence* (Eds. Chen and Edelhoch), Marcel Dekker, New York.
Lewis, C., W. R. Ware, L. J. Doemeny, T. L. Nemzek (1973) 'The measurement of Short lived Fluorescence Decay using the Single Photon Counting Method', *Rev. Sci. Instr.*, **107**, 114.
Wahl, Ph. (1975) 'Decay of Fluorescence Anisotropy in *Concepts in Biochemical Fluorescence* (Eds. R. F. Chen and H. Edelhoch) Marcel Dekker, New York.
Wahl, Ph., J. C. Auchet, A. J. W. G. Visser, F. Muller (1974) 'Time Resolved Fluorescence of Flavin Adenine Dinucleotide', *FEBS Letters*, **44**, 67–70.
Ware, W. R. (1975) 'Time Resolving Spectroscopy and Solvent Relaxation' in *Concepts in Biochemical Fluorescence* (Eds. Chen and Edelhoch), Marcel Dekker, New York.

CHAPTER 8

# The application of carbon-13 NMR spectroscopic techniques to biological problems

J. Feeney
*National Institute for Medical Research*
*Mill Hill, London NW7 1AA*

Over the last ten years we have seen the widespread application of high resolution nuclear magnetic resonance (NMR) techniques to biological problems. Because of the ease with which proton nuclei can be examined by NMR methods it is not surprising that most biological NMR investigations so far reported have involved proton studies. Using these methods several interesting studies of protein unfolding and protein interactions with small molecules have been completed and these have been reviewed by Roberts and Jardetzky (1970) and other workers (Sheard and Bradbury, 1970). The recent advent of facilities for using Fourier transform techniques on commercially available spectrometers has given new impetus to these studies and at the same time opened up the possibility of examining other biologically relevant magnetic nuclei, such as carbon-13 and phosphorus-31 which are inherently less sensitive to NMR detection than proton nuclei. As we shall see later, Fourier transform techniques can reduce the time required to obtain spectra using spectrum accumulation by more than an order of magnitude compared with conventional frequency-swept (continuous-wave) experiments. This allows us to examine proton nuclei at solution concentrations ca. $10^{-5}$ *M* and carbon-13 nuclei in natural abundance at ca. $10^{-2}$ *M*. In addition to the sensitivity improvements, the Fourier transform technique also provides an excellent method for measuring the individual spin–lattice relaxation times for all nuclei with resolved absorption lines in the NMR spectrum. Because such measurements can be interpreted in terms of dynamic processes in the molecules they provide a valuable parameter for probing biological systems.

At the present time NMR measurements have not generally provided conformational information which is as unequivocal as that obtained by X-ray diffraction studies. However, the NMR method has the advantage that it relates to the molecules in aqueous solution and also offers a much more direct method of measuring molecular dynamic processes. Electron spin resonance studies of spin-labelled molecules can also be used to obtain detailed motional information. Clearly a multi-technique approach to solving biological problems is necessary and the results obtained from carbon-13 NMR experiments should be evaluated in this wider context.*

* For a general introduction to proton NMR see, for example, J. Bernstein, J. A. Pople and W. G. Schneider (1959) *High Resolution Nuclear Magnetic Resonance*, Academic Press; J. W. Emsley, J. Feeney and L. H. Sutcliffe (1965) *High Resolution* NMR *Spectroscopy*, Pergamon Press; L. M. Jackman and S. Sternell (1969) *Applications of* NMR *Spectroscopy in Organic Chemistry*, Pergamon Press.

## I. NUCLEAR PROPERTIES OF CARBON-13

The nucleus of carbon-12, the naturally occurring abundant isotope of carbon, does not possess a nuclear magnetic moment, $\mu$, and thus cannot be studied by NMR. Carbon-13 with a natural abundance of only 1·108% does possess a magnetic moment ($\mu_C = 0{\cdot}70220$ nuclear magnetons) and is capable of being examined by this technique. At constant field strength the natural sensitivity to NMR detection is only 1·59% that of proton nuclei because of the small value of the carbon nuclear magnetic moment compared with that for the proton. At natural abundance $^{13}C$ is 5680 times less sensitive than the $^{1}H$ nucleus and it is this serious sensitivity problem which has held back progress in carbon-13 NMR spectroscopy. Various methods of improving carbon-13 sensitivity have been developed and these will be discussed separately in Section IV. The spin number $I$ of carbon-13 is $\frac{1}{2}$ and thus there is no quadruple moment associated with the nucleus: this is favourable for high-resolution studies because quadrupolar nuclei with their short relaxation times can give rise to broad lines. In fact the carbon-13 relaxation times for carbon nuclei without directly bonded protons in small molecules can be very long ($> 50$ sec). This leads to such nuclei being more difficult to examine in Fourier transform experiments (see Section IV.F) where one wishes to examine the spin system repetitively; if insufficient time is allowed between the pulses, nuclei with long relaxation times will not fully recover their magnetization between pulses and will feature in the final spectrum as low-intensity signals. Fortunately carbon nuclei with directly bonded protons have much shorter relaxation times ($< 5$ sec) and for the larger molecules usually examined in biological studies relaxation times of $< 1$ sec are common: these cases are ideal for Fourier transform experiments.

## II. THE NMR EXPERIMENT

When a sample containing magnetic nuclei is introduced into a magnetic field, $B_0$, the nuclei are quantized into $(2I + 1)$ nuclear magnetic energy levels. Thus for carbon-13 nuclei with a spin number $I = \frac{1}{2}$ there are two possible energy levels and the separation between these levels is given by

$$E = \frac{\mu_C B_0}{I} = 2\mu_C B_0$$

As in any spectroscopic method it is possible to induce transitions between these levels by irradiating with energy of the correct frequency $\nu$ (Larmor frequency) to satisfy the Bohr frequency condition

$$2\mu_C B_0 = h\nu$$

For an applied magnetic field of 23,350 gauss the resonance frequency for carbon-13 is 25·0 MHz, i.e. in the radiofrequency range of the electromagnetic spectrum. Because the energy levels are close together there is only a small excess of nuclei in the lower energy level compared with the upper energy level as expressed by the Boltzmann distribution

$$\frac{n_{\text{upper}}}{n_{\text{lower}}} = \exp\left(\frac{-2\mu_C B_0}{kT}\right) \approx 1 - \frac{2\mu_C B_0}{kT}$$

When radiofrequency energy at the resonance frequency is applied, the nuclei in the upper energy level give out energy by stimulated emission while those in the lower level absorb energy; thus the measured absorption of energy corresponds only to that from the small excess of nuclei in the lower energy level. After excitation, the nuclear spins return to the original Boltzmann distribution in a time which depends on the spin–lattice relaxation time $T_1$ of the nuclei.

If the sample under investigation is examined as a liquid, gas or in solution then there is sufficient tumbling and rotational motion of the molecule to average out direct dipole–dipole interactions (such interactions lead to the broad-line NMR spectra observed for solids) and the resulting spectra have sharp absorption bands which can have line widths of less than 0·1 Hz when examined using modern NMR spectrometers of high-field homogeneity and overall system stability. Under these high-resolution conditions two characteristic parameters feature in the NMR spectra, namely, *chemical shifts* and *spin–spin coupling constants*.

## III. THE SPECTRAL PARAMETERS

### A. $^{13}C$ Chemical Shifts

Carbon-13 nuclei in different chemical environments absorb radiofrequency energy at different frequencies when examined in the presence of a constant applied magnetic field. The frequency differences result from the nuclei being shielded to varying extents from the applied magnetic field by their different electronic environments and also from varying magnetic anisotropic and electric field effects from neighbouring groups in the molecule. These frequency differences are referred to as *chemical shifts* and are usually measured with respect to the absorption frequency of a reference material (such as tetramethylsilane, carbon disulphide, benzene or dioxane). It is usual to express the chemical shifts in dimensionless units (ppm) by dividing the observed values in Hz by the operating frequency in MHz. The chemical shifts are characteristic of the particular type of carbon under investigation and when expressed in ppm are independent of the operating magnetic field

**Table 8.1.** The $^{13}C$ chemical shifts in ppm (referred to external $CS_2$) for non-terminal amino acids in peptides

| Amino acid | CO | α-C | β-C | γ-C | δ-C | Others | Reference |
|---|---|---|---|---|---|---|---|
| Gly | 21·4 | 150·5 | | | | | a, c |
| Ala | 19·4 | 143·4 | 177·4 | | | | a, g |
| Leu | 18·3 | 139·4 | 152·3 | 168·1 | 171·5<br>170·2 | | a, c |
| Val | 19·6 | 132·4 | 163·1 | 174·2<br>175·8 | | | a, d |
| Ile | 20·9 | 135·0 | 155·5 | 167·7 | 176·7<br>181.1 | | b |
| Lys | 20·8 | 140·6 | 164·0 | 172·3 | 167·5 | 154·7 (ϵ) | a, g |
| Asp | 20·7 | 142·7 | 157·4 | | | 16·3 (γ-CO) | a, g |
| Glu | 20·6 | 140·2 | 166·6 | 160·2 | | 12·4 (δ-CO) | a, g |
| Asn | 20·9 | 142·4 | 155·0 | | | 20·9 (γ-CO) | b |
| Gln | 21·4 | 140·7 | 167·6 | 163·0 | | 15·9 (δ-CO) | a, g |
| Phe | 21·1 | 138·6 | 157·2 | | | 65·6 (C-4); 58·0 (C-1);<br>63·9 (C-2,3,5,6) | a, g |
| Tyr | | 137.2 | 156.5 | | | 36·3 (C-4); 61·9 (C-2,C-6);<br>64·5 (C-1); 77·2 (C-3, C-5) | b, c |
| Ser | (22·2)<br>(22·8) | 138·5 | 133·4 | | | | a, g |
| Thr | (22·2)<br>(22·8) | 134·4 | 127·6 | 174·2 | | | a, g |
| Arg | 21·0 | 141·0 | 166·2 | 169·8 | 153·2 | 37·2 (ϵ) | a, g |
| Met | 20·9 | 140·5 | 163·8 | 164·7 | 179·5 | | a, g |
| His | 21·3 | 140·8 | 165·8 | | | 75·8 (C-4); 65·6 (C-5);<br>59·4 (C-2) | a, g |
| Trp | | 140·3 | 165·3 | | | 68·3 (C-2); 83·9 (C-3); 70·7 (C-5)<br>74·4 (C-4); 73·4 (C-6);<br>80·8 (C-7); 65·8 (C-8);<br>56·5 (C-9) | c, h |
| Pro | 18·8 | 132·1 | 163·3 | 168·0 | 144·5 | | d |
| Cys-H | 21·9 | 137·9 | 168·1 | | | | e |
| Cys-Cys | 21·9 | 141·0 | 154·9 | | | | e, f |

Positive shifts are to high field of reference.

Reference g data converted from $CH_3I$ reference to $CS_2$ reference using $\delta_{CS_2} = 214{\cdot}2 + \delta_{CH_3I}$.

Reference b data converted from TMS reference to $CS_2$ reference using $\delta_{CS_2} = 193{\cdot}7 + \delta_{TMS}$.

Reference c, d, e data converted from dioxan reference to $CS_2$ reference using $\delta_{CS_2} = 126{\cdot}3 + \delta_{DIOX}$.

a Christl and Roberts, 1972.
b Deslauriers *et al.*, 1972.
c Wessels *et al.*, 1973.
d Zimmer *et al.*, 1972.
e Jung *et al.*, 1972.
f Feeney *et al.*, 1973b.
g Freedman *et al.*, 1973.
h Allerhand *et al.*, 1973.

at which they are measured. Table 8.1 indicates the characteristic $^{13}C$ chemical shifts measured for various amino acids. When these are compared with the $^{1}H$ chemical shifts reported for amino acids it is seen that the carbon-13 chemical shifts are an order of magnitude larger than the proton chemical shifts. Because $^{13}C$ and $^{1}H$ line widths are comparable, $^{13}C$ NMR potentially offers an effectively higher-resolution technique than proton NMR. This possibility has important implications for studying the complex spectra expected from biological systems. By measuring changes in $^{13}C$ chemical shifts which accompany molecular interactions and conformational changes it is sometimes possible to comment on the nature of these effects.

The $^{13}C$ chemical shifts in many biological molecules have been measured and Tables 8.1 to 8.7 summarize some of the more relevant results. Sometimes molecules in different molecular configurations give different $^{13}C$ spectra; thus it is very easy to distinguish *cis*- and *trans*-prolines in peptides because the carbon atoms in the *cis* isomer are shielded differently from the same carbons in the *trans* isomer.

**Table 8.2.** The $^{13}C$ chemical shift changes (ppm) in amino acids resulting from ionizations and incorporation into peptides (Christl and Roberts, 1972)

| Type | $C_0$ | $C_\alpha$ | $C_\beta$ | $C_\gamma$ |
|---|---|---|---|---|
| Peptide shift of an *N*-terminal amino acid[a] | 5·4 ± 1·0 | 1·3 ± 0·4<br>Gly 0·7 ± 0·2 | −0·2 ± 0·5 | 0·4 ± 0·5 |
| Peptide shift of a *C*-terminal amino acid[b] | −3·5 ± 1·0 | −1·0 ± 0·5<br>Gly −2·0 ± 0·2 | −1·2 ± 0·6 | −0·4 ± 0·3 |
| Peptide shift of a non-terminal amino acid[c] | 1·5 ± 1·0 | 0·6 ± 0·4<br>Gly −1·1 ± 0·1 | 0·0 ± 0·5 | 0·0 ± 0·1 |
| Titration shift of *N*-terminal unit[d,e] | −6·9 ± 1·0 | −1·0 ± 0·5<br>Gly −2·6 ± 0·2 | −3·1 ± 1·0 | −0·6 ± 0·4 |
| Titration shift of *C*-terminal unit[f] | 3·0 ± 0·5 | 2·6 ± 0·5<br>Gly 2·2 ± 0·1 | 1·2 ± 0·2 | 0·3 ± 0·2 |

[a] Difference between the chemical shifts of an *N*-terminal amino acid in a zwitterionic peptide and the corresponding free, zwitterionic amino acid.

[b] Difference between the chemical shifts of a *C*-terminal amino acid in a zwitterionic peptide and the corresponding free, zwitterionic amino acid.

[c] Difference between the chemical shifts of a non-terminal amino acid in a zwitterionic peptide and the corresponding free amino acid.

[d] Difference between the chemical shifts of an *N*-terminal amino acid in peptide as anion and zwitterion.

[e] The values for pH 9·5 of the anion solution are probably too small by up to 30% for higher pH values.

[f] Difference between the chemical shifts of a *C*-terminal amino acid in a peptide as cation and zwitterion.

**Table 8.3.** The $^{13}C$ chemical shifts in ppm (referred to external $CS_2$) of the Pro carbon nuclei in some proline-containing peptides

| | | Proline ring carbons | | | | | |
|---|---|---|---|---|---|---|---|
| | | α | β | γ | δ | $CO_2^-$ | CON |
| Proline[a] | | 132·1 | 164·0 | 169·3 | 147·2 | 19·1 | |
| *N*-Ac-Pro[a] | *cis* | 130·9 | 159·5 | 166·1 | 144·5 | 16·6 | 21·7 |
| | *trans* | 132·2 | 161·3 | 166·1 | 143·1 | 16·9 | 22·0 |
| Gly-Pro[a] | *cis* | 129·1 | 159·2 | 168·5 | 143·4 | 12·5 | 25·0 |
| | *trans* | 128·8 | 161·2 | 166·5 | 144·1 | 11·6 | 25·8 |
| Ala-Pro[a] | *cis* | 128·8 | 159·3 | 168·6 | (143·4) | 12·7 | 21·3 |
| | *trans* | 128·8 | 161·5 | 166·1 | (142·5) | 12·0 | 22·2 |
| Val-Pro[a] | *cis* | 128·7 | 159·5 | 168·7 | 143·7 | 12·7 | 22·5 |
| | *trans* | 128·7 | 162·0 | 166·2 | 143·0 | 12·3 | 23·3 |
| *N*-Formyl-Pro[b] | *cis* | 133·1 | 163·4 | 170·3 | 148·5 | | 16·6 |
| | *trans* | 135·8 | 163·4 | 169·1 | 145·5 | | 17·3 |
| *N*-Ac-Pro Me ester[b] | *cis* | 132·1 | 161·8 | 170·4 | 146·1 | | 17·9 |
| | *trans* | 133·8 | 163·6 | 168·4 | 144·4 | | 17·9 |
| *N*-Ac-Pro amide[b] | *cis* | 131·5 | 161·1 | 170·3 | 145·8 | | 15·5 |
| | *trans* | 133·0 | 162·7 | 168·7 | 144·2 | | 15·5 |
| *t*-Butoxycarbonyl-Gly-Pro[b] | *cis* | 133·4 | 161·5 | 170·6 | 146·0 | | 18·1 |
| | *trans* | 133·4 | 163·8 | 168·2 | 146·6 | | 18·1 |

Positive shifts to high field of reference.
Values from reference *a* were converted from a TMS reference scale using $\delta_{CS_2} = 193{\cdot}7 + \delta_{TMS}$.
[a] Thomas and Williams, 1972.
[b] Dorman and Bovey, 1973.

**Table 8.4.** The $^{13}C$ chemical shifts in ppm (referred to external $CS_2$) and assignments of the carbon nuclei in oxytocin in DMSO-$d_6$ solution

| Amino acid | α-C | β-C | γ-C | δ-C | Others |
|---|---|---|---|---|---|
| Gly | 150·1 | | | | |
| Leu | 140·6 | 151·8 | 168·0 | 169·1<br>170·6 | 19·9 (CO) |
| Pro | 132·0 | 163·3 | 167·7 | 145·2 | |
| Cys | (138·8) | 152·1 | | | 24·3 (CO) |
| Asn | 141·6 | 155·2 | | | |
| Gln | (138·1) | 165·3 | 160·5 | | 17·9 (δ-CO) |
| Ile | 133·1 | 156·1 | 167·7 | 176·6<br>180·9 | |
| Tyr | (136·5) | 156·3 | | | C-4 36·3; C-1 64·3;<br>C-2, C-6 62·1; C-3, C-5 77·2 |
| Cys | (148·0) | 152·3 | | | |

Positive shifts to high field of reference.
Measured from TMS external reference and converted to $CS_2$ external reference using $\delta_{CS_2} = 193{\cdot}7 + \delta_{TMS}$.
Assignment uncertain for bracketed values.

**Table 8.5.** $^{13}C$ chemical shifts in ppm (referred to external $CS_2$) for Phe-8 in ribonuclease A S-peptide and its complex with S-protein (Chaiken *et al.*, 1973)

| Carbon | RNase 1–15 peptide | RNase S-peptide/ S-protein complex | Difference (ppm) |
|---|---|---|---|
| α-C | 137·88 | 137·16 | −0·72 |
| β-C | 155·77 | 156·27 | +0·50 |
| CO | 19·60 | 18·27 | −1·33 |

**Table 8.6.** $^{13}C$ chemical shifts in ppm (referred to external $CS_2$) and assignment of some carbons in horse heart cytochrome c (Oldfield and Allerhand, 1973)

| Assignment | Chemical shift (ppm) | |
|---|---|---|
| | reduced | oxidized |
| 2 Arg-δ | 35·7 (1) } | 35·9 (2) |
| Tyr-δ | 36·0 (2) } | |
| Tyr-δ } | 37·2 (3) | { 37·0 (3) |
| Tyr-δ } | | { 38·1 (4) |
| Tyr 67-δ | 40·4 (4) | 31·3 (1) |
| His 18-γ | 71·3 (29) | Not observed |
| Trp 59-γ | 83·6 (30) | 83·9 (16) |

In addition, the porphyrin, Tyr-γ and Trp 59-δ carbons were assigned.
Numbers in parenthesis refer to labels in Figure 8.11.

## B. Spin–Spin Interactions Involving Carbon-13

When the chemically shifted absorptions are examined under high-resolution conditions further multiplet splittings may be observed. These result from interactions between the observed nucleus and other magnetic nuclei in the molecule. The energy of the nucleus is dependent on the orientation of the magnetic moments of neighbouring magnetic nuclei. Thus a carbon directly bonded to a proton gives a $^{13}C$ doublet signal because the carbon nucleus can sense via the bonding electrons the two possible different orientations of the proton nucleus (which also has a spin number $I = \frac{1}{2}$). For a $CH_2$ group the carbon appears as a triplet from interactions of the carbon with the two protons which can have three different values of total spin; similarly a $CH_3$ group gives a quartet splitting on the carbon resonance. These simple splitting patterns are observed when the chemical shift differences between the coupled nuclei are much larger than the spin–spin interactions. For such simple first-order spectra the observed splittings are referred to as *spin coupling constants*, $J_{CH}$, and when measured in cycles/sec (Hz) their magnitude is independent of the field strength (unlike the chemical shifts when these are measured in Hz). For directly bonded carbon and protons the coupling constants are large and depend to some extent on the percentage *s* character in the bonds (e.g. alkanes, $J_{CH} \approx 120$ Hz, alkenes and aromatics, $J_{CH} \approx 170$ Hz, alkynes, $J_{CH} \approx 250$ Hz) (Stothers, 1972). For non-directly bonded carbon and protons much smaller coupling constants are observed ($< 12$ Hz). Three-bond coupling constants are potentially very important because there is a relationship between the dihedral angle and the observed vicinal $J_{^{13}C—C—CH}$ coupling constants. Rodgers and Roberts (1973) have used such coupling constants to determine the conformation of the $C_\alpha$—N bond in *N*-acetyltryptophan when bound to α-chymotrypsin by measuring the coupling constant between the labelled carbonyl carbon and the α-proton (see Section V.G.6).

By studying similar three-bond C—H coupling constants in peptides (e.g. $J_{NH—^{13}C_\beta}$ and $J_{^{13}CONCH}$) important side-chain and backbone conformational information can be obtained (Feeney *et al.*, 1974; Hansen *et al.*, 1975; Bystrov, 1974). At the present time, only limited conformational information of this type is available from studies of proton-proton coupling constants between the NH and α-CH protons. Figure 8.1 shows the $^{13}C$ single-resonance spectrum

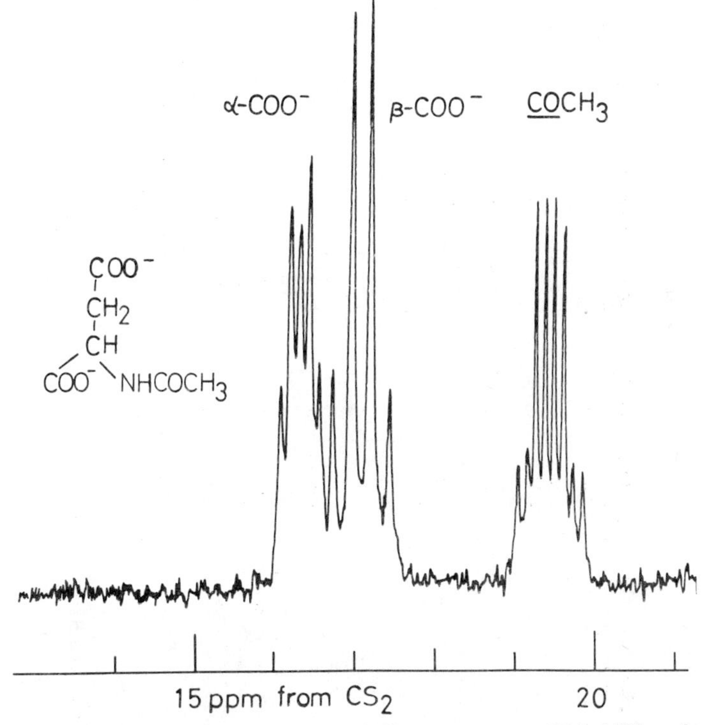


**Figure 8.1.** The single-resonance $^{13}C$ spectrum at 25·2 MHz of *N*-acetylaspartic acid at pD 4·4 in $D_2O$ solution (Feeney, Partington and Roberts, 1973b)

**Table 8.7.** The $^{13}C$ chemical shifts in ppm (referred to external $CS_2$) of some common nucleosides and nucleotides

| Compound | C-2 | C-4 | C-5 | C-6 | C-8 | C-1′ | C-2′ | C-3′ | C-4′ | C-5′ | Reference |
|---|---|---|---|---|---|---|---|---|---|---|---|
| **1** | 42·91 | 31·82 | 72·45 | 59·95 | | 104·94 | 119·33 | 122·18 | 107·66 | 131·17 | a, b |
| **2** | 42·90 | 31·65 | 76·71 | 69·48 | | 104·87 | 119·47 | 122·01 | 107·72 | 130·82 | a, b |
| **3** | 44·09 | 35·94 | 56·15 | — | | 102·88 | 119·95 | 122·11 | 107·83 | 130·53 | a, b |
| **4** | 37·34 | 22·98 | 85·28 | 53·54 | | 100·15 | 112·49 | 108·30 | 105·76 | 131·54 | b |
| **5** | 38·14 | 22·57 | 85·93 | 53·21 | | 98·93 | 112·63 | 110·37 | 107·63 | 131·87 | b |
| **6** See page 297 (compounds 1–11) | 37·85 | 28·66 | 73·75 | 41·36 | | 105·23 | 117·65 | 121·91 | 108·03 | 129·71 | a, b |
| **7** | 38·95 | 26·70 | — | 36·00 | | 102·93 | 118·48 | 123·35 | 108·23 | 132·21 | a, b |
| **8** | 41·40 | 51·11 | 37·65 | 29·16 | | 105·09 | (121·49) | (122·30) | 108·04 | 130·12 | b |
| **9** | 41·51 | 51·55 | 92·25 | 29·33 | | 111·97 | 156·62 | 121·63 | 106·68 | 131·06 | b |
| **10** | 43·24 | 44·16 | — | 43·24 | | 104·14 | (121·12) | (122·43) | 107·90 | 130·17 | b |
| **11** | 26·11 | 25·48 | 152·21 | 41·00 | | 104·03 | (120·46) | (122·19) | 107·75 | 129·78 | b |
| Adenosine | 40·08 | 43·46 | 73·15 | 36·41 | 52·38 | 104·33 | 118·88 | 121·76 | 106·56 | 130·89 | a, c |
| Deoxyadenosine | 39·88 | 43·50 | 73·09 | 36·39 | 52·56 | 107·98 | 153·06 | 121·29 | 104·36 | 130·60 | c |
| Inosine | 44·31 | 46·57 | 68·18 | 35·84 | 53·61 | 104·73 | 118·37 | 122·28 | 106·88 | 131·24 | a, c |
| Guanosine | 38·94 | 41·18 | 75·98 | 35·75 | 56·65 | 106·27 | 118·66 | 121·98 | 107·13 | 131·34 | a, c |
| Uridine-5′-monophosphate | 40·7 | 26·3 | 89·9 | 50·4 | | 104·0 | 118·5 | 122·4 | 108·5 | 129·0 | a, b |
| Thymidine-5′-monophosphate | 41·0 | 26·3 | 81·1 | 55·1 | | 107·7 | 154·0 | 121·3 | 106·6 | 128·5 | d |
| Cytidine-5′-monophosphate | 35·0 | 26·3 | 96·0 | 50·7 | | 103·4 | 118·2 | 122·8 | 109·2 | 129·2 | a, d |
| Inosine-5′-monophosphate | 45·8 | 44·3 | 69·4 | 34·2 | 52·7 | 104·9 | 117·7 | 121·9 | 108·3 | 129·0 | a, d |
| Adenosine-5′-monophosphate | 40·1 | 44·1 | 74·5 | 37·5 | 52·6 | 105·5 | 118·0 | 122·0 | 108·3 | 129·0 | a, d |

a The original assignments for C-2′ and C-3′ have been reversed. Chemical shifts from references b and c were measured from benzene external reference and converted to $CS_2$ external reference using $\delta_{CS_2} = \delta_{C_6H_6} + 65{\cdot}0$ ppm.

b Jones *et al.*, 1970a.

c Jones *et al.*, 1970b.

d Dorman and Roberts, 1970.

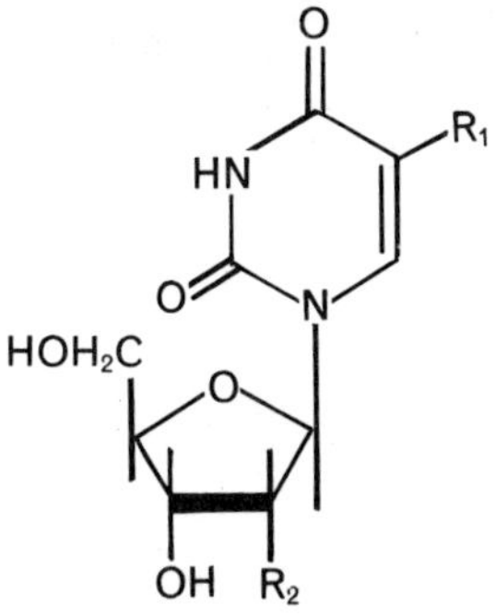

**(1)** $R_1=OH$, $R_2=OH$
**(2)** $R_1=NH_2$, $R_2=OH$

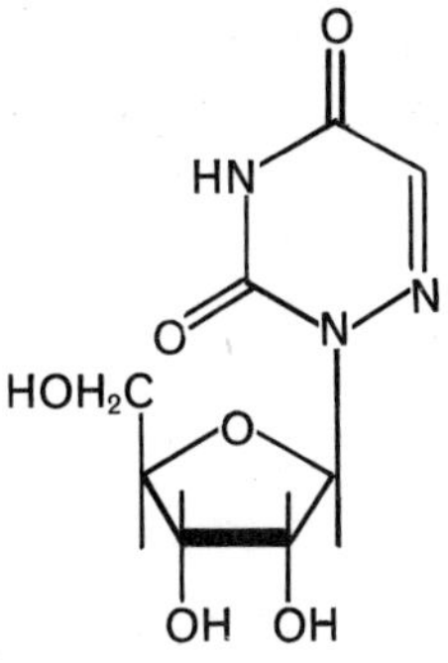

**(3)**

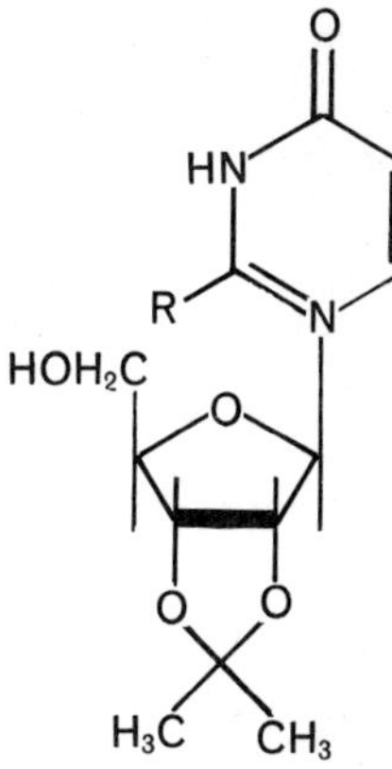

**(4)** $R=OCH_3$
**(5)** $R=NH_2$

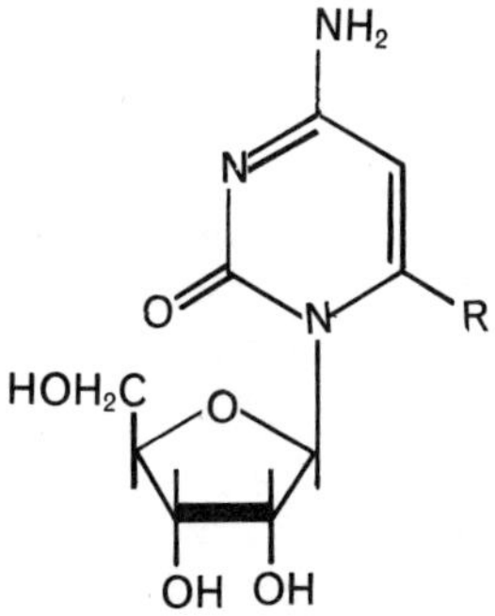

**(6)** $R=OH$

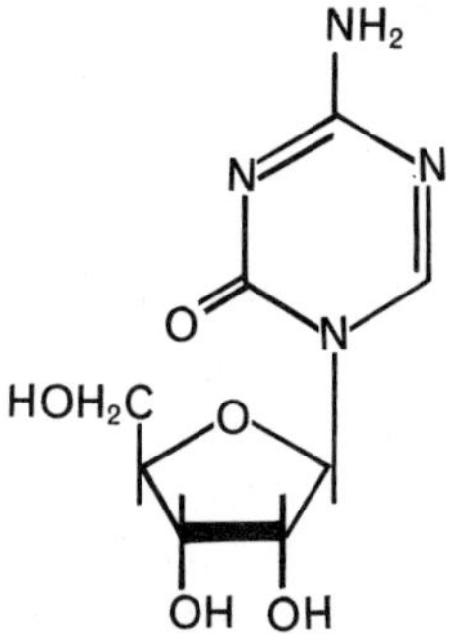

**(7)**

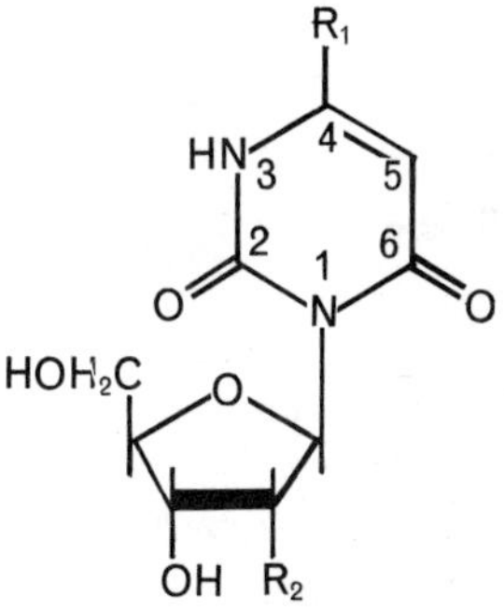

**(8)** $R_1=H$, $R_2=OH$
**(9)** $R_1=H$, $R_2=H$

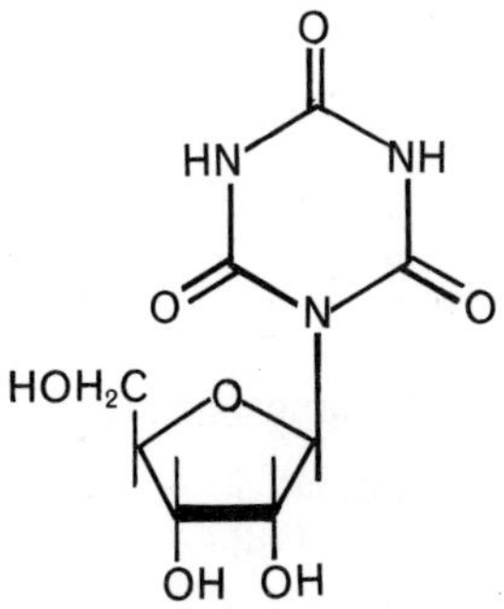

**(10)**

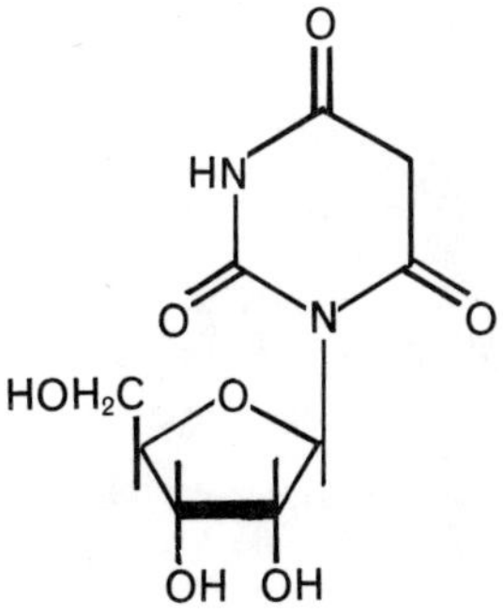

**(11)**

Selected pyrimidine nucleosides

for the carbonyl region of *N*-acetylaspartic acid and this indicates that such coupling constants involving the carbonyl carbon will be measurable with ease, particularly if $^{13}C$-enriched peptides are used. The reliability of this approach will be improved when the Karplus-type relationships* between such coupling constants and the dihedral angles are better characterized (Lemieux *et al.*, 1972).

### C. Carbon-13 Spin–Lattice Relaxation Times ($T_1$)

After the excess nuclei in the lower energy level have absorbed the radio-frequency energy and undergone transition to the upper energy level (Section II) the system will begin to return to the original equilibrium Boltzmann distribution by one of several possible relaxation processes. If the static equilibrium magnetization is $M_0$ and $M_z$ is the magnetization proportional to the excess nuclei in the lower energy level, then $M_z$ approaches $M_0$ exponentially with the characteristic *spin–lattice relaxation time* $T_1$ such that

$$\frac{dM_z}{dt} = \frac{(M_0 - M_z)}{T_1} \tag{8.1}$$

The excess energy is lost to the lattice as a result of interactions with time-dependent fluctuating fields with components at the Larmor frequency capable of stimulating downward transitions of the excited nuclei. Spontaneous emission of energy is of low probability and can be neglected. For carbon nuclei directly bonded to protons in medium-sized and large molecules the usual relaxation mechanism results from modulation of the strong dipole–dipole interaction between the carbon and the directly bonded proton (Allerhand *et al.*, 1971b,; Levy and Nelson, 1972). Because the modulation process results from molecular motions, in favourable cases a study of the relaxation times of carbon nuclei can provide molecular dynamic information. The contribution to the relaxation rate $1/T_1$ from the dipolar mechanism is given by

$$\frac{1}{T_1} = N\hbar^2\gamma_C^2\gamma_H^2\tau_{CH}^{-6}f(\tau_c) \tag{8.2}$$

where $\gamma_C$ and $\gamma_H$ are the carbon and proton magnetogyric ratios,† $N$ is the number of directly bonded protons, $\tau_{CH}$ is the internuclear separation between the carbon and the attached proton, $\hbar$ is (Planck's constant/$2\pi$) and

* In a H—C—C—H fragment the proton–proton three-bond coupling constants $J_{HH}$ depend on the dihedral angle $\phi$ as expressed by a relationship of the form

$$J_{HH} = k_1 + k_2 \cos^2 \phi$$

where $k_1$ and $k_2$ are constants depending on the substituents and other factors.

† The magnetogyric ratio is a characteristic property of a nucleus and is given by $\gamma = \mu/(I\hbar)$ where $\mu$ is the magnetic moment and $I$ is the spin quantum number ($I = 1/2$ for $^{13}C$).

$f(\tau_c)$ is a function of the correlation times and is the only unknown term in the equation. For small molecules where the rotational reorientational correlation time is such that $1/\tau_c$ is much greater than the nuclear resonance frequencies (extreme narrowing condition $\omega^2\tau_c^2 \ll 1$ where $\omega = 2\pi\nu$ radians/sec) then $f(\tau_c)$ is effectively given by $\tau_c$. The rotational correlation time is defined as the time required for the molecule or part of the molecule to rotate through one radian. Thus direct information about such rotational motions in molecules can be obtained from measurements of spin–lattice relaxation times using Equation (8.2).

For short correlation times the contribution to the relaxation rate is seen to increase as the correlation times become longer. Slower motion results in more efficient dipolar relaxation because the intensity of the component at the Larmor frequency in the fluctuating field (i.e. the spectral density) increases as $1/\tau_c$ decreases towards the Larmor frequency.

Fourier transform techniques (see Section IV.F) can be used conveniently to measure the spin–lattice relaxation times (Vold *et al.*, 1968; Farrar and Becker, 1971). In the Fourier transform NMR experiment it is possible to apply a radiofrequency pulse, $H_1$, (180° pulse) along the $x'$ axis for a sufficient length of time to invert completely the magnetization from $M_0$ to $-M_0$ along the magnetic field direction ($z'$ axis). After a time $t$ seconds another pulse (90° pulse) is applied which causes any magnetization along the $z'$ axis to be transferred to the $y'$ axis where it is observed and recorded after Fourier transformation (Figure 8.2). (The axes $x'$, $y'$ and $z'$ refer to the

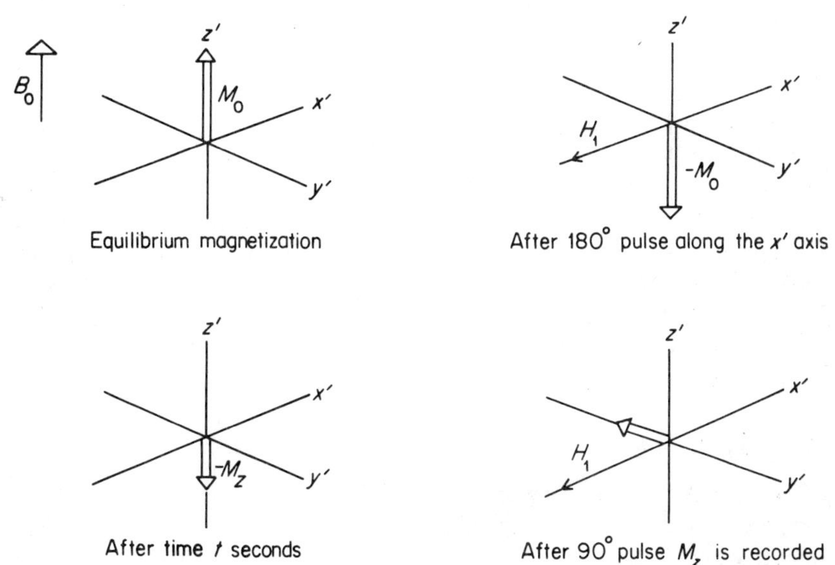


**Figure 8.2.** The effects of applying a 180-$t$-90 pulse sequence to the spin system. The external magnetic field is $B_0$

coordinates in a framework of reference rotating at the Larmor resonance frequency.)

From integration of Equation (8.1) it is seen that

$$M_z = M_0[1 - 2\exp(-t/T_1)] \tag{8.3}$$

As $t$ increases the negative $M_z$ signal decreases in magnitude, goes through a null value at $t = T_1 \ln 2$, then emerges as a positive signal and eventually returns to its equilibrium value, $M_0$. This is illustrated in Figure 8.2. By measuring the resultant magnetization $M_z$ for different values of $t$ it is possible to obtain the value of $T_1$ from Equation (8.3). If a delay of ca. $5T_1$ is allowed after the (180-$t$-90) sequence to allow the Boltzmann distribution to be re-established then the measurement can be repeated and the sensitivity improved by spectrum accumulation techniques. It is clear that for carbon nuclei of similar relaxation times the $T_1$ values of several nuclei with separate absorption bands can be measured in the same experiment. Figure 8.3 shows a set of partially relaxed $^{13}C$ spectra for cholesteryl chloride in carbon tetrachloride (Allerhand *et al.*, 1971) from which it is possible to calculate the $T_1$ values.

In the early $^{13}C$ relaxation time measurements of Allerhand and coworkers (1971b) it was found that in all but very small molecules the carbon-13 relaxation mechanism for protonated carbons is dominated by the dipolar interaction with the directly bonded proton and is largely unaffected by interactions with other protons in the molecule. (Because of the low abundance of $^{13}C$, $^{13}C$–$^{13}C$ dipolar interactions can be neglected.) Anisotropic reorientation of the complete molecule (i.e. unequal rotational rates about different axes) can lead to different relaxation times for different protonated carbons in the same molecule. Furthermore, different carbons in the molecule which are capable of internal reorientation at different rates will also have different relaxation times. Allerhand and Doddrell (1971) were able to assign $^{13}C$ resonances of the two galactopyranose rings in stachyose (Galactose-Galactose-Glucose-Fructose) on the basis of the terminal galactopyranose being expected to have a faster internal reorientation about the —O—$CH_2$— linkage than the other galactopyranose: the galactopyranose carbon resonances could be divided into two sets on the basis of their relaxation times and the terminal galactose was assigned to the lines with the longer carbon-13 relaxation times.

### D. Spin–Spin Relaxation Time ($T_2$)

After a 90° pulse along the $x'$ axis, the $M_z$ component is zero and the magnetization now lies along the $y'$ axis: initially all the precessing magnetic moments are in phase but with the passage of time they lose this coherence in an exponential fashion with a relaxation rate characterized by the spin–spin or transverse relaxation time $T_2$. Any rate process which can cause phase

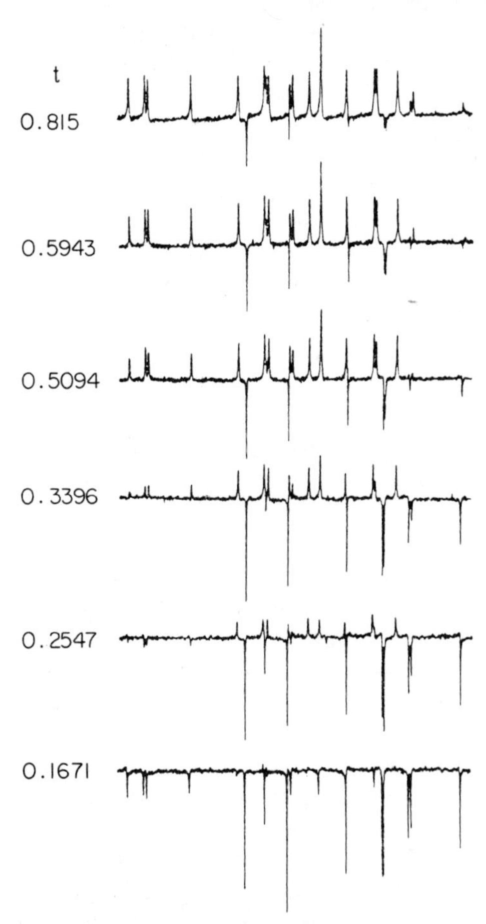


**Figure 8.3.** Partially relaxed $^{13}C$ spectra at 25·2 MHz for cholesteryl chloride in $CCl_4$ observed using a 180-*t*-90 pulse sequence. Time *t* is in seconds (Allerhand, Doddrell and Komoroski, 1971b)

incoherence will make a contribution to $T_2$. In some cases it is possible to measure the rates of chemical exchange processes from $T_2$ measurements.

## IV. METHODS OF IMPROVING SENSITIVITY

We have seen earlier that carbon-13 NMR spectroscopy requires the ultimate in instrumental sensitivity. For large biological molecules where it is difficult to obtain solutions of higher concentration than a few m*M* such studies demand all the available instrumental techniques for improving sensitivity (Ernst, 1966a).

### A. Isotopic Enrichment

This is an obvious method for increasing the sensitivity of any nucleus which has a low natural abundance such as carbon-13. Although carbon-13 enrichment is expensive, both in terms of cost of carbon-13 precursor molecules and also in time required for developing a method for synthesizing the desired compound which efficiently uses the carbon-13 precursor molecule, the advantages of obtaining a specifically enriched carbon-13 compound often make this a worthwhile investment. In addition to simplifying the carbon-13 assignment problem the use of carbon-13-enriched inhibitors, substrates and coenzymes offers a convenient method for studying the interactions of these molecules with proteins, by removing the interference from the background protein natural abundance carbon-13 spectrum. For selective enrichment at a single carbon site it is desirable to aim for the highest enrichment possible. However, for general carbon-13 enrichment, as often encountered in biosynthetic processes, an optimum enrichment level is approximately 20%: at higher enrichments there is an increased probability of having molecules with adjacent carbons both enriched, which will give complicated spectra because of the large $^{13}C$–$^{13}C$ spin couplings between directly bonded carbon atoms.

At the present time the least expensive source of carbon-13 is $^{13}C$ carbon dioxide (£700 for 50 g of 90% $^{13}CO_2$); the widespread use of carbon-13 labelling will result hopefully in substantial cost reductions. In fact, the next predictable advance in biological studies with $^{13}C$ NMR could well result from the availability of inexpensive carbon-13 precuror molecules, particularly C-1 and C-2 labelled sodium acetate for use in biosynthetic work leading to carbon-13 labelled proteins (amino acids), phospholipids and nucleotides.

### B. Use of Large-diameter Sample Tubes

For cases where the size of the sample is not limiting, large volumes of sample can be examined by using large-diameter sample tubes. Typically 12–15 mm diameter tubes are used with 24 Kgauss magnets and no problems are experienced in maintaining a resolution of 0·2 Hz over the large volume element with most of the commercial spectrometers available. However, for cylindrical samples of the same height one might expect a maximum sensitivity improvement proportional to the square of the tube diameter; this is not realized in practice and in fact the sensitivity improvement achieved on examining the same solution in 12 mm and 5 mm tubes (volume ratio 6:1) is no more than 3:1. Thus for experiments where the sample size is a limiting factor and where there is no reason for working at low dilution it is usually better to operate with 5 mm sample tubes. A further effective sensitivity advantage can sometimes result from examining more concentrated solutions since the relaxation times can be shortened in the more viscous solutions which would then allow the Fourier transform spectra to be collected with shorter pulse delays. However, for biological studies the largest-diameter sample tubes available are invariably used because one is usually either limited by low solubility or operating at low concentration to minimize intermolecular effects.

## C. Use of Higher Magnetic Fields

Theoretically the sensitivity increases with the square of the applied magnetic field ($B_0^2$) and, while the observed increase is lower than this (ca. $B_0^{3/2}$), it is usually desirable to operate at the highest magnetic field strength available consistent with the use of large-diameter sample tubes. To obtain fields higher than 25 Kgauss and of the required homogeneity it is necessary to resort to superconducting magnet systems. Generally speaking these instruments cannot accommodate sample tubes greater than 10 mm in diameter and sometimes a higher sensitivity can be achieved by operating at lower fields using iron core electromagnets with larger sample tubes. At very high fields there is also the additional factor for carbon-13 of lower nuclear Overhauser enhancements (Section IV.E) in the proton-noise-decoupled $^{13}C$ spectra and also the considerable problem of obtaining proton-noise-decoupling fields (Section IV.E) of sufficiently high power to cope with the increased proton chemical shift range. However, the chemical shifts increase directly with the magnetic field strength and also one does often achieve better sensitivity, particularly in situations where one is sample limited. For these reasons the trend towards high-field wide-bore superconducting NMR systems will continue.

## D. Spectrum Accumulation

An improvement in sensitivity can be achieved by repeatedly scanning the carbon-13 spectrum and accumulating the spectra in a small computer (see Vol 1, Chapter 8). The spectra are first digitized and then stored in an appropriate number of channels in the computer such that the digitization is not the resolution-limiting factor. Modern NMR spectrometers have such high stability that the signals in the digitized spectra from repeated scans are stored always in the same channels. Because the signals are coherent they add in direct proportionality to the number of scans, $N$, whereas the incoherent noise adds by a square-root relationship proportional to $\sqrt{N}$; thus the overall increase in sensitivity is a signal-to-noise improvement proportional to the square root of the number of scans, $\sqrt{N}$. In Figure 8.4(b) the $^{13}C$ spectrum of pyridine after 16 scans is a factor of 4 better in signal-to-noise ratio than a single-scan spectrum (not shown) (Johnson, personal communication). The spectrum comprises three large doublets ($\alpha$, $\gamma$ and $\beta$ carbons) of multiplets, the doublet splittings (ca. 180 Hz) arising from spin–spin interaction with the directly bonded protons. The smaller splittings result from long-range coupling between the carbons and non-directly bonded protons.

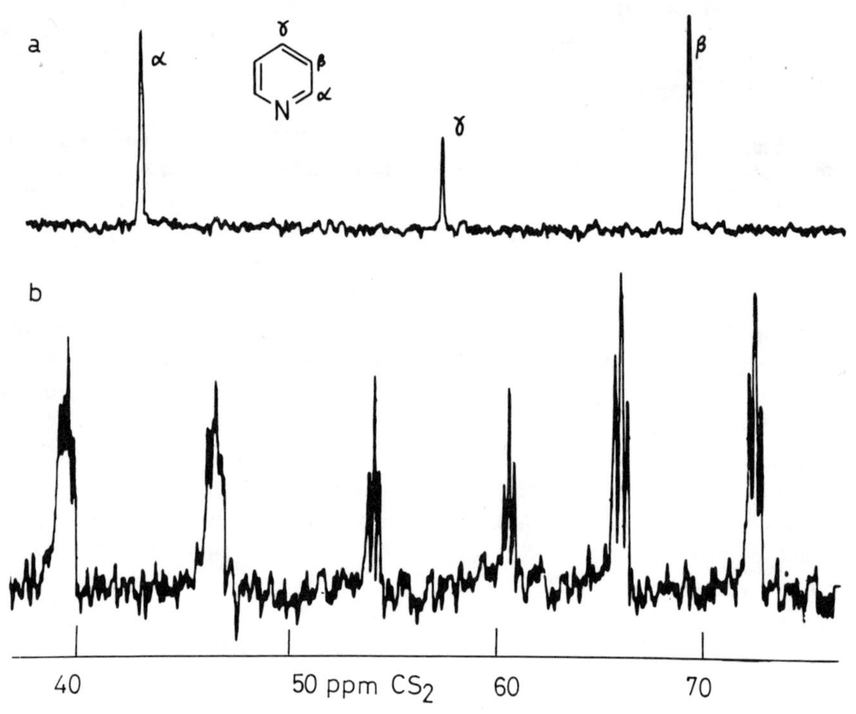


**Figure 8.4.** (a) The $^{13}C$ proton-noise-decoupled spectrum of neat pyridine at 25·0 MHz (single scan). (b) A 16-scan accumulated $^{13}C$ single-resonance spectrum of neat pyridine at 25·0 MHz. (Johnson, personal communication)

## E. Proton Noise Decoupling (Ernst, 1966b)

When the proton nuclei are strongly irradiated at the proton resonance frequency in a double-resonance experiment the C–H spin–spin interactions are removed from the carbon-13 spectrum. Because the intensity of the multiplet is now condensed into a single line of greater height than the multiplet components the signal-to-noise ratio is effectively increased. In fact a further increase in sensitivity often results from such decoupling experiments in that if we have molecules with motions much faster than the Larmor frequency and if the spin–lattice relaxation of the carbon is controlled by dipolar interactions with its proton neighbours then the populations of the energy levels involved in the carbon-13 transitions are changed in a manner which can lead to an increase in sensitivity by almost a factor of three. These so called nuclear Overhauser enhancements are invariably observed for carbon atoms with directly bonded protons (Noggle and Schirmer, 1971). Thus the use of proton decoupling can lead to an overall increase in sensitivity of an order of magnitude which corresponds to a time saving of two orders of magnitude in spectrum accumulation experiments. Figure 8.4(a) shows the sensitivity improvement achieved in the proton-decoupled carbon-13 spectrum of pyridine which has higher sensitivity after a single scan than has the single-resonance spectrum after sixteen scans. Instead of using a coherent radiofrequency field for the proton irradiation it is usual to modulate the decoupling frequency with random noise to produce a broad-band decoupling source which can irradiate simultaneously several nuclei of different chemical shifts (Ernst, 1966b): this *proton-noise-decoupling* technique was used to decouple the differently shielded $\alpha$, $\beta$ and $\gamma$ protons of pyridine coupled to the carbon nuclei shown in Figure 8.4(a). Such methods lead to single sharp absorption bands being observed for each different type of carbon in the spectrum. Of course one loses the potentially valuable C-H spin coupling constant information in a proton-noise-decoupling experiment. By using gated proton decouplers which are switched off during the period that the sample is being examined but irradiate at all other times, undecoupled spectra with some of the nuclear Overhauser enhancement can be obtained (Feeney *et al.*, 1970; Freeman and Hill, 1971). Allerhand and coworkers (Doddrell *et al.*, 1972) have calculated the nuclear Overhauser enhancements expected at various magnetic field strengths for carbon-13 in molecules with the long rotational correlation times that are anticipated in higher molecular weight ($>$ 15,000 M.Wt.) biological compounds. For correlation times of $10^{-8}$ sec, even with carbon nuclei which are completely dipolar relaxed, the nuclear Overhauser enhancements are of the order of $< 20\%$. The nuclear Overhauser enhancements also decrease as the strength of the magnetic field is increased and this can be serious for correlation times longer than $5 \times 10^{-9}$ sec.

Variable nuclear Overhauser enhancements can lead to problems in interpretation of intensity data but there are methods of recording carbon-13 spectra without nuclear Overhauser enhancements which can be realized by using a gated proton-noise-decoupler which is only decoupling during the actual time the sample is being examined and is thus not irradiating the sample for a sufficient length of time to allow the population changes leading to nuclear Overhauser enhancement to be established.

### F. Fourier Transform Technique

When an NMR spectrum is scanned in the normal manner a great deal of time is wasted in that each of the nuclei is being examined only for a small fraction of the total sweep time. Fourier transform techniques overcome this problem by analysing the response to the nuclear system in which all the nuclei are irradiated at their different resonance frequencies at the same time (Gillies and Shaw, 1972; Ernst and Anderson, 1966; Allerhand *et al.*, 1971; Levy and Nelson, 1972). Instead of a continuous radiofrequency source a series of radiofrequency pulses of short pulse duration (e.g. 10–100 $\mu$sec) are used, separated by time intervals much longer than the duration of the pulse (e.g. 1 sec). In this way a distribution of radio-frequencies is achieved which can bring all the carbon nuclei in the sample to resonance at the same time and reduce the 'sweep time' by a factor of up to one hundred. The response to this system is not the conventional NMR spectrum but a beat pattern (time domain spectrum) which is the Fourier transform of the conventional NMR spectrum examined in the frequency domain (see Figure 8.5). By carrying out the Fourier transform on the free induction decay (FID) time domain spectrum (a) the frequency domain spectrum (b) is obtained in a few seconds using a small on-line computer. When the Fourier transform technique is used in conjunction with spectrum accumulation, the impulse responses to many pulses can be added together before eventually transforming the accumulated data, which provides a very powerful method of improving sensitivity. If the relaxation times of the nuclei are long then it is necessary to have adequate time delays between the pulses to allow the nuclei to recover their magnetization. If insufficient time is allowed between the pulses, in the interests of obtaining maximum sensitivity in reasonable times, then the intensity information in such spectra is no longer meaningful. The theoretical signal ($S$)-to-noise ($N$) improvement from a Fourier transform experiment depends on the total width of the spectrum and the required line widths and is given by

$$\frac{S}{N} = \sqrt{\frac{\text{Total width of spectrum}}{\text{Line width}}}$$

Thus the technique is most satisfactory for spectra containing many lines over the spectral region: obviously for a single-line spectrum there would be no signal-to-noise improvement over a conventional frequency-swept NMR experiment.

The resolution in a Fourier transform spectrum is not impaired by the technique if the time for acquisition ($AT$) of data into the computer is such that

$$\text{Instrumental line width} > \frac{1}{AT}$$

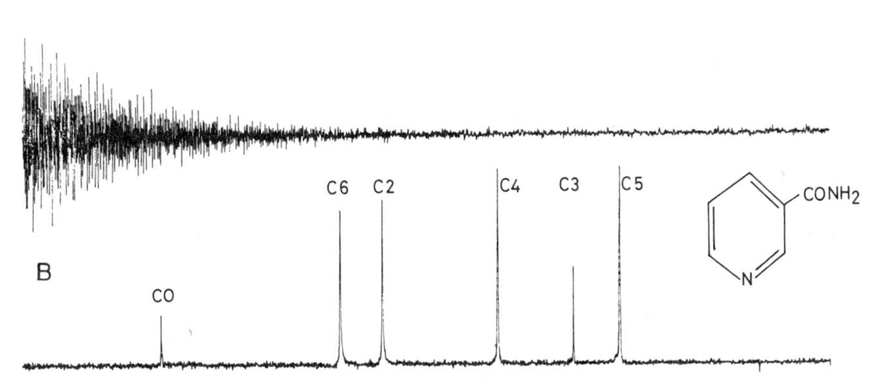


**Figure 8.5.** (a) The $^{13}C$ free induction decay (FID) time domain spectrum of nicotinamide. This is the response of the system as a function of time (s). (b) The frequency domain spectrum of nicotinamide obtained by Fourier transform of the FID in (a). This is the response of the system as a function of frequency ($\omega$)

Under these conditions the line widths are determined by the instrumental inhomogeneities and instabilities as in a conventional frequency-swept experiment. To summarize, Fourier transform techniques (Farrar and Becker, 1971; Levy and Nelson, 1972) can result in an effective saving in time of from one to two orders of magnitude.

However, it is the combination of the many available techniques which make possible the study of carbon-13 NMR at natural abundance, i.e. from Fourier transformation of impulse response spectra collected by spectrum accumulation using proton-noise-decoupling on samples in large-diameter tubes examined at the maximum field strength available.

## V. APPLICATIONS OF $^{13}C$ NMR TO BIOLOGICAL PROBLEMS

In principle NMR can supply detailed information about molecular structure and conformation (from chemical shifts and spin-coupling constants) and molecular dynamic information (from relaxation and line-shape measurements). This information is often difficult to extract from complex spectra of biological molecules due to the extensive overlap of many broadened resonance bands which make it extremely difficult to assign individual nuclei to observed absorption bands. Proton-noise-decoupled $^{13}C$ NMR spectra are intrinsically simpler than $^{1}H$ spectra because of the absence of CH and CC spin coupling and the greater chemical shifts usually found in carbon-13 spectra. Furthermore, by selective carbon-13 enrichment of precursor molecules one can introduce carbon-13 into large complex molecules such as proteins and obtain simplified spectra. In principle, from a series of such spectra, characteristic carbon NMR signals from every carbon in the molecule can be obtained and their chemical shifts and relaxation times monitored. Studies of this kind are now being undertaken and at the present time they have progressed to the point where it has been shown that the experiments are technically possible.

A major problem in $^{13}C$ NMR studies is to obtain quantitative conformational information from $^{13}C$ chemical shift studies. The following review of some of the applications of $^{13}C$ NMR to a range of biologically important molecules will attempt to indicate the present scope and limitations of the technique in this area.

### A. Amino Acids and Peptides

It is not surprising that this important class of compounds has received considerable attention. Several years ago, Horsley and Sternlicht (1968) measured the $^{13}C$ chemical shifts for a series of amino acids, and showed that the range of chemical shifts was much larger than observed for protons (for example $\alpha$-$^{13}C$ chemical shift range of 11·3 ppm compared with $\alpha$-CH proton shift range of 1·4 ppm). More recently, Christl and Roberts (1972) have made a systematic survey of the $^{13}C$ spectra of a large number of small peptides as a function of pH and their findings provide a good starting point for making $^{13}C$ spectral assignments in larger peptides. Table 8.1 gives the predicted $^{13}C$ chemical shifts for amino acid residues in non-terminal positions in a peptide chain and Table 8.2 summarizes the $^{13}C$ chemical shift changes in amino acids resulting from ionizations and incorporation into peptides (Christl and Roberts, 1972). The $^{13}C$ shifts in the peptides show systematic differences from those in the zwitterionic amino acids, the differences depending on whether the residue is *C*-terminal, *N*-terminal or non-terminal. However, the $^{13}C$ chemical shifts are found not to depend on neighbouring residues in the chain unless the neighbour is proline (Christl and Roberts, 1972). A proline residue influences the $^{13}C$ shielding of its *N*-terminal neighbour; thus from comparisons of the $^{13}C$ shifts of phenylalanine in Phe-Pro with those in the Phe-Met and Phe-Arg, the Phe $\alpha$-C is found to be 1 ppm to higher field in the proline analogue but the Phe $\beta$-C shows no effect (Christl and Roberts, 1972).

Useful configuration information can be obtained for prolines from the $^{13}C$ chemical shifts which are sensitive to the *cis*/*trans* configurations. Several workers have pointed out the distinctive chemical shift differences for the $\beta$- and $\gamma$-carbons of *is*- and *trans*-prolines as shown in Table 8.3 (the $\gamma$-carbons providing the most reliable guide) (Smith *et al*, 1972a; Thomas and Williams, 1972; Wuthrich *et al.*, 1972; Deslauriers *et al.*, 1972; Dorman and Bovey, 1973).

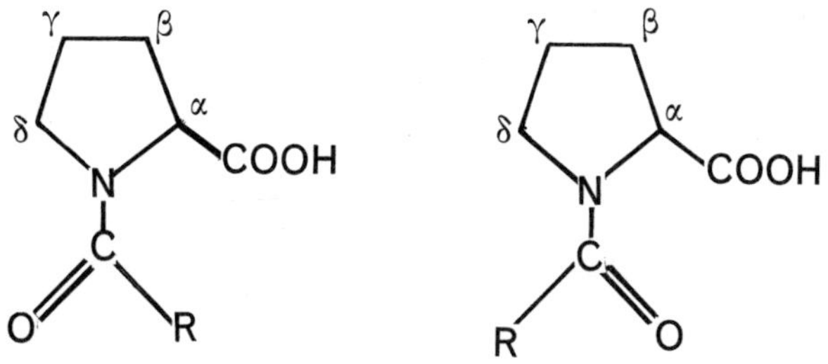

*Cis*-proline

*Trans*-proline

Thomas and Williams (1972) have found these shift differences to be quite insensitive to the bulkiness of the side chain of the preceding amino acid in the sequence and thus they can be used with reasonable confidence to make configurational assignments. For cases where there is a *cis-trans* mixture the ratio of the two isomers can be readily determined. In some of the simple peptides examined substantial amounts of the *cis* isomer were detected (for example in Gly-Pro 39% of the *cis* isomer is present). *Cis-trans* isomerism in naturally occurring peptides has also been observed. From the $^{13}C$ spectrum of thyrotropin releasing factor, PCA-His-ProNH$_2$, in aqueous solution it is found that 14% of this naturally occurring peptide is in a *cis* configuration (Smith *et al.*, 1973). In other solvents such as pyridine only the *trans* form exists.

In the $^{13}C$ spectrum of the oxytocin tail fragment Z-Pro-Leu-GlyNH$_2$ (Figure 8.6)

```
Cys — Tyr — Ile
 |           |
Cys — Asn — Gln
 |
Pro — Leu — GlyNH2
```

Oxytocin

the pro $\alpha$-, $\beta$- and $\delta$-carbons each appear as two signals, clearly reflecting the *cis/trans* isomerism about the Z-Pro bond. In larger component peptides and in oxytocin itself only single bands were observed for the $\alpha$-, $\beta$- and $\delta$-carbons and from the observed chemical shifts it was deduced that the *trans* conformer only is present, (Smith *et al.*, 1972a; Brewster *et al.*, 1973).

For peptides such as oxytocin and lysine vasopressin it is more difficult to assign the carbons simply by reference to the chemical shift data from amino acids and simple peptides because of the effects of cyclization and the possibility of small chemical shift effects from non-adjacent amino acids. However, by examining an extensive series of the component peptides of these molecules nearly all the $^{13}C$ assignments can be made.

Figure 8.7 shows the $^{13}C$ spectrum of oxytocin in fully deuterated dimethyl sulphoxide (DMSO-$d_6$) and Table 8.4 lists the assignments. From their $^{13}C$

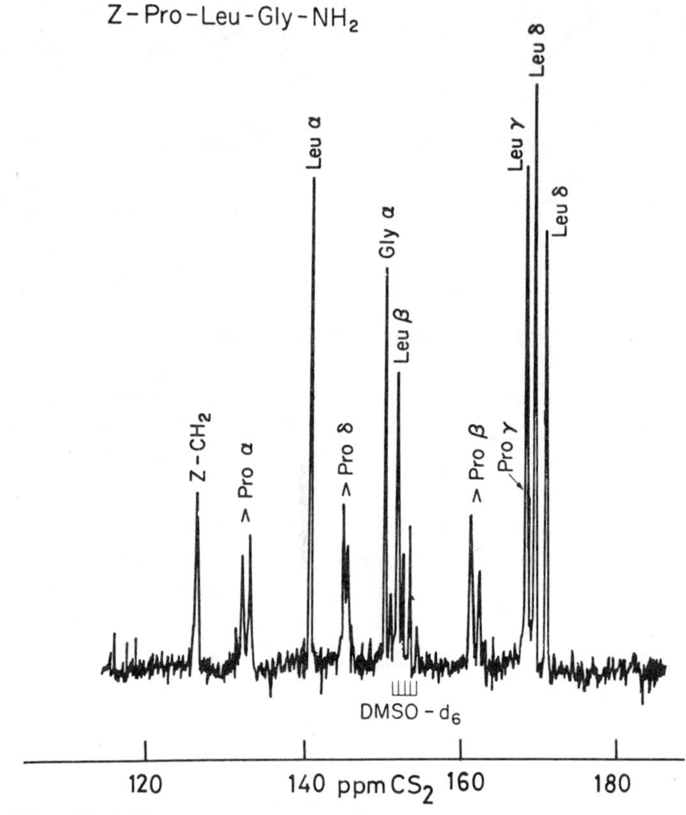


**Figure 8.6.** The $^{13}C$ proton-noise-decoupled spectrum at 25·2 MHz of Z-Pro-Leu-Gly$NH_2$ in DMSO-$d_6$ (Deslauriers, Walter and Smith, 1972)

studies of peptides, Smith and coworkers (Deslauriers, Walter and Smith, 1972) conclude that α-carbons and carbonyl carbons appear to be conformationally sensitive. Several other workers have studied oxytocin and lysine vasopressin and it is found that the oxidized and reduced forms of each peptide show different $^{13}C$ shifts as might be expected from the opening of the six-membered ring. Similar large effects were observed for the NH signals in the proton spectra. Brewster and coworkers (1973) have examined the $^{13}C$ spectra of various oxytocin analogues and found that the $^{13}C$ chemical shift of the Gly-9-$\alpha$ carbon (i.e. in the tail of the molecule) is sensitive to ring closure, substitution of Gln-4 by a Gly, inversion of the $\alpha$-carbon at Pro-7 and removal of the half-cystine-1 $\alpha$-amino group: these effects could possibly provide conformational information about the tail of oxytocin when the shifts are better understood.

Many other small peptides have been examined in this way and for linear peptides such as luteinizing hormone releasing factor (Smith *et al.*, 1973; Wessels *et al.*, 1973) and angiotensin (Zimmer *et al.*, 1972) the assignments can be made with reasonable confidence because the $^{13}C$ spectra of the component peptides are available. From the close agreement between predicted and

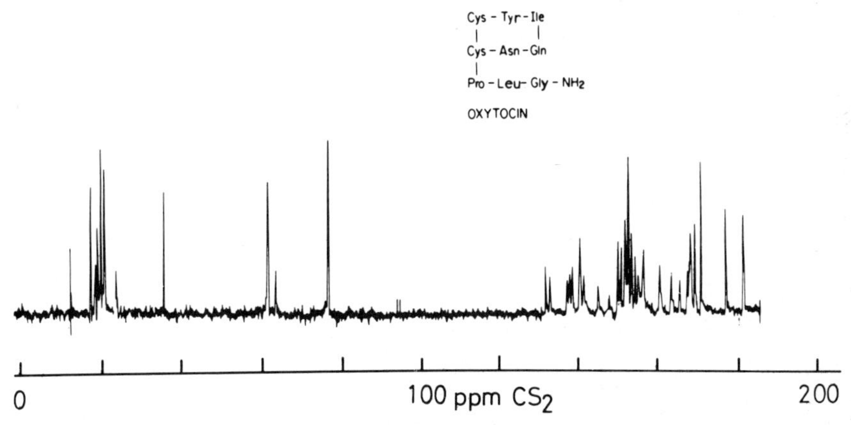


**Figure 8.7.** The $^{13}C$ proton-noise-decoupled spectrum at 25·2 MHz of oxytocin in DMSO-$d_6$ (Deslauriers, Walter and Smith, 1972)

observed $^{13}C$ shifts for linear peptides it is possible to comment on whether or not the peptide is in a random coil or if there are specific interactions present in the hormone which are absent in the component peptides.

Of course it is always difficult to obtain direct conformational information from chemical shift data but in some cases it should be possible to indicate if a conformational change has taken place without being able to specify the nature of the change.

By following the pH dependence of the assigned $^{13}C$ chemical shifts in peptides it is easy to measure the p$K$ values for various ionizable groups in the molecule. For oxidized and reduced glutathione such measurements have been made to attempt to define the microscopic p$K$ values (Jung *et al.*, 1972; Feeney *et al.*, 1973b). If the p$K$ values are known from $^{1}H$ NMR studies then the ionization effects on the $^{13}C$ spectrum can assist with spectral assignments.

This is particularly useful for carbonyl protons which have no directly bonded protons and cannot easily be assigned by proton decoupling experiments.

Zimmer and coworkers (1972) have studied the ionization behaviour of angiotensin-II (Asn-Arg-Val-Tyr-Val-His-Pro-Phe) by $^{13}$C NMR and found that there is no indication of preferential interactions between any of the side chains with the charged groups in angiotensin-II. Thus in aqueous solution there is no evidence for any ion–ion interaction between the phenylalanine carboxylate group and the protonated imidazole ring which has been proposed to exist in non-aqueous solvents: the His C-2 and C-4 carbons show no chemical shift changes when the Phe carboxylate group is protonated.

The cyclic membrane-active antibiotic valinomycin and its alkali metal complexes have been studied by $^{13}$C NMR by several workers (Bystrov *et al.*, 1972; Ohnishi *et al.*, 1972; Grell *et al.*, 1973; Patel, 1973). From the carbonyl chemical shifts of the stable complexes in methanol with $K^+$, $Rb^+$ and $Cs^+$ it was concluded that the ligand conformation is similar in all cases. The carbonyl signals are shifted 4–5 ppm downfield when the valinomycin forms its complex with these cations. The much weaker complex with $Na^+$ has a very different ligand conformation from that observed for the other alkali metal complexes.

For the $K^+$, $Rb^+$ and $Cs^+$ complexes the $^{13}$C shifts are consistent with all the ester carbonyl groups being involved in metal coordination ($\sim$ 4 ppm shifts on metal coordination) and formation of intramolecular hydrogen bonds between NH protons and the amide groups ($\sim$ 1 ppm shifts on formation of such hydrogen bonds). It is likely that temperature dependencies of $^{13}$CO chemical shifts will be useful for monitoring those groups involved in intramolecular hydrogen bonding.

## B. Proteins

It is obvious that $^{13}$C NMR can provide the necessary spectral resolution to enable most of the carbon nuclei in medium-sized peptides (up to decapeptides) to be assigned to their separated absorption bands. We should now consider the extent to which carbon-13 NMR can give useful information about proteins. The $^{13}$C spectra of the 1-13, 1-15 and 1-20 (S-peptide) *N*-terminal peptides of bovine pancreatic ribonuclease A have been examined by Freedman and coworkers (Chaiken *et al.*, 1973; Freedman *et al.*, 1973) and even in these large fragments many of the $^{13}$C spectral assignments could be made. By enriching the Phe-8 with $^{13}$C (15% uniformly labelled) the Phe assignments in the S-peptide could be made unequivocally. By combining the labelled S-peptide with S-protein they were able to see the $^{13}$C-labelled nuclei with ease in the protein $^{13}$C spectrum. However, disappointingly small $^{13}$C chemical shift differences were found between the $^{13}$C resonances in the S-peptide and S-peptide/S-protein mixture. Table 8.5 gives the carbon chemical shifts for the $^{13}$C-labelled Phe-8 group in 1-15 terminal S-peptide and in its complex with ribonuclease S-protein at pH 5·5.

Interestingly, the observed trend in $^{13}$C chemical shifts on complex formation is noted to be in the same direction as those observed for poly-L-glutamic acid (PGA) when it goes through its coil-to-helix transition although somewhat larger $^{13}$C shift differences are observed in the latter case ($C_\alpha = -1{\cdot}81$, $C_\beta = +2{\cdot}01$, $CO = -1{\cdot}79$ ppm for coil-to-helix transition in PGA) (Lyerla *et al.*, 1973).

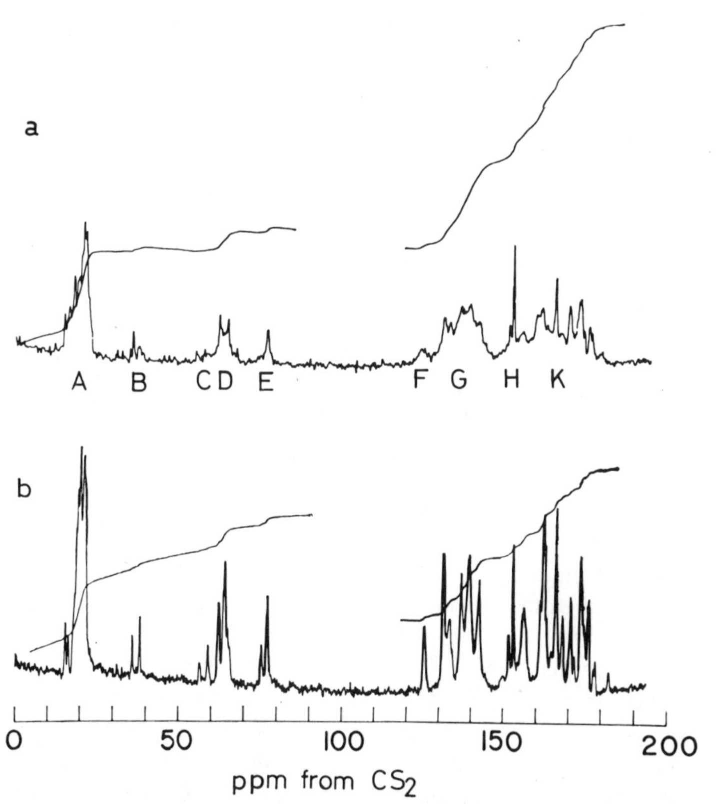


**Figure 8.8.** The $^{13}$C proton-noise-decoupled spectrum at 15·08 MHz of (a) ribonuclease A: 0·017 *M* native protein at pH 4.12, 32,768 transients in 6 h. Assignments: A, carbonyl carbons; B, C-6 carbons of tyrosine and ε-carbons of arginine; C, imidazole C-2 carbons of histidines and phenylalanine quaternary carbons; D, tyrosine ring C-3 and C-4 carbons, phenylalanine C-4, C-5, C-6 carbons and histidine C-5 carbons; E, histidine C-4 carbons and tyrosine C-5 carbons; F, β-carbons of threonine; G, β-carbons of serine and all α-carbons except those of glycine; H, ε-carbons from lysine: K, remaining aliphatic carbons (Allerhand *et al.*, 1970, 1971a). (b) Denatured ribonuclease A

Bovine pancreatic ribonuclease A had previously been studied by Allerhand and coworkers (1970) in one of the earliest $^{13}C$ studies of proteins. By using the chemical shifts of the 124 component amino acids of the enzyme they were able to present a partial assignment of the $^{13}C$ spectrum as indicated in Figure 8.8.

It is sometimes possible to make $^{13}C$ assignments in protein $^{13}C$ spectra if the carbon signals are in a region of the spectrum which does not overlap too badly with the spectral region of other carbons. Thus Packer and his coworkers (1972) have been able to assign the 2′,6′-ring carbons (*meta* to OH) in the two tyrosyl residues in ferredoxin (from *Clostridium acidi-urici*) and observe large downfield shifts compared with those normally observed for Tyr residues (2·7 ppm in the oxidized form and 7·3 ppm in the reduced form). They deduced from this that both Tyr residues are in equivalent magnetic environments in close proximity to the two iron–sulphur clusters in the oxidized and reduced forms of ferredoxin.

The fact that $^{13}C$ resonance spectra of proteins are very complicated has not prevented some workers from extracting useful information by noting changes in the $^{13}C$ spectra of a series of related proteins. Thus Moon and Richards (1972a) examined variously liganded human haemoglobin molecules and observed that the $\epsilon$-carbons of lysine and the $\beta$-carbon signals of alanine have different intensities in the different systems (see Figure 8.9). Along the series oxyhaemoglobin ($HbO_2$), acid methaemoglobin (Hi), deoxyhaemoglobin (Hb) and (Hb + 2,3-diphosphoglycerate), the $\epsilon$-C (Lys) intensities decreased and they interpreted this to indicate that immobilization of the lysine side chains as the protein becomes tighter results in excessive line broadening which effectively removes them from the spectra. On the other hand, the alanine methyl groups show an increase in intensity along the series which could result from the low saturation suffered by the less mobile methyl carbons with shorter relaxation times. However, one cannot completely rule out the possibility that chemical shift changes between the $\epsilon$-C (Lys) carbons and $\beta$-C (Ala) carbons could also be contributing to these intensity effects.

The most promising $^{13}C$ results on proteins so far reported have come from studies on native hen egg-white lysozyme by Allerhand and coworkers (1972, 1973). Examination of the unsaturated carbon region of the spectrum (Figure 8.10) revealed that most of the bands numbered 1–22 arise from single resonances from the 28 non-protonated carbons in lysozyme. Bands 18 to 22 correspond to the $\gamma$-carbons of the six tryptophans in lysozyme and it is encouraging that there is a total spread of chemical shifts of almost 5 ppm which is removed by denaturation. Thus, the hope that carbon-13 chemical shifts will be very sensitive to conformational effects seems to be well founded. When a detailed assignment of the resonance bands has been achieved many interesting unfolding and inhibitor interactions can be studied by monitoring the assigned resonance bands on the protein. To obtain the excellent $^{13}C$

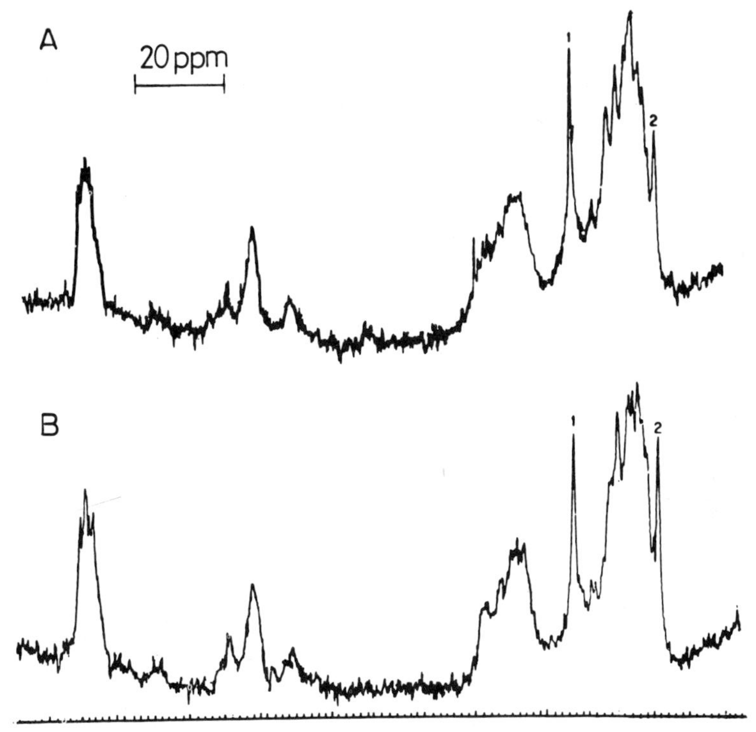


**Figure 8.9.** The $^{13}C$ proton-noise-decoupled spectrum at 25·2 MHz of (a) oxyhaemoglobin ($HbO_2$) at pH 7·25 and (b) deoxyhaemoglobin (Hb) at pH 7·41. The ε-carbon signals are labelled 1 and the alanine β-carbon signals are labelled 2 in the spectra (Moon and Richards, 1972a)

spectra shown in Figure 8.10 required 40 h of Fourier transform spectrum accumulation using a 20% (w/v) solution of lysozyme in 20 mm tubes.

Similar studies of concentrated aqueous solutions (~ 14 m*M*) of ferrocytochrome c and ferricytochrome c have been made (Oldfield and Allerhand, 1973). Because of rapid electron transfer between the two redox states, mixtures of the two species give an average $^{13}C$ spectrum and it is thus possible to establish a 1:1 correspondence between the resonances in the two states by varying the amounts of the two species present in the equilibrium. Many single-carbon resonances can be detected and assigned in the aromatic region of the $^{13}C$ spectrum corresponding to unprotonated carbons (see Table 8.6). These carbons retain their sharp appearance even when the proton-noise-decoupling field is shifted several ppm off the proton spectrum whereas protonated carbons are not observed under these conditions (Oldfield and Allerhand, 1973).

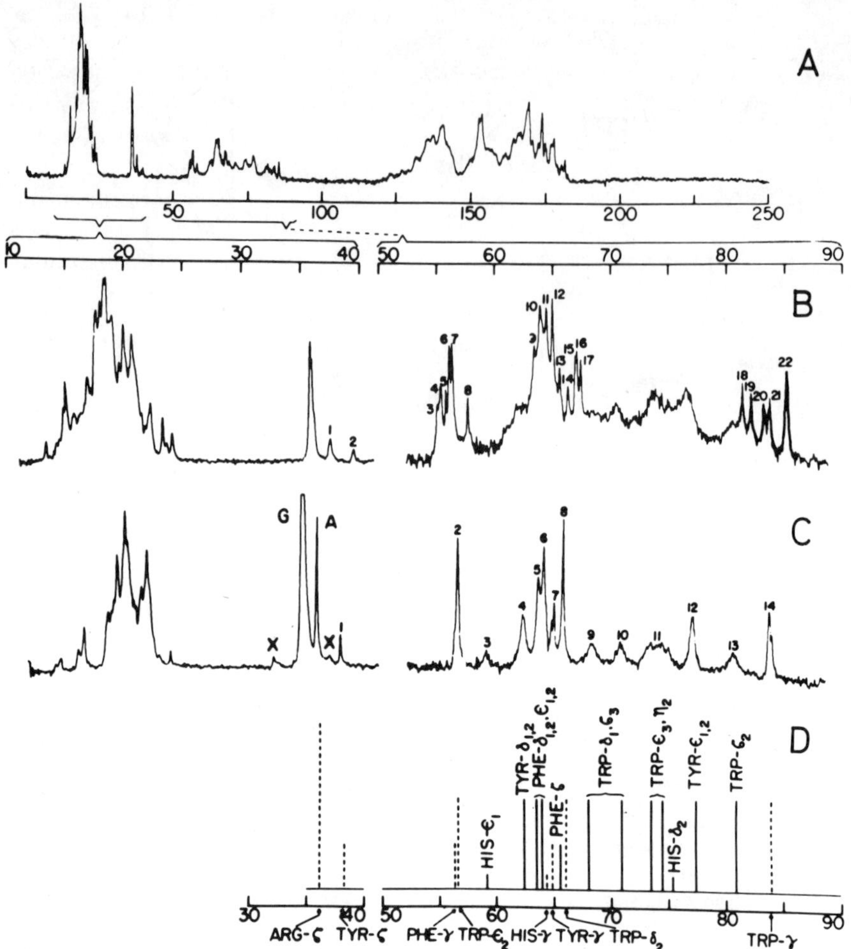


**Figure 8.10.** The $^{13}C$ proton-noise-decoupled spectrum at 15·18 MHz of native hen egg-white lysozyme (Allerhand *et al.*, 1973). (A) Complete spectrum. (B) Unsaturated carbon region of the spectrum. (C) Guanidine denatured sample of HEW lysozyme. (D) Simulated $^{13}C$ spectrum of unsaturated carbons in simple peptides: differences in (B) and (D) reflect conformational effects present in the native enzyme. For assignments, see text

By observing the resonances which broaden and undergo paramagnetic shifts as the amount of ferricytochrome c increases one can assign the carbons which are near to the iron atom. Many of the signals which disappear on going from ferro- to ferricytochrome c (compare Figures 8.11(a) and (b)) arise from the haem carbons in the region 46 to 55 ppm. It is interesting that the signal at 83·6 ppm assigned to the $\gamma$-C of the single trytophan (Trp 59) is essentially unchanged in the two states (Oldfield and Allerhand, 1973).

For less soluble proteins available in smaller quantities this approach using large sample tubes with concentrated solutions will be less useful. An alternative method possible for bacterial enzymes will be to examine specifically $^{13}C$-labelled compounds obtained by providing $^{13}C$-enriched amino acids in the growing medium. By using only one enriched amino acid not only will the sensitivity problem be overcome but the resulting $^{13}C$ spectra will be far less complex. Of course, one will still need to undertake the considerable task of assigning the observed resonances to their respective amino acids in the sequence before significant information can be obtained from the system. In principle, most of the carbon atoms in bacterial enzymes of molecular weight $< 20{,}000$ could be monitored in this manner.

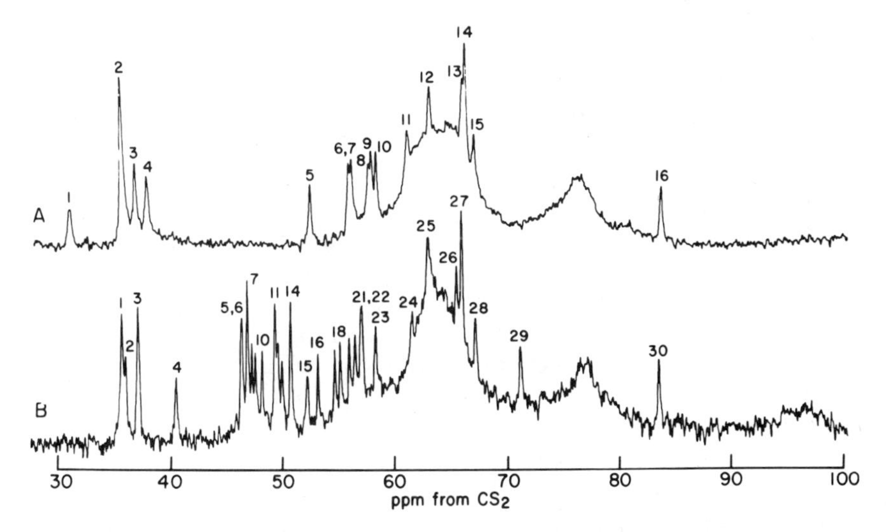


**Figure 8.11.** The aromatic region of the noise-modulated off-resonance proton-decoupled $^{13}C$ spectrum at 15·18 MHz of (a) 14·4 m*M* horse heart ferricytochrome c and (b) 11·5 m*M* horse heart ferrocytochrome c (Oldfield and Allerhand, 1973)

Browne and coworkers (1973) have extracted $^{13}C$-labelled tryptophan synthetase from *E. Coli* grown on an amino acid culture containing $^{13}C$-2 labelled histidine as one of the constituents. The C-2 carbons of the four His residues are clearly visible in the spectrum but unfortunately do not show any chemical shift differences detectable within the line width of the $^{13}C$ resonance ($\sim$ 2 ppm).

Hunkapiller and coworkers (1974) have reported on important method of measuring the ionisation state of His residues. By examining α-lytic protease from *Myxobacter* 495 with incorporated $^{13}C$-2 labelled His they showed that the single His has a very low pK value ($<$4) as indicated by the observed C-H spin coupling constant between the C-2 carbon and the directly bonded proton.

Chemical modification can also be used to introduce $^{13}C$ into proteins (Nigen *et al.*, 1973).

Carbon-13 enriched amino acids for incorporation into bacterial enzymes can be made either synthetically or biosynthetically (for example from the protein fraction of *E. Coli* grown on either $^{13}$C-1 or $^{13}$C-2 labelled sodium acetate as the sole carbon source).

Undoubtedly, this area of study will be one of major activity over the next few years.

## C. Protein Unfolding Studies using $^{13}$C NMR

Proton NMR studies of unfolding of proteins have proved that the NMR technique can make unique contributions to the understanding of this problem. Detailed unfolding studies by proton NMR can only be made on proteins which have resonances assignable to single protons such as the C–2H protons of the histidines of ribonuclease (Roberts and Jardetzky, 1970) or the tyrosine ring protons in partially deuterated staphylococcus nuclease (Jardetzky *et al.*, 1971). Thus the success of $^{13}$C resonance studies will depend largely on whether or not one can make assignments of single carbons to their respective absorptions particularly in $^{13}$C specifically enriched bacterial proteins. While the potential of $^{13}$C resonance studies in this area remains to be demonstrated it seems likely that they will prove to be most valuable. When unfolding occurs one can expect changes in both the chemical shifts and the relaxation times of the carbons in the unfolded fragments.

$^{13}$C studies so far reported provide only general information about the unfolding process because one is dealing with spectral changes in either unassigned single-carbon resonances or in absorptions corresponding to several carbon nuclei in different positions in the molecule.

From the studies on lysozyme by Allerhand and coworkers (1973) it is clear that large shift changes (as much as 4 ppm) in single-carbon resonances are observed on denaturation (see Figure 8.10). Similarly, large shift changes are apparent in the $^{13}$C spectra of native and denatured ribonuclease A (Figure 8.8) (Allerhand *et al.*, 1971a). Thus when we have carbon-13 assignments for single resonances in proteins it should be possible to use $^{13}$C NMR to present a detailed sequence of the events in the unfolding process.

Relaxation time changes on unfolding have been demonstrated to be substantial, the relaxation becoming less efficient as the carbon nuclei undergo increased molecular motion in the denatured enzyme. Allerhand and coworkers (1971a) have demonstrated this in their study of ribonuclease unfolding.

## D. Nucleosides and Nucleotides

Grant (Jones *et al.*, 1970a, 1970b) and Roberts (Dorman and Roberts, 1970) and their coworkers have examined an extensive series of nucleosides and nucleotides and some of their results are given in Table 8.7.

The sugar carbon resonances are in a separate region of the spectrum (98–157 ppm from $CS_2$) from that of the pyrimidine and purine base carbons 22–96 ppm) (Dorman and Roberts, 1970).

It was found that for both nucleosides and nucleotides the chemical shifts of the sugar carbon resonances appear to be independent of the heterocyclic base present except for the C–1′ resonances which can vary by as much as ca. 6 ppm. Of course different sugars such as ribosides and deoxyribosides have different $^{13}C$ spectra and can be distinguished with ease particularly by noting the C-2′ chemical shifts. To make $^{13}C$ assignments for the heterocyclic base carbons in compounds of this type one often has to rely on chemical shift changes induced by various chemical modifications. This approach is less reliable than assignment methods using heteronuclear $\{^1H\}$–$^{13}C$ spin decoupling to relate carbons to assigned protons. However, when a large number of chemically modified bases are used to obtain the assignments they can be accepted with some confidence (Jones *et al.*, 1970a).

Base pairing effects and stacking interactions were not detected in the nucleotide $^{13}C$ spectra by Roberts and coworkers (Dorman and Roberts, 1970) over the limited concentration range available to them (greater than 0·25 *M*). More recently, Smith and coworkers (Smith *et al.*, 1972b) have observed concentration-dependent intermolecular association effects on the $^{13}C$ spectrum of 5′-AMP over the concentration range 1·2 to 0·04 *M* in $D_2O$.

The high charge density calculated for the C-5 carbons in compounds such as uridine and cytidine correlate well with the large observed upfield values of C-5 $^{13}C$ chemical shifts, and reflects the high electrophilic character of this position in pyrimidine nucleosides (Jones *et al.*, 1970b).

$^{13}C$ NMR provides an excellent method of determining p$K$ values for various ionizable groups in nucleotides and nucleotide-containing coenzymes. For example, in the $^{13}C$ spectrum of nicotinamide adenine dinucleotide ($NAD^+$) (Figure 8.12) one can assign all the carbons and observe the protonation of the adenine ring at the N-1 position (p$K = 4{\cdot}0$) from the pH titration curves of the adenine ring carbons (Figure 8.13) (Birdsall and Feeney, 1972). At lower pH values the diphosphate group is protonated (p$K < 1{\cdot}0$) and this can be followed by observing the titration curves for the ribose C-4′ and C-5′ carbons.

When the adenine ring of $NAD^+$ is protonated it is thought that molecules present in a folded conformation become unfolded (Sarma *et al.*, 1968). This effect is manifested in the $^1H$ and $^{13}C$ chemical shift changes in the nicotinamide ring when one protonates the adenine ring. Several studies on the conformation of nucleotides in solution made by measuring pseudocontact shifts arising from paramagnetic lanthanide ions bound to the phosphate groups have given promising preliminary results (Barry *et al.*, 1971; Birdsall *et al.*, 1975).

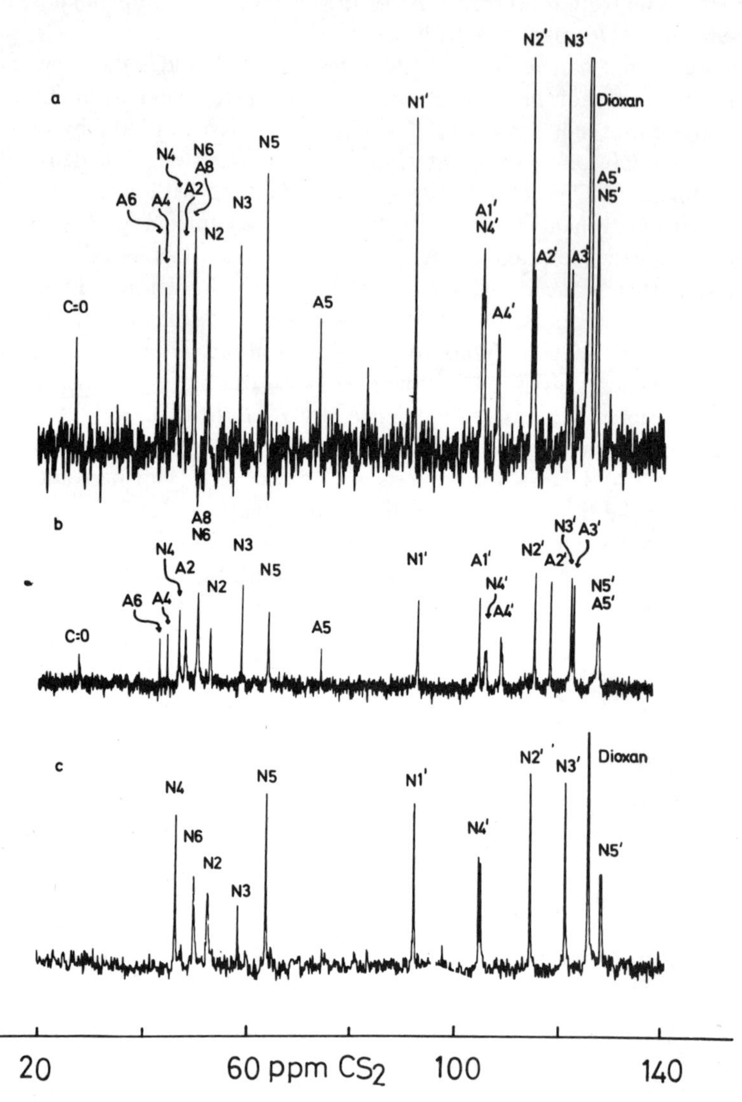


**Figure 8.12.** The $^{13}C$ proton-noise-decoupled spectrum at 25·2 MHz of (a) nicotinamide adenine dinucleotide phosphate, $NADP^+$, aqueous solution at pH 2·0; (b) nicotinamide adenine dinucleotide, $NAD^+$, pH 1·0; (c) nicotinamide mononucleotide, $NMN^+$, pH 2·8 (Birdsall and Feeney, 1972)

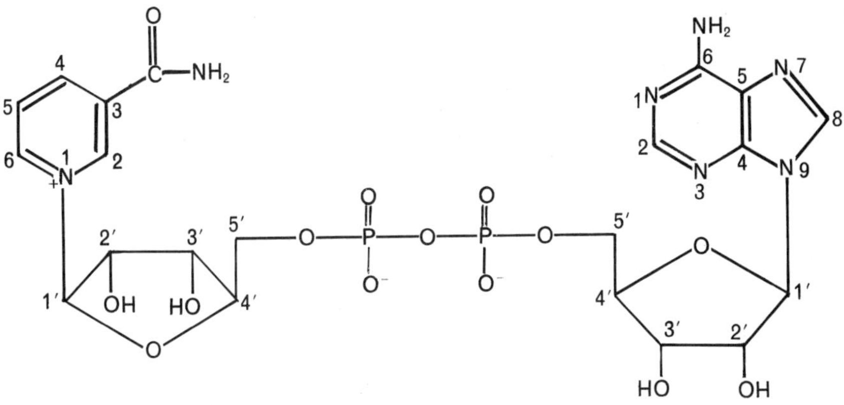

Nicotinamide adenine dinucleotide, $NAD^+$

In the early studies on nucleotides, doublet splittings were observed on the C-4′ and C-5′ ribose carbons in the proton-noise-decoupled $^{13}C$ spectra. These splittings arise from spin–spin interaction with the phosphorus nucleus ($I = \frac{1}{2}$). The three-bond $J_{C4'-P}$ values (1–10 Hz) can be larger than the two-bond $J_{C5'-P}$ values ($\approx$ 4 Hz). Recently Smith and coworkers (Lapper *et al.*, 1973) have shown how the $H_{C4'-P}$ three-bond couplings can be used to give conformational information about the dihedral angle between the planes $^{31}P—O—C'$ and $O—C'—^{13}C'$). From conformationally rigid cyclic nucleotides such as thymidine 3′,5′-cyclic phosphate and adenosine 3′,5′-phosphate Smith and coworkers have deduced that the *trans* and *gauche* coupling constants are 8 and 2 Hz respectively.

The same authors (Lapper and Smith, 1973; Smith *et al.*, 1972b) have used these values to estimate the conformational preference of the phosphate group in non-cyclized nucleotides such as 3′-UMP and 5′-UMP. Polyuridylic acid (M.Wt. 130,000) gives a remarkably well-resolved $^{13}C$ spectrum in which three bond $^{31}P–^{13}C$ coupling constants can be detected. The observed values are consistent with the backbone conformation being predominantly 2′-*trans*-4′-*gauche* about the $C_{3'}O$ bond and 4′-*trans* about the $OC_{5'}$ bond (Lapper and Smith, 1973; Smith *et al.*, 1972b).

## E. Phospholipids in Model Systems and Natural Membranes

The complex dynamic behaviour of the phospholipid hydrocarbon chains in bilayer model compounds and natural membranes has recently received much attention. By examining the electron spin resonance spectra of phospholipid chains selectively spin labelled at different chain positions Hubbell and McConnell (1968) established the pattern of motion in the hydrocarbon

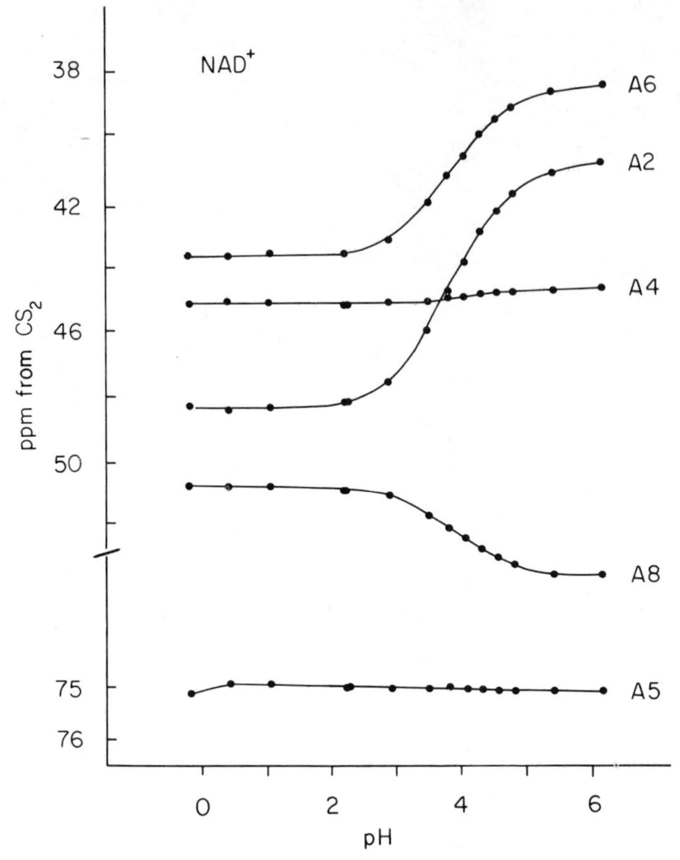


**Figure 8.13.** The $^{13}C$ chemical shifts of the adenine carbons of $NAD^+$ as a function of pH (Birdsall and Feeney, unpublished results)

chains of phospholipid bilayers. The hydrophobic interiors of the membranes studied were shown to be in a highly fluid state and other spin-label experiments provided evidence for rapid lateral diffusion of the lipid molecules (Trauble and Sackmann, 1972). A problem associated with the spin-labelling technique is that the introduction of the bulky spin labels could significantly perturb the system. Metcalfe and coworkers (Metcalfe *et al.*, 1971; Levine *et al.*, 1972a; Metcalfe *et al.*, 1972b; Robinson *et al.*, 1972) have shown how carbon-13 spin lattice relaxation time studies can also be used to obtain dynamic information in membrane systems. An obvious advantage of the carbon-13 studies is that they involve insignificant perturbations of the system under study.

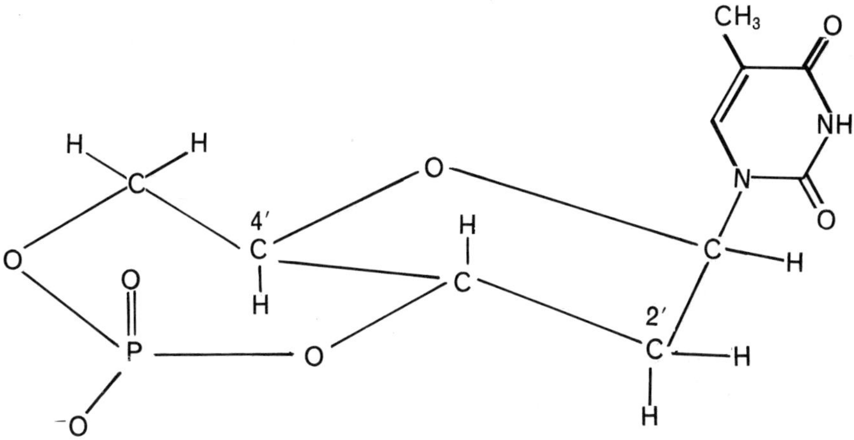

Thymidine 3′, 5′—cyclic phosphate

A further possible advantage of using carbon-13 enriched rather than spin-labelled phosphlipids is that the former would be expected to present fewer problems when these are being incorporated into natural membrane systems. Of course, the major disadvantage of the $^{13}C$ approach is that of sensitivity and even with specifically enriched carbon-13 phospholipids incorporated into the membranes the NMR technique in its present state of development is fully extended when examining such systems.

The usefulness of $^{13}C$ spectroscopy in this area can best be assessed by considering some of the results of Metcalfe and coworkers (Metcalfe *et al.*, 1971, 1972b; Levine *et al.*, 1972a, 1972b; Robinson *et al.*, 1972; Birdsall *et al.*, 1972). The $^{13}C$ spectrum of dipalmitoyl lecithin in deuterochloroform is shown in Figure 8.14 (Birdsall *et al.*, 1972). In the alkyl carbon region of the spectrum, separate $^{13}C$ resonance signals are observed for five of the hydrocarbon chain carbon atoms (C-4 to C-13 given an unresolved band but specific isotopic enrichment can be used to examine the individual carbons in this part of the chain). It is possible to use the Fourier transform technique to measure the $^{13}C$ relaxation times for the resolved carbon signals of phospholipids in various solvents and Figure 8.15 gives the $T_1$ values for sonicated dipalmitoyl lecithin in $D_2O$ solution. Under these conditions the molecules exist as vesicles with bilayer structures with the charged quarternary onium headgroups pointing into the aqueous solvent on the outside and inside of the spherical vesicles. Qualitatively it is seen that the shortest relaxation

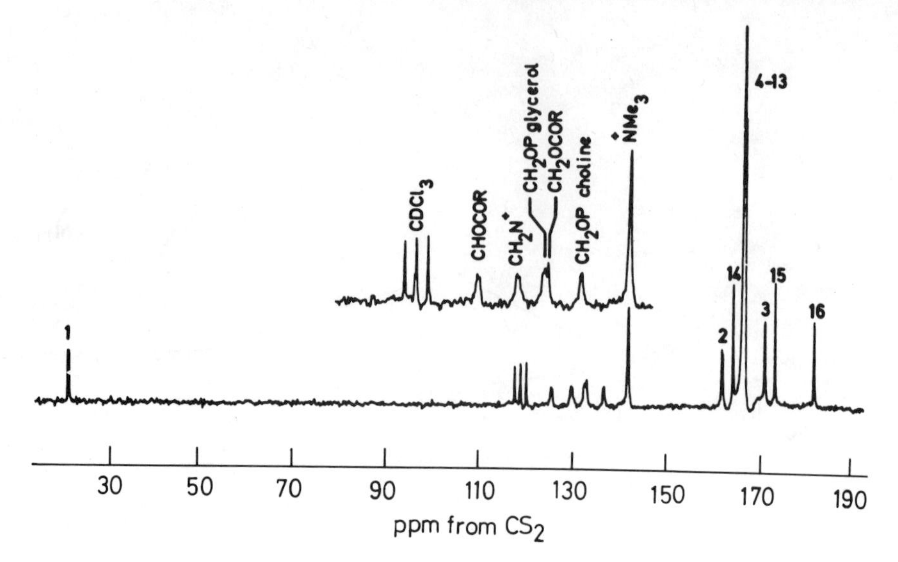


**Figure 8.14.** The [13]C proton-noise-decoupled spectrum at 25·2 MHz of dipalmitoyl lecithin in deuterochloroform (Metcalfe *et al.*, 1971). Numbers refer to hydrocarbon chain carbon atoms: for structure see Figure 8.15

times (reflecting the slowest molecular motion) are in the glycerol carbon atoms while the relaxation times progressively increase towards the methyl end of the hydrocarbon chains and towards the $N(CH_3)_3^+$ end of the choline headgroups indicating increasing molecular motion. A more quantitative understanding of the observed relaxation times can be reached by considering the component anisotropic motions in the system (Levine *et al.*, 1972b, 1973);

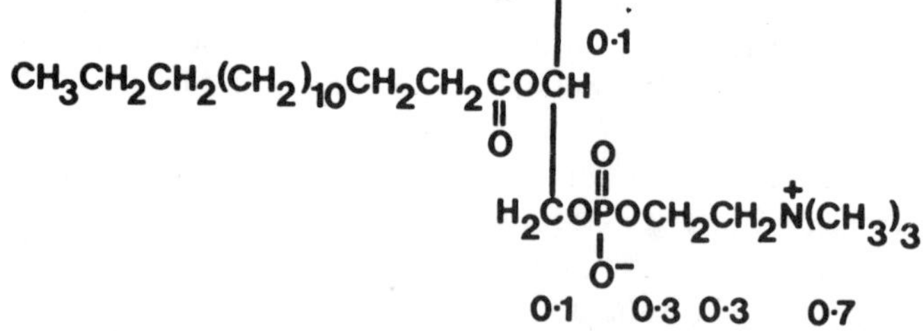

**Relaxation Times $T_1$ in secs.**
**in $D_2O$ at 52°C**

3·3 1·8 1·1 0·6 0·2 0·1 2·3 0·1
$CH_3CH_2CH_2(CH_2)_{10}CH_2CH_2\overset{\|}{C}OCH_2$ (C=O)

0·1
$CH_3CH_2CH_2(CH_2)_{10}CH_2CH_2\overset{\|}{C}OCH$ (C=O)

$H_2COP(=O)(O^-)OCH_2CH_2\overset{+}{N}(CH_3)_3$

0·1 0·3 0·3 0·7

**Figure 8.15.** The carbon-13 spin-lattice relaxation times ($T_1$) for sonicated dipalmitoyl lecithin in $D_2O$ (Levine *et al.*, 1972a)

such an analysis indicates that there is an upsurge in molecular motion over the last few C—C bonds in the alkyl chains which was not noted in the spin-label ESR experiments.

The measured $^{13}C$ relaxation times in systems where the phospholipids are not in a bilayer structure indicate that the average effective correlation times are somewhat longer than in the bilayer structure. Metcalfe and coworkers (1972b) explored the effects on $^{13}C$ relaxation times of cholesterol addition to the bilayer structures (decrease in $T_1$ values), of using shorter lengths of the lecithin chains (increase in $T_1$ values) and of using unsaturated chains (increase in $T_1$ values). They obtained a $^{13}C$ spectrum for sarcoplasmic reticulum membranes in $D_2O$ (Figure 8.16) and were able to show that the hydrocarbon chain $^{13}C$ relaxation times in the intact membrane are similar to those in the vesicles of the lipids extracted from the membranes,

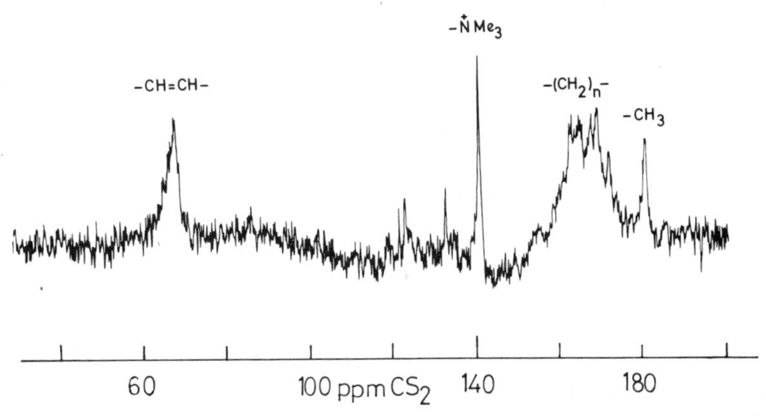


**Figure 8.16.** The $^{13}C$ proton-noise-decoupled spectrum at 25·2 MHz of sarcoplasmic reticulum membranes in $D_2O$ (Metcalfe *et al.*, 1972b)

but that the relaxation times for the $N^+Me_3$ carbons are considerably shorter in the membrane. This partial immobilization of the $N^+Me_3$ headgroup could possibly result from protein/lipid interactions in the membrane.

Whether or not $^{13}C$ studies in membranes will lead to useful structural and dynamic information relevant to an understanding of membrane function will depend largely on how successfully one can introduce specifically enriched $^{13}C$ phospholipids into functional membranes and on the availability of adequate $^{13}C$ instrumental sensitivity. Promising results have been obtained in initial experiments based on (i) biosynthetic incorporation of $^{13}C$-labelled palmitic acid into the *acholeplasma laidlawaii* membranes (Metcalfe *et al.*, 1972a); (ii) fusion of highly sonicated labelled phospholipids into functional membranes (sarcoplasmic reticulum membranes) and (iii) reconstitution experiments in which functional membrane systems are reconstituted using defined phospholipid components (Warren *et al.*, 1974). Thus, ATPase can be extracted from sarcoplasmic reticulum membranes and then reconstituted with selected phospholipids to give a system which has recovered its full $Ca^{2+}$ pump activity. By using suitably labelled phospholipids in the reconstitution process it might be possible (depending on sensitivity limitations) to obtain useful results about protein-phospholipid interactions in such systems (Warren *et al.*, 1974).

### F. Biosynthetic Studies

Carbon-14 studies have been used for several years to elucidate the pathways of biosynthesis of natural products. The technique involves the use of C-14 labelled precursor molecules in the biosynthetic process followed by isolation and selective degradation of the natural product prior to radioactive counting to determine the specific activity of the carbon in the degradation fragments. With the widespread availability of NMR spectrometers capable of examining $^{13}C$ nuclei it is not surprising that similar biosynthetic studies are now being made using $^{13}C$-labelled precursor molecules (Tanabe and Jankowski, 1972). This approach has the obvious advantage that it is no longer necessary to degrade the labelled natural product: it is possible to see in the $^{13}C$ spectrum the sites containing the enriched nuclei even at quite low $^{13}C$ enrichments (2–5 %). Other advantages of the technique are apparent when one considers an actual example of such a study. Tanabe and coworkers (1973) have prepared the antifungal agent avenaciolide by using, in separate experiments, C-1 (90 %) and C-2 (60 %) labelled sodium acetate as the carbon source for the organism which produces avenaciolide. Figure 8.17 shows the $^{13}C$ spectrum obtained for one of the labelled avenaciolides. If the $^{13}C$ spectral assignments are known, then the $^{13}C$ signals of high intensity clearly can be identified with specific carbons in the molecule. When two enriched carbons are incorporated in adjacent carbon sites, if the enrichment factor is high then the adjacent carbon-13 nuclei will be conspicuous because of the large C–C coupling constants on the absorptions. In avenaciolide, the olefinic carbons C-11 and C-15 from the C-2 sodium acetate culture are both $^{13}C$ nuclei and a coupling constant of 75 Hz is clearly visible on their absorption bands (Figure 8.17). From the carbon-13 spectrum it was also possible for Tanabe to conclude (Tanabe *et al.*, 1973) that the 3-ketodecanoic acid derived portion of avenaciolide had a higher level of carbon-13 enrichment than the succinic acid portion: this result clearly implicates these two separate biogenetic units in the formation of avenaciolide.

It should be mentioned that carbon-13 biosynthetic studies are most convenient for those organisms which make efficient use of the precursor molecules. However, the many carbon-13 biosynthetic studies already reported establish that this technique is of some general importance. The interesting

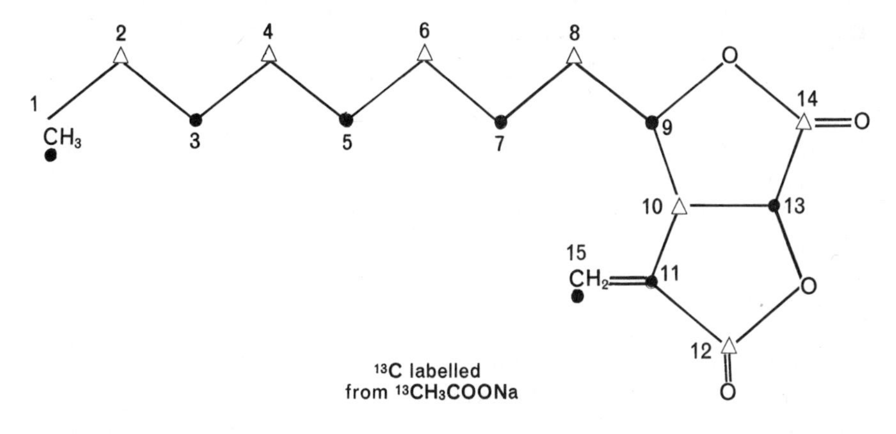


systems examined include protoporphyrins (Battersby *et al.*, 1972a, 1972b), rifamycin (White *et al.*, 1973), radicinin (Tanabe *et al.*, 1970a), asperlin (Tanabe *et al.*, 1971), sterigmatocystin (Tanabe *et al.*, 1970b), cephalosporin C (Neuss, *et al.*, 1971), prodigiosin (Cushley *et al.*, 1971), antibiotic X-537A (Westley *et al.*, 1972), vitamin $B_{12}$ (Scott *et al.*, 1972; Brown *et al.*, 1972), virescenosides (Polonsky *et al.*, 1972), aureothin (Yamasaki *et al.*, 1972b), ochratoxin (Yamasaki *et al.*, 1972a), sepedonin (McInnes *et al.*, 1971) and pyrrolnitrin (Martin *et al.*, 1972).

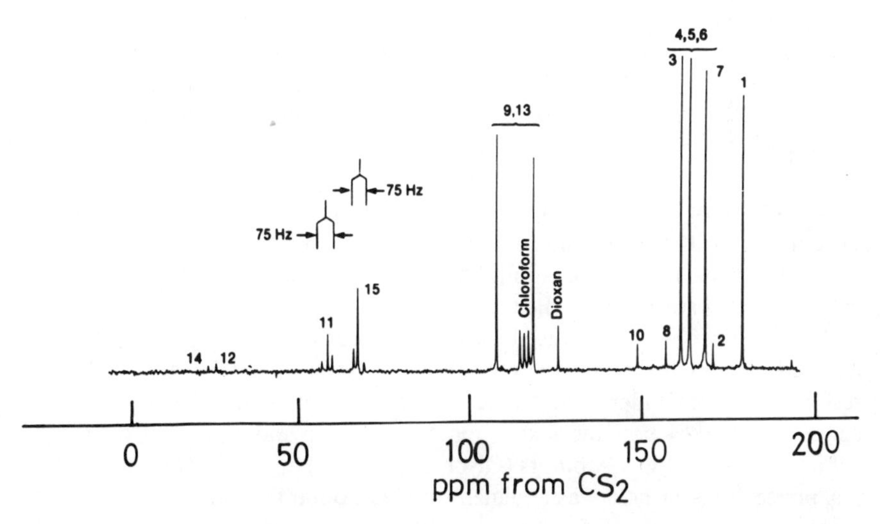


**Figure 8.17.** The $^{13}C$ proton-noise-decoupled spectrum at 25·2 MHz of avenaciolide grown on $^{13}CH_3COONa$ (43 mg/0·7 ml $CDCl_3$, 4000 transients of 0·8 sec acquisition time (0·89 h) (Tanabe and Jankowski, 1972)

## G. Interactions of Small Molecules with Proteins

There have been numerous proton NMR studies of interactions of this type for enzyme–inhibitor, enzyme–coenzyme and hapten–antibody complexes (Sheard and Bradbury, 1970). For strongly bound 1:1 complexes it is difficult to observe the small molecule proton spectrum because of the large background signals from the protein. To overcome the background overlap problem it is necessary to resort to either protein deuteration or to difference spectroscopy techniques. Carbon-13 NMR offers a convenient method of overcoming the protein background problem because by enriching the small interacting molecule it can be made easier to monitor. This is illustrated in Section V.G.2 where carbon-13 enriched cyanide ions are examined when strongly bound to carbonic anhydrase (Feeney *et al.*, 1973a). A further factor assisting such studies is that the protein carbon spectrum is more extended in its shift range and contains more signal-free regions than the proton spectrum. For example, the $^{13}C$ spectral region containing ribose and hexose signals is a signal-free region in protein spectra and this will facilitate the study of interactions between proteins and ribose-containing coenzymes, or between proteins such as concanavalin A and $^{13}C$-enriched sugars (Brewer *et al.*, 1973) (Section V.G.5).

### *1. Carbon monoxide binding to haemoglobins*

By observing the $^{13}C$ resonance signals from $^{13}CO$ (90% enriched) bound to various haemoglobins it has been possible to comment on the extent to which the haemoglobin subunits interact differently with carbon monoxide (Moon and Richards, 1972b). Figure 8.18 shows the $^{13}C$ spectrum of carbon monoxide bound to sperm whale myoglobin and for this monomeric protein only a single carbon monoxide $^{13}C$ resonance signal is observed. When carbon monoxide binds to human adult haemoglobin, two CO signals are observed (Figure 8.18(b)) corresponding to the carbon monoxide bound to the $\alpha$-subunits being in a different environment from that on the $\beta$-subunits. An experiment with foetal human haemoglobin binding to carbon monoxide gave similar results to the adult human haemoglobin despite 39 different amino acid substitutions in the two enzymes.

When oxygen was introduced into the haemoglobin–CO systems the low-field CO signal decreased in intensity at a faster rate than the high-field resonance. From independent kinetic measurements it had been previously suggested that carbon monoxide replaced oxygen more rapidly on the $\beta$-subunits than on the $\alpha$-subunits (Olson *et al.*, 1971) and based on this the $^{13}C$ resonance at lower field was assigned to $^{13}CO$ bound to $\alpha$-subunits in haemoglobin.

Antonini and coworkers (1973) have obtained the $^{13}C$ spectra of $^{13}CO$ complexes of the isolated $\alpha$- and $\beta$-chains of haemoglobin and their results

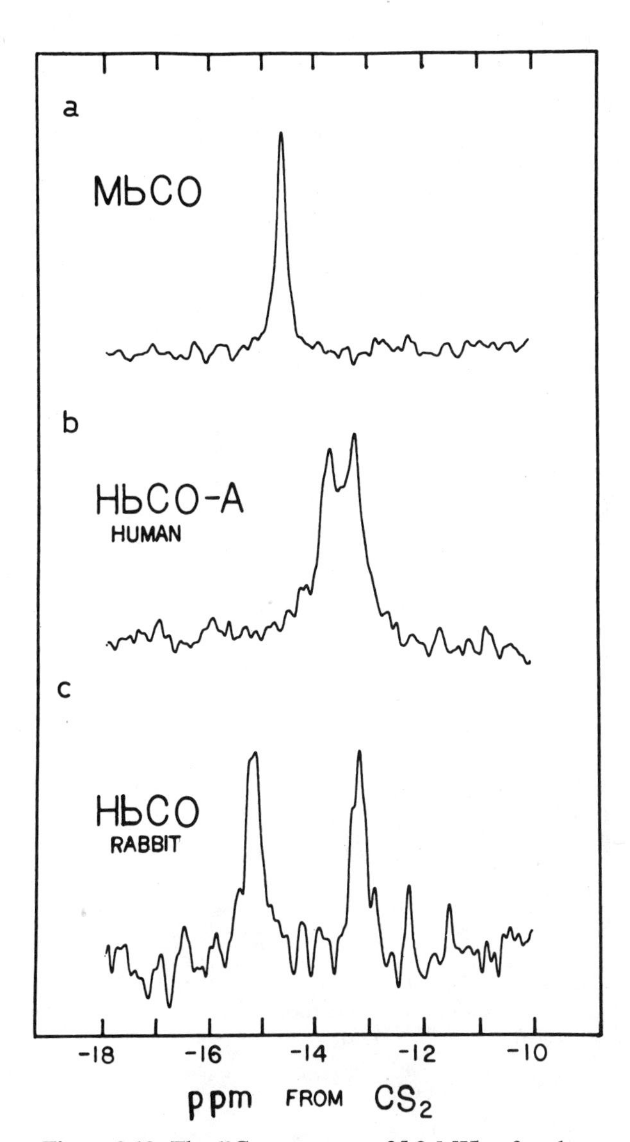


**Figure 8.18.** The $^{13}C$ spectrum at 25·2 MHz of carbon monoxide bound to (a) sperm whale myoglobin, (b) adult human haemoglobin, (c) rabbit haemoglobin (Moon and Richards, 1972b)

lead to assignments for the signals observed in the haemoglobin–CO complexes. The low-field signal corresponds to CO bound to $\alpha$-chains (−13·71 ppm from $CS_2$ reference for haemoglobin, −13·8 ppm for the isolated $\alpha$-chain), while the high-field signal corresponds to CO bound to the $\beta$-chains (−13·25 ppm in haemoglobin, −13·02 ppm for the isolated $\beta$-chain). The small difference observed in the shielding of the CO bound to the $\beta$-chains probably reflects the effects of specific chain–chain interactions in the tetramer.

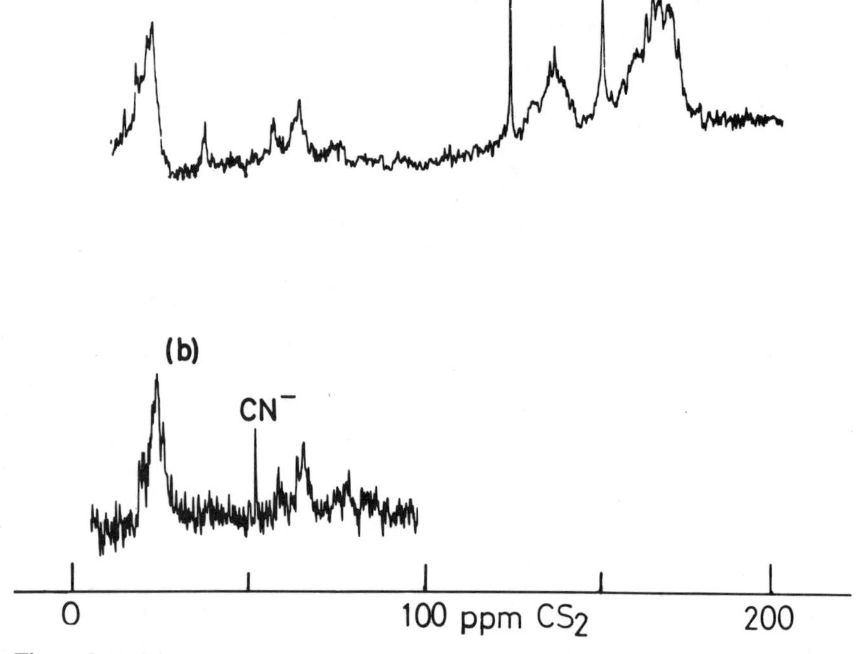


**Figure 8.19.** The $^{13}C$ proton-noise-decoupled spectrum at 25·2 MHz of (a) bovine carbonic anhydrase B and (b) bovine carbonic anhydrase in the presence of equimolar $^{13}CN^-$ ions (Feeney *et al.*, 1973a)

## 2. *Cyanide binding to carbonic anhydrase*

At millimolar concentrations, cyanide ions form 1:1 complexes with carbonic anhydrase B enzymes (human and bovine) and the $^{13}C$ signals from the bound cyanide (70 % enriched) can be easily detected in the presence of the protein spectrum (see Figure 8.19). Because of the similarity of the measured $CN^-$ shift in the enzyme complexes with that found for cyanide in inorganic zinc cyanide complexes, it appears that the cyanide is bound directly in its non-protonated form to the Zn ion in carbonic anhydrase (Feeney *et al.*, 1973a). ($^{13}$C chemical shifts: $^{13}$CN-human carbonic anhydrase B 49·9 ppm $\pm$ 1 ppm; $^{13}$CN-bovine carbonic anhydrase B 51·9 ppm; Zn $(CN)_4^{2-}$ 47·7 ppm from $CS_2$ external reference.) The cyanide shifts in the enzyme complexes are pH independent over the range pH 7·6 to 8·9 indicating that no protolytic reactions involving the cyanide ions are taking place over this pH range.

### 3. *Interactions of $^{13}CO_2$ and bicarbonate with human haemoglobin preparations*

When $^{13}CO_2$ is equilibrated with glycine and other amino acids and peptides, carbamino derivatives are formed with the characteristic $^{13}$C chemical shift of 29 $\pm$ 1 ppm upfield of $CS_2$. Similarly shifted signals are observed when $^{13}CO_2$ is equilibrated with deoxyhaemoglobin at alkaline pH values and these have been attributed to carbamino group formation (Morrow *et al.*, 1973). This process is inhibited by the addition of 2,3-diphosphoglycerate. A broad $^{13}$C absorption (+33 ppm from $CS_2$) arising from the bicarbonate/carbonate rapid exchange system was also observed and the line width of this becomes very narrow when acetazolamide, a carbonic anhydrase inhibitor, is added. Clearly the preparations used contain some carbonic anhydrase. One can confidently anticipate some useful $^{13}$C studies of the $^{13}CO_2$-carbonic anhydrase system to study the hydration of carbon dioxide and the nature of the $CO_2$ binding site on the enzyme.

### 4. *Interaction of pyruvate ions with pyruvate carboxylase and pyruvate kinase*

Fung and coworkers (1973) have reported a very elegant study of the interaction of pyruvate ions with Mn(II)-pyruvate carboxylase and Mn(II)-pyruvate kinase by measuring the $^{13}$C relaxation times of pyruvate-1-$^{13}$C and pyruvate-2-$^{13}$C ions in the presence of the enzymes. Pyruvate-1-$^{13}$C provides

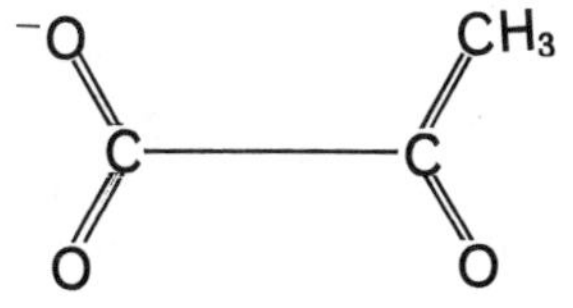

Pyruvate

an unambiguous measurement of the carboxyl $^{13}$C chemical shift (+22·65 ppm from $CS_2$) and similarly the pyruvate-2-$^{13}$C gives the carbonyl $^{13}$C chemical shift (−12·25 ppm from $CS_2$). By using the singly labelled pyruvate ions the problem of direct $^{13}$C–$^{13}$C coupling was avoided. The $T_1$ values for both carbons are very similar for the free pyruvate ions. However, when 40 $\mu M$ $MnCl_2$ is added to an 82·3 m$M$ solution of pyruvate ($^{13}$C-1 and $^{13}$C-2 labelled) the normalized paramagnetic contribution to the relaxation rate, $1/(fT_{1p})$, of the carboxyl carbon is twice that of the carbonyl carbon of the pyruvate and, furthermore, the carboxyl carbon resonance is significantly broadened. ($f =$ [Paramagnetic species]/[Pyruvate].)

When the pyruvate is examined in the presence of the Mn(II)-pyruvate carboxylase the paramagnetic contribution to the relaxation rate of the carbonyl carbon is now 3·5-fold greater than that to the carboxyl carbon. The carbonyl $^{13}C$ line width is also substantially broader than that of the carboxyl signal. Control experiments were carried out using the non-paramagnetic enzyme Mg(II)-pyruvate carboxylase. Thus the results for pyruvate in the presence of the Mn(II) enzyme are in marked contrast to those observed in the presence of Mn(II) and clearly indicate which groups are nearest to the metal.

The value of $1/(fT_{1p})$ can be used to calculate the relaxation time of the bound ligand $T_{1M}$ from which one can obtain the distance of the paramagnetic ion to the magnetic nucleus being relaxed if the relaxation rate is not limited by chemical exchange. The dipolar contribution to the relaxation is given (Dwek, 1972) by

$$\frac{1}{T_{1M}} = \frac{2g^2\beta^2\gamma_I^2 S(S+1)}{15r^6}\left(\frac{3\tau_c}{1+\omega_I^2\tau_c^2} + \frac{7\tau_c}{1+\omega_S^2\tau_c^2}\right) \tag{8.4}$$

where $g$ is the electron $g$ factor, $\beta$ the Bohr magneton, $S$ the electron spin quantum number of the paramagnetic ion, $\omega_I$ and $\omega_S$ are the nuclear and electron precession frequencies respectively, $r$ is the metal–carbon internuclear distance and $\tau_c$ the correlation time for the electron–nuclear interactions. The correlation time $\tau_c$ is given by

$$\frac{1}{\tau_c} = \frac{1}{\tau_r} + \frac{1}{\tau_m} + \frac{1}{\tau_S} \tag{8.5}$$

where $\tau_r$ = effective rotational correlation time
$\tau_m$ = residence time of ligand bound to protein
$\tau_S$ = electron spin lattice relaxation time at 23·4 KG.

The most difficult part of the analysis is obtaining the correct value of $\tau_c$ which can be obtained from a study of the frequency dependence of the relaxation times. The $1/(fT_{1p})$ value for the Mn(II)-pryuvate complex is independent of frequency as would be expected if the correlation time is that for the molecular tumbling of the complex; for the enzyme Mn(II)-pyruvate complex $1/(fT_{1p})$ depends inversely on the operating frequency and it can be shown that the correlation time is now determined by the electron spin relaxation time of the bound Mn(II).

The results show that for the Mn(II)-pyruvate system the pyruvate is binding simply as a monodentate carboxyl ligand. In contrast to this result, the Mn(II)-pyruvate carboxylase–pyruvate complex has the pyruvate carbonyl carbon much nearer to the metal than the carboxyl or methyl protons. However the Mn(II)-carbonyl carbon distance is fairly large (7·1 Å) which clearly rules out the possibility of the pyruvate forming an inner-sphere complex as had been suggested by earlier proton NMR studies.

### 5. *α-Methyl-D-glucopyranoside binding to concanavalin A*

Concanavalin A (Con A) is a sugar-binding protein which can agglutinate selectively cells transformed by oncogenic viruses. There are two metal-binding sites on the protein, one which binds $Ca^{2+}$ and the other which binds transition metal ions and both metal-binding sites need to be occupied for the protein to bind sugars. By examining carbon-13 enriched α-methyl-D-gluco-pyranoside bound to Con A (zinc and manganese derivatives) (Brewer *et al.*, 1973) it has been possible to measure the $^{13}C$ relaxation times of the bound sugar carbons and to assess the contribution $T_{1p}^{-1}$, from the paramagnetic manganese ions to the relaxation rates. There is no scalar contribution to this relaxation because no unpaired spin density is delocalized from the manganese ion on to the bound sugar carbons. Thus the paramagnetic contribution comes only from the dipolar mechanism given by Equation (8.4).

For Mn-Con A the $S$ value is 5/2 and $g = 2.0$ and the rotational correlation time is the dominant one in Equation (8.5) at 23·4 KG; the value of $\tau_c$ ($8 \times 10^{-8}$ sec) was determined from the Zn-Con A-sugar $^{13}C$ relaxation times.

From the observed paramagnetic contribution to the sugar carbon-13 relaxation rates, values of the $r$ distances can be estimated using Equation (8.4) and the effective rotational correlation time $\tau_c$. The sugar is found to bind in the C1 chair conformation with the 3 and 4 ring carbons being closest to the manganese at a mean distance of approximately 10 Å: this contrasts markedly with the value of 20 Å deduced from the X-ray crystallographic studies (Edelman *et al.*, 1972). For sugar binding at 20 Å from the manganese no paramagnetic contributions to the relaxation rates would have been observed. More recent measurements have indicated the presence of two separate sugar binding sites.

### 6. N-*Acetyltryptophan binding to α-chymotrypsin*

Roberts and Rodgers (1973) have used three-bond $^{13}C$–$^{1}H$ coupling constants to determine the conformation of the $C_\alpha$—N bond in *N*-acetyl-tryptophan when bound to α-chymotrypsin. They measured the coupling constant between the labelled carbonyl carbon and the α-proton in the bound state. The bound *N*-acetyltryptophan exists in rapid exchange with the free *N*-acetyltryptophan in solution and thus the observed coupling constants are weighted averages of the values in the free (2·3 Hz) and bound states. Knowing the binding constant and the concentrations of the enzyme and *N*-acetyl-tryptophan, it is possible to calculate the coupling constant in the bound state

($J = -0{\cdot}8 \pm 0{\cdot}8$ Hz). To facilitate these measurements, the acetyl carbonyl carbon of *N*-acetyltryptophan was enriched to 90% with carbon-13 and the spin–spin interactions with the acetyl methyl protons were removed from the carbonyl $^{13}C$ resonance signal by selectively irradiating the acetyl methyl protons at their resonance frequency. The only splitting remaining on the carbonyl signal is a doublet from interaction with the α-proton which is three bonds away. The calculated bound coupling constant corresponds to a dihedral angle about the $C_{\alpha}$—N bond of (±) 90° in the complex which is in reasonable agreement with the crystallographic findings for *N*-formyltryptophan binding to α-chymotrypsin.

## VI. CONCLUSIONS

It is now clear that $^{13}C$ NMR spectroscopy can reveal unique information about biological molecules in solution. Although many of the applications have been used previously in studies using the much more sensitive proton nucleus, the capability of being able to extend these methods to study carbon-13 has opened up new possibilities. Proton-noise-decoupled $^{13}C$ spectra are simpler than proton spectra and extend over a wider chemical shift range which increases the chance of being able to detect single resonance bands for specific nuclei in large molecules. Further spectrum simplification can be achieved by isotopic enrichment with $^{13}C$. Protein unfolding studies and conformational changes occurring when small molecules bind to proteins can be monitored by examining the $^{13}C$ spectra of assigned single-carbon resonance bands.

When $^{13}C$ chemical shifts are better understood, important conformational information in proteins might become available from their $^{13}C$ spectra. Already useful configurational information concerning *cis/trans*-prolines is available from $^{13}C$ chemical shift studies. More detailed temperature-dependent studies of carbonyl chemical shifts potentially could provide intramolecular hydrogen bonding information analogous to that obtained from the temperature dependence of the chemical shifts of NH protons. In these studies NH protons involved in intramolecular hydrogen bonds have low temperature coefficients for their chemical shifts.

Three-bond coupling constants ($^{13}C$–X–Y–H and $^{13}C$–X–Y–P) are conformationally dependent and can be usefully studied in small molecules of biological interest (peptides and nucleotides). Such studies are more difficult in large molecules where the short relaxation times of the large molecules often lead to line broadening thus preventing the detection of these coupling constants.

Relaxation time studies involving $^{13}C$ are attractive in that the relaxation mechanism for medium and large size diamagnetic molecules is always dipolar in origin and is not influenced appreciably by intermolecular dipolar relaxation effects (in contrast to proton relaxation times). Thus one can obtain dynamic information with ease from the measured relaxation times as illustrated in the phospholipid studies (Section V.E).

Overshadowing all the attractive features of carbon-13 is the outstanding problem of sensitivity which prevents a much wider application of such studies to biological problems. Undoubtedly the full potential of $^{13}C$ NMR in this area will only be realized when increased sensitivity is available either through improvements in instrumentation or more likely in improved carbon-13 enrichment procedures.

## VII. ACKNOWLEDGEMENTS

I would like to thank Drs. G. C. K. Roberts and P. J. Sadler for useful discussions and the many who supplied me with preprints of their work, in particular Drs. A. Allerhand, A. S. Mildvan, J. D. Roberts, G. C. K. Roberts, I. C. P. Smith and W. A. Thomas.

## VIII. REFERENCES

Allerhand, A., D. W. Cochran and D. Doddrell (1970) 'Carbon-13 Fourier transform NMR. II. Ribonuclease', *Proc. Natl. Acad. Sci. U.S.*, **67**, 1093.

Allerhand, A. and D. Doddrell (1971) 'Strategies in the application of partially relaxed Fourier transform NMR spectroscopy in assignments of $^{13}C$ resonances of complex molecules. Stachyose', *J. Amer. Chem. Soc.*, **93**, 2777–9.

Allerhand, A., D. Doddrell, V. Glushko, D. W. Cochran, E. Wenkert, P. J. Lawson and F. R. N. Gurd (1971a) 'Conformation and segmental motion of native and denatured ribonuclease A in solution. Application of natural-abundance $^{13}C$ partially relaxed Fourier-transform NMR', *J. Amer. Chem. Soc.*, **93**, 544–6.

Allerhand, A., D. Doddrell and R. Komoroski (1971b) 'Natural abundance carbon-13 partially relaxed Fourier transform nuclear magnetic resonance spectra of complex molecules', *J. Chem. Phys.*, **55**, 189–98.

Allerhand, A., R. F. Childers, R. Coodman, E. Oldfield and X. Ysern (1972) 'Increased sensitivity in carbon-13 FT (Fourier transform) NMR using 20 mm sample tubes'. *Amer. Lab.*, **4**, 19.

Allerhand, A., R. F. Childers and E. Oldfield (1973) 'Natural abundance $^{13}C$ NMR studies in 20 mm sample tubes. Observation of numerous single-carbon resonances of hen egg white lysozyme', *Biochemistry*, **12**, 1335–41.

Antonini, E., M. Brunori, F. Conti and G. Geraci (1973) 'NMR studies of $^{13}CO$-Hemoglobin $\alpha$ and $\beta$ chain identification', *FEBS Letters*, **34**, 69–70.

Barry, C. D., A. C. T. North, J. A. Glasel, R. J. P. Williams and A. V. Xavier (1971) 'Quantitative determination of mononucleotide conformations in solution using lanthanide ion shift and broadening NMR probes', *Nature*, **232**, 236–45.

Battersby, A. R., J. Moron, E. McDonald and J. Feeney (1972a) 'Studies of porphyrin biosynthesis by $^{13}C$ NMR: synthesis of ($^{13}C$) porphobilinogen and its incorporation into protoporphyrin-IX', *J. Chem. Soc. Chem. Commun.*, 920–1.

Battersby, A. R., G. L. Hodgson, M. Ihara, E. McDonald and J. Saunders (1972b), *J. Chem. Soc. Chem. Commun.*, 441.

Birdsall, B. and J. Feeney, unpublished results.

Birdsall, B. and J. Feeney (1972) 'The $^{13}C$ and $^{1}H$ NMR spectra and methods of their assignment for nucleotides related to dihydronicotinamide adenine dinucleotide phosphate (NADPH)', *J. Chem. Soc. Perkin II*, 1643–9.

Birdsall, B., N. J. M. Birdsall, J. Feeney and J. Thornton (1975) 'An NMR study of the conformation of nicotinamide mononucleotide in aqueous solution', *J. Amer. Chem. Soc.*

Birdsall, N. J. M., J. Feeney, A. G. Lee, Y. K. Levine and J. C. Metcalfe (1972) 'Dipalmitoyl-lecithin: Assignment of the $^{1}H$ and $^{13}C$ NMR spectra, and conformation studies', *J. Chem. Soc. Perkin II*, 1441–5.

Brewer, C. F., H. Sternlicht, D. M. Marcus and A. P. Grollman (1973) 'Binding of $^{13}C$ enriched α-methyl-D-glucopyranoside to Concanavalin A as studied by carbon magnetic resonance', *Proc. Natl. Acad. Sci. U.S.*, **70**, 1007–11.

Brewster, A. I. R., V. J. Hruby, A. F. Spatola and F. A. Bovey (1973) '$^{13}C$-NMR spectroscopy of oxytocin, related oligopeptides and selected analogs', *Biochemistry*, **12**, 1643–9.

Brown, C. E., J. J. Katz and D. Shemin (1972) 'The biosynthesis of vitamin $B_{12}$: a study by $^{13}C$ magnetic resonance spectroscopy', *Proc. Natl. Acad. Sci. U.S.*, **69**, 2585–8.

Browne, D. T., G. L. Kenyon, E. L. Packer, H. Sternlicht and D. M. Wilson (1973) 'Studies of macromolecular structure by $^{13}C$ NMR. II. $^{13}C$-specific labeling approach to the study of histidine residues in proteins', *J. Amer. Chem. Soc.*, **95**, 1316–23.

Bystrov, V. F., V. T. Ivanov, S. A. Kozmin, I. I. Mikhaleva, R. Kh. Khalilulina, Yu. A. Ovchinnikov, E. I. Fedin and P. V. Petrovskii (1972) 'Biologically active alkali metal complexones. A $^{13}C$-NMR study of ion–dipole interaction', *FEBS Letters*, **21**, 34–8.

Bystrov, V. F. (1974). Private communication.

Chaiken, I. M., M. H. Freedman, J. R. Lyerla, Jr. and J. S. Cohen (1973) 'Preparation and studies of $^{19}F$-labeled and enriched $^{13}C$-labeled semisynthetic ribonuclease-$S^{1}$ analogues', *J. Biol. Chem.*, **248**, 881–91.

Christl, M. and J. D. Roberts (1972) 'Nuclear magnetic resonance spectroscopy. C–13 chemical shifts of small peptides as a function of pH', *J. Amer. Chem. Soc.*, **94**, 4565–73.

Cushley, R. J., D. R. Anderson, S. R. Lipsky, R. J. Sykes and H. H. Wasserman (1971) '$^{13}C$ Fourier transform NMR spectroscopy. II. The pattern of biosynthetic incorporation of (1-$^{13}C$) and (2-$^{13}C$) acetate into prodigiosin', *J. Amer. Chem. Soc.*, **93**, 6284–6.

Deslauriers, R., R. Walter and I. C. P. Smith (1972) 'A $^{13}C$ NMR study of oxytocin and its oligopeptides', *Biochem. Biophys. Res. Comm.*, **48**, 854–9.

Doddrell, D., V. Glushko and A. Allerhand (1972) 'Theory of Nuclear Overhauser Enhancement and $^{13}C$-$^{1}H$ dipolar relaxation in proton-decoupled carbon-13 spectra of macromolecules', *Chem. Phys.*, **56**, 3683–89.

Dorman, D. E. and J. D. Roberts (1970) 'NMR spectroscopy $^{13}C$ spectra of some common nucleotides', *Proc. Natl. Acad. Sci. U.S.*, **65**, 19–26.

Dorman, D. E. and F. A. Bovey (1973) 'Proton-coupled $^{13}C$ magnetic resonance spectra. The simple amides', *J. Org. Chem.*, **38**, 1719–22.

Dwek, R. A. (1972) 'Proton relaxation enhancement probes', *Advan. in Molecular Relaxation Processes*, **4**, 1.

Edelman, G. M., B. A. Cunningham, G. N. Reeke, Jr., J. W. Becker, M. J. Waxdal and J. L. Wang (1972) 'The covalent and three dimensional structure of Concanavalin A', *Proc. Natl. Acad. Sci. U.S.*, **69**, 2580–4.

Ernst, R. R. (1966a) *Advances in Magnetic Resonance*, Vol. 2 (Ed. J. S. Waugh), Academic Press Inc., New York.

Ernst, R. R. (1966b) 'Nuclear magnetic double resonance with an incoherent radio-frequence field', *J. Chem. Phys.*, **45**, 3845–61.

Ernst, R. R. and W. A. Anderson (1966) 'Application of Fourier transform spectroscopy to magnetic resonance', *Rev. Sci. Instr.*, **37**, 93–102.

Farrar, T. C. and E. C. Becker (1971) *Pulse and Fourier Transform NMR*, Academic Press, New York.

Feeney, J., P. J. S. Pauwels and D. Shaw (1970) 'A method of increasing sensitivity in $^{13}$C. NMR spectroscopy via a heteronuclear Overhauser effect without loss of spin–spin coupling information', *Chem. Commun.*, 554–5.

Feeney, J., A. S. V. Burgen and E. Grell (1973a) 'Cyanide binding to carbonic anhydrase—a $^{13}$C NMR study', *Eur. J. Biochem.*, **34**, 107–11.

Feeney, J., P. Partington and G. C. K. Roberts (1973b) 'The assignment of $^{13}$C resonances from carbonyl groups in peptides', *J. Mag. Res.*, **13**, 268–74.

Feeney, J., P. E. Hansen and G. C. K. Roberts (1974) 'Use of $^{13}$C-$^{1}$H coupling constants in the determination of side-chain conformations of amino acids', *J. Chem. Soc., Chem. Commun.*, 465–6.

Freedman, M. H., J. R. Lyerla, Jr., I. M. Chaiken and J. S. Cohen (1973) '$^{13}$C-NMR studies on selected amino-acids, peptides and proteins', *Eur. J. Biochem.*, **32**, 215–26.

Freeman, R. and H. D. W. Hill (1971) 'Nuclear Overhauser effect in undecoupled NMR spectra of $^{13}$C', *J. Mag. Res.*, **5**, 278–9.

Fung, C. H., A. S. Mildvan, A. Allerhand, R. Komoroski and M. C. Scrutton 1973) 'Interaction of pyruvate with pyruvate carboxylase and pyruvate kinase as studied by paramagnetic effects on $^{13}$C relaxation rates', *Biochemistry*, **12**, 620–9.

Gillies, D. G. and D. Shaw (1972) 'The application of Fourier Transform to high resolution NMR Spectroscopy', *Annual Reports on NMR Spectroscopy*, Vol. 5 (Ed. E. F. Mooney), Academic Press.

Grell, E., T. Funck and H. Sauter (1973) '$^{13}$C NMR and infra red absorption spectroscopy of valinomycin and its alkali-ion complexes', *Eur. J. Biochem.*, **34**, 415–24.

Hansen, P. E., J. Feeney and G. C. K. Roberts (1975) 'Long-range $^{13}$C-$^{1}$H spin coupling constants in amino acids and peptides: conformational applications', *J. Mag. Res.*

Horsley, W. J. and H. Sternlicht (1968) '$^{13}$C magnetic resonance studies of amino acids and peptides', *J. Amer. Chem. Soc.*, **90**, 3738–48.

Hubbell, W. L. and H. M. McConnell (1968) 'Spin-label studies of the excitable membranes of nerve and muscle', *Proc. Natl. Acad. Sci. U.S.*, **61**, 12–16.

Hunkapiller, M. W., S. H. Smallcombe, D. R. Whitaker and J. H. Richards (1973) 'Carbon NMR Studies of the Histidine Residue in $\alpha$-Lytic Protease', *Biochemistry*, **12**, 4732–43.

Jardetzky, O., J. L. Markely, H. Thielmann, Y. Arata and M. N. Williams (1971) 'Tentative sequential model for the unfolding and refolding of Staphylococcal nuclease at high pH', *Cold Spring Harbor Symposium on Quantitative Biology*, **36**, 257.

Johnson, L. F., personal communication.

Jones, A. J., D. M. Grant, M. W. Winkley and R. K. Robins (1970a) '$^{13}$C magnetic resonance. XVIII. Selected nucleotides', *J. Phys. Chem.*, **74**, 2684–9.

Jones, A. J., M. W. Winkley, D. M. Grant and R. K. Robins (1970b) '$^{13}$C NMR: naturally occurring nucleosides', *Proc. Natl. Acad. Sci. U.S.*, **65**, 27–30.

Jung, G., R. Breitmaier and W. Voelter (1972) 'Dissoziationsgleichgewitchle von Glutathion. Eine Fourier-Transform $^{13}$C NMR specktroskopische Untersuchung der pH—Abhangigkeit der Ladungsverteilung', *Eur. J. Biochem.*, **24**, 438–45.

Lapper, R. D. and I. C. P. Smith (1973) 'A $^{13}$C and $^{1}$H NMR study of the conformations of 2′, 3′-cyclic nucleotides', *J. Amer. Chem. Soc.*, **95**, 2880–4.

Lapper, R. D., H. H. Mantsch and I. C. P. Smith (1973) 'A $^{13}$C and $^{1}$H NMR study of conformation of 3′, 5′-cyclic nucleotides. A demonstration of the angular dependence of 3 bond spin–spin couplings between carbon and phosphorus', *J. Amer. Chem. Soc.*, **94**, 6243–4; **95**, 2878–80.

Lemieux, R. U., T. L. Nagabhushan and B. Paul (1972) 'Relationship of $^{13}$C to vicinal $^{1}$H coupling to the torsion angle in uridine and related structures', *Canad. J. Chem.*, **50**, 773–6.

Levine, Y. K., N. J. M. Birdsall, A. G. Lee and J. C. Metcalfe (1972a) '$^{13}$C NMR relaxation measurements of synthetic lecithins and the effect of spin labeled lipids', *Biochemistry*, **11**, 1416–21.

Levine, V. K., P. Partington, G. C. K. Roberts, N. J. M. Birdsall, A. G. Lee and J. C. Metcalfe (1972b) '$^{13}$C nuclear magnetic relaxation times and models for chain motion in lecithin vesicles', *FEBS Letters*, **23**, 203–7.

Levine, Y. K., P. Partington and G. C. K. Roberts (1973) 'Calculation of dipolar nuclear magnetic relaxation times in molecules with multiple internal rotations. 1. Isotropic overall motion of the molecule', *Mol. Phys.*, **25**, 497–514.

Levy, G. C. and G. L. Nelson (1972) *Carbon-13 Nuclear Magnetic Resonance for Organic Chemists*, Wiley–Interscience, New York.

Lyerla, Jr., J. R., B. H. Barber and M. H. Freedman (1973) 'Carbon-13 chemical shifts accompanying helix formation', *Canad. J. Biochem.*, **51**, 460–4.

Martin, L., C. Chang, H. Floss, J. Mabe, E. Hagaman and E. Wenkert (1972) 'A $^{13}$C NMR resonance study on the biosynthesis of pyrrolnitrin from tryptophan by *Pseudomonas*', *J. Amer. Chem. Soc.*, **94**, 8942–4.

McInnes, A. G., D. G. Smith, L. C. Vining and L. F. Johnson (1971) 'Use of $^{13}$C in biosynthetic studies. Location of isotope from labelled acetate and formate in fungal tropolone, sepedonin, by $^{13}$C nuclear magnetic resonance spectroscopy', *Chem. Commun.*, 325–6.

Metcalfe, J. C., N. J. M. Birdsall, J. Feeney, A. G. Lee, Y. K. Levine and P. Partington (1971) '$^{13}$C NMR spectra of lecithin vesicles and erythrocyte membranes', *Nature*, **233**, 199.

Metcalfe, J. C., N. J. M. Birdsall and A. G. Lee (1972a) '$^{13}$C NMR spectra of acholeplasma membranes containing $^{13}$C labelled phospholipids', *FEBS Letters*, **21**, 335–40.

Metcalfe, J. C., N. J. M. Birdsall and A. G. Lee (1972b), *Mitochondria/ Biomembranes*, North Holland, Amsterdam.

Moon, R. B. and J. H. Richards (1972a) 'Conformational studies of various hemoglobins by natural abundance $^{13}$C spectroscopy', *Proc. Natl. Acad. Sci. U.S.*, **69**, 2193–7.

Moon, R. B. and J. H. Richards (1972b) 'NMR resonance studies of $^{13}$CO binding to various hemoglobins', *J. Amer. Chem. Soc.*, **94**, 5093–5.

Morrow, J. S., P. Kein, R. B. Visscher, R. C. Marshall and F. R. N. Gurd (1973) 'Interaction of $^{13}CO_2$ and bicarbonate with human hemoglobin preparation', *Proc. Natl. Acad. Sci. U.S.*, **70**, 1414–18.

Neuss, N., C. H. Nash, P. A. Lemke and J. B. Grutzner (1971) 'The use of $^{13}C$ NMR (CMR) spectroscopy in biosynthetic studies. Incorporation of carboxyl and methyl $^{13}C$ labeled acetates into cephalosporin C', *J. Amer. Chem. Soc.*, **93**, 2337–9.

Nigen, A. M., P. Keim, R. C. Marshall, J. S. Morrow, R. A. Vigna and F. R. N. Gurd (1973) '$^{13}C$ NMR spectroscopy of myoglobins carboxymethylated with enriched (2-$^{13}C$) bromoacetate', *J. Biol. Chem.*, **248**, 3724–32.

Noggle, J. H. and R. E. Schirmer (1971), *The Nuclear Overhauser Effect, Chemical Applications*, Academic Press, New York.

Ohnishi, M., M. C. Fedarco, J. D. Baldeschwieler and L. F. Johnson (1972) 'Fourier transform $^{13}C$ NMR analysis of some free and potassium-ion complexed antibiotics', *Biochem. Biophys. Res. Comm.*, **46**, 312–27.

Oldfield, E. and A. Allerhand (1973) 'Cytochrome c. Observation of numerous single-carbon sites of the reduced and oxidised species by means of natural abundance $^{13}C$ NMR spectroscopy', *Proc. Natl. Acad. U.S.*, **70**, 3531.

Olson, J. S., M. E. Andersen and Q. H. Gibson (1971) 'The dissociation of the first oxygen molecule from some mammalian oxyhemoglobins', *J. Biol. Chem.*, **246**, 5919–23.

Packer, E. L., H. Sternlicht and J. C. Rabinowitz (1972) 'The possible role of aromatic residues of *Clostridium acidi-urici* ferrodoxin in electron transport', *Proc. Natl. Acad. Sci. U.S.*, **69**, 3278–82.

Patel, D. J. (1973) 'Carbon framework of valinomycin and its metal ion complex in solution', *Biochemistry*, **12**(3), 496–501.

Polonsky, J., Z. Baskevitch, N. Cagnoli-Bellavita, P. Ceccherelli, B. L. Buckwalter and E. Wenkert (1972) '$^{13}C$ NMR spectroscopy of naturally occurring substances. XI. Biosynthesis of the virescenosides', *J. Amer. Chem. Soc.*, **94**, 4369–70.

Roberts, G. C. K. and O. Jardetzky (1970) 'Nuclear magnetic resonance spectroscopy of amino acids, peptides and proteins', *Advan. Protein Chem.*, **24**, 447–545.

Roberts, G. C. K. and P. Rodgers (1973) '$^{13}C$ NMR studies of the conformation of N-acetyl-L-Tryptophan in its complex with chymotrypsin', *FEBS Letters*, **36**, 330–3.

Robinson, J. D., N. J. M. Birdsall, A. G. Lee and J. C. Metcalfe (1972) '$^{13}C$ and $^{1}H$ NMR relaxation measurements of the lipids of sarcoplasmic reticulum membranes', *Biochemistry*, **11**, 2903–9.

Sarma, R. H., V. Ross and N. O. Kaplan (1968) 'Investigation of the conformation of $\beta$-diphosphopyridine nucleotide ($\beta$-nicotinamide-adenine dinucleotide) and pyridine dinucleotide analogs by proton magnetic resonance', *Biochemistry*, **9**, 3052–62.

Scott, A. I., C. A. Townsend, K. Okada, M. Kajiwara, P. J. Whitman and R. J. Cushley (1972) 'Biosynthesis of corrinoids. Concerning the origin of the methyl groups in vitamin $B_{12}$', *J. Amer. Chem. Soc.*, **94**, 8267–69.

Sheard, B. and E. M. Bradbury (1970) 'Nuclear magnetic resonance in the study of biopolymers and their interaction with ions and small molecules', *Progr. Biophys.*, **20**, 187–246.

Smith, I. C. P., R. Deslauriers and R. Walter (1972a) *Chemistry and Biology of Peptides* (Ed. J. Meienhofer), Ann Arbor Science Publ.

Smith, I. C. P., H. H. Mantch, R. D. Lapper, R. Deslauriers and T. Schleich (1972b) *Proceedings Fifth Jerusalem Symposium on Quantum Chemistry and Biochemistry* (Eds. E. Bergmann and A. Pullmann), Academic Press, New York.

Smith, I. C. P., R. Deslauriers, H. Saito, R. Walter, C. Carrigau-Lagrange, H. McGregor and D. Sarantakis (1973) 'Carbon-13 NMR studies of peptide hormones and their components', *Ann. N.Y. Acad. Sci.*, **222**, 597–627.

Stothers, J. B. (1972), *Carbon-13 NMR Spectroscopy*, Academic Press.

Tanabe, M., H. Seto and L. F. Johnson (1970a) 'Biosynthetic studies with $^{13}$C. $^{13}$C NMR spectra of radicinin', *J. Amer. Chem. Soc.*, **92**, 2157–8.

Tanabe, M., T. Hamasaki. H. Seto and L. F. Johnson (1970b) 'Biosynthetic studies with $^{13}$C. $^{13}$C NMR spectra of the metabolite sterigmatocystin', *Chem. Commun.*, 1539–40.

Tanabe, M., T. Hamasaki, D. Thomas and L. F. Johnson (1971) 'Biosynthetic studies with $^{13}$C asperlin', *J. Amer. Chem. Soc.*, **93**, 273–4.

Tanabe, M. and W. C. Jankowski (1972) *Varian Instrument Application*, Vol. 7, p. 2.

Tanabe, M., T. Hamasaki, Y. Suzuki and L. F. Johnson (1973) 'Biosynthetic studies with $^{13}$C: Fourier transform NMR spectra of the metabolite avenaciolide', *J. Chem. Soc. Chem. Comm.*, 212–13.

Thomas, W. A. and M. K. Williams (1972) '$^{13}$C NMR spectroscopy and *cis/trans* isomerism in dipeptides containing proline', *Chem. Commun.*, 994.

Trauble, H. and E. Sackmann (1972) 'Studies of crystalline–liquid phase transition of lipid model membranes. III. Structure of steroid-lecithin system below and above lipid-phase transition', *J. Amer. Chem. Soc.*, **94**, 4499–510.

Vold, R. L., J. S. Waugh, M. P. Klein and D. E. Phelps (1968) 'Measurement of spin relaxation in complex systems', *J. Chem. Phys.*, **48**, 3831–2.

Warren, G., P. Toon, N. J. M. Birdsall, A. G. Lee and J. C. Metcalfe (1974) 'Reconstitution of a calcium pump using defined membrane components', *Proc. Natl. Acad. Sci., U.S.*, **71**, 622–6.

Wessels, P. L., J. Feeney, H. Gregory and J. J. Gormley (1973) 'High resolution NMR studies of the conformation of LH-RH and its component peptides', *J. Chem. Soc. Perkin II*, 1691–8.

Westley, J. W., D. L. Pruess and R. G. Pitcher (1972) 'Incorporation of (1-$^{13}$C) butyrate into antibiotic X-537A. $^{13}$C Nuclear magnetic resonance study', *Chem. Commun.*, 161–2.

White, R. J., E. Martinelli, G. G. Gallo, G. Lancini and P. Beynon (1973) 'Rifamycin biosynthesis studies with $^{13}$C enriched precursors and carbon magnetic resonance', *Nature*, **243**, 273–7.

Wüthrich, K., A. Tun-Kyi and R. Schwyzer (1972) 'Manifestation in the $^{13}$C-NMR spectra of two difference molecular conformations of a cyclic pentapeptide', *FEBS Letters*, **25**, 104–8.

Yamasaki, M., Y. Macbayashi and K. Miyaki (1972a) 'Application of $^{13}$C NMR to biosynthetic investigations. I. Biosynthesis of ochratoxin A', *Chem. Pharm. Bull. Tokyo*, **20**, 2172–5.

Yamasaki, M., F. Katoh, J. Ohichi and Y. Koyama (1972b) 'Study on biosynthesis of aurethin, a nitro-containing metabolite from *Streptomyces luteoreticuli*, using $^{13}$C-NMR spectroscopy', *Tetrahedron Letters*, No. 26, 2701–4.

Zimmer, S., W. Haar, W. Maurer, H. Ruterjans, S. Fermandjian and P. Fromageot (1972) 'Investigation of the structure of angiotensin II using $^{13}$C NMR spectra', *Eur. J. Biochem.*, **29**, 80–7.

CHAPTER 9

# The Mössbauer effect and its applications in biology

R. Cammack
*University of London King's College,*
*Department of Plant Sciences,*
*68 Half Moon Lane, London SE24 9JF*

## I. INTRODUCTION

It is often the case when a new spectroscopic technique is discovered, that it is applied first as a tool of physics, then later of chemistry, and it is only much later that it can be applied to the more intractable problems of biology. In the case of Mössbauer spectroscopy this development has been exceptionally rapid. The phenomenon of recoilless gamma-ray emmission was discovered by R. L. Mössbauer in 1957; almost immediately it found applications in solid-state physics and in verifying Einstein's theory of Relativity. By the early 1960's the potential of the technique as a unique means of studying the chemistry of iron, tin and other elements was being realized, and it was soon found that the method could be applied to the study of iron-containing proteins. Since then a considerable number of biological materials have been examined by this technique, and in favourable cases it has been possible to obtain information about the iron atoms in them which could not have been obtained in any other way.

Mössbauer spectroscopy is basically a form of gamma-ray spectroscopy which is concerned with the transition of the nucleus of an atom to an excited state. Small changes in the energy of this nuclear transition can be measured with extraordinary precision—of the order of 1 part in $10^{14}$ for $^{57}Fe$. At this level of precision it is possible to measure minute changes due to interaction of the nucleus with those electrons that are involved in chemical bonding. The nucleus thus acts as a probe of the chemical state of the atom. Mössbauer spectroscopy is completely specific for the type of nucleus being studied; if one is examining $^{57}Fe$, no other nuclei will be observed. It is a very local probe; it is capable of giving detailed information about the chemical state of one atom, but comparatively little about other atoms in its vicinity.

The Mössbauer effect is only applicable to certain nuclei, the requirement being that they have an excited state which is fairly close in energy to the ground state. The effect has been observed in nearly 80 isotopes which fulfil this condition, though, for various reasons, spectroscopy has been found to be technically feasible in relatively few cases. Most of these isotopes are of the heavier elements; of these isotopes, $^{57}Fe$ is the one which is of most interest biologically. It is a naturally occurring stable isotope, and fortunately it turns out to be a very good Mössbauer isotope. Moreover the variety of magnetic and valence states which the iron atom can take up mean that there is a considerable amount of useful information to be obtained. As a result, more work has been done on the Mössbauer effect in $^{57}Fe$ than in any other isotope, so that the theory and techniques are comparatively well understood. The type of questions we would expect to answer would be, in the case of an iron compound, whether the iron is ferrous or ferric, high spin or low spin, the type of ligands present and their arrangement about the iron atom. In practice the solution of such problems is often best achieved by a concerted approach using Mössbauer spectroscopy supplemented by data from other physicochemical techniques such as electron paramagnetic resonance (EPR) and magnetic susceptibility.

This article is intended to give a description of Mössbauer spectroscopy as applied to biological problems, in particular to the study of iron-containing proteins. It is hoped to give an idea of the scope of the method, with particular emphasis on special aspects of technique concerned with handling of these materials. However it is not intended to give an account of the fundamental principles involved in interpretation of Mössbauer spectra, which requires a more rigorous mathematical approach based on wave mechanics. For further reading the authoritative review by Lang (1970) on Mössbauer spectroscopy of haem proteins can be recommended. For more general reading on the Mössbauer Effect see Wertheim (1964), and Goldanskii and Herber (1968). The rapidly increasing literature on Mössbauer spectroscopy is surveyed by a considerable number of reviews which are published at intervals, including the *Mossbauer Effect Data Index*, of which the latest edition, covering the literature of 1970, is edited by J. G. and V. E. Stevens (1972) and a series in *Analytical Chemistry Reviews*, of which the latest is by Stevens and Bowen (1974), which give useful comments on trends in methods and instrumentation, and in which references to other reviews may be found.

## II. PRINCIPLES OF $^{57}$Fe MÖSSBAUER SPECTROSCOPY

The basic process of Mössbauer spectroscopy is illustrated in Figure 9.1 for the case of $^{57}$Fe. A $^{57}$Fe nucleus in an excited state (nuclear spin $I = \frac{3}{2}$) is produced as a result of radioactive decay of $^{57}$Co. The excited $^{57}$Fe nucleus decays to the ground state, with the release of a $\gamma$-ray quantum of energy 14·4 keV. The gamma ray produced by this source can be absorbed by another $^{57}$Fe nucleus, which then becomes excited to the $I = \frac{3}{2}$ state. This process is called *resonant absorption;* it is analogous to optical absorption, where the absorption of a quantum of light results in the excitation of an electron to a higher energy state. The excited absorber $^{57}$Fe nucleus subsequently decays to the ground state and the $\gamma$-ray is re-radiated in a random direction.

The energy of the $\gamma$-ray is defined extremely closely, and as a result, if the Fe atoms in the source and absorber are in different chemical states, resonant absorption may not take place. Then in order for the $\gamma$-ray to be absorbed, its energy must be altered by a certain amount. This can be achieved by means of the Doppler shift, by moving the source with velocity $v$ relative to the absorber. The frequency of the $\gamma$-ray will be changed by a factor $v/c$, and since $E = h\nu$ its energy will also be changed by this factor. Thus we have the basis of a spectrometer (Figure 9.2). $\gamma$-rays from a $^{57}$Co source are passed

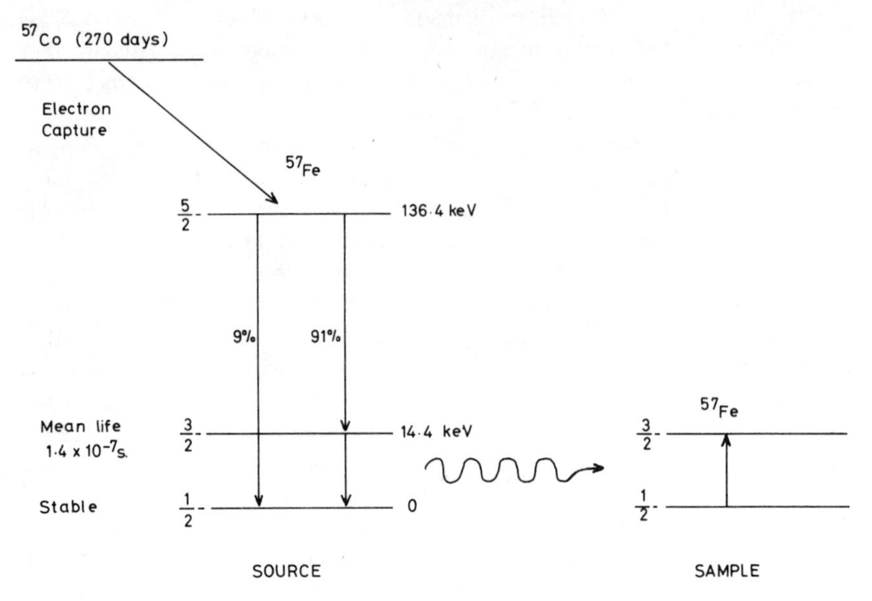


**Figure 9.1.** Nuclear transitions involved in the $^{57}Fe$ Mössbauer effect

through a sample containing the $^{57}Fe$ atoms of interest, and into a suitable counter. The source is moved at a certain velocity relative to the sample, and the intensity of 14·4 keV radiation transmitted through the sample is measured. This process is repeated at a number of different values of velocity, and the results can be plotted as an absorption spectrum. By convention, positive values of velocity are *towards* the absorber. The velocity at which maximum absorption takes place is then a measure of the difference in nuclear transition energy between source and absorber $^{57}Fe$ nuclei, due to the difference in chemical states between the two atoms. It turns out that the sort of velocity required for absorption to take place is literally a snail's pace—of the order of a few millimetres per second. Thus we are observing very small differences in energy; 1 mm/sec is equivalent to an energy difference of $4{\cdot}8 \times 10^{-8}$ eV. All the same, the graph which we obtain by this type of experiment, with the somewhat bizarre units of counts per second *versus* millimetres per second, is analogous to the more familiar forms of absorption spectrum.

In practice, the velocity range is usually scanned continuously by moving the source with a parabolic oscillating motion, and the counts over each small velocity range are summed by means of a multichannel analyser. The absorption by the sample is usually only a few percent so that a large number of counts must be obtained for good statistical accuracy and it is normal to measure Mössbauer spectra over the course of several hours.

According to theory based on free atoms, this spectrometer should not

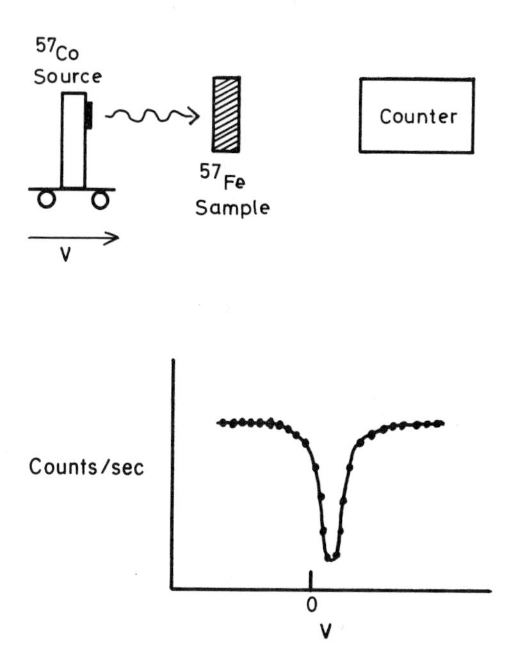


**Figure 9.2.** A simple Mössbauer spectrometer and spectrum

work. The emission of the $\gamma$-ray should cause the emitting nucleus to recoil, and the energy of this recoil should be subtracted from that of the $\gamma$-ray. This recoil energy is a small fraction of $\gamma$-ray energy, but is large compared with the very small energy differences we are trying to observe, so that combined effects of recoil on emission and absorption would cause a shift in the spectrum of about $10^5$ mm/sec. It was not until 1957, when R. L. Mössbauer discovered the phenomenon of *recoil-free emission* that it was realized that this situation does not necessarily apply to a solid. If the atoms are bound in a lattice of a solid, their vibrational energy will be quantized as phonons. The energy which can be transferred to the lattice by the $\gamma$-ray recoil must be an integral number of phonons. The phonon energy depends on the strength of the lattice binding forces, but in most compounds is greater than the recoil energy of the 14·4 keV $\gamma$-rays. As a consequence there is a probability that a given $\gamma$-ray will be emitted without recoil. It may be noted that this does not contravene the laws of conservation of energy and momentum, since other $\gamma$-rays will be emitted with recoil energy equal to one (or even more) phonons, so that when an average is taken over a large number of emission processes, the energy transferred per emission is exactly the free-atom recoil energy. Therefore we can consider that a fraction $f$ of the $\gamma$-rays will be emitted without any alteration of their energy by recoil, and it is these $\gamma$-rays which have the precisely defined energy which is used in Mössbauer spectroscopy. Similarly a fraction $f'$ of these $\gamma$-rays will be absorbed without recoil by an atom in the absorber solid lattice, so that the number of $\gamma$-rays absorbed will be proportional to $ff'$. The value of $f$ places a theoretical limit on the sensitivity of Mössbauer spectroscopy; it is large if the lattice binding forces are strong and the temperature is low. Fortunately $f$ is fairly large for typical $^{57}Fe$ compounds, being about 0·8 at temperatures below 77°K, and about 0·5 even at 300°K.

For other types of nuclei, $f$ will only be appreciable when the $\gamma$-ray recoil energy is small in relation to the lattice binding forces and hence, if an isotope is to be suitable for Mössbauer spectroscopy, the energy of the $\gamma$-ray produced by a nuclear transition should be fairly small—less than 150 keV. For high recoil energies, it may be necessary to cool both source and sample to very low temperatures to obtain a significant recoil-free fraction.

A second requirement for a useful isotope for Mössbauer spectroscopy is that the width of absorption lines in the spectrum should be of the right magnitude. This is determined by the linewidth of the $\gamma$-ray, which in turn is determined by the average lifetime of the excited state—the longer the lifetime, the narrower the line width of the $\gamma$-ray. This is a consequence of the Heisenberg uncertainty principle of energy and time. The average lifetime of the excited state of the $^{57}Fe$ nucleus is about 140 nsec and the resulting natural line width is 0·19 mm/sec, which is the minimum line width which can be observed in an $^{57}Fe$ Mössbauer spectrum, (though the lines are usually broader than this for a number of reasons).

## A. States of the Iron Atom

The iron atom in chemical compounds normally exists in either the trivalent, ferric, form or the divalent, ferrous, form. In addition, as a consequence of interactions with its ligands, the electronic configuration can be either high spin or low spin. These states of the iron atom can be understood in terms of the ligand field theory (see, for example, Griffith, 1961). The ferric ion contains twenty-three electrons, distributed in orbitals $1s^2$, $2s^2$, $2p^6$, $3s^2$, $3p^6$, $3d^5$; that is, there are five valence electrons in the $3d$ shell, and the remaining electrons form a stable core. The ferrous ion is similar, but contains six electrons in the $3d$ shell. The $3d$ shell is capable of accommodating ten electrons in five orbitals: the $d_{xy}$, $d_{yz}$ and $d_{xz}$ orbitals, collectively known as $d_\epsilon$ or $t_{2g}$, and the $d_{z^2}$ (sometimes called $d_{3z^2-r^2}$) and $d_{x^2-y^2}$ collectively known as the $d_\gamma$ or $e_g$. In a free ion all these orbitals are of equal energy, but the presence of ligands causes them to become separated in energy (Figure 9.3). The ligands are usually arranged about the iron atom in either octahedral or tetrahedral symmetry. With six equivalent ligands in octahedral symmetry the $e_g$ orbitals are higher in energy than the $t_{2g}$; the separation in energy is

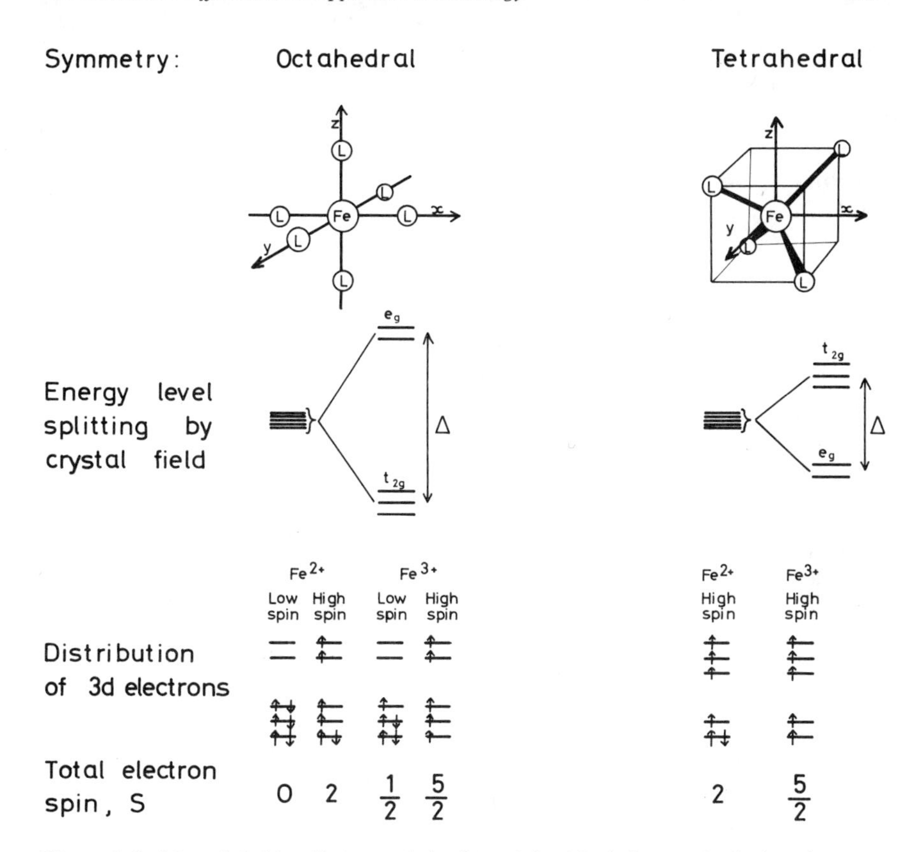


**Figure 9.3.** Ligand field splittings of the iron 3*d* orbitals in octahedral and tetrahedral symmetry

called the crystal field splitting $\Delta$. With four equivalent ligands in tetrahedral symmetry the $t_{2g}$ levels will be higher than the $e_g$. Normally the symmetry will not be perfect, particularly in iron–protein complexes where the ligands are not all equivalent and the iron atom is distorted. This will cause further separations of energy of the orbitals within the $t_{2g}$ and $e_g$ levels, though these separations will be small compared with $\Delta$ if the distortion is not excessive.

The way in which the *d* electrons fill the orbitals depends on the magnitude of $\Delta$, which in turn depends on the type of ligands present. Each orbital can contain two electrons, but to place two electrons in an orbital requires pairing energy. Therefore if $\Delta$ is small the electrons will fill the orbitals singly as far as possible, and the complex will be high-spin. On the other hand if $\Delta$ is large the electrons will fill the lower energy levels and the complex will be low-spin. $\Delta$ is smaller for tetrahedral complexes than for octahedral with the same type of ligands, so that in practice low-spin tetrahedral complexes are unknown.

It is possible to determine whether iron is high-spin or low-spin by measurements of magnetic susceptibility, since the unpaired electrons have a magnetic moment and give rise to paramagnetism. As will be seen later, the magnetic effects of these electrons on the nucleus adds considerably to the amount of information which can be obtained by Mössbauer spectroscopy.

The 'core' electrons are not directly involved in chemical bonding, but it will be seen later that since the *s* electrons have a certain charge density, i.e. probability of being present at the nucleus, whereas the 3*d* valence electrons do not, many of the consequences of chemical bond formation are transmitted to the nucleus by indirect processes through the *s* electrons.

## B. Chemical Shift

The chemical shift (or isomer shift) is caused by interaction of the nucleus with the *s* electrons, and is observed in Mössbauer spectra of all $^{57}$Fe atoms regardless of symmetry or spin state. The effect arises because in the excited nuclear state ($I = \frac{3}{2}$) the $^{57}$Fe nucleus has a slightly smaller radius than in the ground state ($I = \frac{1}{2}$). This means that the energy of the nuclear transition

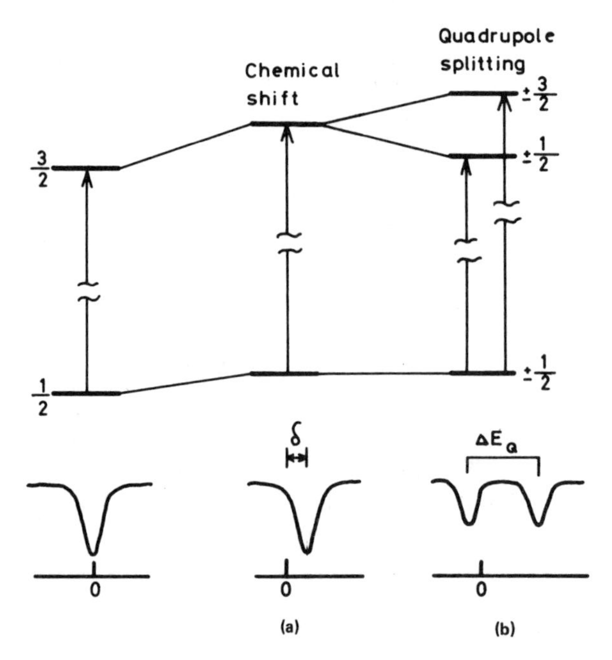


**Figure 9.4.** The chemical shift and quadrupole splitting of the nuclear energy levels, and their effects on the Mössbauer spectrum

is affected by the electron density at the nucleus. There will be a positive shift in the spectrum, as shown in Figure 9.4(a) corresponding to an increase in energy, if the absorber nucleus has a decreased electron density at the nucleus relative to the source. If in addition the spectrum shows other types of structure such as quadrupole splitting (see next section), the 'centre of gravity' of the whole spectrum will be shifted by the chemical shift. The chemical shift is expressed relative to a standard which is taken as the zero on the velocity scale. This may be defined as zero velocity of a standard source, such as $^{57}Co$ in Pd, Cr or Cu, or the centre of the spectrum of a standard absorber such as sodium nitroprusside or metallic iron. Since the observed chemical shift depends on the standard used, this must be specified when results are to be compared. Except where otherwise stated, the spectra in this paper are calibrated relative to metallic iron.

Chemical shifts usually lie within the range $-0{\cdot}3$ to $+1{\cdot}4$ mm/sec. In terms of the energy of the nuclear transition this is very small—about 1 part in $10^{12}$—but the precision of the Mössbauer spectroscopic technique is such that it can readily be measured.

It has proved very difficult to predict chemical shifts from first principles. As already noted, the 3*d* electrons which are involved in formation of chemical bonds have negligible charge density at the nucleus. However they can have an indirect effect by shielding the 3*s* electrons from the nucleus. Thus in ionic compounds, $Fe^{2+}$ atoms, $3d^6$, have a more positive shift than $Fe^{3+}$, $3d^5$. If however covalent bonds are formed there is a mixing of 3*d* electrons with 4*s* orbitals and this produces an increased electron density at the nucleus together with a negative chemical shift. Therefore in compounds where the bonding shows covalency the two effects tend to cancel out, and both $Fe^{2+}$ and $Fe^{3+}$ show small chemical shifts.

Where a series of closely related iron compounds of known structure is available it is possible to use the chemical shift to estimate valence and spin states. However this is rarely possible in molecules of biological interest, where the iron atom is often contained in unusual, highly strained configurations for which there are few good chemical models, so that the chemical shift has only a limited usefulness. A possible exception is high-spin $Fe^{2+}$, which tends to have large positive chemical shifts—about 1 mm/sec.

## C. Quadrupole Splitting

The chemical shift occurs as a result of electrostatic interaction between the nucleus and electrons, and occurs even when the charge distribution due to the electrons is spherically symmetrical (such as a metal where all the atoms have cubic symmetry). Usually however, the electrical charge distribution about the nucleus is asymmetrical, giving rise to an *electric field gradient* (EFG). The $^{57}Fe$ nucleus in its ground state ($I = \frac{1}{2}$) has a symmetrical charge distribution so that it is unaffected by the EFG. However the excited ($I = \frac{3}{2}$) state has an electric quadrupole moment which interacts with the EFG to split the excited energy states (Figure 9.4b). As a result there are two possible nuclear transitions and the Mössbauer spectrum consists of two lines. The splitting between these lines is the quadrupole splitting $\Delta E_Q$.

As with the chemical shift, the quadrupole splitting is difficult to calculate theoretically, since it is the resultant of a number of competing effects. The EFG arises from two sources—charges on distant ions, and electrons in incompletely filled shells of the atom itself. These effects cause distortion of the wave functions of the core electrons, which usually results in an amplified EFG (a phenomenon known as antishielding).

High-spin $Fe^{3+}$, as seen in Figure 9.3, has five electrons distributed singly in the five $3d$ orbitals. This arrangement is spherically symmetrical and produces no EFG, so that the quadrupole splitting due to the valence electrons should be very small. High-spin $Fe^{2+}$ contains one more electron in addition to this spherically symmetrical subshell, and this gives rise to a large EFG. Because this electron can populate several orbitals of slightly greater energy the quadrupole splitting of high-spin $Fe^{2+}$ may vary considerably with temperature even down to 4·2°K. The temperature dependence of the quadrupole splitting in this case can give an indication of the ground-state orbital, and thus of the configuration of the ligands about the Fe atom.

In practice, high-spin $Fe^{2+}$ is the only form which can readily be distinguished in the Mössbauer spectra by its quadrupole splitting, often more than 3 mm/sec. By contrast, low-spin $Fe^{2+}$ and high- and low-spin $Fe^{3+}$ all tend to have small quadrupole splittings.

The EFG is a tensor, which in a suitable coordinate system can be reduced to three components, $V_{xx}$, $V_{yy}$ and $V_{zz}$ (or $\partial^2 V/\partial x^2$, $\partial^2 V/\partial y^2$ and $\partial^2 V/\partial z^2$). These components are not independent and can be reduced to two components, $V_{zz}$ and the asymmetry parameter $\eta$, defined by $\eta = (V_{xx} - V_{yy})/V_{zz}$. Axes are usually chosen so that $V_{zz} > V_{xx} \geqslant V_{yy}$ so that $\eta$ can take values between 0 and 1. When $\eta = 0$ the EFG is axially symmetrical and is defined by $V_{zz}$. The sign of $V_{zz}$ in the principal axis system is called the *sign of the field gradient* (or sign of the EFG) and can be determined from studies of magnetic splittings. The sign of the field can be useful in identifying the ground-state orbital.

### D. Magnetic Hyperfine Splitting

The nuclear hyperfine interaction is the interaction of a nuclear spin with a magnetic field. In an iron atom, this field can arise from unpaired electrons of the atom itself. In addition it is possible to apply an external magnetic field to the sample. The hyperfine interaction causes a hyperfine splitting of the nuclear energy levels which considerably increases the complexity of the

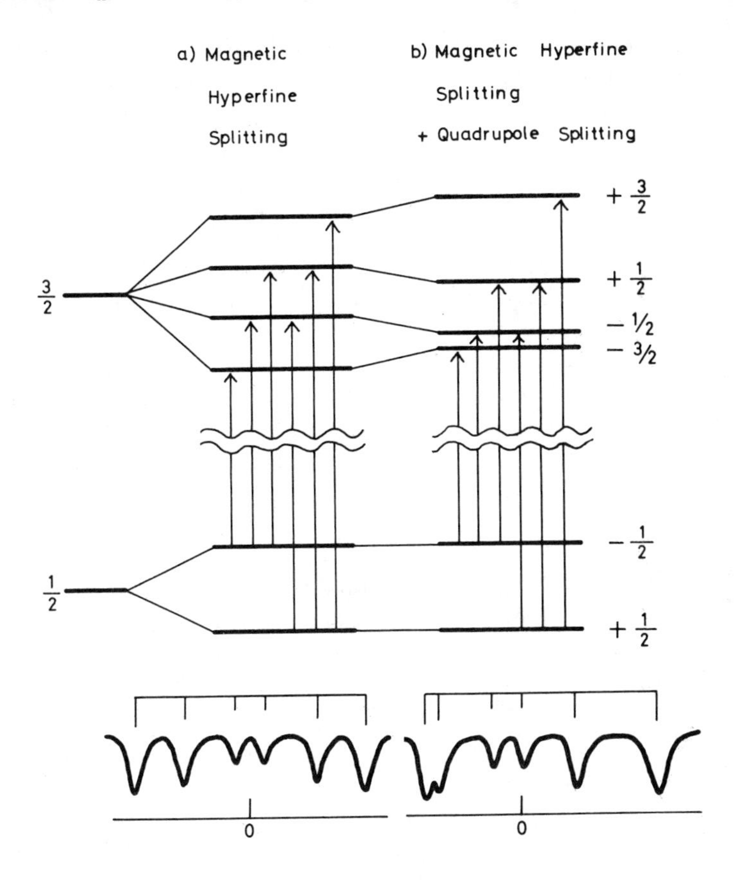


**Figure 9.5.** Effects of magnetic hyperfine splitting and quadrupole splitting. A magnetic field is assumed to lie parallel to the principal axis of the EFG

Mössbauer spectra, and the amount of information that can be obtained from them.

A magnetic field applied to a nucleus causes it to take up $(2I + 1)$ energy levels, i.e. 2 for the ground ($I = \frac{1}{2}$) state and 4 for the excited ($I = \frac{3}{2}$) state. When the $\gamma$-ray is absorbed by the sample, energy transitions are permitted between these levels with selection rules $\Delta m_I = 0, \pm 1$. Thus of the eight possible transitions, the six shown in Figure 9.5 are permitted, and in the absence of quadrupole splitting the spectrum will be a symmetrical pattern as shown in Figure 9.5(a). This effect is sometimes called *Zeeman splitting.* The width of the pattern is proportional to the magnetic field at the nucleus, which can be calculated from it.

In most cases the magnetic hyperfine splitting pattern is modified by the effect of the quadrupole splitting. Figure 9.5(b) shows the effect of a small quadrupole splitting on the energy levels, and the resulting shift in the six-line hyperfine pattern. The two outer lines of the spectrum arising from transitions to the $I = \pm \frac{3}{2}$ states of the excited nucleus are shifted in one direction, and the four inner lines from transitions to the $I = \pm \frac{1}{2}$ states are shifted in the other direction. The sigh of the EFG can be determined from this spectrum; if the $I = \pm \frac{3}{2}$ states are increased in energy relative to the $I = \pm \frac{1}{2}$ states, as in this case, the sign of the EFG is positive.

In samples where the quadrupole splitting is large the pattern may change completely and the lines broaden out. This is because the iron ions in the sample are randomly oriented, so that in each atom the applied field is in a different direction relative to the EFG. Even in such cases relatively sharp lines are often observed, since the probability $p(\theta)$ that the field makes an angle $\theta$ with the principal axis of the EFG is proportional to $\sin \theta$, which is a maximum for $\theta = 90°$. It is possible to simulate the spectra by computer calculation of the average interaction over all possible orientations of the Fe atom.

The above treatment is concerned with the interaction of a nucleus with an externally applied field, or the internal magnetic field due to a single unpaired electron. In high-spin iron complexes there are several unpaired electrons, and these are capable of giving further splitting. In the general case, interaction with an electron spin $S$ can split the nuclear energy levels into $(2I + 1)(2S + 1)$ states, but there are simplifying factors which make the situation less complicated than this. For example the high-spin $Fe^{3+}$ in oxidized rubredoxin has five unpaired electrons, which will couple together to give three possible states, with resultant spin $\pm \frac{5}{2}$, $\pm \frac{3}{2}$ and $\pm \frac{1}{2}$, (Note that this is not the same as low-spin $Fe^{3+}$; there are still five unpaired electrons.) These states differ in energy, so that at low temperatures the lowest state, with spin $\pm \frac{5}{2}$ is principally populated. The interaction of the nucleus with the spin $\pm \frac{5}{2}$ state of the electrons gives rise to the six-line pattern shown in the low-temperature spectrum of oxidized rubredoxin (Figure 9.12).

The magnetic field at the nucleus due to the atom's own electrons is zero in low-spin $Fe^{2+}$, which is non-magnetic, and can be very large in other states of the atom—up to 500 kG in high-spin $Fe^{3+}$. For example, the field at the nucleus of $Fe^{3+}$ in rubredoxin (Figure 9.12) is calculated to be 370 kG.

Often, magnetic hyperfine splitting is only observed at low temperatures. At higher temperatures only quadrupole splitting and the chemical shift are seen. This is because the electron-spin relaxation time must be long; if the electron is changing its spin direction too rapidly the magnetic hyperfine interaction will average to zero. This is analogous to the situation in electron paramagnetic resonance (EPR), where the electron-spin relaxation must be slow for signals to be observed. Rapid electron-spin relaxation is due mainly to two processes: spin–spin relaxation, which in iron-containing proteins is usually insignificant as they are magnetically dilute, and spin–lattice relaxation, which can be decreased by lowering the temperature. As a result many studies of the hyperfine interaction are made by measuring the spectra at liquid helium temperature, 4·2°K. EPR spectroscopy at these very low temperatures is often complicated by power saturation problems because the electron-spin relaxation is *too* slow, but in Mössbauer spectroscopy this problem does not arise. An additional advantage of using these very low temperatures is that it simplifies the use of superconducting magnets for the application of external magnetic fields to the sample.

The hyperfine interaction with the atom's own electrons is sometimes not observed in zero field. When an external field is applied, an induced hyperfine splitting is observed which is much greater than would be expected from the direct effect of the applied field on the nucleus, and the direction of this induced field is such that it opposes the applied field. The example considered by Johnson (1971), which has bearing on the zero-field spectra of the two-iron ferredoxins, is of an iron ion with electron spin $S = \frac{1}{2}$. When no external field is applied, the electron spin $S$ and the nuclear spin $I$ will couple together to give a total spin angular momentum $F = S + I$, which results in an asymmetrical three-line spectrum. When a small field of about 0·1 kG is applied, the spins become decoupled from each other and align with the external field, so that the normal six-line hyperfine pattern is resolved. The direction of the applied field has an effect on the relative intensities of the lines in this spectrum; if the field is perpendicular to the direction of the $\gamma$-rays, the lines are in the ratio 3:4:1:1:4:3; if it is parallel to the $\gamma$-ray, the ratio is 6:0:2:2:0:6.

In high-spin $Fe^{2+}$ the electrons are often coupled in such a way that the ground state is a singlet, which can have no magnetic moment in the absence of an applied magnetic field. In order to see hyperfine interaction, an external field must be applied to cause a mixing of the electronic states to produce states with a magnetic moment. In contrast to the previous case, where a small field is sufficient, the fields needed in this case are of the order of several kilogauss.

There is a close relationship between the magnetic hyperfine splittings observed in Mössbauer spectroscopy and EPR spectroscopy. The magnetic moment of the electrons causes a field at the nucleus given by

$$H_{\mathrm{n}} = \frac{A}{2g_{\mathrm{n}}\beta_{\mathrm{n}}}$$

which causes the splitting observed in the Mössbauer spectrum, while the magnetic moment of the nucleus causes a field at the electrons given by

$$H_{\mathrm{e}} = \frac{A}{2g\beta}$$

which causes the EPR lines to be split by a factor $2H_e$ (g and $g_n$ are the electron and nuclear $g$-values respectively, $\beta_n$ the nuclear magneton and $\beta$ the Bohr magneton). Generally $2H_e$ is smaller than $H_n$ by a factor of about $10^4$ for $^{57}Fe$, so that hyperfine splittings in the EPR spectra are of the order of 20 G. $A$, the magnetic hyperfine coupling constant, which is common to both expressions, is a tensor, which can be reduced to three components in a suitable coordinate system. These components cannot readily be measured in EPR spectra of randomly orientated samples, but in favourable circumstances can be measured by electron-nuclear double resonance (ENDOR). (Fur further reading on EPR and ENDOR techniques, see, e.g. Ingram, 1969).

It is possible to describe the magnetic hyperfine interaction between the nucleus, electrons and applied magnetic field in the form of a spin hamiltonian (see Lang, 1970). The advantage of this approach is that it is possible to correlate experimentally derived parameters from other techniques such as EPR, ENDOR and magnetic susceptibility, with the Mössbauer spectroscopic data.

## III. EXPERIMENTAL TECHNIQUES

Figure 9.6 shows the layout of a Mössbauer spectrometer suitable for work with $^{57}Fe$. The various components will now be discussed, with special reference to the requirements of biological work. The previous section has indicated that it is desirable to be able to measure Mössbauer spectra over a wide range of temperature from liquid helium to room temperature, and in variable applied magnetic fields of up to 50 kG. Since biological samples are normally very dilute in iron by chemical standards, the apparatus must be designed for maximum sensitivity—for example, the $\gamma$-ray counter should have a high efficiency, and the electronics should be stable to permit long periods of counting.

### A. Source

The $\gamma$-ray source should contain $^{57}Co$ in a suitable environment so as to give an unsplit emission line, with line width as close as possible to the natural line width for maximal resolution of the spectral lines. The recoil-free fraction $f$ should be as high as possible. Since the error in the observed Mössbauer absorption is determined by the number of counts, the source should also be of high activity. A suitable source is provided by $^{57}Co$ diffused into a foil of a diamagnetic metal in which Co enters substitutionally, such as Pd, Cu or Cr. The source should be very thin to prevent reabsorption of $\gamma$-rays. Strong Pd sources are relatively easy to prepare, and have narrow line widths; an example is the 100 mCi source in Pd foil 0·00625 mm thick (line width 0·21 mm/sec) produced by the Radiochemical Centre, Amersham,

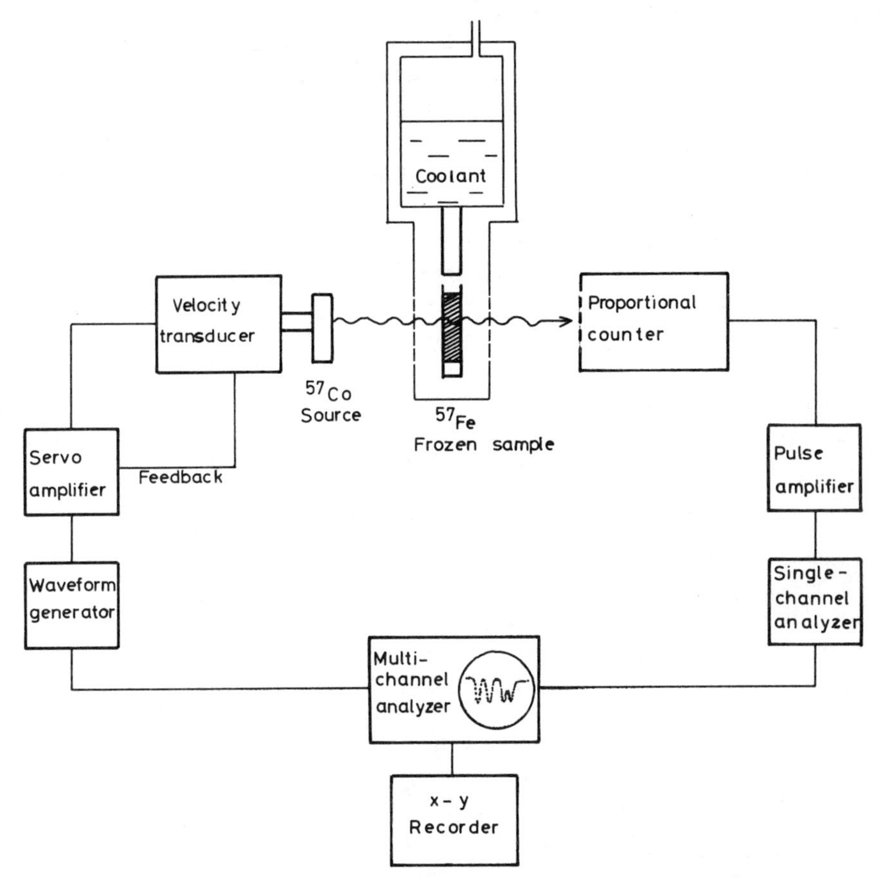


**Figure 9.6.** Diagram showing the major components of a typical Mössbauer spectrometer

Bucks, U.K. The cost of such high-activity sources may be as much as £1000. The frequency with which the source must be replaced is determined by the 270-day half-life of $^{57}$Co. A disadvantage of Pd sources is the emission of a 21 keV characteristic X-ray. In this respect Cr is a better host metal, though such sources are more difficult to prepare.

## B. Source Drive Unit and Multichannel Analyser

An unusual feature of the Mössbauer spectrometer is the system for modulating the velocity of the source in a reproducible fashion, while the $\gamma$-ray counts corresponding to each velocity range are added up in a multichannel analyser. Because of the high reproducibility required over counting times of hours or days, mechanical systems using constant velocity drives or cam-driven parabolic motion have now mostly been superseded by electromechanical systems somewhat similar to a high-fidelity loudspeaker system, in which an electronic drive unit is linked to an electromechanical transducer, which causes the vibration of the source. Incorporated in the transducer is a second coil which monitors the velocity of the source, and provides feedback to the drive unit.

In one type of drive unit, pulses from a crystal oscillator are used to trigger a parabolic motion of the transducer, so that the velocity of the source varies linearly with time over each velocity sweep. The pulses are also used to synchronize a multichannel analyser, which transfers counts to a number of channels (usually 100–512) in sequence throughout the velocity sweep. Thus each channel will accumulate the counts corresponding to a particular velocity. Another method is to incorporate the velocity drive into the multichannel analyser; the channels are switched at rapid intervals, and simultaneously small increments are made in the velocity of the transducer. In this way the address of a channel effectively determines the velocity at which the counts are taken.

During accumulation of counts the Mössbauer spectrum can be displayed on an oscilloscope as a plot of counts *versus* channel number. After a sufficient number of counts has been accumulated the spectrum can be plotted on an $x$–$y$ recorder, or alternatively transferred to punched tape for computer analysis. The computer can process the spectrum, for example converting counts into % absorption, correcting for baselines and velocity calibration, etc. There is an increasing use of minicomputers in these systems, which besides acting as a multichannel analyser can carry out subsequent operations on the spectral data. (See, for example, Chapter 8 in volume 1 of this series).

## C. Sample Holder and Cryostat

The technique of sample preparation will be described in a later section. In order to make measurements at low temperatures the sample holder is mounted in a Dewar system or cryostat. Suitable coolants are 'Freon' (for temperatures above 200°K), solid $CO_2$ (198°K), liquid nitrogen (boiling point 77°K) or liquid helium (boiling point 4·2°K). Temperatures down to about 1·2°K can be obtained by reducing the pressure over liquid helium with a vacuum pump. In the arrangement shown in Figure 9.6, the sample is in a vacuum, mounted on a brass or copper rod in thermal contact with the coolant vessel. For work at liquid helium temperatures the cryostat usually contains several radiation shields. Such a system is very economical of coolant. The temperature of the sample may be maintained above that of the coolant by means of a small electric heater. An alternative system which allows greater variability in temperature is to place the sample in a gas stream which is cooled by a suitable coolant then heated to the desired temperature by an electric heater controlled by a thermocouple.

The $\gamma$-rays pass through the sample *via* windows of Mylar and heat shields of aluminium foil. To minimize attenuation of the beam the Dewar should allow the source and counter to be mounted close to the sample. Other important practical considerations are that the cryostat system be designed to minimize vibration of the sample, which causes line broadening, and that it should be easy to insert and remove frozen samples without danger of the solution thawing out.

To apply small magnetic fields to the sample, small permanent magnets can be used. At liquid helium temperatures small fields, of the order of 100 G, can be obtained by means of Pb washers mounted near the sample. These are superconducting at 4·2°K and can be magnetized by passing a permanent magnet over them.

Since small magnetic fields can have a considerable effect on the Mössbauer spectrum, it is important to avoid them when they are not wanted. For example, the Dewar near to the sample should be of brass rather than stainless steel, which can develop ferromagnetic regions at low temperature. Soft solder joints should also be avoided where they can become superconducting and develop ring currents.

Superconducting magnets for the application of large fields may be either external to the sample cryostat, or built in. Normally the magnet is mounted in a fixed direction, though systems have been designed using split-coil solenoids around the sample so that the $\gamma$-rays can be passed either perpendicular or parallel to the magnetic field. Where the magnet is mounted inside the Dewar there is less flexibility of temperature though the sample may be contained in a separate internal Dewar system so that its temperature can be varied relative to the liquid helium temperature of the magnet.

## D. Counting System

To detect the 14·4 keV $\gamma$-rays transmitted by the sample, proportional counters containing 10% methane and 90% argon or xenon are normally used. The characteristics required of a suitable counter are discussed by Lang (1970). The $^{57}$Co emits a flux of 122 keV $\gamma$-rays (see Figure 9.1) which is 20–40 times more intense at the counter than the 14·4 keV $\gamma$-rays. These are of too high an energy to be detected directly by the counter, but they Compton scatter in the counter gas and produce a background of radiation in the 0–40 keV region which, if counted, would considerably decrease the apparent Mössbauer absorption and increase the statistical error of measurement. The purpose of the single-channel analyser is to select the 14·4 gammas from this background as accurately as possible, with high efficiency and at high counting rates. The system described by Lang (1970) has an overall efficiency of about

20% with a non-14·4 keV fraction of about 20%. The 14·4 keV fraction still contains about 30% of $\gamma$-rays which were emitted with recoil and cannot contribute to the Mössbauer absorption. The 14·4 keV recoil-free fraction can be detected specifically by resonant detectors, in which the stopping material is $^{57}Fe$, but these have an efficiency of only a few percent. Obviously in terms of increasing the sensitivity of Mössbauer spectrometers there is scope for improvement in the detecting and counting systems.

## E. Availability of Equipment

In the early days of Mössbauer spectroscopy the apparatus used was normally constructed by the experimenter, often specially designed for a particular purpose. A number of the components of a spectrometer such as counters, cryostats and multichannel analysers are fairly standard pieces of equipment. In recent years components specially designed for Mössbauer spectroscopy, and some complete systems, have become available commercially. A list, which is probably not exhaustive, of suppliers of such equipment is given in Appendix 1. It is important to determine that the apparatus is suitable for the special requirements of biological work, of which the most important is a high sensitivity for $^{57}Fe$.

The cost of Mössbauer spectrometers is not excessive when compared, say, with magnetic resonance spectrometers. In round figures, the transducer and drive unit may cost about £1000; the counter, amplifiers and a fairly basic multichannel analyser, about £4000, and a cryostat system about £1000. A separate superconducting magnet for fields of up to 60 kG can be obtained for about £5000. A strong source, as already mentioned, may cost up to £1000, and must be renewed every year. Thus the initial cost of a Mössbauer system, excluding peripheral equipment such as computers, is in the region of £6000–£12,000. (*N.B. All prices refer to 1972*).

## F. Computer Analysis of Data

The raw Mössbauer spectrum produced by a multichannel analyser usually appears as a series of dots, each one corresponding to the counts in a single channel. Quite often the dots are joined by a smooth curve which is sometimes simply drawn in as a guide to the eye but which is usually a computer-fitted curve.

The simplest way to fit a Mössbauer spectrum is to represent it as the sum of $n$ Lorentzian curves. The computer is then programmed to find the curve which gives the best fit to the points, by adjusting the $3n$ variables—the position, depth and width of each of the Lorentzians. This method is useful in simple quadrupole-split spectra, in order to determine the values of the chemical shift $\delta$ and quadrupole splitting $\Delta E_Q$.

In the more complex case of spectra of paramagnetic complexes in which there is hyperfine splitting, as already mentioned, it is useful to describe the interactions in the form of a spin hamiltonian, so that data from other techniques can be used in simulating the spectrum. Then one can assume a model for the state of the Fe in the molecule, and, using the spin hamiltonian, attempt to fit the observed Mössbauer spectra. There are many variables—often more than twenty—which can be adjusted, so it is necessary to determine the values of as many of these as possible by some other means, either by experiment or theory. Then if the computer-fitted spectrum fits the observed data well, it is evidence for the correctness of the model. Further confirmation of the model is achieved if the variables derived to fit one spectrum give a good fit to another spectrum measured under different conditions of temperature and applied magnetic field. All the same, the difficulty lies in proving that a computer fit is unique—all it can do is demonstrate that a model is *consistent* with the observed results, though there may be another model which fits the data equally well.

## G. Sample Preparation

The most careful Mössbauer spectroscopic measurements and interpretation are useless if the sample being studied is of poor quality. It is therefore essential to give careful consideration to the special problems involved in preparation of samples.

Many of the problems arise from the low concentration of $^{57}$Fe in most biological materials, both because the iron atom is usually contained in a large molecule, and because natural iron contains only 2·2% $^{57}$Fe. The statistics of radioactive counting mean that the time required for a measurement of a certain degree of accuracy is inversely proportional to the *square* of the $^{57}$Fe concentration.

As an example, Lang and Marshall (1966) considered the comparatively favourable case of haemoglobin, which contains one iron atom per subunit of molecular weight 17,000. A sample of packed oxygenated red cells contains about 30% haemoglobin by weight—a higher concentration than is possible in ordinary solution. The spectrum (Figure 9.7) consists of a quadrupole-split doublet, and calculations showed that at natural abundance of $^{57}$Fe the expected peak absorption would be about 2·5%; the experimental value was 2·2%. To measure the spectrum with an accuracy of 5% would typically require counting for about 6 hours, which is not unreasonable. However in the less favourable case of methaemoglobin, where the spectrum is decreased in depth by a factor of about 20 by magnetic hyperfine interaction, and the most concentrated solution conveniently obtainable is 10%, to obtain a spectrum of the same accuracy would take 2·5 years. As these authors concluded, the need for enrichment is obvious.

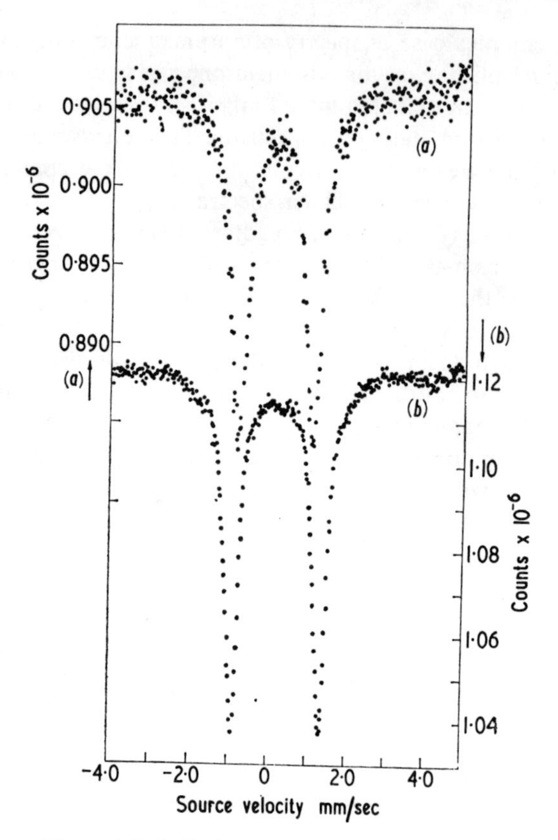


**Figure 9.7.** Mössbauer spectra of rat oxyhaemoglobin (a) at 195°K; (b) at 1·2°K, showing the temperature dependence of the quadrupole splitting. (Reproduced, with permission, from Lang and Marshall, *Proc, Phys. Soc.*, **87**, 3 (1966))

Although work has been done on a range of biological materials containing $^{57}Fe$ at natural abundance, in most cases it is well worth trying to enrich the sample with $^{57}Fe$ isotope. Replacement of natural iron by 90% enriched $^{57}Fe$ would result in a decrease in running time by a factor of about 1700, and considerably decrease the difficulties due to iron contamination in the Dewar windows, heat shields and counter windows.

The methods which can be used for $^{57}Fe$ enrichment are limited by the high cost of the isotope—about £3 per milligram for the metal at 90% enrichment. The most satisfactory method of incorporating the iron is by growing an organism under conditions where the natural iron is replaced by enriched $^{57}Fe$ isotope, then extracting the required material. This method has been very successful with other isotopes; for example, to obtain milk xanthine oxidase enriched with $^{95}$Mo, Bray and Meriwether (1966) injected a cow with a solution of the isotope, in sufficient quantity to swamp the animal's store of molybdenum, milked the cow and were then able to extract xanthine oxidase approximately 75% enriched in $^{95}$Mo. This method could not be used to enrich the xanthine oxidase in $^{57}$Fe, since the cow probably contained about 30 g of iron, which would dilute out any injected $^{57}$Fe to an impossible degree. The problems, then, are to ensure that the natural $^{57}$Fe in the organism is diluted out by the $^{57}$Fe, and that a large proportion of the $^{57}$Fe is converted into the desired form.

The process of incorporating $^{57}$Fe is easiest with microorganisms, where the isotope can be introduced in the growth medium, and large quantities of material can be grown from a small quantity of cells, so that the natural iron in the initial inoculum can be diluted out until it is negligible. Animals are more difficult, as they contain considerable reserves of iron, and it is necessary to deplete these before introducing $^{57}$Fe. The enriched haemoglobin studied by Lang and Marshall (1966) was from rats that had been made anaemic by growth on an iron-free diet, and were then periodically injected with ferric citrate containing 80% enriched $^{57}$Fe. The enrichment of the iron in the haemoglobin was estimated at about 50%. In a similar way plants may be made chlorotic by growth in water culture in a nutrient solution lacking iron; on introduction of $^{57}$Fe to the medium, new growth often takes place of leaves and root systems in which a high proportion of the iron is $^{57}$Fe. In all work of this type it is very important to exclude natural iron, which would cause isotopic dilution of the $^{57}$Fe. Therefore precautions must be taken to remove iron from all media, glassware and other apparatus. The extent of enrichment with $^{57}$Fe may be estimated by introducing $^{59}$Fe as a radioactive tracer with the $^{57}$Fe, then measuring the specific activity of the isolated protein.

The second problem with introducing $^{57}$Fe into an organism is that the organism may convert most of it into unwanted forms. For example, of the total iron in an animal, about two-thirds is in the form of haemoglobin which is therefore a good subject for this method. By contrast, all enzymes, cytochromes and iron–sulphur proteins total less than 2% of the iron. Many iron-containing enzymes require kilogram quantities of starting material such as liver for their extraction, so that growth on $^{57}$Fe is plainly out of the question. The alternative is to isolate the protein containing natural iron, then enrich it chemically by removal of the natural iron and replacement with $^{57}$Fe in a suitable form. It is possible to enrich many iron–sulphur proteins by removing the iron and labile sulphide by precipitation with trichloroacetic acid, then reconstituting it with $^{57}FeSO_4$ and sulphide under anaerobic conditions in the presence of a thiol. After such drastic treatments it is important to check that the protein has not been irreversibly altered in its properties, by means of spectroscopic properties such as optical absorption, circular dichroism and EPR, and preferably by some kind of enzymic assay. Another problem arises from the fact that the proteins contain other groups with chelating properties, so that they are often very 'sticky' for extra atoms of $^{57}$Fe. These extra iron atoms will contribute proportionately to the Mössbauer spectrum and cause confusion. After repurification of the reconstituted protein its iron content should therefore be measured by colorimetric analysis or atomic absorption spectrometry.

In order for the Mössbauer spectrum to be observed the sample must be solid, either as a frozen solution or a freeze-dried powder. The assumption must be made that the structure of the sample molecules in the solid is the same as in solution, and in most cases there is no way of demonstrating that this is true. When the samples are thawed or redissolved afterwards they are usually found to be unchanged in their properties, but that is not the same thing. Evidence with haem proteins at least indicates that freezing of the sample causes less distortion of the protein molecules than freeze drying. Therefore frozen solutions are to be preferred, though they result in more dilute samples.

The protein sample must be very concentrated. Depending on the width of the spectrum a concentration of 0·5–5 m*M* of each species of $^{57}$Fe is required for accurate results. For a protein containing one $^{57}$Fe atom per 10,000 molecular weight this represents 5–50 mg protein in a sample cell of volume 1 ml. Concentrations of this magnitude are commonly required for other techniques such as magnetic resonance, but are several orders of magnitude higher than those used in most enzyme studies. It may be unwise to assume that the behaviour of the protein is the same as that in dilute solution; for example the catalytic behaviour of an enzyme may be different when the enzyme concentration is comparable with that of the substrate; aggregation of protein molecules may also take place. Protein solutions of suitable concentration may be obtained in many ways, including ultrafiltration, drying with a stream of dry gas and salt precipitation. A simple point which is sometimes overlooked is to ensure that the pH is adequately buffered to prevent denaturation. Salts and buffers should not contain heavy atoms, which have a strong electronic absorption of $\gamma$-rays; ammonium is a better cation than potassium for example. Bromide should be particularly avoided, as the Br nucleus has a strong K absorption of X-rays at 14·4 keV.

A suitable holder for frozen samples is a polyethylene cell of 3–5 mm thickness, holding 0·7–1 ml of solution, as illustrated in Figure 9.8. The sample should be clearly labelled. Solutions can be frozen in liquid nitrogen, and it is advisable to store the cells at liquid nitrogen temperature if the sample is an unstable protein. When transferring to the Dewar of the Mössbauer apparatus some quick work may be necessary to avoid melting the sample.

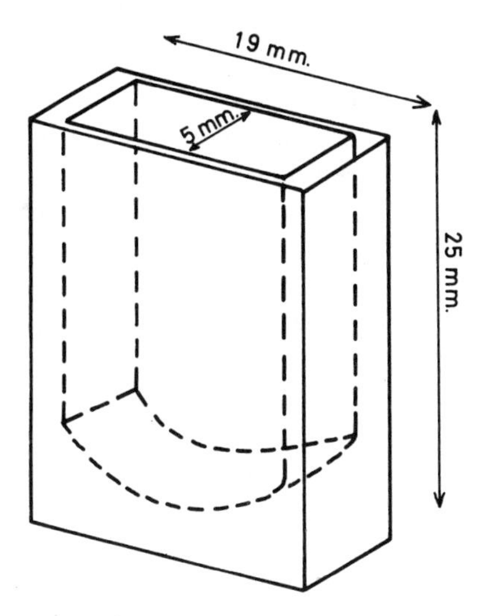


**Figure 9.8.** A suitable holder for frozen protein samples, made of polythene or acrylic plastic

## IV. PRACTICAL APPLICATIONS

This section is not intended to be a review of all biological Mössbauer work, but to give some examples of work on compounds which are of direct biological interest, and which have been selected to give an idea of the scope of the method as it has so far been applied. Possible ways in which the method may be extended in the future will also be indicated.

### A. Measurement of Small Vibrations

Not all biological applications of the Mössbauer effect are concerned with chemistry. Another ingenious application has been to measure vibrations of small amplitude. For example Rhode (1971) used the $^{57}$Fe Mössbauer effect to measure vibrations in the basilar membrane of the ear of the squirrel monkey. To do this, a tiny $^{57}$Co source was mounted surgically on the membrane. 14·4 keV $\gamma$-rays from this source were counted by a proportional counter through a $^{57}$Fe absorber, mounted externally to the ear. When the source was at rest, most of the $\gamma$-rays were absorbed, but when the source was moving, their energy was shifted and fewer were absorbed, so that vibration was accompanied by an increase in count rate. The counts were measured while sound of a particular frequency was applied to the ear, and by the use of a computer it was possible to determine both the amplitude and phase of the vibration of the membrane. Because of the inaccessibility of the vibrating system, this type of measurement on a living system would have been almost impossible by other means.

Bonchev *et al.* (1968) investigated small abdominal movements of the ant using the $^{119}Sn$ Mössbauer effect. In contrast to the experiments just described, the absorber rather than the source was attached to the vibrating system; $^{119}Sn_2O_3$ powder was stuck to the abdomens of the ants and fifty of them were mounted in cages in a Mössbauer spectrometer. In this case the vibration resulted in a broadening of the absorption line, which was a function of the average speed of vibration. The broadening observed (of the order of 3 mm/sec) was attributed to the ants' breathing, and flickering movements of the abdomen.

## B. Haem Proteins

In these compounds the Fe atom is bound to four planar nitrogen ligands of a porphyrin ring, and two ligands perpendicular to the haem plane. The symmetry of the Fe atom is thus perforce octahedral, with varying degrees of tetragonal distortion. This can be considered as a stretching or squashing of the ligands along the *z* direction (Figure 9.3), and gives rise to characteristic splittings of the orbital states. It is a feature of the haem compounds that the state of the Fe atom is strongly influenced by the nature of the 5th and 6th ligands. Thus in haemoglobin the $Fe^{2+}$ atom changes from high-spin to low-spin when the $H_2O$ in the 6th ligand position in deoxyhaemoglobin is replaced by $O_2$. This conversion may have considerable effects on the properties of the protein, including the well-known cooperativity between subunits.

Much of the ensuing discussion is based on the review by Lang (1970) and the references therein.

### 1. *Haemoglobin*

This protein, being of obvious vital importance, has probably been the subject of more intensive study than any other metalloprotein, and the results have frequently been unexpected. The Mössbauer effect has also shown that the molecule has some unusual properties.

Oxyhaemoglobin, the oxygen-carrying protein of arterial blood, contains low-spin $Fe^{2+}$. Magnetic susceptibility shows that the molecule is diamagnetic, and the Mössbauer spectra show no magnetic hyperfine structure. Contrary to expectation from simple theory, the quadrupole splitting is large and temperature dependent, varying from 2·2 mm/sec at 1·2°K to 1·9 mm/sec at 195°K (Figure 9.7). Lang and Marshall (1966) showed that this is consistent with the $O_2$ molecule lying parallel to the haem plane, with $\pi$-bonding to the iron. With the small amount of data available, however, this explanation would not be expected to be unequivocal, and Lang (1970) has pointed out

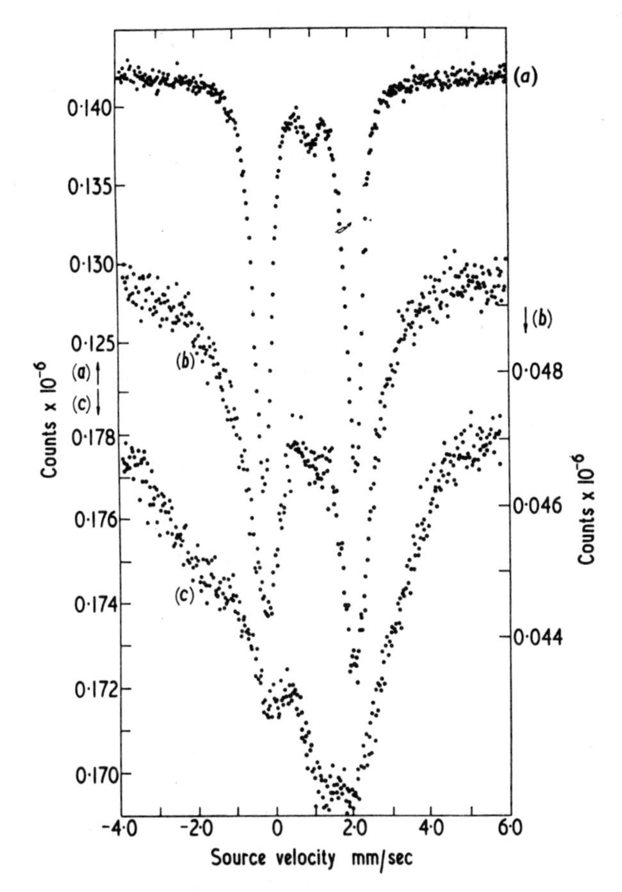


**Figure 9.9.** Mössbauer spectra of rat deoxyhaemoglobin in dithionite-reduced red cells, at 4°K, (a) in zero magnetic field; (b) in a field of 7·5 kG applied perpendicular to the $\gamma$-rays; (c) in a field of 30 kG applied perpendicular to the $\gamma$-rays. (Reproduced, with permission, from Lang and Marshall, *Proc, Phys. Soc.*, **87**, 3 (1966))

that the spectra could also be explained by the oxygen atom standing on end, perpendicular to the haem plane, with strong $\sigma$-bonding to the iron atom.

Haemoglobin carbon monoxide also contains low-spin $Fe^{2+}$, but its Mössbauer spectra conform more closely to expectation; the quadrupole splitting is small (0·4 mm/sec) and shows little temperature dependence.

In deoxyhaemoglobin, the protein of venous blood, the iron is high-spin $Fe^{2+}$, and, as expected, the quadrupole splitting is both large and temperature dependent (Figure 9.9). No magnetic hyperfine splitting is seen in zero field, indicating that the ground state is a singlet, and hence non-magnetic. Application of a field of 30 kG produces a broadening of the spectrum, indicating an internal field of about 100 kG, in agreement with rough theoretical estimates.

The haemoglobin molecule can form a number of derivatives which, though not normally present in nature, are of interest from the point of view of the chemistry of haem proteins generally. The iron atom may be oxidized to $Fe^{3+}$, forming methaemoglobin, and the 6th ligand position normally occupied by $O_2$ replaced by $CN^-$, $N_3^-$, NO, $F^-$ or other ligands. The nature of this 6th ligand causes a considerable variation in the spin state of the iron atom (Winter *et al.*, 1972), and some of these derivatives can be used as models for other haem proteins. For example, data obtained from low-spin $Fe^{3+}$ in haemoglobin cyanide were used in the interpretation of the spectra of cytochrome c (Lang *et al.*, 1968). The ferric compounds all show magnetic hyperfine splitting, and with the increase in complexity the spectra become potentially much more informative. Haemoglobin nitric oxide represents an interesting case, where the state of the Fe is in some doubt. Lang and Marshall (1966) suggested that the covalent bonding in this case may be so strong that the distinction between $Fe^{2+}$ and $Fe^{3+}$ assignments becomes blurred.

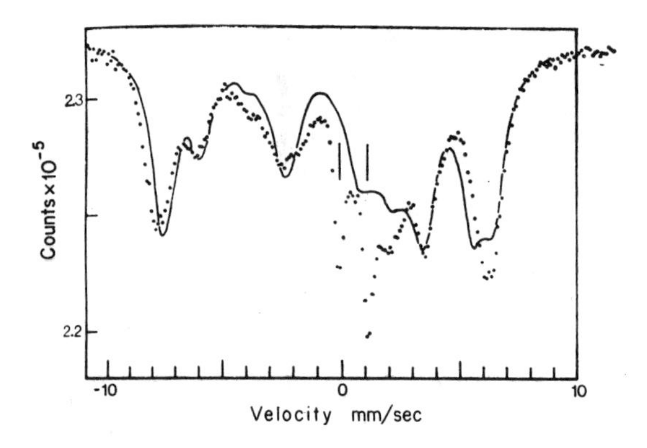


**Figure 9.10.** Mössbauer spectrum of oxidized cytochrome c of *C. utilis* at 1·8°K. Vertical lines mark the positions of absorption lines caused by a small amount of reduced cytochrome. The solid curve is a calculated spectrum containing no free parameters except line width, which was taken as 1·2 mm/sec in order to represent relaxation broadening. The *g*-values of horse heart cytochrome c were assumed to apply. (Reproduced, with permission, from Lang, Herbert and Yonetani, *J. Chem. Phys.*, **49**, 944 (1968))

## 2. *Cytochrome c*

The oxidation and reduction of this protein in the mitochondrial respiratory chain involve interconversions between the low-spin $Fe^{3+}$ and $Fe^{2+}$ states. Mössbauer spectra were recorded by Lang *et al.* (1968) on the protein isolated from *Candida utilis* grown on a medium containing 90%-enriched $^{57}Fe$.

The spectrum of oxidized cytochrome c shows magnetic hyperfine splitting at low temperatures in the absence of an applied field (Figure 9.10). This spectrum and others in applied fields were computer fitted (solid line in Figure 9.10), and it was found that the positions of lines in the simulated spectra were strongly influenced by the *g*-values. The *g*-values which gave the best fit were closely similar to those obtained from EPR spectra of horse heart cytochrome c (see Lang *et al.*, 1968).

Reduced cytochrome c contains low-spin $Fe^{2+}$ so the spectrum shows no hyperfine splitting due to unpaired electrons. The spectrum in zero field contains two sharp lines, and the quadrupole splitting is small and temperature

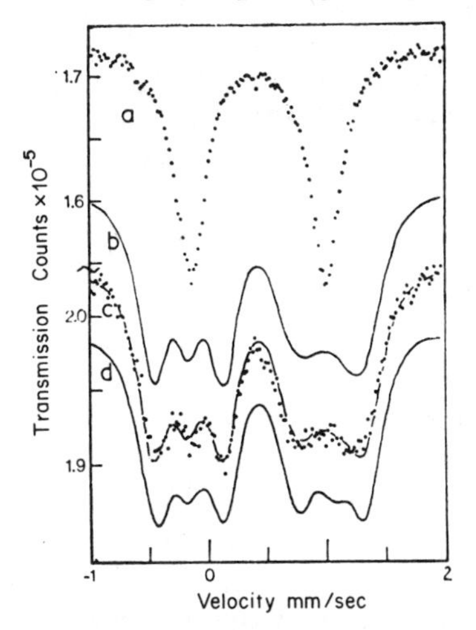


**Figure 9.11.** Mössbauer spectra of reduced cytochrome c of *C. utilis*, at 4·2°K, (a) in zero applied field and (c) with a field of 25·7 kG applied perpendicular to the $\gamma$-rays. The solid curves are calculated spectra with the asymmetry parameter set to (b) 0·0, (c) 0·5 and (d) 0·8. (Reproduced, with permission, from Lang, Herbert and Yonetani, *J. Chem. Phys.*, **49**, 944 (1968))

insensitive. The splitting on applying a large magnetic field (Figure 9.11) is due to the direct effect of the applied field on the nucleus. The computer-simulated curves show that it is possible to determine from the shape of the spectrum the sign of $V_{zz}$ and the value of $\eta$, the asymmetry parameter. From the observed spectrum the sign of the EFG is positive, and the value of $\eta$ is about 0·5, though this value is subject to a considerable degree of uncertainty because of statistical error in the spectrum.

## C. Iron–Sulphur Proteins

These are a diverse group of proteins, characterized by the presence of iron and (with the exception of rubredoxin) 'labile sulphide' which are released upon mild acidification. The proteins generally have highly reducing redox potentials and are involved in electron transport processes in mitochondria, chloroplasts and in many bacterial enzyme systems such as nitrogen fixation. The relative instability of these proteins, and the lack of distinct prosthetic groups such as haem which are amenable to chemical analysis have meant that the structure of the iron–sulphur group and the state of the iron atom are not readily determined. The proteins are difficult to crystallize, so that X-ray Fourier analysis has so far only been possible in a few cases, when it has been found that each iron atom is coordinated to four sulphur atoms in approximately tetrahedral symmetry.

### *1. Rubredoxin*

This protein is found in a number of bacteria, mostly anaerobes. It is the simplest of the iron–sulphur proteins, and in many ways the best understood chemically. Unfortunately its function is unknown. Normally the rubredoxin molecule consists of a protein of molecular weight about 6000 containing one iron atom, which undergoes oxidation and reduction. The amino acid sequences of several of the proteins are known. X-ray crystallography of *Micrococcus aerogenes* rubredoxin (Watenpaugh *et al.*, 1971) showed that the Fe atom is coordinated approximately tetrahedrally to four sulphur atoms from cysteine residues in the protein. Phillips *et al.* (1970) showed by magnetic susceptibility that the iron atom is high-spin $Fe^{3+}$ and $Fe^{2+}$ in the oxidized and reduced forms respectively, as would be expected from tetrahedral symmetry. The oxidized rubredoxin shows an EPR signal at $g = 4{\cdot}3$ which again is indicative of $Fe^{3+}$ in a strongly distorted tetrahedral ligand field (Peisach *et al.*, 1971).

The Mössbauer spectrum of oxidized rubredoxin at 4·2°K shows an almost symmetrical six-line pattern (Figure 9.12), indicating a hyperfine field at the nucleus of $370 \pm 3$ kG, and a small quadrupole splitting. This is typical of high-spin $Fe^{3+}$, as previously noted.

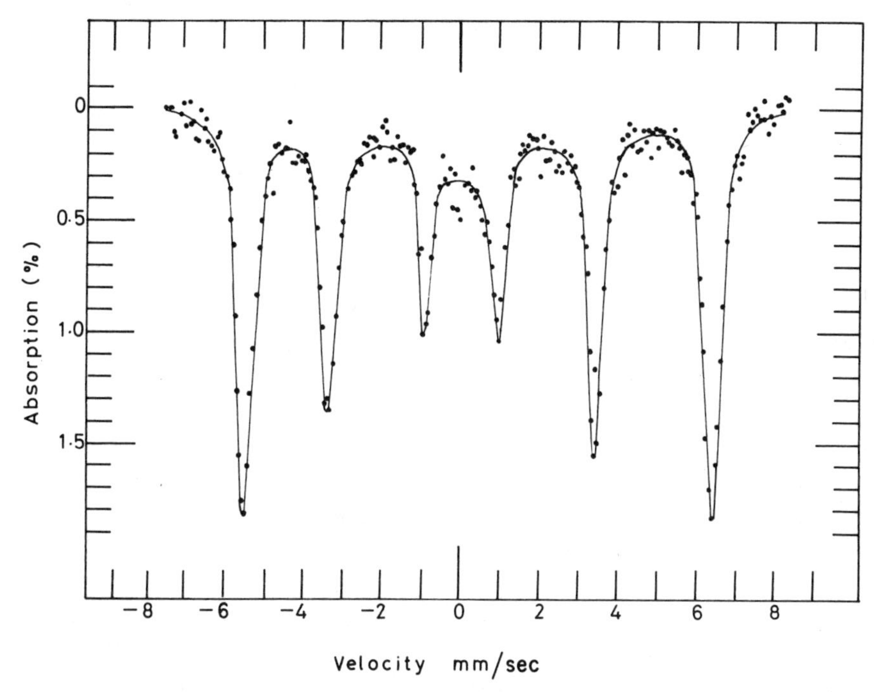


**Figure 9.12.** Mössbauer spectrum of oxidized rubredoxin from *Chloropseudomonas ethylica*, at 4·2°K. (After Rao *et al.*, 1972)

The spectrum of reduced rubredoxin consists of two lines with chemical shift 0·65 mm/sec and quadrupole splitting 3·1 mm/sec and is almost independent of temperature between 1·4 and 195°K, indicating that the ground state of the electrons is pure and lies well below the level of the first excited state. No magnetic hyperfine splitting is seen in zero applied field, but on application of a field of 30 kG an internal field is induced (see p. 353). The lines in the spectrum, Figure 9.13, are fairly sharp because, as previously noted, there is a tendency in a randomly oriented sample for the effective field to be directed perpendicular to the axis of symmetry of the ion. The effective field at the nucleus in Figure 9.13 determined from the width of the spectrum is — 133 kG. This means that since the applied field of 30 kG will be opposed by the field due to the unpaired electrons, the value of the latter is −163 kG. The value of the internal field increases with applied field, and the saturation value of the field $H_n$ perpendicular to the axis of symmetry is estimated to be −210 kG. From the behaviour of the lines in the field the sign of $V_{zz}$ is negative, which identifies the ground state orbital as $d_{z^2}$. This would be the case if the effective distortion of the four sulphur ligands from

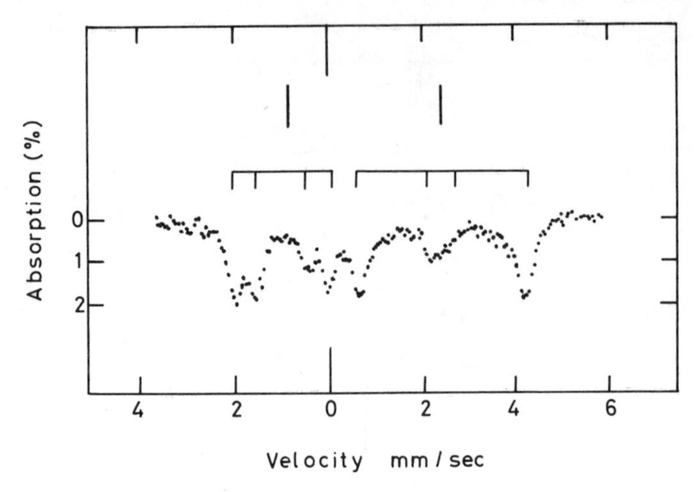


**Figure 9.13.** Mössbauer spectrum of reduced rubredoxin from *Ch. ethylica*, at 4·2°K with a field of 30 kG applied perpendicular to the $\gamma$-rays. Vertical lines indicate the position of the two lines in the zero-field spectrum. (After Rao *et al.*, 1972)

tetrahedral symmetry were equivalent to a squashing about a tetragonal direction (Rao *et al.*, 1972).

As will be seen later, the results obtained from Mössbauer spectroscopy of oxidized and reduced rubredoxin—the chemical shifts, quadrupole splittings and hyperfine fields—are valuable because they provide basic data on these states of the iron atom, with which the results from more complex iron–sulphur proteins may be compared.

*2. The two-iron ferredoxins*

This group of proteins form a widely distributed class; they are sometimes called 'plant type' since the first one isolated (spinach ferredoxin) is a component of the photosynthetic electron transport chain. Subsequently similar proteins were isolated from other sources, including adrenodoxin from adrenal glands and putidaredoxin from *Pseudomonas putida*, both involved in hydroxylation reactions. These proteins contain two atoms each of iron and labile sulphide per molecule, and on reduction one electron is added per two-iron unit.

The proteins show no EPR signal in the oxidized state, indicating that they are non-magnetic, and in the reduced state show a signal with $g$-values centred around 1·94, corresponding to spin $S = \frac{1}{2}$. To account for these properties, Gibson *et al.* (1966) and Thornley *et al.* (1966) proposed the model shown in Figure 9.14. The two iron atoms are assumed to be bound to sulphur atoms with tetrahedral symmetry, the labile sulphide atoms acting as bridging

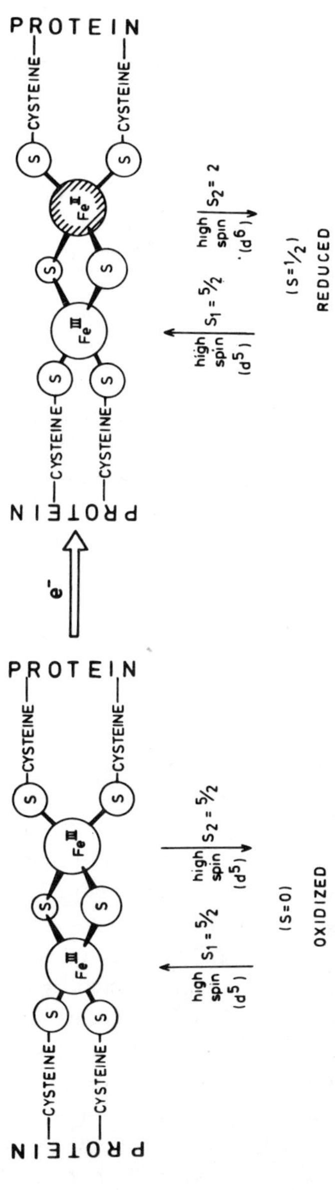


**Figure 9.14.** Model of the iron–sulphur group in the two-iron ferredoxins. (After Rao *et al.*, 1971)

ligands. The indirect evidence for this structural arrangement has been summarized by Rao *et al.* (1971). In the oxidized state, the two iron atoms are high-spin $Fe^{3+}$, and on reduction, one of them is reduced to high-spin $Fe^{2+}$. To explain the magnetic properties, the iron atoms are antiferromagnetically coupled, i.e. spin coupled *via* the sulphur ligands. Thus in the oxidized state, the spins on the $Fe^{3+}$ atoms effectively cancel out; in the reduced state, the excess spin on the $Fe^{3+}$ means that the coupled system has a net spin $S = \frac{1}{2}$.

The Mössbauer effect provided a method of confirming the model, since it provided a potential means of resolving the chemical state of the individual Fe atoms. The problem was a challenging one, because little information was available on the structure of the active site, so that the symmetry of the ligands was unknown, and no inorganic compounds of similar properties were known. It was by no means certain that the different types of two-iron proteins have the same type of active group. Moreover the Mössbauer spectra are complicated by the presence of two types of iron atom, which give two superimposed sets of spectra.

The plant ferredoxins were the first of this class to be studied. Mössbauer spectra were recorded on the proteins containing $^{57}Fe$ at natural enrichment, from cells of the green alga *Euglena* grown on enriched $^{57}Fe$ (Johnson *et al.*, 1968) and from the ferredoxins enriched in $^{57}Fe$ by chemical substitution. All three proteins gave essentially the same spectra, so the chemically reconstituted protein was used. Some early difficulties were experienced due to the presence of contaminant Fe which caused two narrow lines at the centre of the spectra. This was prevented by more careful sample preparation (see Rao *et al.*, 1971).

One of the conclusions from Mössbauer spectroscopy of the two-iron ferredoxins was that the structure of the iron–sulphur group is essentially similar in all of them, though they showed differences in the temperature at which magnetic hyperfine splitting in the reduced form was observed, because of differences in the electron-spin relaxation rates.

Figure 9.15 shows the spectra of oxidized and reduced ferredoxin from the green alga *Scenedesmus*, at 195°K. The oxidized spectrum shows two lines of slightly different width, which were interpreted by Rao *et al.* (1971) as due to two high-spin $Fe^{3+}$ atoms in slightly different environments so that they had slightly different values of chemical shift and quadrupole splitting. The reduced spectrum at this high temperature shows no hyperfine splitting because the electron-spin relaxation is too rapid. Four lines are now seen; two in the same position as those in the oxidized spectrum, corresponding to an unchanged $Fe^{3+}$ atom, and two corresponding to an atom with large chemical shift and quadrupole splitting, typical of high-spin $Fe^{2+}$. These spectra are therefore in agreement with the proposal of Gibson *et al.* (1966) for the valence states of the iron atoms.

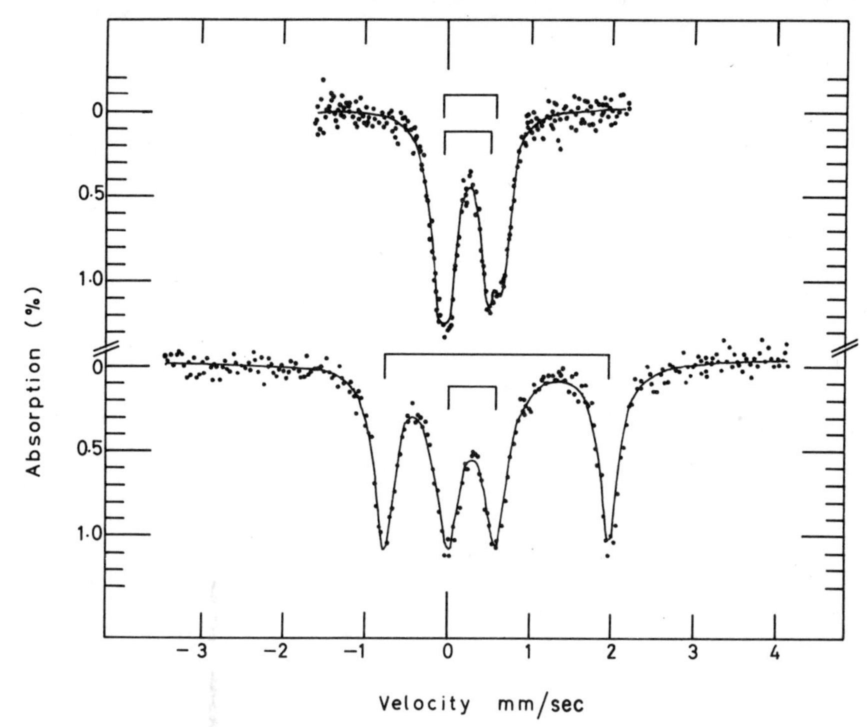


**Figure 9.15.** Mössbauer spectra of ferredoxin from *Scenedesmus obliquus* at 195°K, upper spectrum, oxidized, and lower spectrum, reduced. (From Rao *et al.*, 1971)

The spectra of the reduced ferredoxins at 4·2°K in zero field are complex, but contain at least three lines (Johnson *et al.*, 1968), in agreement with the proposal of Johnson (1971) for the spectrum of an iron atom interacting with a single spin $S = \frac{1}{2}$, as described in a previous section. The hyperfine splitting can be resolved by applying a small field. Figure 9.16 shows the spectra of reduced adrenodoxin. The various components of these spectra are perhaps best understood by means of 'stick' spectra representing the contributions of the individual iron atoms (Cammack *et al.*, 1971). The contribution of the $Fe^{3+}$ atom can be represented by a six-line Zeeman pattern since its quadrupole splitting is small. It can be seen that there are more than six lines in the spectrum of Figure 9.16(a); the other lines are attributed to the $Fe^{2+}$ atom, which will give broader lines because it is anisotropic (i.e. the field at the nucleus varies with direction) and the molecules of the sample are randomly oriented. Since quadrupole splitting of the $Fe^{2+}$ is large, its spectrum can be approximated by a five-line spectrum as shown.

The application of a large field perpendicular to the $\gamma$-rays causes these lines to become split, as the spins of the iron atoms can align either with the field ('spin up') or against the field ('spin down'). These two forms represent

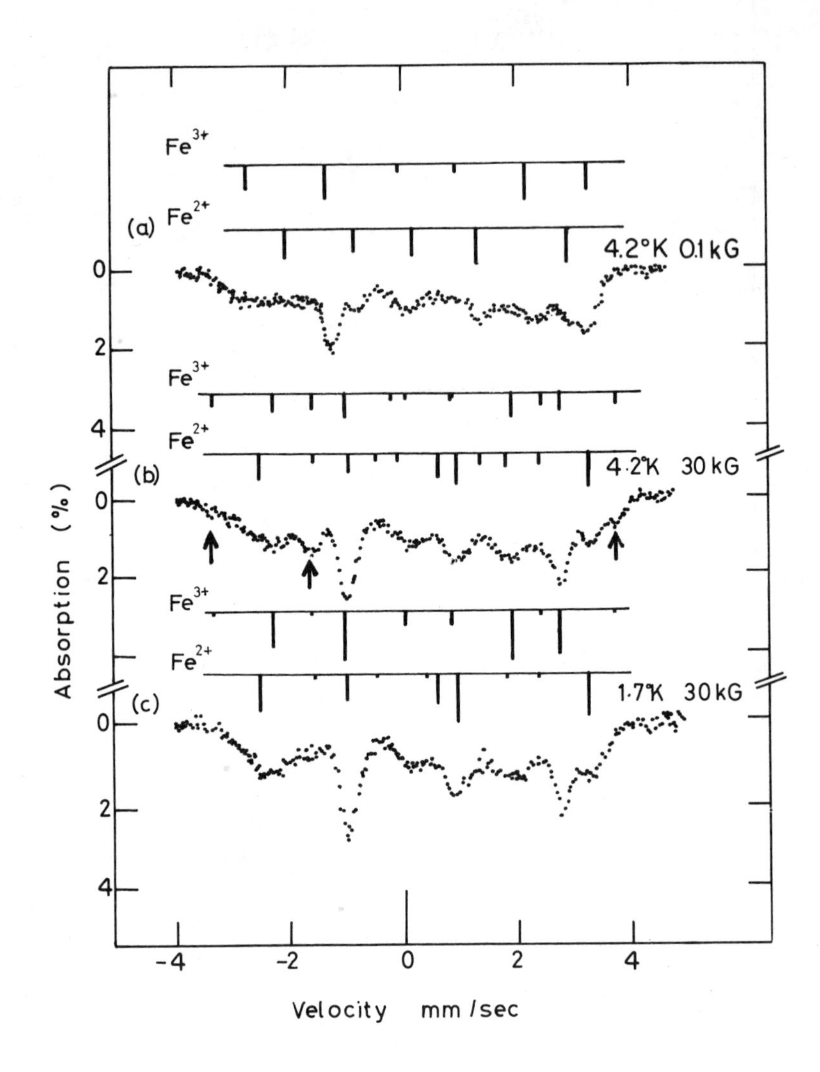


**Figure 9.16.** Mössbauer spectra of reduced porcine adrenodoxin, (a) at 4·2°K, in a field of 0·1 kG; (b) at 4·2°K, in a field of 30 kG; (c) at 1·7°K in a field of 30 kG. The fields were applied perpendicular to the $\gamma$-rays. The proposed assignments of the $Fe^{3+}$ and $Fe^{2+}$ spectra are described in the text. The arrows under (b) indicate lines visible in the spectrum at 4·2°K and not at 1·7°K. (After Cammack *et al.*, 1971)

different energy levels, and for a single Fe atom the spin-down state would be more populated than the spin-up state. Consider first the spectrum at 1·7°K (Figure 9.16c). At this temperature in a field of 30 kG it would be expected that the distribution would be 91% in the spin-down state and 9% in the spin-up state. In the spin-down state the applied field opposes the internal field. Therefore the $Fe^{3+}$ spectrum consists mainly of another six-line pattern corresponding to a somewhat smaller field at the nucleus, together with six much smaller lines corresponding to an increased field. However the lines due to $Fe^{2+}$ do not seem to have shifted the same way. Consider the right-hand line in Figure 9.16(a), which splits to give a new line of *increased* energy; this could not be due to a spin-up $Fe^{3+}$ line, as the intensity of these is too weak. The most likely explanation is that the line is due to $Fe^{2+}$, with an increased effective field at the nucleus. Other features of the spectrum also support the proposal that the magnetic field at the $Fe^{2+}$ atom is in the opposite direction to that of the $Fe^{3+}$, and therefore the two iron atoms must be spin coupled. The spectrum in a field of 30 kG at 4·2°K (Figure 9.16b) provides further evidence for the assignments of the stick spectra, since at this temperature the spin-up and spin-down states will tend to be more equally populated, (28% of one form, 72% of the other), and lines due to the less populated states will become more prominent; places where this can be seen are indicated by the arrows. From this interpretation it is clear that the fields at the two iron atoms are pointed in opposite directions, and therefore must be spin coupled.

This interpretation of the spectra of the reduced ferredoxins is instructive as an example of the amount of information that can be obtained just from examination of the Mössbauer spectra recorded over a range of temperatures and magnetic fields. More detailed information can be obtained from a quantitative approach using computer-fitting methods as already described.

Dunham *et al.* (1971) measured the Mössbauer spectra of six different two-iron ferredoxins, and Münck *et al.* (1972) of putidaredoxin, and using the values for the components of the $A$ tensor from ENDOR results and $g$-values from EPR, fitted the spectra by computer. Figure 9.17 shows the computer-fitted spectrum of reduced spinach ferredoxin at 4·2°K in a large field applied parallel to the $\gamma$-rays; these results are not directly comparable with Figure 9.16, where the field is applied perpendicular to the $\gamma$-rays. For example, in a parallel field, the $Fe^{3+}$ spectrum has lines with intensities 6:0:2:2:0:6, i.e. four lines are seen, as opposed to six in the perpendicular field. The computer-fitted spectra for $Fe^{2+}$ still give discernible lines, so the assumption that they can be approximated by 'stick' spectra is valid, (though the stick spectra can obviously give no information about line shapes).

From the computer-fitted spectra it was possible to obtain more detailed information about the $Fe^{2+}$ atom in reduced ferredoxin. The sign of the $A$-values for $Fe^{2+}$ is positive, in contrast with the free-atom values which are

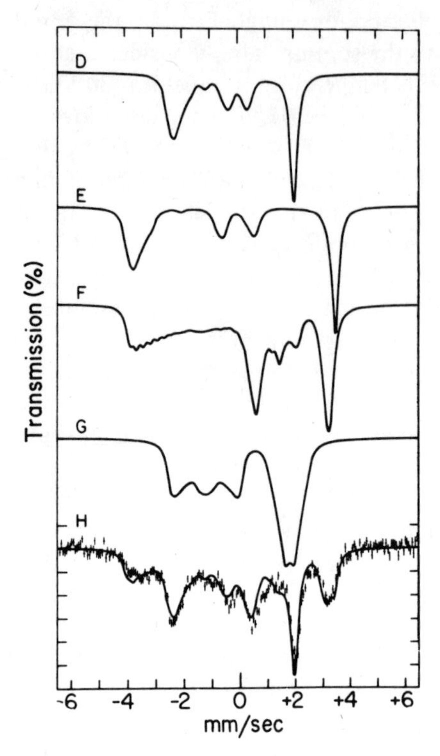


**Figure 9.17.** Experimental (strokes) and computer-fitted (solid lines) Mössbauer spectra for reduced spinach ferredoxin (lyophilysed) at 4·2°K and with a field of 46 kG applied parallel to the $\gamma$-ray direction. The zero of velocity is for a $^{57}Co$ in Pt source. (*D*) $Fe^{3+}$, $m_S = -\frac{1}{2}$; (*E*) $Fe^{3+}$, $m_S = +\frac{1}{2}$; (F) $Fe^{2+}$, $m_S = -\frac{1}{2}$; (G) $Fe^{2+}$, $m_S = +\frac{1}{2}$; (H) weighted sum of D, E, F and G superimposed over experimental spectrum (E and G are weighted 0·3 times the weights of D and F). (Reproduced, with permission, from Dunham *et al.*, *Biochem, Biophys. Acta*, **253**, 134 (1971))

always negative. This has the result that the internal field will add to the external field, as already noted, and must be due to spin coupling to an atom of even greater magnetic moment, i.e. high-spin $Fe^{3+}$. The quadrupole splitting of the $Fe^{2+}$ is large and as determined from the computer analysis, temperature insensitive; this and the *g*-values being near to 2 indicate that, as in reduced rubredoxin, the $Fe^{2+}$ is high spin and the ground-state orbital is $d_{z^2}$, lying well below the next orbital state. This indicates that the symmetry of the iron atom is not octahedral, but either distorted square planar or tetrahedral.

It is interesting to compare the Mössbauer parameters for the iron atoms in the plant-type ferredoxins with those in rubredoxin. The values of chemical shift in reduced *Scenedesmus* ferredoxin (0·22 mm/sec for $Fe^{3+}$, 0·56 mm/sec for $Fe^{2+}$) agree reasonably well with those for rubredoxin at 77°K (0·25 mm/sec for $Fe^{3+}$, 0·65 mm/sec for $Fe^{2+}$). Moreover the hyperfine field at the $Fe^{3+}$ nucleus in reduced *Scenedesmus* ferredoxin is about 180 kG. The effects of antiferromagnetic coupling would be expected from theory to reduce the value of the hyperfine field by a factor of (7/6)/(5/2) of the value in a free atom (Gibson *et al.*, 1966; Rao *et al.*, 1971). The resulting expected value, 386 kG, compares well with the experimental value of 370 kG observed in oxidized rubredoxin (Rao *et al.*, 1971). This agreement in experimental parameters indicates a chemical similarity between the iron atom in rubredoxins and those in the plant-type ferredoxins.

The two-iron ferredoxins therefore represent a case where the Mössbauer effect, used in conjunction with magnetic resonance techniques, has made it possible to obtain not only chemical information, about the valence and spin states of the iron atoms, but also some insight into their structure which could not be obtained by other methods.

## D. Applications in Medical Research

The fact that Mössbauer spectroscopy offers a non-destructive and completely specific test for iron and also gives a good indication of its chemical state, suggests that, if the problems of sensitivity can be resolved, it may have many medical applications.

Johnson (1971) has measured the Mössbauer spectra of material from human lungs infected with hemosiderosis, a fatal disease contracted among coal-miners (Figure 9.18). Spectrum (a) is from a normal lung for comparison, and shows a weak signal due to haemoglobin (cf. Figure 9.7). Spectrum (b) is from a lung infected with hemosiderosis, and shows a much more intense absorption (in fact the iron content of the material was about 2%) due to the presence of large quantities of iron in a magnetic state.

The behaviour of the hyperfine splitting is typical of compounds in which the iron is in a finely divided physical state, such as the iron-storage protein

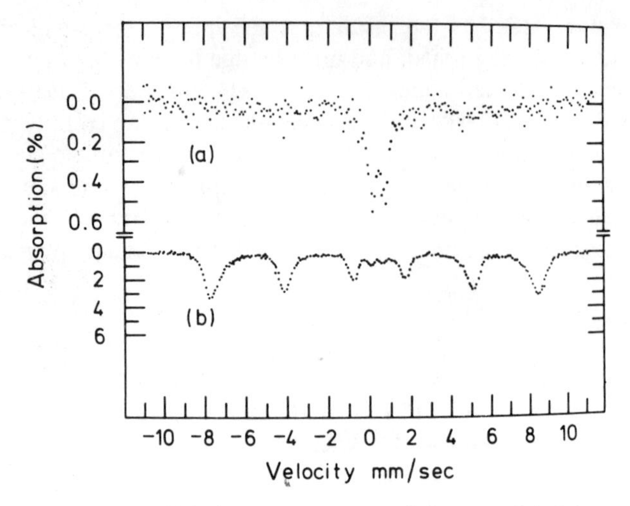


**Figure 9.18.** Mössbauer spectrum of human dried lung material at 4·2°K. (a) from a healthy lung; (b) from a hemosiderosis victim. (Reproduced, with permission, from Johnson, *J. Appl. Phys.*, **42**, 1325 (1971))

ferritin (Boas and Window, 1966). This agrees with the known structure of hemosiderin, which contains small particles of $Fe_2O_3$ enveloped in a protein molecule.

## E. Isotopes Other than $^{57}$Fe

It is possible to study cobalt complexes by carrying out the $^{57}$Fe Mössbauer spectroscopy in reverse—labelling the compound with $^{57}$Co and using this as the source, and a $^{57}$Fe absorber. Studies of this type on vitamin $B_{12}$ were reported by Nath *et al.* (1968). It is not immediately obvious which element one is looking at in these experiments, since the chemical state of an element depends on its electrons rather than the nucleus. The decay of the $^{57}$Co nucleus involves the capture of an electron from the *K* shell, and is followed by a complex series of electronic rearrangements involving multiple-charge states of $^{57}$Fe, finally resulting in a conversion of the electronic configuration to that of $Fe^{2+}$. All of this takes place in a time considerably less than $10^{-7}$ seconds, so that in the Mössbauer experiment one is effectively looking at an $Fe^{2+}$ atom in a site normally occupied by Co, and all the usual types of interpretation of the structure will apply. It is therefore possible to obtain information about the environment of the metal in these complexes, though one must be cautious in extrapolating the results to the condition in

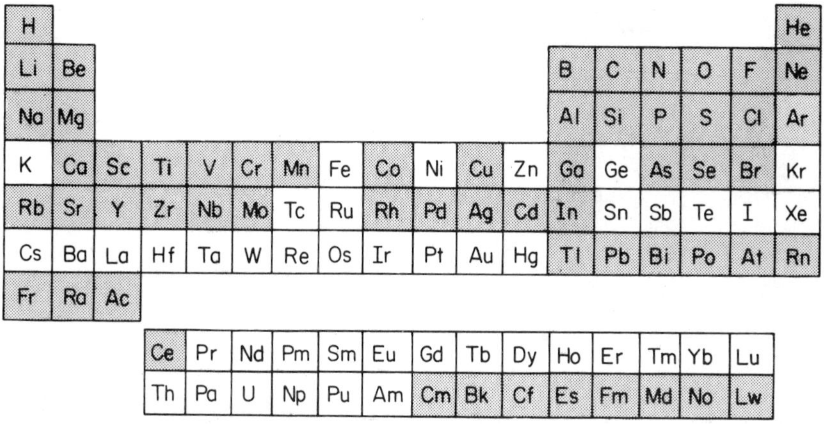

| H | | | | | | | | | | | | | | | | | He |
|---|---|---|---|---|---|---|---|---|---|---|---|---|---|---|---|---|---|
| Li | Be | | | | | | | | | | | B | C | N | O | F | Ne |
| Na | Mg | | | | | | | | | | | Al | Si | P | S | Cl | Ar |
| K | Ca | Sc | Ti | V | Cr | Mn | Fe | Co | Ni | Cu | Zn | Ga | Ge | As | Se | Br | Kr |
| Rb | Sr | Y | Zr | Nb | Mo | Tc | Ru | Rh | Pd | Ag | Cd | In | Sn | Sb | Te | I | Xe |
| Cs | Ba | La | Hf | Ta | W | Re | Os | Ir | Pt | Au | Hg | Tl | Pb | Bi | Po | At | Rn |
| Fr | Ra | Ac | | | | | | | | | | | | | | | |

| Ce | Pr | Nd | Pm | Sm | Eu | Gd | Tb | Dy | Ho | Er | Tm | Yb | Lu |
|---|---|---|---|---|---|---|---|---|---|---|---|---|---|
| Th | Pa | U | Np | Pu | Am | Cm | Bk | Cf | Es | Fm | Md | No | Lw |

**Figure 9.19.** The periodic table. Elements in which the Mössbauer effect has been observed are unshaded. (After Stevens and Bowen, 1974)

the natural compound. The aromatic ring system of vitamin $B_{12}$ seems to survive these violent processes, but some molecules fly apart.

Figure 9.19 lists the elements in which the Mössbauer effect has been observed so far, though by no means all of them are suitable for spectroscopy. It can be seen that few of the lighter elements are represented. Next to $^{57}$Fe, the isotope which is of greatest potential interest in the study of metalloproteins is $^{67}$Zn which is found in a number of enzymes. However this is a much less suitable Mössbauer isotope. It has a high $\gamma$-ray energy (93 KeV), so that both source and sample must be cooled to liquid helium temperatures to get a significant recoil-free fraction. More important, its natural line width is 100 times smaller than for $^{57}$Fe, so that the type of spectrometer described in this article would be unsuitable, because of relative vibrational motion of source and absorber. Some success has recently been achieved by holding the source and sample close together in a Dewar, and vibrating the source by means of the piezoelectric effect. It seems likely that with improvements in techniques it may be possible to study the Mössbauer effect of $^{67}$Zn in biological materials.

$^{40}$K is a difficult isotope because the line width is too broad, and the recoil-free fraction is very small. $^{129}$I is a promising subject; iodine is present in compounds such as thyroxine, and could also be used as a label for aromatic groups such as tyrosine in proteins. Since it is a radioactive isotope, it must be produced artificially and is therefore rather scarce. Studies of haem iodide and haem-histidine iodide have been reported by Pasternak *et al.* (1970). The spectrum of haem-histidine-$^{129}$I-iodide (Figure 9.20) shows a more complex quadrupole splitting than $^{57}$Fe, because of the larger nuclear spins involved. It was concluded that in this compound the iodine is covalently bonded to the

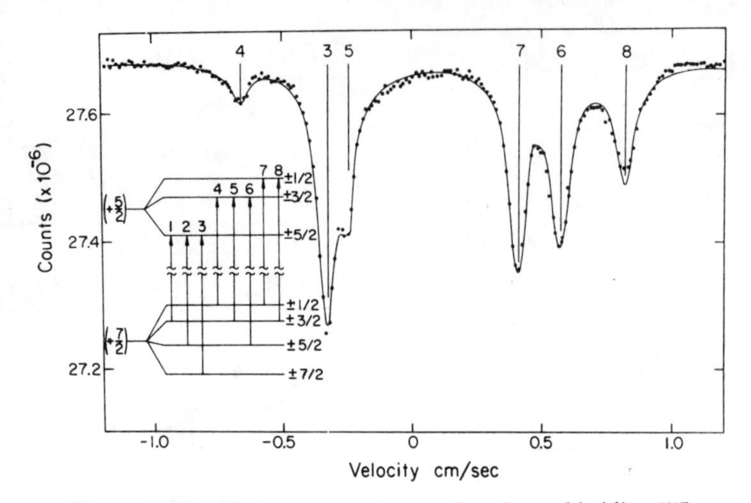


**Figure 9.20.** Mössbauer spectrum of a haem-histidine-$^{129}$I absorber and a Zn$^{129}$Te source at 4·2°K, showing the complex quadrupole splitting pattern. Lines 1 and 2 are out of the velocity scale. (Reproduced, with permission, from Pasternak *et al.*, *Proc. Natl. Acad. Sci. U.S.*, **66**, 1142 (1970))

iron atom of haem. These workers also investigated the spectrum of the corresponding telluride, by preparing the $^{125}$I derivative and using this as the *source* in a Mössbauer spectrometer with an absorber of Zn$^{125}$Te. The Mössbauer effect in this case involves decay of the $^{125}$I to $^{125}$Te, so that this is analogous to the experiments with $^{57}$Co, giving a quadrupole-split spectrum of Te in the haem-histidine complex.

Other Mössbauer isotopes are of no direct biological interest, but some of them might possibly substitute for elements which are. For example, Williams (1971) has pointed out the eligibility of $^{151}$Eu as a substitute for calcium. This rare earth metal is a good Mössbauer isotope, and can take up a magnetic divalent state, so that it is suitable for studies of magnetic hyperfine interaction and also EPR. It might therefore provide a probe for the active centre of calcium-binding proteins. These proteins appear to be widespread. Although they are mostly associated with insoluble membrane systems, some are now being obtained in a purified state suitable for physicochemical studies. $^{151}$Eu might also act as a probe of the structure of bone, teeth and related materials. Other possible substitutions are $^{125}$Te for sulphur (though this might present some problems of preparative chemistry), $^{61}$Ni for magnesium in complexes with nucleotides, and $^{182}$W for molybdenum. There is obviously considerable scope for further work.

## V. APPENDIX 1—SUPPLIERS OF APPARATUS FOR MÖSSBAUER SPECTROSCOPY

|  |  |
|---|---|
| Austin Science Associates Inc.<br>5902 W. Bee Caves Road<br>Austin, Tex. 78746<br>U.S.A. | Mössbauer spectrometers<br>Cryostats<br>Superconducting magnets<br>Sources |
| Ealing Scientific<br>2225 Massachusetts Avenue<br>Cambridge, Mass. 02140, U.S.A. | Mössbauer spectrometers |
| Elscint Ltd.<br>P.O. Box 5258<br>Haifa, Israel | Mössbauer spectrometers |
| Frieseke and Hoepfner GmbH<br>D-7500 Karlsruhe-Durbach<br>Bergwaldstrasse 30, Postfach 410220<br>W. Germany | Mössbauer spectrometers |
| Harshaw Chemical Co.<br>6801 Cochran Road<br>Solon, Ohio 44139, U.S.A. | Counting systems |
| Harwell Mössbauer Group<br>Building 8, AERE Harwell<br>Didcot, Berks., U.K. | Mössbauer spectrometers<br>Cryostats<br>Iron-57<br>Sources |
| J. & P. Engineering Ltd.<br>Portman House, Cardiff Road<br>Reading RG1 8JF, U.K. | Mössbauer spectrometers |
| L.N.D. Inc.<br>3230 Lawson Boulevard<br>Oceanside, N.Y. 11572<br>U.S.A. | Proportional counters |
| New England Nuclear Corporation<br>575 Albany Street<br>Boston, Mass. 02118, U.S.A. | Sources<br>Iron-57 |
| Numelec<br>2 Petite Place<br>78000 Versailles,<br>France | Mössbauer spectrometers |

| | |
|---|---|
| Ortec Inc.<br>100 Midland Road, Oak Ridge<br>Tennessee 37830, U.S.A. | Counting electronics<br>Multichannel analysers |
| Oxford Instruments Co. Ltd.<br>Osney Mead<br>Oxford OX2 0DX, U.K. | Mössbauer spectrometers<br>Cryostats<br>Superconducting magnets |
| Radiochemical Centre<br>Amersham, Bucks, U.K. | Sources |
| Ranger Electronics Corporation<br>1718 Oklahoma Blvd.<br>P.O. Box 863<br>Alva, Oklahoma 73717, U.S.A. | Mössbauer spectrometers<br>Cryostats |

## VI. APPENDIX 2—GLOSSARY OF TERMS USED

*Bohr magneton*, $\beta$: The atomic unit of magnetic moment, defined as the moment of a single electron spin.

*g-value:* A term in the expression for the energy of an electron spin in a magnetic field. On applying a field $H$ to an electron, its energy levels will be split into two, separated by a quantity $g\beta H$. These levels correspond to the electron spin being aligned parallel or antiparallel to the field. For a free electron, $g$ is 2·0023, but for an electron associated with a transition metal atom, $g$ differs from this value because of orbital angular momentum. $g$-values are determined experimentally by EPR spectroscopy.

*Hamiltonian:* A form of equation used in quantum mechanics to describe the total energy of a system.

*Lorentzian lineshape:* For a Mössbauer spectrum this is defined by the formula

$$\sigma(E) = \sigma_0 \left[1 + 4\left(\frac{E - E_0}{\Gamma}\right)^2\right]^{-1}$$

$\sigma$ is the intensity of absorption at energy $E$, $\sigma_0$ and $E_0$ being the values at peak absorption. $\Gamma$ is the width of the spectrum at half maximal absorption. The shape of the curve is similar to the more familar gaussian curve, but is narrower at the peak and broader at the base.

*Multichannel analyser*, or computer of average transients (CAT) is used to average out spectra which must be scanned repeatedly to improve the signal to noise ratio. The spectrum is divided into a large number of equal sections and during each scan the signal in each section is added up digitally in a particular channel.

*Nuclear g-value*, $g_n$ is a property of any nucleus which has a spin, and is analagous to the $g$-value of an electron. In a magnetic field $H$ the nuclear energy levels are split by $g_n\beta_n H$.

*Nuclear magneton*, $\beta_n$: The magnetic moment of a single proton. Its value is 1/1836 of the Bohr magneton.

*Quadrupole moment* of a nucleus: arises from the distribution of electric charge in a nucleus. Spherically symmetrical nuclei ($I = 0$ or $\frac{1}{2}$) have no quadrupole moment, but nuclei with spin $I > \frac{1}{2}$ have a quadrupole moment. An oblate (flattened) nucleus has a negative quadrupole moment while a prolate (elongaged) one has a positive moment.

*Tensor:* a mathematical function, usually expressed in the form of a matrix, which is used in quantum mechanics to describe functions which vary in multidimensional space.

## VII. ACKNOWLEDGEMENTS

I would like to express my gratitude to Professor C. E. Johnson for introducing me to biological Mössbauer spectroscopy and for much helpful advice, and to numerous other workers in this field for valuable discussions. Several of the examples cited in this article were experiments carried out in collaboration with Dr D. O. Hall, Dr K. K. Rao, Dr M. C. W. Evans and Prof. C. E. Johnson, and supported by the Science Research Council.

## VIII. REFERENCES

Boas, J. F. and B. Window (1966) 'Mössbauer effect in ferritin', *Austr. J. Phys.*, **19**, 573.

Bonchev, T., I. Vassilev, T. Sapundzhiev and K. Evtimov (1968) 'Possibility of investigating movement in a group of ants by the Mössbauer Effect', *Nature*, **217**, 96.

Bray, R. C. and L. S. Meriwether (1966) 'Electron spin resonance of xanthine oxidase substituted with molybdenum-95', *Nature*, **212**, 467.

Cammack, R., K. K. Rao, D. O. Hall and C. E. Johnson (1971) 'Mössbauer studies of adrenodoxin. The mechanism of electron transfer in a hydroxylase iron-sulphur protein', *Biochem. J.*, **125**, 849.

Dunham, W. R., A. J. Bearden, I. T. Salmeen, G. Palmer, R. H. Sands, W. H. Orme-Johnson and H. Beinert (1971) 'The two-iron ferredoxins in spinach, parsley, pig adrenal cortex, *Azotobacter vinelandii* and *Clostridium pasteurianum*: studies by magnetic field Mössbauer spectroscopy', *Biochim. Biophys. Acta*, **253**, 134.

Gibson, J. F., D. O. Hall, J. H. M. Thornley and F. R. Whatley (1966) 'The iron complex in spinach ferredoxin', *Proc. Natl. Acad. Sci. U.S.*, **56**, 987.

Goldanskii, V. I. and R. H. Herber (1968) *Chemical Applications of Mössbauer Spectroscopy*, Academic Press, New York.

Griffith, J. S. (1961) *Theory of Transition Metal Ions*, Cambridge University Press.

Ingram, D. J. E. (1969) *Biological and Biochemical Applications of Electron Spin Resonance*, Adam Hilger Ltd., London.

Johnson, C. E. (1971) 'Applications of the Mössbauer effect in biophysics', *J. Appl. Phys.*, **42**, 1325.

Johnson, C. E., E. Elstner, J. F. Gibson, G. L. Benfield, M. C. W. Evans and D. O. Hall (1968) 'Mössbauer effect in the ferredoxin of *Euglena*', *Nature*, **220**, 5174.

Lang, G. (1970) 'Mössbauer spectroscopy of haem proteins', *Quart. Rev. Biophys.*, **3**, 1.

Lang, G., D. Herbert and T. Yonetani (1968) 'Mössbauer spectroscopy of cytochrome c', *J. Chem. Phys.*, **49**, 944.

Lang, G. and W. Marshall (1966) 'Mössbauer effect in some haemoglobin compounds', *Proc. Phys. Soc.*, **87**, 3.

Münck, E., P. G. Debrunner, J. C. M. Tsibris and I. C. Gunsalus (1972) 'Mössbauer parameters of putidaredoxin and its selenium analogue', *Biochemistry*, **11**, 855.

Nath, A., M. Harpold, M. P. Klein and W. Kündig (1968) 'Emission Mössbauer spectroscopy for biologically important molecules. Vitamin $B_{12}$ and its analogues and cobalt phthalocyanine', *Chem. Phys. Lett.*, **2**, 471.

Pasternak, M., P. G. Debrunner, G. DePasquali, L. P. Hager and L. Yeoman (1970) 'Application of $^{129}I$ Mössbauer effect to biological systems: studies with heme models', *Proc. Natl. Acad. Sci. U.S.*, **66**, 1142.

Peisach, J., W. E. Blumberg, E. T. Lode and M. J. Coon (1971) 'An analysis of the electron paramagnetic resonance spectrum of *Pseudomonas oleovorans* rubredoxin. A method for determination of the ligands of ferric iron in completely rhombic sites', *J. Biol. Chem.*, **246**, 5877.

Phillips, W. D., M. Poe, J. F. Weiher, C. C. McDonald and W. Lovenberg (1970) 'Proton magnetic resonance, magnetic susceptibility and Mossbauer studies of *Clostridium pasteurianum* rubredoxin', *Nature*, **227**, 574.

Rao, K. K., R. Cammack, D. O. Hall and C. E. Johnson (1971) 'Mössbauer effect in *Scenedesmus* and spinach ferredoxins; the mechanism of electron transfer in a plant-type iron–sulphur protein', *Biochem. J.*, **122**, 257.

Rao, K. K., M. C. W. Evans, R. Cammack, D. O. Hall, C. C. Thompson, P. J. Jackson and C. E. Johnson (1972) 'Mössbauer effect in rubredoxin. Determination of the hyperfine field in a simple iron–sulphur protein', *Biochem. J.*, **129,** 1063.

Rhode, W. S. (1971) 'Observations of the vibration of the basilar membrane in squirrel monkeys using the Mössbauer technique', *J. Acoust. Soc. Amer.*, **49**, 1218.

Stevens, J. G. and V. E. Stevens (1972) *Mössbauer Effect Data Index, Covering the 1970 Literature*, Plenum Press, New York.

Stevens, J. G., and L. H. Bowen (1974) 'Mössbauer spectrometry', *Anal. Chem.*, **46**, 287R.

Thornley, J. H. M., J. F. Gibson, F. R. Whatley and D. O. Hall (1966) 'Comment on a recent model of the iron complex in spinach ferredoxin', *Biochem. Biophys. Res. Comm.*, **24**, 877.

Watenpaugh, K. D., L. C. Sieker, J. R. Herriott and L. H. Jensen (1971) 'The structure of a non-heme iron protein: rubredoxin at 1·5 Å resolution', *Cold Spring Harbor Symp. Quant. Biol.*, **36**, 359.

Wertheim, G. K. (1964) *Mossbauer Effect: Principles and Applications*, Academic Press, New York.

Williams, R. J. P. (1971) in *Bioinorganic Chemistry* (Ed. R. F. Gould), American Chemical Society.

Winter, M R. C. C. E. Johnson, G. Lang and R. J. P. Williams (1972) 'Mössbauer spectroscopy of haemoglobin derivatives' *Biochim. Biophys. Acta*, **263**, 515.

# Biological Glossary

*Assimilate*—Formation of organic substances from $CO_2$ and water by green plant cells in light.
*Auxin*—A plant growth hormone.
*Chimeral tissue*—Tissue made up of cells of more than one genetic constitution.
*Coleoptile*—Sheath surrounding the young shoot of grass seedlings.
*Cytokinin*—A six-substituted purine which elicits a growth response in plant cells.
*Fusion-body*—Product of the fusion of two or more plant protoplasts.
*H.A.T. selection*—A selection system which can be used to isolate hybrid animal cells. The selection is dependent upon the blocking of DNA synthesis from sugars and amino acids by aminopterin. This requires cells to use an alternative pathway involving thymidine kinase and hypoxanthine guanine phosphoribosyl transferase. Parent cell lines possessing only one of these enzymes are unable to grow. Hybrid cells containing both enzymes can grow in medium containing aminopterin and nucleosides.
*Homokaryon*—A multinucleate cell in which the nuclei have the same genetic constitution.
*Meristematic*—Refers to plant cells which undergo division giving rise to the plant body.
*Mesophyll*—Tissue of the leaf differentiated into palisade cells and spongy mesophyll cells.
*Palisade cells*—Elongated cells containing numerous chloroplasts located beneath the epidermis of the leaf with the long axis perpendicular to the leaf surface.
*Parenchymatous*—refers to thin-walled plant cells which in general are not highly differentiated.
*Plantlet*—A small developing plant.
*Plasmalemma*—Lipoprotein membrane bounding the cytoplasm of plant cells.
*Plasmodesmata*—Fine protoplasmic threads passing through the cell walls that separate plant protoplasts.
*Symplastic*—refers to the cellular interconnecting nature of the cytoplasmic component of the plant body.
*Virion*—The term used for an intact virus particle.

# Index